T0174744

Introduction to Renewable Power Systems and the Environment with R

Introduction to Renewable Power Systems and the Environment with R

Miguel F. Acevedo

CRC Press
Taylor & Francis Group
Boca Raton London New York

CRC Press is an imprint of the
Taylor & Francis Group, an **informa** business

MATLAB® is a trademark of The MathWorks, Inc. and is used with permission. The MathWorks does not warrant the accuracy of the text or exercises in this book. This book's use or discussion of MATLAB® software or related products does not constitute endorsement or sponsorship by The MathWorks of a particular pedagogical approach or particular use of the MATLAB® software.

CRC Press
Taylor & Francis Group
6000 Broken Sound Parkway NW, Suite 300
Boca Raton, FL 33487-2742

First issued in paperback 2020

© 2019 by Taylor & Francis Group, LLC
CRC Press is an imprint of Taylor & Francis Group, an Informa business

No claim to original U.S. Government works

ISBN 13: 978-0-367-57130-6 (pbk)
ISBN 13: 978-1-138-19734-3 (hbk)

This book contains information obtained from authentic and highly regarded sources. Reasonable efforts have been made to publish reliable data and information, but the author and publisher cannot assume responsibility for the validity of all materials or the consequences of their use. The authors and publishers have attempted to trace the copyright holders of all material reproduced in this publication and apologize to copyright holders if permission to publish in this form has not been obtained. If any copyright material has not been acknowledged, please write and let us know so we may rectify in any future reprint.

Except as permitted under U.S. Copyright Law, no part of this book may be reprinted, reproduced, transmitted, or utilized in any form by any electronic, mechanical, or other means, now known or hereafter invented, including photocopying, microfilming, and recording, or in any information storage or retrieval system, without written permission from the publishers.

For permission to photocopy or use material electronically from this work, please access www.copyright.com (http://www .copyright.com/) or contact the Copyright Clearance Center, Inc. (CCC), 222 Rosewood Drive, Danvers, MA 01923, 978-750-8400. CCC is a not-for-profit organization that provides licenses and registration for a variety of users. For organizations that have been granted a photocopy license by the CCC, a separate system of payment has been arranged.

Trademark Notice: Product or corporate names may be trademarks or registered trademarks, and are used only for identification and explanation without intent to infringe.

Library of Congress Cataloging–in–Publication Data
Names: Acevedo, Miguel F., author.
Title: Introduction to renewable power systems and the environment with R / Miguel F. Acevedo.
Description: First edition. | Boca Raton, FL : CRC Press/Taylor & Francis Group, 2018. | "A CRC title, part of the Taylor & Francis imprint, a member of the Taylor & Francis Group, the academic division of T&F Informa plc." | Includes bibliographical references.
Identifiers: LCCN 2017060764 | ISBN 9781138197343 (hardback : acid-free paper)
Subjects: LCSH: Electric power production–Data processing. | Renewable energy sources. | R (Computer program language)
Classification: LCC TK1005 .A28 2018 | DDC 621.31/2102855133–dc23
LC record available at https://lccn.loc.gov/2017060764

Visit the Taylor & Francis Web site at
http://www.taylorandfrancis.com

and the CRC Press Web site at
http://www.crcpress.com

Contents

Preface

This aim of this book is to introduce the fundamentals of electrical power systems while examining their relationships with the environment. In particular, I want the book to help students understand those aspects of environmental systems that motivate the development and utilization of renewable power systems technology. Importantly, I cover conventional power systems and opportunities for increased efficiencies and friendlier environmental interactions. My approach is to explain the basics of each energy conversion system, emphasizing what limits their efficiency, both theoretically and practically.

On purpose, I intertwine chapters on thermodynamics, electricity, and environmental systems, instead of separating these topics into parts. This may seem unconventional, but I have successfully employed this approach for several years when teaching a course in renewable electrical power systems for undergraduate and graduate students in electrical engineering and mechanical and energy engineering. This approach allows the student to feel more comfortable moving across disciplines, while at the same time avoids consecutive weeks of a course focusing on topics of a particular discipline. However, the instructors may choose to vary the order in which the chapters are covered according to the particular disciplines of the students enrolled in their classes.

Naturally, some chapters may be more familiar to some students according to their background. For example, electrical engineering students would be very familiar with the circuit chapters but would feel more challenged by other topics such as thermodynamics and environmental science, which are not taught in the typical electrical engineering curricula. Mechanical engineers would know more about thermodynamics and thermal conversion processes. Chemical engineers are better prepared to understand combustion processes, energy balances, and process control. Civil and environmental engineers are better able to understand hydrological and hydraulic analysis. Typically, students of all these engineering disciplines have less background on natural resources, ecosystems, and the environment. Principles of biology, ecology, geology, and atmospheric science are not common in engineering curricula, except agricultural and environmental engineering. An environmental science and engineering reader may be more familiar with the environmental systems concepts but less familiar with electrical circuits and thermodynamics; using this book, they will learn more about the basis for energy conversion and production of electrical power.

As you will note, there is broad range of interrelated problems that come together when producing electricity, which calls for contributions of engineering disciplines together with applied science, such as environmental science, geography, and ecology. All these disciplines contribute to understanding the basic interactions of a power system technology with its environment.

Although my target is a textbook, I have also structured the material in such a way that could serve as a reference book. The material is organized in 14 chapters; therefore when used as a textbook, it can be covered on a chapter-per-week basis in a typical semester. I include example and exercise problems that can serve as the basis of homework assignments.

The computer examples use the R system, which is open source, and it is very simple to download, install, and run. As some authors put it, R has evolved into the *lingua franca* of scientific computing; currently there are tens of thousands of packages to perform a tremendous variety of analysis. Even students with no prior knowledge of programming are quickly acquainted with the basics of programming in R. Engineering students are typically familiar with MATLAB® by their junior year and would find R easy to understand. Students can execute the computer examples and exercises in the classroom environment and at their own pace using their computers. I have written the R package renpow to support this textbook. My intention is to bundle many lengthy calculations into functions requiring a single call or at most a sequence of few lines of code. Functions of renpow have been described and used in the textbook in the context of specific examples. More options and details are available from the help of renpow.

The first two chapters are introductory and motivational. Chapter 1 covers basic physical and chemical principles underlying the concepts of energy and power, ranging from the basics of mechanical and thermal energy to other forms that will be described later in the book. A large part of this chapter is devoted to understanding how most electricity is produced around the world and the need to shift from carbon-based energy conversion to other forms, including renewables. In the second chapter we examine the basics of environmental systems emphasizing global cycles, particularly carbon, and ecosystems. We briefly review the geologic history of Earth to place the contemporary issue of climate change and carbon-based energy conversion on a broader perspective.

I have included three circuit analysis chapters to allow all readers to either learn or review these concepts. Chapters 3, 5, and 8 introduce electrical circuits from the simplest circumstances of steady voltage and current (Direct Current or DC) to more complicated time-varying voltage and current because of either transients or periodic variations (Alternating Current or AC). In addition, these chapters present conversion of DC to DC, AC to DC, and DC to AC that are central to the use of renewable power. These chapters would probably be a review for many engineering students; however, this material allows introducing electrical technology to many readers who need this background to work with electrical power systems. Later chapters (10 and 11) offer more details on electrical power systems, particularly some topics not covered in the typical circuit analysis courses.

Fundamental concepts of thermodynamics and heat engines are covered in Chapters 4 and 6, which are also introductory and look at the simplest laws that underlie generation of power. Mechanical and chemical engineering students are likely familiar with this material. However, it is typically new subject for other students. Chapters 7 and 9 cover the bulk of conventional power generation using thermal processes and relying heavily on the basis laid out by Chapters 4 and 6.

Chapters 12, 13, and 14 constitute most of the renewable energy conversion focus of this book. They are devoted to hydroelectrical, wind, and solar, because these currently represent the bulk of electrical production by renewables around the world. However, as a reader will notice, I have made an effort to include other existing and promising technologies as they relate to the topic under discussion in earlier chapters.

Acknowledgments

I thank the many individuals and institutions that have contributed to my experience on the subject matter of this book. I feel fortunate to have accumulated this experience from many projects on environmental systems, climate change, oil and gas, watershed and reservoirs, hydroelectric power generation, and more recently off-grid solar and wind systems for desalination. I have learned from many colleagues, in academia, government agencies, and industry, with whom I have worked throughout the years. To all of them I express my sincere gratitude.

Many thanks go to the numerous students of electrical engineering and mechanical and energy engineering at the University of North Texas (UNT) who have taken successive versions of my renewable power systems courses for the past several years. All these students, undergraduate and graduate alike, have contributed to shaping my approach to this subject and the structure of this book, including the use of examples and exercises for homework and exams.

I also thank senior design, and MS and PhD students of geography, environmental science, and electrical engineering at UNT whom I have supervised through the years. I am grateful that they centered their projects and research on topics related to this subject matter.

Special thanks go to Breana Smithers who has helped for several years deploying and installing systems for off-grid wind and solar power systems, as well as proofreading and drafting figures for this book.

My son Andrés Acevedo-Cross has been a source of inspiration, as he worked on many large and complex projects of utility-scale solar power plants. His professional commitment to renewable power has encouraged me through the efforts of writing this book.

Finally, but not least, I am very grateful to Irma Shagla-Britton, editor for Environmental Science and Engineering at CRC Press, for her enthusiasm and her patience as I completed this project.

Author

Miguel F. Acevedo, PhD, has vast interdisciplinary experience, especially at the interface of science and engineering. He has served the University of North Texas (UNT) since 1992 in the Department of Geography, the graduate program in environmental sciences, the Electrical Engineering Department, and the Advanced Environmental Research Institute. He earned his PhD in biophysics from the University of California, Berkeley (1980) and master's degrees in electrical engineering and computer science from Berkeley (ME, 1978) and the University of Texas at Austin (MS, 1972). Before joining UNT, he was at the Universidad de Los Andes, Merida, Venezuela, where he taught since 1973 in the School of Systems Engineering, the graduate program in tropical ecology, and the Center for Simulation and Modeling. He has served on the Science Advisory Board of the US Environmental Protection Agency and on many review panels of the US National Science Foundation. He has received many research grants, and written several textbooks and numerous journal articles, as well as many book chapters and proceeding articles. UNT has recognized him with the Regents' Professor rank, the Citation for Distinguished Service to International Education, and the Regents' Faculty Lectureship.

1 Introduction

As we humans developed our capacity to modify and exploit our *environment* for food and shelter, along with using various forms of *energy*, we have become more and more aware of the importance of managing and preserving the quantity and quality of natural resources upon which we depend to sustain our livelihood. *Electricity* has become one of the major uses of energy in the world today contributing greatly to prosperity and many forms of advancements in industry, technology, and information. However, the major sources used to generate electricity today across the world are *fossil fuels*, which cause at least two complicated issues: These resources are finite and cannot be realistically replenished once depleted, and their consumption enhances Earth's greenhouse effect causing *global climate change*. Hence, shifting the sources of electricity generation to more lasting resources that at the same time reduce the impact on climate change has become an important and challenging endeavor.

This chapter introduces the needs for intertwining the two major themes of this book: the environment and electrical power systems. Being an introductory chapter, we first review basic concepts of energy and *power*, using simple and intuitive mechanical, thermal, radiant, and chemical energy as our first few examples. Second, we examine the sources of electricity generation, emphasizing carbon-based sources and the need for renewable power systems. Important distinctions are made among several terms commonly employed to qualify energy conversion systems. Several textbooks [1–10] serve as reference and supplementary readings to the material presented in this chapter.

1.1 ENERGY AND POWER

1.1.1 BASICS OF MECHANICAL POWER AND ENERGY

We start by reviewing some basic notions from elementary mechanics as applied to a moving object. *Velocity* is speed of motion or distance traveled per unit time; it is given, for example, in meters per second (m/s or ms^{-1}). The rate of change of velocity is *acceleration*, or how fast the velocity changes; it is then ms^{-1} per second, which is meters per square second (ms^{-2}). Newton's second law of motion states that *force* is a measure of how much acceleration you can give to a moving mass and it is given by $F = ma$, where m is mass and a is acceleration. One newton (N) is the force required to give a mass of one kilogram (1 kg) an acceleration of 1 ms^{-2}.

A force does *work* when it is applied to an object of a given mass moving it along a certain distance or, in other words, displacing its position (Figure 1.1). This work is a form of *energy* provided by the force and given to the object. Therefore, we can think of work as a transfer of energy. One *joule* (J) of energy corresponds to the work done by a 1 N force acting for one meter (1 J = 1 Nm).

Besides the total amount of work done, the time taken to move the object is important and determines *power*, which is the rate of change of work or energy. Thus, power is energy per unit time, say, joules per hour or joules per second (s). In fact, the latter is the unit of power called *watt* (W), which is equal to 1 J/s. From this perspective, energy is an accumulation of power over time, thus a joule would be W × s.

Both energy and power may vary with time. Mathematically, we can calculate *instantaneous power* as a derivative of energy with respect to time t

$$p(t) = \frac{de(t)}{dt} \quad \text{and} \quad p(t) = \frac{dw(t)}{dt} \tag{1.1}$$

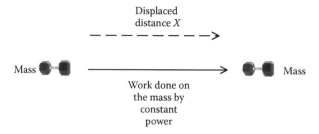

FIGURE 1.1   Displacing a mass.

where $e(t)$ is energy, $w(t)$ is work, and $p(t)$ is power. Conversely, we can calculate energy as an integral of power up to that time t starting at some arbitrary reference time t_0:

$$e(t) = \int_{t_0}^{t} p(x)dx \quad \text{and} \quad w(t) = \int_{t_0}^{t} p(x)dx \qquad (1.2)$$

In many cases, it is also relevant to think of the value of energy or work that has been integrated up to time t, as *total energy* or *work* accumulated up to a final time t_f. Typically, these total quantities will be denoted by capitalized E and W, respectively. Therefore,

$$E = \int_{t_0}^{t_f} p(x)dx \quad \text{and} \quad W = \int_{t_0}^{t_f} p(x)dx \qquad (1.3)$$

Please pay close attention to notation. Total work is W (italicized) and power units watts is W (not italicized). It is customary to use lowercase letters to denote time-varying quantities (instantaneous value at t) and uppercase letters to denote constant or total quantities.

Example 1.1

Suppose we exert 10 N of force on a mass and we displace it 12 m at constant speed with no friction in 1 min. What are the total work and energy? What are the instantaneous work, power, and energy? Answer: Total work is $W = 10 \times 12$ Nm = 120 J. Assuming there was no loss due to other factors (e.g., overcoming friction) this work is the same as total energy E (which came from us who exerted the force). Because this work was done at constant speed, instantaneous power is actually a constant $p(t) = \dfrac{W}{t_f} = \dfrac{120 J}{60 s} = 2$ W. Again, be careful distinguishing W for work and W for watts. Assuming initial time is 0 and integrating over time $w(t) = \int_0^t p(x)dx = \int_0^t 2dx = 2\int_0^t dx = 2t$ J in the time interval $0 \le t \le 60$ s. After 1 min, the work is $w(t = 60 \text{ s}) = 2$ J/s $\times$ 60 s = 120 J as expected because this amount is total work and remains constant at this value. Figure 1.2 illustrates these results.

It is convenient to use programs to visualize plots of varying power and work as in Figure 1.2. For this purpose, we can use functions pow.work and pow.work.plot, which are part of the R functions of package renpow that I have written to accompany this book. A brief introduction to R is given in the Appendix. A call to pow.work requires three arguments: a time sequence, a type of power

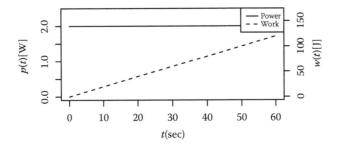

FIGURE 1.2   Displacing a mass at constant power requires linearly increasing work.

function such as constant or linear, and a value of power for the parameter of the function. In this case, we use time from 0 to 60 s in intervals of 0.1 s, and constant power with value 2W:

```
t <- seq(0,60,0.1)
x <- pow.work(t,pow="const",p=2)
```

The output contains the values of power and work as a function of time. Then we call the plot function using the output x as argument to the pow.work.plot function:

```
pow.work.plot(x)
```

This statement produces Figure 1.2.

Of special interest in this book is the mechanical power involved in *rotational* movement because currently most electricity worldwide is produced by rotating an electrical generator using a prime mover (Figure 1.3). *Torque* is force F multiplied by radius r of the circle along which we are considering the rotational movement, that is to say, $\tau = Fr$. Because force is in newtons (N) and radius in meters (m), torque units are in newton meters (Nm). This is dimensionally the same as a joule, but torque is not energy! For rotational movement the linear distance $l(t)$ that the force is applied for angular displacement of $\theta(t)$ is $w(t) = Fl(t) = Fr\theta(t) = \tau\theta(t)$ or $w(t) = \tau\theta(t)$, which can be rewritten as

$$\tau = \frac{w(t)}{\theta(t)} \tag{1.4}$$

In other words, torque is the same as work per unit angle, and thus torque can be expressed as joule per radian (J/rad).

Now, calculate power as rate of change of work, $P = \dfrac{dw(t)}{dt} = \tau\dfrac{d\theta(t)}{dt}$. The last term is the *angular speed* $\omega_{mech} = \dfrac{d\theta(t)}{dt}$. Then,

$$P = \tau\omega_{mech} \tag{1.5}$$

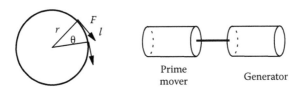

FIGURE 1.3   Rotational mechanical energy. Basic relations between force, torque, and angle. A prime mover provides rotational mechanical power to an electrical generator.

Put differently, power is torque multiplied by angular speed; with ω_{mech} in rad/s, and torque in J/rad, we get P in J/s = W. We have used the subscript *mech* in ω_{mech} so that we can reserve ω for electrical angular frequency later in the textbook.

In practical terms, angular speed is often given in revolutions per minute (rpm) and torque in Nm. It is common then, to convert rpm to revolutions per seconds and multiply by 2π to get meters per rad, that is to say $P = \tau \times 2\pi \times \dfrac{\omega_{mech}}{60\,\text{s/min}}$. Therefore, with torque in Nm and speed in rpm we can calculate power from

$$P = \tau\omega_{mech} \times \frac{\pi}{30}\ \text{W} \qquad (1.6)$$

Example 1.2

Consider a device delivering 3 kW of mechanical power at 1800 rpm. What is the torque in Nm? Answer: Use Equation 1.6 and solve for torque $\tau = \dfrac{P}{\omega_{mech} \times \dfrac{\pi}{30}}$; evaluate $\tau = \dfrac{3000}{1800 \times \dfrac{\pi}{30}}\text{Nm} =$ 15.92 Nm.

Two cases are relevant to our future discussion in this book. In one case, the force increases the object's elevation thereby giving it *potential energy*, and in the other case, it increases the object's velocity and gives it *kinetic energy*. We will discuss these two forms of energy in the following two sections.

1.1.2 POTENTIAL ENERGY

Think about the work done in lifting a weight of mass m (in kg) against the pull of gravity to a given height h (in m) (Figure 1.4). The mechanical force is mg, where g is acceleration of gravity 9.8 ms^{-2}. Work is then force times height, $W = mgh$. Potential energy becomes available to do work and can be stored to perform this work at a different time as needed, for example, letting the object fall and pull some other mass (Figure 1.4). Upon doing this, the mass loses this potential energy after it drops, transferring to the other object. Power relates to the speed at which we lift that weight or the speed at which we release the mass.

Example 1.3

Consider lifting a mass of 10 kg to a height of 10 m during 5 min at constant speed. What is the work required? What is the power required? What is the stored potential energy? What is the instantaneous potential energy? Answer: Total work is $W = mgh = 10 \times 9.8 \times 10$ N $\times$ m = 980 J. Assuming no loss, this work would be the potential energy E. If the power is constant, we can

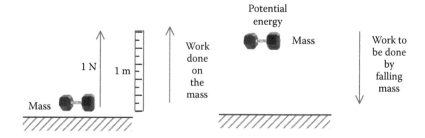

FIGURE 1.4 Mechanical work and potential energy: lifting a mass.

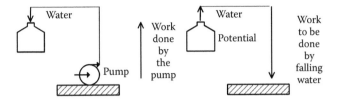

FIGURE 1.5 Pumping water to store potential energy. Release water to perform work.

calculate as total work divided by the interval duration $p(t) = \dfrac{W}{t_f} = \dfrac{980\,\text{J}}{300\,\text{s}} = 3.27\,\text{W}$. Assuming

initial time is 0 and integrating over time $w(t) = \int_0^t p(x)dx = \int_0^t 3.27 dx = 3.27 \int_0^t dx = 3.27t\,\text{J}$ in the time

interval $0 \le t \le 300$ s.

Potential energy of water stored at a substantial height is the basis of one system of hydroelectric power generation, which is a form of renewable energy we will discuss later in Chapter 12. For now, to give a very simple example of the aforementioned concepts, water can be stored in an elevated tank by pumping; work is done by the pump and is converted to potential energy of the water in the tank (Figure 1.5). Power relates to the rate of performing work by the pump, or how fast we lift water. Later, when needed, water is released to perform work by moving the blades of a turbine, which in turn spins the rotor of an electrical generator. We will talk about generators in Chapter 5 and water turbines in Chapter 12. Water flow rate in a pipe draining the tank will determine power (Figure 1.5). In this trivial example, if there were no losses, the work performed by the stored water would be the same as the work done to lift it. However, there are losses due to inefficiencies of the pump, the turbine, and the generator.

1.1.3 KINETIC ENERGY

A force can also impress kinetic energy upon an object by changing its velocity, and the object would keep this energy unless the velocity changes. Mathematically, velocity is given by the derivative of distance x with respect to time $v(t) = \dfrac{dx}{dt}$ and acceleration by the derivative of velocity $a(t) = \dfrac{dv}{dt}$. Thus, Newton's second law $F = ma$ can be rewritten as $F = m\dfrac{dv}{dt}$. The quantity mv or product of mass and velocity is called *momentum*. If the mass m is constant, the derivative of momentum is $\dfrac{d(mv)}{dt} = m\dfrac{dv}{dt}$. Therefore, we can state Newton's second law saying that force is the product of constant mass and rate of change of momentum.

Many times it is convenient to use $\dot{x} = \dfrac{dx}{dt}$ or a dot on top of the symbol to denote time derivative of the variable. It just helps to abbreviate mathematical writing when you have many time derivatives. We will use this convention $\dot{x}$ when convenient to do so, and keep using $\dfrac{dx}{dt}$ when it helps clarify the equations.

To derive an expression for kinetic energy you may think of the infinitesimal work $dw = Fdx$ needed to impress velocity to an object while it travels an infinitesimal distance dx. Divide both sides by dt to get $\dot{w} = F\dot{x} = Fv(t)$. Because of Equation 1.1, we can then say that the power required to move an object at a given velocity is the force times the velocity. Using $F = m\dot{v}$ and dx from the definition of velocity $dx = v(t)dt$, we rewrite work as $dw = Fdx = m\dot{v}v(t)dt = mv(t)dv$ resulting in the simple expression $dw = mv(t)dv$. Now, integrate dw up to time t to get the work that has taken up to

that time $w(t) = \int_0^t dw = \int_0^t mv(t)dv = \frac{1}{2}mv^2(t)$, where we have assumed that mass is constant. Finally,

assume that all this work is converted to kinetic energy and state that

$$e(t) = \frac{1}{2}mv^2(t) \tag{1.7}$$

where $e(t)$ is the instantaneous kinetic energy of mass m moving at velocity $v(t)$.

Example 1.4

Suppose a constant force moves a 10 kg mass from rest to a velocity of 3 m/s for 1 min at constant acceleration (Figure 1.6). What is the kinetic energy of the object after 1 min? What is the power required? What is the work done by the force? Answer: The kinetic energy at the final velocity $e(t) = \frac{1}{2} \times 10 \times 3^2 \frac{\text{kg} \times \text{m}^2}{\text{s}^2} = 45$ Nm = 45 J. Acceleration is $a = \frac{3\text{m/s}}{60\text{s}} = 0.05$ m/s^2 in the interval $0 \le t \le 60$ s. Therefore, force is $F = 10$ kg $\times 0.05$ m/s$^2 = 0.5$ N in this interval. Velocity is a linear function of time $v(t) = at = 0.05t$ m/s in the same interval and power is proportional to $v(t)$ and linear with time $p(t) = Fv(t) = 0.5$ N $\times 0.05t$ m/s $= 0.025t$ W in the interval (Figure 1.6). The work is the integral of power over the interval $w(t) = \int_0^t p(x)dx = 0.025 \int_0^t xdx = 0.025\frac{t^2}{2}$ J a parabolic function of time. At the end of interval, total work is $W = 0.025 \times \frac{60^2}{2} = 45$ J, which is equal to the kinetic energy calculated using the final velocity. These results can be visualized using renpow as we explain next.

We can visualize plots of varying power and work using functions pow.work and pow.work.plot of package renpow. A call to pow.work involves three arguments: a time sequence, a type of power function (linear in this case), and a value of power, which is the slope in this case:

```
t <- seq(0,60,0.1)
x <- pow.work(t,pow="linear",p=0.025)
```

The output contains the values of power and work as a function of time. Then we call the plot function with this output as argument:

```
pow.work.plot(x)
```

This produces Figure 1.7.

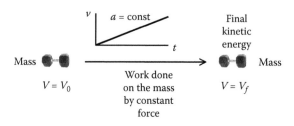

FIGURE 1.6 Moving a mass with constant force produces a linearly increasing velocity that represents an increase in kinetic energy.

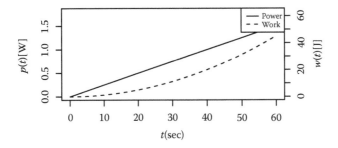

FIGURE 1.7 Work is parabolic for linearly increasing power.

It is useful at this point to mention that we can send graphics organized by pages and panels to pdf as explained in the Appendix (Sections A.3.6 and A.5.5). For instance, the plots of Figure 1.2 and Figure 1.7 can be combined in one pdf with two pages using

```
wd=6; ht=3
outfile <- paste(path,"mech-power.pdf",sep="")
pdf(outfile,wd,ht)
panels(wd,ht,1,1,pty="m")
t <- seq(0,60,0.1)
x <- pow.work(t,pow="const",p=2)
pow.work.plot(x)
x <- pow.work(t,pow="linear",p=0.025)
pow.work.plot(x)
dev.off()
```

where *path* is the path of a directory to store your pdf (say, "output/"). Alternatively, we can produce one pdf with only one page composed of two panels (as shown in Figure 1.8).

```
wd=6; ht=6
outfile <- paste(path,"mech-power-2panels.pdf",sep="")
pdf(outfile,wd,ht)
panels(wd,ht,2,1,pty="m")
t <- seq(0,60,0.1)
x <- pow.work(t,pow="const",p=2)
pow.work.plot(x)
x <- pow.work(t,pow="linear",p=0.025)
pow.work.plot(x)
dev.off()
```

As mentioned in the Appendix (Section A.5.5), you can use this method to send the graphics output of most of the examples we will work with in this book to pdf, and organize it by pages and panels, as you deem necessary and useful.

Kinetic energy can also be converted to other forms of energy and can be used to perform work as long as there is force sustaining the momentum. A simple scenario would be to give sustained velocity to a fluid and use the kinetic energy to move something else. Examples of renewable power systems that use this principle are hydroelectric power generation using the flow of water as in "run-of-river" (Chapter 12) and wind power generation using the kinetic energy in the wind (air is the fluid) (Chapter 13). As a simple example, suppose velocity of the fluid is constant v, then the kinetic energy is $e(t) = \frac{1}{2}mv^2$ and the power would be its rate of change:

$$p(t) = \frac{d}{dt}\left(\frac{1}{2}mv^2\right) = \frac{1}{2}\frac{dm}{dt}v^2 = \frac{1}{2}\dot{m}v^2 \tag{1.8}$$

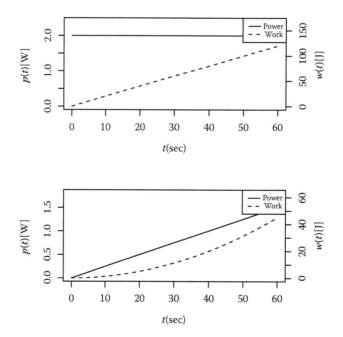

FIGURE 1.8 One-page figure composed of two panels.

The term $\dot{m}$ is the fluid *mass flow rate*. We can determine it by thinking of the *density* ρ of the fluid and a simple geometry as a pipe of cross-sectional area A (Figure 1.9). Density is mass per unit volume, therefore the mass of fluid in an element of volume dV given by area times distance dx is the density times this volume $dm = \rho dV = \rho A dx$ (Figure 1.9). Now use $dx = vdt$ from the definition of velocity, and then $dm = \rho A v dt$. Move dt to the left and obtain mass flow rate as $\dfrac{dm}{dt} = \dot{m} = \rho A v$. Let us now substitute this in Equation 1.8 to calculate power available in this moving fluid $p(t) = \dfrac{1}{2}\rho A v \times v^2 = \dfrac{1}{2}\rho A v^3$, which shows that power available is proportional to the cube of velocity of the fluid:

$$p(t) = \frac{1}{2}\rho A v^3 \tag{1.9}$$

This equation is used often for calculations of wind power generation after adjusting air density according to elevation above sea level, as we will see in Chapter 13.

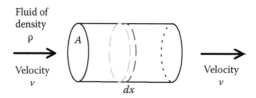

FIGURE 1.9 Kinetic energy and power available for a fluid in motion as wind or running water.

1.1.4 THE WH ENERGY UNIT

As we have discussed, energy is an accumulation of power over time, thus a joule would be W × s. This concept is equally applicable by changing the units of time. Indeed, it is common in electrical power systems to use Watts × Hours = Wh as energy units. We refer to this unit as "watt hour." Please note that it would be incorrect to say "watts per hour" when referring to the units of energy, since the term *per* implies a rate quantity and what we really mean is "watts multiplied by hours." Since J and Wh are both units for energy, let us see how many J are in 1 Wh. Simply take the number of seconds in one hour $60\frac{\text{s}}{\text{min}} \times 60\frac{\text{min}}{\text{h}} = 3600\frac{\text{s}}{\text{h}}$ and convert $3.27 \text{ W} \times \frac{5\,\text{min}}{60\,\text{min}\,/\text{h}} = 0.272$ Wh. Therefore, 1 Wh = 3600 J, a convenient number to remember.

Example 1.5

Calculate work in Example 1.3 giving the value in Wh. Answer: We know that work is 980 J. Converting $W = \frac{980\ \text{J}}{3600\ \text{J/Wh}} = 0.272$ Wh. Alternatively, we know the power is 3.27 W and if we provided it for 5 min, $3.27 \text{ W} \times \frac{5\,\text{min}}{60\,\text{min}\,/\text{h}} = 0.272$ Wh.

For larger amounts, we use kWh, MWh, and GWh. Examples: A 100 W device used for 10 h, would have consumed 100 × 10 Wh = 1 kWh of energy. A 1 MW electric power plant producing power constantly for one year would produce annual energy of $E = 1 \text{ MW} \times 365 \text{ d} \times 24 \text{ h/d} = 1 \text{ MW} \times 8760 \text{ h} = 8.76$ GWh. The number 8760 of hours in a year is a number worth remembering and we will use it frequently.

1.1.5 ELECTROMAGNETIC RADIATION

An *electric field* emanates from electric charges (e.g., an electron) and exerts a force on other charges, whereas a *magnetic field* results from moving electric charges and moments of particles of magnetic materials. The propagation of fluctuating electric and magnetic field at the speed of light carry *radiant energy*, which can be modeled as an *electromagnetic (EM) wave* or as a flux of particles called *photons*. We will refer to the waves or photons as *EM radiation*, indistinctly from the model employed to analyze it. Atoms emit EM radiation due to the rearrangement of previously excited electrons. Radiant energy carried by the EM radiation can in turn be transferred as some other form of energy upon interaction with other matter. For example, energy from *fusion* (a form of nuclear energy) in the sun is transferred to EM radiation emitted at its surface, which when reaching the Earth's surface can transfer energy to the land, ocean water, and plants. EM radiation from the sun is referred to as *solar radiation* or commonly as *sunlight*.

Using the wave model, we employ the frequency ν of the EM fluctuation (in Hz or s^{-1}), which when multiplied by the wavelength λ (length of the wave, in m, Figure 1.10) equals the speed of wave propagation, which is the speed of light c (3×10^8 m/s). The relationship is

$$c = \nu\lambda \tag{1.10}$$

In other words, the frequency would be the number of cycles that go through a fixed point in space every second; wavelength is the distance between a point of the wave and a similar point in the next cycle, say, between peaks (Figure 1.10).

The energy E of a photon of an EM wave is directly related to the frequency ν of the EM wave by the Planck's constant $h = 6.6 \times 10^{-34}$ Js

$$E = h\nu \tag{1.11}$$

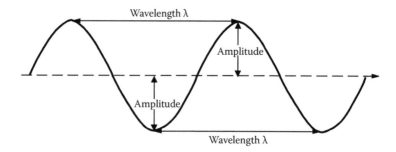

FIGURE 1.10   Wavelength for wave representation of EM radiation.

Using Equation 1.10, we can also write the energy of a photon as

$$E = \frac{hc}{\lambda} \tag{1.12}$$

Thus, photon energy is inversely proportional to wavelength. The shorter the wave, the more energetic the photons. We can express radiant energy as *moles* of photons and scale it per unit area and time to obtain photon *flux* as moles per m^2 per second. Recall that a mole of a substance corresponds to a standard number of elemental units of that substance; this standard number is *Avogadro's number*, which is 6.022×10^{23}.

Example 1.6

What is the frequency and energy of a photon of solar radiation of 1100 nm wavelength? What is the energy of a μmole of these photons? Here n is nano (10^{-9}) and μ is micro (10^{-6}). What is the power per m^2 of a photon flux of one μmole of these photons per second? Answer: The frequency is $\nu = \frac{c}{\lambda} = \frac{3 \times 10^8 \ \text{m/s}}{1100 \times 10^{-9} \ \text{m}} = 272.72$ THz. Here we are using tera (10^{12}). The energy is $E = h\nu = 6.6 \times 10^{-34} \times 272.72 \times 10^{12}$ Js/s = 179.9 zJ. Here we have used zepto (z), which is 10^{-21}. Now using Avogadro's number $10^{-6} \times 6.022 \times 10^{23} \times 179.9 \times 10^{-21}$ J = 0.1083 J. The power of the flux of 1 μmol/(m^2s) would be 0.1083 J/(m^2s) = 0.1083 W/m^2.

In general, EM radiation is a continuous mixture of many wavelengths according to the *EM spectrum*. From short to long waves, or high to low energy, the spectrum includes gamma rays (less than 0.01 nm), x-rays (0.01–1.10 nm), ultraviolet (10–400 nm), visible light (400–700 nm), infrared (700 nm–1 mm), and radio waves (1 mm–100 km). For example, the various wavelength ranges of solar radiation determine important types of radiation in regard to Earth's climate. *Incoming shortwave* solar radiation reaching the Earth's surface includes ultraviolet (UV, 10–400 nm), visible (VIS, ~380–750 nm), and near infrared (NIR, 750–1400 nm). The Earth's surface absorbs incoming radiation and increases its temperature, thereby emitting radiation but at longer wavelengths in what is considered the thermal part of EM radiation (or heat). We will discuss heat and temperature in the next section. This *outgoing longwave* (low energy) emitted by the Earth's surface is in the infrared (IR) (e.g., 4–50 μm). The balance of incoming and outgoing radiation determines Earth's temperature (Chapter 2).

EM radiation interacts with matter according to wavelength. For example, vegetation is able to use radiation within a band of wavelengths or photosynthetically active radiation (PAR), which is radiation in the ~400 to 700 nm range (Chapter 2). A *photovoltaic (PV)* cell, commonly called a solar cell, which is a source of renewable electric power generation, requires that the wavelength of radiation be such that the photon energy overcomes the band-gap energy of *silicon* (Chapter 14).

1.1.6 THERMAL ENERGY: HEAT AND TEMPERATURE

Two fundamental quantities in thermal processes relate to energy, *temperature* and *heat*. Temperature T is a measure of average kinetic energy of molecules in a sample of matter, say, of water in a cup or gas in a tank. It is measured in several scales; among them, we will use absolute temperature in kelvins (K) and degrees Celsius (°C). Absolute temperature is calculated by adding 273 (reference) to the temperature in Celsius. For example, 25°C is $273 + 25 = 298$ K. Temperature measures the ability of that sample to transfer its energy to another object. The actual energy transfer is heat and occurs from a higher temperature (hot) object to a lower temperature (cold). Heat transfer can occur by several processes, for instance, heat *conduction* and *radiation*. Conduction requires solids to touch each other; kinetic energy is transferred from atoms of one solid to the other. Radiation does not require that the objects touch. EM radiates from one object in the IR part of the EM spectrum and is absorbed by the other object. Heat Q is measured in units of energy (J). Upon heat transfer, the hot object's temperature drops (as it loses energy) and the cold object's temperature rises (as it gains energy). Once they achieve the same temperature, heat transfer ceases, and the two objects are at thermal equilibrium.

Heat capacity is a property of material indicating the heat Q required to change the temperature one degree (Celsius or K) and thus is expressed in J/K. For a sample changing its temperature by an amount ΔT we can write $C = \dfrac{Q}{\Delta T}$, where C is heat capacity.

Temperature T is considered an *intensive* property because it does not depend on the amount of the sample under consideration as long as it has the same value throughout the sample. However, C is an *extensive* property because it is proportional to the amount of the sample. A larger amount of hot water can transfer more heat than a small amount. To make an extensive property independent of the amount, we divide by the mass m. Thus, we have *specific* extensive properties, typically denoted by lowercase letters. We then define *specific heat capacity* c as the heat required to change the temperature one degree per g of the sample. We can write

$$c = \frac{C}{m} = \frac{Q}{m\Delta T} \tag{1.13}$$

in units of J/gK. Note that the change of temperature in degrees Celsius (°C) is the same as the change in K, because $\Delta T(K) = (273 + T_f(°C)) - (273 + T_i(°C)) = \Delta T(°C)$. Therefore, c can also be given using temperature in degrees Celsius.

Example 1.7

The specific heat capacity of liquid water is 4.187 J/gK and water vapor is 1.996 J/gK. Compare the heat required to change the temperature of 1 kg of liquid water and water vapor by 2°C. Answer: For liquid water $Q = c \times m \times \Delta T = 4.184 \dfrac{J}{gK} \times 1000\ g \times 2\ K = 8.368$ kJ and water vapor $Q = c \times m \times \Delta T = 1.996 \times 1000 \times 2 = 3.992$ kJ.

1.1.7 CHEMICAL ENERGY

Chemical compounds have internal energy represented by the *bonds* required for the arrangement of atoms and molecules making the compounds. A chemical reaction involves a change of energy related to the differences in bonding energy between *reactants* and *products* (Figure 1.11). When energy is *released*, the change of energy is negative because the reactants have more energy than the products (Figure 1.11, left-hand side). Conversely, when energy is *absorbed*, the change of energy is positive because now the products have more energy than the reactants (Figure 1.11, right-hand side). Therefore, it takes energy to form a compound from others and this energy can be released by

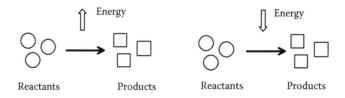

FIGURE 1.11 Chemical reaction. Left: Exothermic, change of energy is negative (released). Right: Endothermic, change of energy is positive (absorbed).

chemical reactions to form other compounds. In Chapter 6, we will explain the concept of *enthalpy* (*H*). For now, think of enthalpy as total energy contained in a compound that can be converted to other forms of energy, such as work, heat, and EM radiation.

To illustrate these concepts, we use the simple example of the formation of liquid water H_2O (l) from hydrogen gas H_2 (g) and oxygen O_2 gas:

$$H_2 \text{ (g)} + \frac{1}{2}O_2(g) \rightarrow H_2O \text{ (l)} \tag{1.14}$$

Here the (g) denotes gas and (l) denotes liquid. This reaction releases energy, and the change of enthalpy (total energy) per mole of hydrogen is $dH = -285.8$ kJ/mol. Therefore, liquid water has less energy than the summation of the energy of hydrogen gas and of oxygen. Because it releases energy, this is an example of an *exothermic* reaction; part of the energy released by this reaction can perform work and the rest is heat. In reaction 1.14, hydrogen is *oxidized* by oxygen. Arranged together with electrodes and electrolytes, this reaction is the basis for a hydrogen *fuel cell*, which we will cover with more detail in Chapter 6.

Recall that a mole of a substance corresponds to 6.022×10^{23} atoms of that substance. A mole is used as a unit in chemistry to convert atom counting to a macroscopic property as mass. For example, one mole of water weighs 18.015 g. Careful with notation! Here g denotes grams and it is different from (g) that denotes gas.

The reverse reaction to reaction 1.14

$$H_2O \text{ (l)} \rightarrow H_2 \text{ (g)} + \frac{1}{2}O_2(g) \tag{1.15}$$

has the same change of enthalpy, but of opposite sign $dH = +285.8$ kJ/mol. This reaction absorbs energy and is an example of an *endothermic* reaction. In reaction 1.15, a water molecule is split into hydrogen and oxygen; it is the *hydrolysis* reaction and requires energy.

Example 1.8

What is the energy released by oxidizing two moles of hydrogen according to reaction 1.14? How much energy is required to hydrolyze 72.06 g of liquid water? Answer: For oxidation of 2 moles the reaction is $2H_2 + O_2 \rightarrow 2H_2O$, which is the typical form of the *combustion* reaction of hydrogen in presence of oxygen (Figure 1.12). In this combustion reaction, hydrogen is the fuel, and water is the combustion product. The energy released is $dH = -2\text{mol} \times 285.8$ kJ/mol $= -571.6$ kJ. Now, for the second question, convert grams to moles $\frac{72.06g}{18.015g/mol} = 4\,\text{mol}$. Thus, hydrolysis of 4 moles is $4H_2O$ (l) $\rightarrow 4H_2$ (g) $+ 2O_2$ and requires $dH = +4\text{mol} \times 285.8$ kJ/mol $= +1143.2$ kJ.

This example shows that increasing the amount of reactants, increases the energy released or absorbed. Molecular weight is used to convert energy per mol to energy per unit mass, typically

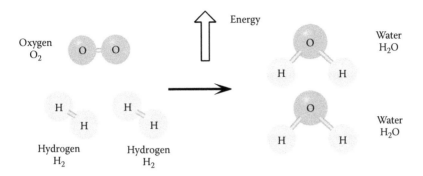

FIGURE 1.12   A simple example of an exothermic reaction. Combustion of hydrogen. Ball-and-stick models are used for illustration only.

kJ/mol to MJ/kg. For instance, one kg of H_2 yields $\dfrac{285.8\,\text{kJ/mol}}{1\,\text{g/mol}} = 285.8\,\text{MJ/kg}$. When using hydrogen to obtain electricity in a fuel cell, the $H_2(g)$ flow rate in g/s will determine the power output. First convert g/s to mol/s, then calculate power taking into account that only part of the energy released can be converted to work.

Example 1.9

Assume a $H_2(g)$ flow rate of 1 mg/s and that 40% of the energy can be converted to electricity. What is the power output? Answer: The flow rate in mol/s is $\dfrac{1\,\text{mg/s}}{1\,\text{g/mol}} \simeq 0.001\,\text{mol/s}$. Then, $dH/dt =$ -0.001 mol/s $\times$ 285.8 kJ/mol $= -0.2858$ kJ/s $= 285.8$ W. But the power output is only 40% of this total energy released or $P \simeq 285.8 \times 0.4 = 114.32$ W.

1.1.8   NUCLEAR ENERGY

Nuclear energy is derived for the binding force that holds the nucleons (subatomic particles) of the atomic nucleus [2,4]. For our purposes, we will only consider protons and neutrons as nucleons. Elements have an *atomic number* Z (number of protons) and *atomic mass* A (sum of number of protons and neutrons). Each proton or neutron has 1 atomic mass unit (1 amu). One amu is 1.66×10^{-27} kg or 1/12 of the mass of the most common carbon atom, with 6 protons and 6 neutrons. An element's nucleus corresponds to the number of protons Z, for instance, carbon has 6 protons. But many elements have several forms or *isotopes* according to the atomic mass [11]. For instance, ^{12}C has 6 neutrons ($A = 6 + 6 = 12$), but ^{14}C has 8 neutrons ($Z = 6 + 8 = 14$).

Example 1.10

Uranium has 92 protons. How many neutrons are in the ^{235}U isotope and in the ^{238}U isotope? Answer: $235 - 92 = 143$ neutrons in the ^{235}U isotope and $238 - 92 = 146$ neutrons in the ^{238}U isotope.

Binding energy per nucleon is greater for the middle value of atomic mass; it decreases for larger or smaller Z. Binding energy increases when lighter nuclei undergo *fusion* or when heavier nuclei undergo *fission*. Currently, nuclear-fueled power plants are based on fission (Chapter 7). To initiate a fission *chain reaction* using uranium, ^{235}U is bombarded by a neutron producing fission into two lighter and unstable isotopes (which decay later to stable forms emitting β and γ rays), and two or three neutrons that can bombard other nuclei and sustain a chain reaction. Several different fission reactions could occur.

For instance, one possible fission reaction of ^{235}U is

$$^{235}U + {}^1n \rightarrow {}^{144}Ba + {}^{89}Kr + 3{}^1n \tag{1.16}$$

Here ^1n denotes one neutron. Mass is not conserved in this reaction, with the mass deficit converted to energy according to $E = mc^2$. Reaction 1.16 produces 73×10^6 MJ per kg of ^{235}U. This is a great amount of energy, for instance, compared to combustion of hydrogen 285 MJ/kg (see previous section), the fission reaction releases 0.25 million times more energy.

Example 1.11

Calculate the energy released by reaction 1.16 [2]. Answer: Calculate atomic mass deficit; the left side is 235.04 + 1.00 = 236.04 amu and the right-hand side is 143.92 + 88.91 + 3.02 = 235.85 amu. The mass deficit is 0.19 amu. Now convert to kg $m = 0.19 \times 1.66 \times 10^{-27}$ kg = 0.315 $\times 10^{-27}$ kg and use $E = mc^2 = 0.315 \times 10^{-27}$ kg $\times (3 \times 10^8$ m/s$)^2 = 2.84 \times 10^{-11}$ J per atom of ^{235}U. Multiply by Avogadro's number to get moles $2.84 \times 10^{-11} \times 6.023 \times 10^{23} = 17.10 \times 10^{12}$ J/mole and now use molar mass to $\dfrac{17.10 \times 10^{12} \text{ J/mol}}{235 \text{ g/mol}} = 73 \times 10^6$ MJ/kg.

1.2 CARBON-BASED POWER SYSTEMS

1.2.1 ENERGY FROM HYDROCARBON COMBUSTION

As we will see shortly, the major source of today's generation of electricity is burning *fossil fuels*. These fuels are derived from oil, coal, and gas that formed from algae, trees, and other living organisms in the geologic past (we will study this with detail in Chapter 2). The fossil matter contains energy stored in chemical form that can be converted to heat by combustion. The common type of compound found in fossil fuels are *hydrocarbons*, which are organic compounds formed entirely by *carbon* C and hydrogen H. For example, *methane* CH_4 has one carbon atom and four hydrogen atoms; *ethane* C_2H_6 has two carbon atoms and six hydrogen atoms (Figure 1.13). Note that as you increase the number of C atoms and keep all bonds single and occupied with hydrogen (a *saturated* hydrocarbon), you form chains with the general formula C_nH_{2n+2}, because there are two H atoms per C atom and one at each end of the chain. These are called *alkanes*.

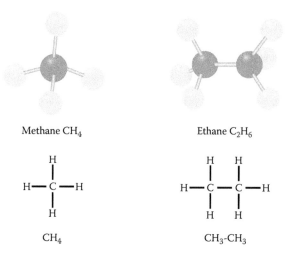

FIGURE 1.13   Examples of simple hydrocarbons: methane and ethane. Ball-and-stick molecular model, name and formula, structure, and condensed structure.

When burned in the presence of sufficient oxygen, the *combustion* products are water vapor H_2O, carbon dioxide CO_2, and energy. As the simplest illustration of hydrocarbon combustion, consider burning methane CH_4 with sufficient oxygen. The reaction is exothermic and given by

$$CH_4 + 2O_2 \rightarrow 2H_2O + CO_2$$

It releases $dH = -802.3$ kJ/mol (Figure 1.14). If there is not enough oxygen, the reaction yields carbon monoxide CO rather than CO_2.

Example 1.12

What is the energy produced by burning one gram of methane? What is the mass of CO_2 produced? Answer: First, convert grams of methane to moles using the molecular formula for methane. Use ≈ 12 g/mol for C and ≈ 1 g/mol for H. Then $\dfrac{1\,\text{g}}{(1 \times 12 + 4 \times 1)\text{g/mol}} = \dfrac{1}{16}$ mol. Then multiply by $dH = -802.3$ kJ/mol to get $dH = -802.3$ kJ/mol $\times \dfrac{1}{16}$ mol $= -50.14$ kJ/g. Now, convert moles of CO_2 to mass, use ≈ 12 g/mol for C and ≈ 16 g/mol for O, so that $\dfrac{1}{16}$ mol $\times (1 \times 12 + 2 \times 16)$g/mol $= \dfrac{44}{16}$ g $= 2.75$ g.

This example illustrates the importance of the relationship between energy released and the mass ratio of CO_2 to fuel combusted. In this simple case, there is a ratio of 2.75:1 of CO_2 to fuel. You can build an intuitive reasoning for this ratio by realizing that in an alkane the C atoms are bonded with H, and when combusted the C atoms are oxidized (bonded with O). Since oxygen weighs more than H (16 to 1), then the mass of CO_2 emitted is larger than the mass of methane combusted. The ratio of CO_2 to fuel allows an estimate of how much CO_2 is emitted per J of energy. Since it takes 1 g of methane to produce 50.14 kJ while releasing 2.75 g of CO_2, we can conclude that there is ~0.055 g CO_2 produced per kJ of energy. Converting to kg and kWh, we obtain ~0.20 kg CO_2 per kWh.

This is an idealized estimate not applicable to real electricity production since we are considering the simplest alkane and ignoring inefficiencies. In general, the combustion of an alkane containing n C atoms is

$$C_nH_{2n+2} + \frac{3n+1}{2}O_2 \rightarrow (n+1)H_2O + nCO_2$$

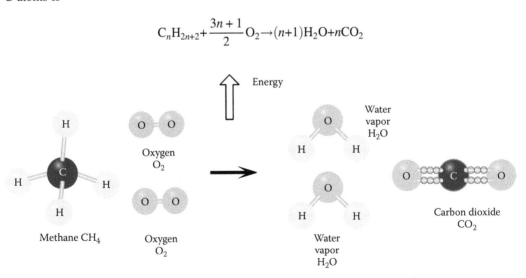

FIGURE 1.14 Simple example of hydrocarbon combustion: methane combustion with sufficient oxygen. Ball-and-stick models are used for illustration only.

The energy released can be estimated from the number of C-C bonds (347 kJ/mol) and C-H bonds (413 kJ/mol). You can see that this combustion will produce n molecules of CO_2 that are typically emitted to the Earth's atmosphere. In Chapters 7 and 9, we will consider coal, natural gas, and oil, which have alkanes with n C atoms and other compounds. Realistic estimates of CO_2 emission from fossil fuel combustion are higher. Typical values are ~0.55 kg/kWh for natural gas, ~0.7 kg/kWh for oil, and ~0.8 kg/kWh for coal.

1.2.2 CARBON, ELECTRICITY, AND CLIMATE

As we just studied, combustion of hydrocarbon (the major source of electricity generation) leads to CO_2 emissions; and as we shall see in Chapter 2, increased atmospheric CO_2 leads to enhanced greenhouse effect and global climatic change. Undoubtedly, Earth's climate system is complex and we still do not fully understand many aspects of how it works. However, our examination of carbon dynamics, atmospheric CO_2, and global temperature in Chapter 2 indicates that at the very least we should proceed with caution and decrease CO_2 emissions by reducing electricity generation from fossil fuels. That is to say, we may follow the "precautionary principle," meaning that in the absence of complete knowledge we are better served by reducing CO_2 emissions in order to avoid potential enhancement of global warming.

Therefore, a strong motivation to develop non-fossil-based power systems is to reduce the consumption of fossil fuels because of the contribution of CO_2 to the atmosphere leading to enhanced greenhouse effect and thus global climate change. In Figure 1.15, we compare two contrasting options to generate electricity. A key aspect of this diagram is that both options are ultimately derived from solar radiation. The path counterclockwise from the sun illustrates the option of burning fossil fuels. As we will discuss further in Chapter 2, these are derived from oil, coal, and gas that formed from biomass in the geologic past. The fossil matter contains energy stored in chemical form that can be converted to heat by burning it. This heat is used to produce mechanical work using a heat engine (Chapter 4), and this mechanical work is used to run an electrical generator (Chapter 5).

This counterclockwise path is then an option that is nonrenewable (depletes the resource used for energy conversion), generates electricity indirectly from the sun, and produces CO_2 emissions. In contrast, in the path clockwise from the sun, a photovoltaic (PV) cell (Chapter 14) provides a renewable source, provides electricity directly from the sun, and does not produce CO_2 emissions. This would seem a perfect option; however, as we will see, there are other environmental issues to this option. As PV-based electric power plants become larger, requirements for land increase with a concomitant impact on land use. Resolving environmental conflicts is not easy and leads to a variety of trade-offs.

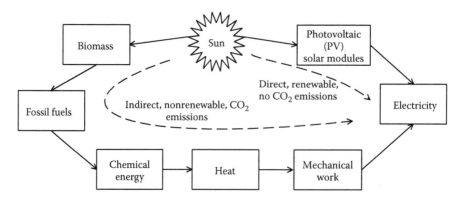

FIGURE 1.15 Comparing two options to generate electricity.

1.2.3 CARBON-BASED AND FOSSIL FUEL–BASED POWER SYSTEMS

World electricity production in 2014 amounted to ~23,815 TWh in a year [12] and represented 15% of total world energy consumption. Electricity production by source [12] shows a reliance on carbon-based sources (Figure 1.16, top panel) with a total contribution of 68.8% of world production by four sources: coal, natural gas, oil, biomass, and waste. As shown in the top panel, the two most important sources are coal (40.8%) and natural gas (21.6%) contributing on their own 62.4% of world's electricity production. They are followed by two major non-carbon-based sources (nuclear and hydroelectric), which combined represent 27% of production. Therefore, nearly 90% of production is by these four major sources, and the carbon based (coal and natural gas) is more than double the noncarbon sources (nuclear and hydroelectric). Out of the remaining ~10%, more than half is also carbon based (oils and biomass and waste, adding up to 6.4%), leaving only 4.2% for other renewables different from hydro (e.g., solar, wind, geothermal). When taking the carbon-based sources only (which represent nearly 70% of the total production) and looking at contributions of each source (Figure 1.16, bottom panel), we see that coal and natural gas alone represent 90% of carbon-based production.

Electricity production by source varies from country to country. For example, in the United States, production by natural gas (33.8%) has surpassed that by coal (30.4%), but their combined contribution 64.2% is similar to the world's value of 62.4% [13]. The two major non-carbon-based sources (nuclear 19.7% and hydroelectric 6.5%) combined represent 26.5% of production. So the percent contribution of these four major sources is lower (80.4%) than the worldwide contribution (~90%). Carbon-based contribution from these four major sources is 2.5 times the amount of noncarbon based. Biomass and waste together with oil amounts to about 2.4%, so carbon-based generation totals about 66.6%, lower than the world total of ~70%. Other noncarbon renewables different from hydroelectric (e.g., solar, wind, geothermal) amount to 6.9%, which is higher than the world total of 4.2% [12].

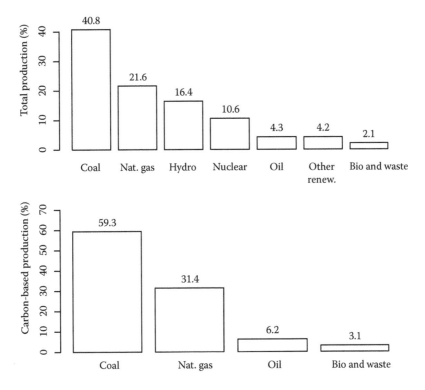

FIGURE 1.16 World electricity production by source. Top: Percent of world production by all sources. Bottom: Percent of world production by source for carbon-based only sources.

1.2.4 Terminology: Clean, Alternative, Renewable, Green, or Sustainable

Being this an introductory chapter, it is worthwhile to consider terms commonly used to describe energy conversion technologies that are more favorable to the environment than conventional technologies [14]. Frequently used terms are "clean," "alternative," "renewable," "green," and "sustainable." These terms are sometimes used loosely speaking to mean the same thing, namely, environmentally friendly, but they are actually different. How do we distinguish among them? Which one of these adjectives best applies to a particular technology?

The method described by Traum [14] to distinguish these terms is based on three properties: (1) the energy resource serving as input to the energy conversion process, (2) the relative consumption to regeneration rates of the resource, and (3) environmental effluents from the energy conversion process. You can think about these properties in terms of their potential reduction of negative effects on the environment: (1) a nonfossil resource reduces carbon emissions and thus helps curb climate change effects; (2) lower consumption rate implies resource conservation or longer time to depletion of the resource; and (3) effluent controls to reduce accumulated impacts on the environment. You could argue that property 1 is subsumed in property 3 because, after all, climate change is an environmental effect. However, separating them accounts for how these terms have evolved over the years.

By *conventional* we mean the processes have low values for all three properties, for example, a coal-fired power plant without emission controls. This process is fossil-fuel based, depletes the resource, and pollutes air and water. A first step-up level is a *clean* process, which is still low in regard to properties 1 and 2, but attempts to curb nonclimate environmental effects; for example, a coal-fired power plant with emission controls for sulfur and particulate matter reduces air quality effects (we will discuss this at length in Chapter 7). *Alternative* is a term applied to fossil-free resource conversion but without improving parameters 2 and 3. The major example is a nuclear power plant, which is non-fossil fuel based, but depletes the resource and has environmental risks.

Next in being more favorable to the environment are *renewable* conversion processes, which are high for characteristics 1 and 2 (non-fossil based and low resource depletion), but not necessarily so for characteristic 3. Examples are hydroelectric power plants (Chapter 12), wind turbines (Chapter 13), and solar panels (Chapter 14). Environmental protection is not necessarily high for these technologies for various reasons; for instance, large PV-based solar plants, wind farms, and hydroelectric power plants have land-use impacts.

The term *green* is mostly a popular term and used almost as synonymous to renewable. However, it is useful to think of it as an additional effort to improve on property 3; for example, PV-based deployments using available space and limiting land-use change, and low-head hydroelectric generation. Finally, we have the term *sustainable*, which would score high on all three properties. Examples are renewable systems that include further efforts to reduce environmental effects, improve efficiency, and repurpose energy waste.

Now that we understand these terms, it is easier to discuss the pros and cons of various energy conversion technologies or processes. We can say that this book consists of a review of clean power systems, plus an introduction of renewable power systems, with some indications of the potential for moving toward green and sustainable processes. The latter is the desirable system where we provide power for prosperity without degrading the environment nor depleting resources, and accounting for equitable availability of power and accommodating for a growing population. A feature of sustainable systems is to form closed loop or self-sustaining interacting high-efficiency processes, where waste from one process becomes a resource for another.

1.2.5 Electrical Power Sources and Conversion

As shown in Figure 1.17 most energy resources are ultimately derived from the sun. We already saw in the last section the difference between fossil and nonfossil resources and this is repeated in the

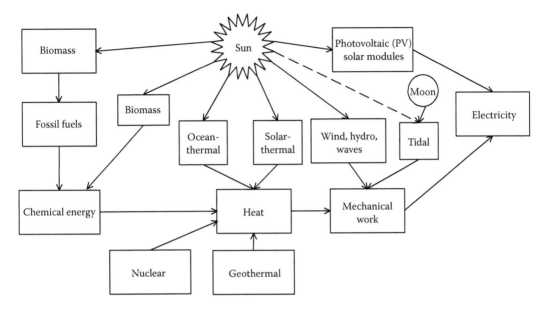

FIGURE 1.17 Energy resources. (Adapted from Tester, J.W. et al., *Sustainable Energy: Choosing Among Options*, second edition, 2012: MIT Press.)

diagram. Take wind for example: solar radiation reaching the Earth's surface creates differentials in atmospheric pressure, which are one of the major drivers of wind. Likewise, hydroelectric power is possible because solar radiation drives the hydrologic cycle; water from the Earth's surface converts to atmospheric water vapor, which condenses and falls as rain. The runoff becomes streamflow, which can be stored in reservoirs and used to generate electricity. Both examples, wind and hydroelectric, represent renewable options that are indirectly derived from the sun and do not generate CO_2 emissions.

As we strive to develop environmentally friendly energy conversion processes, it is imperative that we understand the environmental variables related to the resource (solar radiation, wind speed, and rainfall) and particularly their spatial and temporal variation. As renewable components become part of the electric power grid, there is a need for long-term continuous weather measurements in order to forecast potential production and support grid operation and management. Moreover, understanding environmental effects of various conversion processes is facilitated by environmental monitoring and modeling. This support can be provided by quantitative data analysis [15], as well as simulation and modeling [16], and long-term continuous environmental monitoring [17].

1.2.6 QUANTIFICATION: ANALYSIS AND MODELING

A model is a simplified representation of reality based on concepts, hypotheses, and theories of how the real system works. Some models represent reality as a set of mathematical equations based on the *processes* at work [16], for example, a differential equation representing pollutant transported over time based on pollutant discharge to a stream and streamflow. For this purpose, we use the concept that mass transport is the product of flow and concentration. This method is in contrast to *empirical* models that build a quantitative relationship between variables based on data without an explicit consideration of the process yielding that relation [15]. For example, using regression, we can derive a predictor of pollutant transport as a function of rainfall events of different intensities and durations.

Let us relate these concepts to what we have done in this chapter and what we will do in Chapter 2. Our equations to calculate potential and kinetic energy in the first section of this chapter, as well as the carbon cycle dynamics model and the Earth's radiation balance of the second chapter, are examples of process-based models. The regression models based on time series of atmospheric CO_2 and global temperature of the second chapter are examples of empirical models.

However, we also use empirical models to estimate parameters of the process-based models using data from field and laboratory experiments, as well as monitoring programs. For example, we can use a mechanistic model to calculate flow of a stream using water velocity and cross-sectional area, but we estimate velocity using an empirical relation of velocity to water depth. In addition, we will use empirical models to convert output variables of process-based models to other variables. For example, we could predict tree diameter increments from a process-based model of tree growth, and then convert diameter to height using an empirical relation of height versus diameter. Temporal and spatial dynamics are of paramount importance in renewable power systems in relationship to the environment, and are integrated in mathematical and simulation models to predict future behavior of the environmental system.

Understanding and designing renewable power systems in conjunction with the environment requires interdisciplinary work among scientists and engineers with various backgrounds and training. Electrical engineers would be very familiar with electrical circuit analysis and electric technology. Mechanical engineers would know more about thermodynamics, thermal conversion processes, and machinery. Chemical engineers are better prepared to analyze combustion processes, energy balances, and process control. Civil and environmental engineers are better able to understand hydrological and hydraulic analysis, soils, and structures.

However, all these engineering disciplines have less background on natural resources, ecosystems, and the environment. For example, principles of biology, ecology, geology, and atmospheric science are not typical in engineering curricula, except agricultural and environmental engineering. In this book, I strive to present a broad range of interrelated problems that call for contributions of engineering disciplines together with applied science, such as environmental science, geography, and ecology. All these disciplines contribute to understanding the basic interactions of a power system technology with its environment.

EXERCISES

1.1. Suppose we used 10 W to displace a mass 10 m at constant speed with no friction in 2 min. How much work was consumed? What was the force applied?

1.2. Suppose we can let a 10 kg mass fall at a controlled constant speed from a height of 10 m for 5 min. What is the force? What is the instantaneous power it can deliver? What is the instantaneous work it can perform?

1.3. A prime mover rotates at 3600 rpm to provide enough mechanical power to a 10 MW electrical generator. What is the torque?

1.4. Convert world electricity production in 2014 to J. Hint: For numbers this high, it is convenient to use EJ (E or Exa is 10^{18}). Convert the energy released by combustion of one gram of methane to Wh.

1.5. Assume a photon flux of 1 mmol/(s m^2) composed of equal proportions of 460 nm and 480 nm radiation (in the blue band of visible radiation). Calculate the flux in kW/m^2.

1.6. Suppose all energy from burning one gram of methane is converted to heat and used to raise the temperature of 10 kg of water. What is the change of water temperature?

1.7. What is the energy produced by burning one gram of octane? How many grams of CO_2 are produced? What is the mass ratio of CO_2 to fuel?

1.8. Countries in the Organisation for Economic Cooperation and Development (OECD) produced ~11,000 TWh electricity in 2014 out of the world total [12]. What was the percent production in non-OECD countries with respect to the world total? Non-OECD electricity production in 2014 by source was coal 47.8%, natural gas 19.5%, hydro 19.1%, oil 5.7%, nuclear 4.3%, other renewable 2.3%, biomass, and waste 1.3%. Develop a bar graph. Calculate percent contribution of all carbon-based sources. Compare to the world total and to the United States.

1.9. Suppose we were able to "sequester" all CO_2 produced from fossil fuel combustion to generate electricity thus avoiding CO_2 emission to the atmosphere. Would a coal-fired power plant with this technology be a clean process? Would it be an alternative process? Would it be a renewable process?

REFERENCES

1. Boyle, G., *Renewable Energy: Power for a Sustainable Future*. Third edition. 2017: Oxford University Press. 566 pp.
2. Fay, J.A., and D.S. Golomb, *Energy and the Environment. Scientific and Technological Principles*. Second edition. MIT-Pappalardo Series in Mechanical Engineering, M.C. Boyce and G.E. Kendall. 2012: Oxford University Press. 366 pp.
3. Tester, J.W. et al., *Sustainable Energy: Choosing Among Options*. Second edition. 2012: MIT Press. 1021 pp.
4. Vanek, F., L. Albright, and L. Angenent, *Energy Systems Engineering: Evaluation and Implementation*. Third edition. 2016: McGraw-Hill. 704 pp.
5. Demirel, Y., *Energy: Production, Conversion, Storage, Conservation, and Coupling*. Second edition. Green Energy and Technology. 2016: Springer. 616 pp.
6. Sørensen, B., *Renewable Energy: Physics, Engineering, Environmental Impacts, Economics and Planning*. Fourth edition. 2011: Academic Press. 954 pp.
7. Nelson, V., and K. Starcher, *Introduction to Renewable Energy*. Second edition.Energy and the Environment, A. Ghassemi. 2016: CRC Press. 423 pp.
8. da Rosa, A., *Fundamentals of Renewable Energy Processes*. Third edition. 2013: Elsevier, Academic Press. 884 pp.
9. Goswami, Y., and F. Kreith, editors. *Energy Conversion*. Second edition. Mechanical and Aerospace Engineering, ed. F. Kreith and E.E. Michaelides. 2017, Boca Raton, Florida: CRC Press. 1193 pp.
10. Masters, G.M., *Renewable and Efficient Electric Power Systems*. Second edition. 2013: Wiley-IEEE Press. 690 pp.
11. Graedel, T.E., and P.J. Crutzen, *Atmospheric Change: An Earth System Perspective*. 1993, Freeman: New York. 446 pp.
12. International Energy Agency. *Key Electricity Trends*. 2017. Accessed June 2107. Available from: https://www.iea.org/publications/freepublications/publication/KeyElectricityTrends.pdf.
13. US EIA. What Is U.S. Electricity Generation by Energy Source? 2017. Accessed July 2017. Available from: https://www.eia.gov/tools/faqs/faq.php?id=427&t=3.
14. Traum, M., *Conventional, Clean, Alternative, Renewable, Green, or Sustainable: Formalizing Definitions for Energy Conversion Technology Descriptors*, 2005, University of North Texas.
15. Acevedo, M.F., *Data Analysis and Statistics for Geography, Environmental Science and Engineering: Applications to Sustainability*. 2013: CRC Press, Taylor & Francis Group. 535 pp.
16. Acevedo, M.F., *Simulation of Ecological and Environmental Models*. 2012, Boca Raton, Florida: CRC Press, Taylor & Francis Group. 464 pp.
17. Acevedo, M.F., *Real-Time Environmental Monitoring: Sensors and Systems*. 2015: CRC Press. 356 pp.

2 Environmental Systems, the Carbon Cycle, and Fossil Fuels

This chapter introduces one of the three major themes of this book: environmental systems, emphasizing global cycles, particularly carbon, and ecosystems. This will provide a motivation of understanding carbon-based power generation. We first explain ecosystems concepts, particularly energy flow, and relate these to the global carbon cycle. Then, from the entire carbon cycle, we focus on the atmospheric content of carbon dioxide and its change during the last decades and its relation to global temperature. We demonstrate that these increases are faster than the ones predicted by a simple exponential model. In order to strengthen the reader's background on environmental systems, we provide a brief review of geologic history and of fuels found in sedimentary rocks.

2.1 ECOSYSTEMS AND THE CARBON CYCLE

Often, when we refer to environmental systems we not only consider our surrounding air, water, and soils but also living entities sharing these resources with us. Technically, we could distinguish between the purely nonliving components as our environment, independently from the living components. Both components are, however, interrelated since living creatures interact with their environment. Thus, for practical reasons, ecological interactions become part of our working concept of environmental systems [1].

2.1.1 ECOSYSTEMS

A concept that provides a framework to ecological interactions is that of *ecosystem*, which emphasizes relations of *biotic* components (living) such as animals and plants with *abiotic* factors (nonliving) such as air, light, soil, and water. Key aspects of ecosystems are functional relationships among species focusing on the transfer of material and energy among them, and interactions with the abiotic factors. As a generalization, we are concerned with how materials *cycle* among components of the system and how energy *flows* from one component to another (Figure 2.1). We see energy passage through plants (primary producers), herbivore animals, and carnivore animals. In other words, it flows through *food chains* or *trophic chains*. For simplicity, and to aid modeling and quantification, we have abstracted each one of the energy flow levels in a *compartment* shown in Figure 2.1 as a box. A useful concept is *biomass*, or the mass in these compartments of living entities, and an important component of biomass is carbon (C).

Biomass in these compartments of living entities eventually become dead organic matter, which can return to minerals through decomposition and made available to build plant biomass, as shown on the left-hand side of Figure 2.1. Solar input drives the top part of the diagram on the right-hand side. Using *photosynthesis*, primary producers are able to capture solar radiation to make carbohydrates. There is energy loss at each step, due to respiration, and actually, the energy capture efficiency is small at each step.

Two important concepts that allow quantification of ecosystem dynamics are *stock*, how much we have of that variable at a particular time, and *rate*, how much does the variable change per unit time. For example, plant biomass is a stock, as mass per unit area, say, tons/ha, whereas primary productivity is a rate, say, (tons/ha) yr^{-1}.

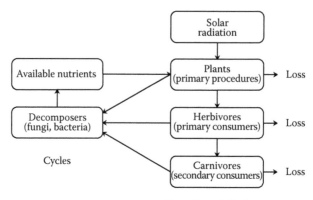

FIGURE 2.1   Nutrient cycle (left) and energy flow (right) in ecosystems.

Armed with the concepts of stock and rate we can establish models of the dynamics of the ecosystem through time. When the rate depends on the stock, we can describe the relationship with a *rate coefficient*, k in yr^{-1}. This rate coefficient multiplies the current stock to obtain the rate in $(tons/ha) yr^{-1}$.

Mathematically, denoting the stock at time t by $x(t)$ and the rate by dx/dt we can write a differential equation $\dfrac{dx}{dt} = kx$ that has a solution $x(t) = x(t_0) \exp(kt)$ at a given time t starting at time t_0.

Example 2.1

Suppose a stock this year, call it year t_0, is 10 tons/ha and it changes with a rate coefficient of 0.01 yr^{-1}. What is the rate at t_0? What is the stock 5 years from now? What is the rate 10 years from now? Answer: Rate at t_0 is $0.01 \times 10 = 0.10$ (tons/ha) yr^{-1}. Projection 5 years from now, say, $t = 5$ years, is $x(t) = 10 \exp(0.01 \times 5) = 10 \exp(0.05) = 10 \times 1.051 = 10.51$ (tons/ha). The rate 5 years from now is $10.51 \times 0.01 \approx 0.11$ (tons/ha) yr^{-1}.

2.1.2   PRIMARY PRODUCTIVITY AND RESPIRATION

Using photosynthesis, primary producers convert solar energy to chemical energy stored in the form of carbohydrates. This process requires carbon dioxide CO_2 and water H_2O and produces oxygen O_2. The chemical reaction is

$$6CO_2 + 6H_2O \xrightarrow{\text{light}} C_6H_{12}O_6 + 6O_2 \tag{2.1}$$

To convert one mole of CO_2 to chemical energy requires eight moles of photons. In terms of joules, it takes 0.472 MJ to convert one mole of CO_2 or 0.00004 MJ to fix 1 mg of C.

One important concept that we will employ in this book discussing various forms of energy conversion is *efficiency*, that is, how much in percent of one form of energy is converted to another. This concept is used in two ways: as its theoretical maximum calculated from first principles, and as a lower practical value taking into account a variety of other considerations.

The aforementioned basic ratio of sunlight energy to chemical energy by photosynthesis (eight moles of photons to one mole of CO_2), leads to a theoretical efficiency of about 30%. However, photosynthesis is wavelength dependent and only occurs for *photosynthetically active radiation* (PAR), which is radiation in the ~400 to 700 nm range. Therefore, only part of the total solar radiation (~45% is PAR) can be converted to carbohydrates. Thus, theoretical efficiency with respect to total solar radiation is only about $0.3 \times 0.45 = 0.13$ or ~13%.

Furthermore, primary producers lose some of the captured energy to *respiration*, which in an opposite manner to photosynthesis, requires oxygen O_2 and produces water and CO_2. This is represented defining *net primary productivity* (NPP) rate as the gross productivity minus respiration. In other words, some of the carbon captured by photosynthesis goes back to atmospheric CO_2 by respiration; the rest is stored, consumed, and decomposed. Once accounting for respiration and other factors (e.g., reflection), efficiency is reduced further to about 1% to 6% of total radiation.

2.1.3 Secondary and Tertiary Producers: Food Chains

Secondary producers are herbivores, or consumers of biomass made by primary producers. They derive energy and make biomass from the compounds contained in the primary producer's biomass. On land, or terrestrial ecosystems, examples include mammals (e.g., deer) that graze on plants. In water, or aquatic ecosystems, examples include zooplankton (typically microscopic, e.g., crustaceans) feeding on algae. Tertiary producers, usually carnivores, are consumers of secondary producers or herbivores. On land, examples include wild cats, snakes, and spiders. In water, examples include predatory fish (e.g., sharks, tuna). A *food or trophic chain* is the system of flow of energy from primary producer, to secondary producer, to tertiary producer. When many species are involved at each level of this chain (i.e., a trophic level) and multiple relationships exist, then we call this a *food web*.

2.1.4 Global Carbon Cycle

Ecosystem concepts are important at the planetary scale as well. We typically group biotic components in the *biosphere* and abiotic components in the *geosphere* comprised of *atmosphere* (gaseous envelope of the planet), *hydrosphere* (water in oceans, lakes, rivers, glaciers), and *lithosphere* (rocks and mineral matter). Cycles and flows are then *global* or occurring at planetary scale, and include relations of the geosphere and biosphere. There are multiple systems in the Earth's biosphere and geosphere, with multiple interactions among its components. For example, the hydrological cycle is of great importance; atmospheric water as vapor is condensed and falls as rain, feeding the soil with water that can be stored for use by the vegetation. Transpiration by plants and evaporation from the soil returns water vapor to the atmosphere.

Importantly, the global carbon cycle involves net primary productivity and other ecosystem processes (Figure 2.2) [2,3]. Carbon dioxide, CO_2, in the atmosphere (X_1) is used by primary producers (e.g., terrestrial plants and algae) to make carbohydrates by photosynthesis, utilizing sunlight.

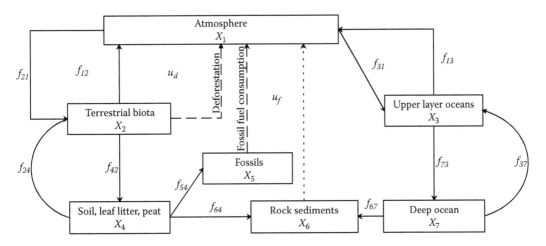

FIGURE 2.2 A simple model of the carbon cycle. Variable X_i represents stock in compartment i and f_{ij} is a transfer rate from compartment j to compartment i.

We have two compartments—one for terrestrial biota X_2, another for the upper layer of the ocean X_3—and two fluxes from the atmosphere to these compartments f_{21}, f_{31}. We denote f_{ij} as flux from compartment j to i. Some of the carbon goes back to atmospheric carbon (CO_2) by respiration (fluxes f_{12}, f_{13}); the rest is stored, consumed, and decomposed (fluxes f_{42}, f_{73}) and part is recycled (fluxes f_{24}, f_{37}). At slower rates, carbon transfers to fossil organic matter, sediments, and sedimentary rocks (fluxes f_{54}, f_{64}, f_{67}). Sedimentary deposits contain most of the global carbon. Human action accelerates release of CO_2 by burning fossil fuels (coal, oil, gas) at rate u_f and from the terrestrial biota by deforestation at rate u_d. The time scale is mixed: Some processes occur rapidly as the exchanges between atmosphere and *biota*, and others slowly, such as sedimentation. We will use a time scale of years and gigatons or Gt for carbon amount in compartments. The fluxes are then in Gt/yr. One Gt is one peta g or 10^{15} g.

Estimated values for the compartments X_1, X_2, ..., X_7 are 740, 1760, 1000, 1500, 10000, 20 million, and 39000 Gt, respectively. These values are just approximations.

Example 2.2

Let us calculate atmospheric carbon using net rates of inflow and outflow assuming that these rates do not vary with the stocks in the compartment. Assume that the flow from rocks to atmosphere is negligible. What is the total flow of C into the atmosphere and what is the total flow of C out of the atmosphere? What would be the C content in the atmosphere 50 years from now? Answer: Calculate flux out of the atmosphere as uptake from land and oceans $F_{out} + F_{to\_land} + F_{to\_oceans} = 102 + 92 = 194$ Gt/yr. Calculate flux into the atmosphere as emissions from land biota, ocean biota, deforestation, and burning fossil fuels: $F_{in} + F_{from\_land} + F_{from\_oceans} + F_{from\_def} + F_{from\_fossil} = 100 + 90 + 2 + 5.3 = 197$ Gt/yr.

The difference in flux into and out gives us net rate of change: $\Delta F = F_{in} - F_{out} = 3.3$ Gt/yr. Content 50 years from now: $C_{50yr} = 740$ Gt $+ 3.3$ Gt/yr $\times 50$ yr $= 905$ Gt.

Atmospheric carbon or carbon dioxide in the atmosphere contributes to determine the greenhouse effect, which in turns affects global air temperature, as we will discuss in a couple of sections. Let us first examine existing data on changes of atmospheric carbon.

2.2 CARBON DIOXIDE IN THE ATMOSPHERE AND GLOBAL TEMPERATURE

2.2.1 Increasing Atmospheric CO_2 Concentration

An important piece of our knowledge of planetary carbon dynamics comes from the measurement of atmospheric CO_2 concentrations recorded in Mauna Loa, Hawaii [4,5]. A visit to the website of NOAA's Global Monitoring Division [6] will inform us of recent values of monthly average of CO_2 concentration in parts per million (ppm). For example, at the time of this writing, the most recent value was for May 2017 and was 409.65 ppm, ~2 ppm up from 407.70 ppm for the same month the previous year (May 2016).

Concentration in ppm express dry air mole fraction defined as the number of molecules of carbon dioxide divided by the number of all molecules in air, including CO_2 itself, after water vapor has been removed [6]. The May 2017 value of 409.65 ppm represents a mole fraction of 0.000409. On the website, we can see a graph of CO_2 in ppm as monthly average and its trend (seasonal correction) for the last 5 years of record. The trend is calculated by a moving average of seven (an odd number) adjacent seasonal cycles centered on the month to be corrected [6]. The trend changes from 395 to 406 ppm in 5 years, which is an average increase of ~2 ppm/year.

Besides the graph, the website offers the data for download. Figure 2.3 illustrates the CO_2 trajectory for the entire record of measurement (since March 1958) using the data downloaded from this website. From the data set we plot the same two lines shown at the website. The dashed line represents the monthly average values (centered on the middle of each month), which fluctuates up

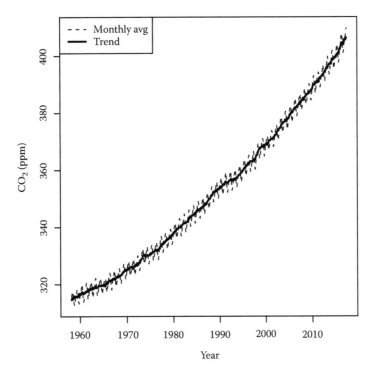

FIGURE 2.3 Monthly mean CO_2 at Mauna Loa, entire record from 1958 to 2017. (Data downloaded from NOAA, Trends in Atmospheric Carbon Dioxide, 2017. NOAA, Earth System Research Laboratory, Global Monitoring Division, accessed June 2017, available from: http://www.esrl.noaa.gov/gmd/ccgg/trends/.)

and down during the year according to the seasons. Removing the average of this seasonal cycle yields the solid line that shows a clear accelerating increase during the entire record. Figure 2.4 uses a time window of the most recent 10 years (2007–2017) so that we can visualize the graphs more clearly.

Example 2.3

The plots in Figures 2.3 and 2.4 were produced using R as shown in this example. We will learn how to read data using R, convert to time series, and plot. First, look at a segment of the downloaded file:

```
CO2 expressed as a mole fraction in dry air, micromol/mol, abbreviated as ppm
#
# (-99.99 missing data; -1 no data for #daily means in month)
#
#      decimal   average  interpolated  trend   #days
#        date               (season corr)
1958  3  1958.208   315.71    315.71      314.62   -1
1958  4  1958.292   317.45    317.45      315.29   -1
1958  5  1958.375   317.50    317.50      314.71   -1
1958  6  1958.458   -99.99    317.10      314.85   -1
```

The file has 72 lines of useful information before the first record (March 1958). We will use columns 5 and 6 that have the completed values by interpolation and the seasonal corrected trend. We read the downloaded data file into a dataset or data frame, which is organized by columns and can be manipulated by R functions. Use function read.table described in the Appendix. Assuming you copied folder extdata in your working folder as described in the Appendix you could type.

```
CO2.mo <- read.table("extdata/CO2monthly.txt",skip=72)[,5:6]
names(CO2.mo) <- c("Monthly Avg","Trend")
```

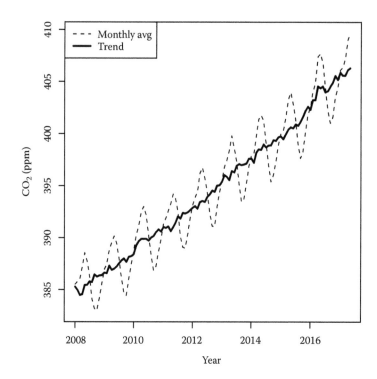

FIGURE 2.4 Monthly mean CO_2 at Mauna Loa, 2008–2017. (Data downloaded from NOAA, Trends in Atmospheric Carbon Dioxide, 2017. NOAA, Earth System Research Laboratory, Global Monitoring Division, accessed June 2017, available from: http://www.esrl.noaa.gov/gmd/ccgg/trends/.)

or alternatively

```
CO2.mo <- read.table(system.file("extdata","CO2monthly.txt",package="renpow"),
skip=72)[,5:6]
names(CO2.mo) <- c("Monthly Avg","Trend")
```

Notice that we skip the first 72 lines of the file and only use columns 5 and 6. We named the two variables accordingly. The first 10 records show us that we have correctly imported the data into a data set:

```
> CO2.mo[1:10,]
   Monthly Avg Trend
1     315.71 314.62
2     317.45 315.29
3     317.50 314.71
4     317.10 314.85
5     315.86 314.98
6     314.93 315.94
7     313.20 315.91
8     312.66 315.61
9     313.33 315.31
10    314.67 315.61
>
```

As explained in the Appendix, many of data files used in the book are included as data of the renpow R package. For instance, the CO2 monthly data can be easily accessed by simply typing CO2monthly at the console. A shortcut then to build the data frame used here would be `CO2.mo <- CO2monthly[,5:6]`.

We now convert to a time series starting month 3 of 1958 with monthly frequency:

```
CO2.mo.ts <- ts(CO2.mo, start=c(1958,3), frequency=12)
```

The first few lines show that we now have the data stamped by a month and year:

```
> CO2.mo.ts
      Monthly Avg Trend
Mar 1958    315.71 314.62
Apr 1958    317.45 315.29
May 1958    317.50 314.71
Jun 1958    317.10 314.85
Jul 1958    315.86 314.98
Aug 1958    314.93 315.94
Sep 1958    313.20 315.91
Oct 1958    312.66 315.61
Nov 1958    313.33 315.31
```

Now we plot and obtain Figure 2.3.

```
ts.plot(CO2.mo.ts,type="l",lty=2:1,lwd=1:2,xlab="Year",ylab="CO2 (ppm)")
legend('topleft',leg=names(CO2.mo),lty=2:1,lwd=c(1,2))
```

Figure 2.4 shows the same variables but over the most recent 10 years. It was obtained by a window of the time series:

```
recent10yrs <- window(CO2.mo.ts,start=c(2007,5), end=c(2017,5))
ts.plot(recent,type="l",lty=2:1,lwd=1:2,xlab="Year",ylab="CO2 (ppm)")
legend('topleft',leg=names(CO2.mo),lty=2:1,lwd=c(1,2))
```

2.2.2 EXPONENTIAL INCREASE

An interesting feature of Figure 2.4 is that the trend is nonlinear, that is, it displays an increase of the rate of change over time. What is the nature of this accelerated growth? A first thought is that the rate is itself proportional to the concentration $X(t)$, so that as concentration increases, so does the rate. This is modeled as a linear ordinary differential equation (ODE). Using the derivative of X with respect to time t for the rate of change of X we write

$$\frac{dX(t)}{dt} = kX(t) \tag{2.2}$$

where the coefficient k is a *per unit* rate of change or *rate coefficient*.

The solution of an ODE, like Equation 2.2, is a function X that satisfies the ODE. A solution is found by separating terms in X and t in different sides of the equation and integrating both sides. Start by dividing both sides by X to obtain the per unit rate of change $\frac{1}{X}\frac{dX}{dt} = k$. Then move dt to the right-hand side (multiply both sides by dt) and integrate both sides between the initial time t_0 and final time t_f, $\int_{x(t_0)}^{x(t_f)}\frac{1}{X}dX = \int_{t_o}^{t_f}kdt$. The integral on the left is the natural log

$$\ln\left[X(t_f)\right] - \ln[X(t_0)] = \ln\left[\frac{X(t_f)}{X(t_0)}\right] = k(t_f - t_0) \tag{2.3}$$

and now invert the log to obtain the exponential function for X at the final time t_f

$$X(t_f) = X(t_o)\exp\left[k(t_f - t_0)\right] \tag{2.4}$$

which can be calculated once we have the initial condition $X(t_0)$. It is common practice to let $t_0 = 0$, and t_f as variable t. In this manner, we get

$$X(t) = X(0)\exp(kt) \tag{2.5}$$

as the solution to the ODE. This a commonly used model for many processes, and we will refer to it simply as the exponential model.

Look at Equation 2.3 again and focus on the right-hand side equality. It says that the log of the ratio of final and initial values is proportional to the time interval. In other words, the log of the ratio should plot as a straight line with respect to time. Solving for the rate coefficient k, we see that the slope of this line is the rate coefficient:

$$k = \frac{\ln\left[\dfrac{X(t_f)}{X(t_0)}\right]}{t_f - t_0} \tag{2.6}$$

An interesting derived concept from an exponential model is the doubling time t_d. By making $X(t_d) = 2X(t_0)$ and $t_d = t_f - t_0$ we can write $k = \dfrac{\ln(2)}{t_d} = \dfrac{0.693}{t_d}$ or $t_d = \dfrac{0.693}{k}$.

Example 2.4

Let us see how to apply this concept to the CO_2 data using annual data, which is also available on the NOAA website [6]. A file downloaded from the website is available as CO2annual.txt in folder extdata. Then we read the data file into a data frame as in Example 2.3, from your local extdata or from the package by using system.file. Then, calculate time as a difference in years, and calculate the natural logarithm of the ratio of CO_2 concentration at year t to CO_2 concentration at $t = 0$ (initial value at 1959). Alternatively, simply do CO2.yr <- CO2annual since CO2annual is a data set in renpow.

```
CO2.yr <- read.table("extdata/CO2annual.txt",skip=57)[,1:2]
names(CO2.mo) <- c("Year","Annual Avg")
t <- CO2.yr[,1]-CO2.yr[1,1]
y <- log(CO2.yr[,2]/CO2.yr[1,2])
```

and run a linear regression with zero intercept to obtain the slope

```
lm(y ~ 0+ t)
```

The results indicate that the slope is 0.0039 per year and that this is a good estimate (R^2 is close to 1, and p-value is negligible).

```
> summary(lm(y ~ 0+ t))

Call:
lm(formula = y ~ 0 + t)

Residuals:
     Min        1Q    Median        3Q       Max
-0.0167316 -0.0119660 -0.0071152 -0.0002329 0.0238539

Coefficients:
  Estimate Std. Error t value Pr(>|t|)
t 0.0039023 0.0000433   90.12  <2e-16 ***
—
Signif. codes:  0 '***' 0.001 '**' 0.01 '*' 0.05 '.' 0.1 ' ' 1

Residual standard error: 0.0109 on 57 degrees of freedom
Multiple R-squared: 0.993,   Adjusted R-squared: 0.9929
F-statistic: 8121 on 1 and 57 DF, p-value: < 2.2e-16

>
```

This means a 0.39% per year rate coefficient, which translates to a doubling time of 0.693/0.0039 ≈ 177 years. A double CO_2 concentration with respect to a reference year is often used as the scenario for climate change modeling.

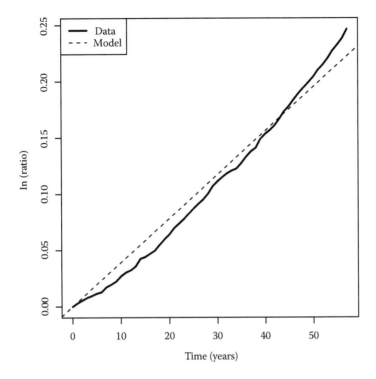

FIGURE 2.5   Determining possible exponential increasing trend of CO_2.

However, plotting this log of ratios,

```
plot(t,y, type="l",ylab="ln(ratio)",xlab="Time (Years)",lwd=c(2,1))
abline(lm(y ~ 0+ t),lty=2)
legend('topleft',leg=c("Data", "Model"),lty=1:2,lwd=c(2,1))
```

we realize that it cannot be approximated to a straight line (Figure 2.5)! Therefore, our first guess of a linear rate of change as proposed by Equation 2.2 is not a good enough approximation, and the rate may be a nonlinear function of the concentration. This implies that prediction of doubling time would be off and the doubling time could be shorter. The trajectory before $t \sim 46$ years (~2005) indicates underprediction and it seems that thereafter the concentration increases faster.

We will have more to say later about carbon cycling, its relation to atmospheric carbon dioxide, and a better estimate of doubling time, but for now let us focus on global temperature.

2.2.3   Earth's Atmosphere and Energy Balance

Earth's atmosphere is a gaseous envelope, held by gravitation, that extends for 10,000 km from the surface of the planet, but it is concentrated near the surface of the Earth. In addition to *gases*, it contains small particles in suspension termed *aerosols*. Almost all (99%) of the gases are nitrogen (78%) and oxygen (21%) with concentrations that are relatively constant. The rest includes argon (~0.9%), water vapor and *trace gases* such as methane (CH_4), carbon dioxide (CO_2), and ozone (O_3); these are present in small quantities, are more variable, and affect weather and climate. Out of the trace gas total, CO_2 is the most abundant representing ~94%, while methane is only 0.43%. However, methane is a more powerful absorber and emitter of longwave EM radiation and therefore an important contributor to climate. Particles come from volcanic eruptions, salt spray, fires, and dust storms. Particles serve as condensation nuclei for clouds and can absorb or reflect sunlight. Water vapor stays mostly near the surface, is spatially variable, and can form clouds.

The most abundant trace gas, CO_2, is important in photosynthesis as we described in the previous two sections and it is an emission product of fossil fuel combustion. CO_2 absorbs infrared radiation thus preventing heat to escape to space, which is the essence of the *greenhouse* (GH) effect. A good absorber at a certain wavelength is also a good emitter at that wavelength, therefore heat absorbed by the GH gases (infrared absorbers) is emitted back as heat and can reradiate back to the Earth's surface, which warms the surface, producing more heat release from the surface, therefore leading to warming.

Methane, CH_4, less abundant than CO_2, can absorb and emit longwave EM radiation ~30 times more effectively than CO_2 and therefore is a much more powerful greenhouse gas and has potentially greater impact on warming [7]. Methane is released to the atmosphere by a variety of natural processes occurring on land (e.g., termites), in the oceans (e.g., microorganisms in the seafloor), and in inundated ecosystems (e.g., decomposition in wetlands). These emissions are mostly offset by natural uptake processes. However, CH_4 atmospheric emissions due to human activities have increased. These correspond, for instance, to cultivating rice under inundated conditions and decay of solid waste in municipal landfills. CH_4 is the main component of natural gas, which is one of the most important fossil fuels for electricity production (Chapter 9).

The GH effect is a natural process that occurs on other planets as well. Mars has very little and thus its average temperature is low, but Venus has too much and therefore has a hot surface. On Earth, human activities enhance the GH effect; these activities are primarily emission of CO_2 from fossil fuel burning and deforestation, and methane from agricultural activities and landfills.

Incoming solar radiation reaching Earth is distributed by wavelength; increasing for short wavelengths from ultraviolet (UV) to visible, reaching a peak in the visible range, and then decreasing as wavelength increases. As Earth's surface warms, it emits *outgoing radiation* in longer infrared (IR) waves. Earth's average temperature results as a balance of incoming and outgoing radiation. A fraction of the incoming shortwave radiation reflects back by reflective surfaces like clouds, snow, and particles. The coefficient representing this fraction is termed *albedo*. A fraction of the outgoing longwave radiation is reradiated to Earth due to the GH effect. Thus, we can reduce warming by lowering of GH atmospheric concentrations by energy conservation and reforestation.

An extremely simple model of energy balance for planet Earth is given by the equation for incoming and outgoing radiation (Figure 2.6) [8]:

$$\frac{P_{solar}}{4}(1-\alpha) = \sigma T^4(1-f) \tag{2.7}$$

P_{solar} is solar irradiance received by Earth 1380 W m^{-2}, which is a power density or power per unit area. Coefficient α is albedo or reflectivity (incoming reflected loss), f is the factor due to GH (fraction of outgoing radiation trapped by GH effect), σ is the Stefan-Boltzmann constant 5.6704×10^{-8} W m^{-2} K^{-4}, and T is temperature of the Earth in kelvins. Outgoing radiation is calculated here as σT^4 using blackbody radiation.

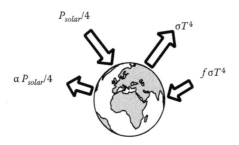

FIGURE 2.6   A very simple model of energy balance for Earth.

Example 2.5

Assume albedo is α = 0.28 and f = 0.39. What is the Earth's temperature at equilibrium in °C?
Answer: Solve for T in Equation 2.7

$$T = \left(\frac{\frac{P_{solar}}{4}(1 - \alpha)}{\sigma(1 - f)} \right)^{1/4} = \left(\frac{\frac{1380 \frac{W}{m^2}}{4}(1 - 0.28)}{5.67 \times 10^{-8} \frac{W}{m^2} K \times (1 - 0.39)} \right)^{1/4} = \left(\frac{248.4}{3.45 \times 10^{-8}} \right)^{1/4} K = 291.3 \text{ K}$$

Convert to °C: $T = 291.3 - 273 = 18.3°C$.

Please realize that this is an extremely simple calculation and cannot be used to predict changes. Also, it is not intended to prove or disprove the GH effect on planetary temperature. In the next section, we will examine the trends in global temperature from existing data.

2.2.4   GLOBAL TEMPERATURE: INCREASING TREND

Now let us visit NASA's Global Climate Change website [9]. You can see a graph of the change in global surface temperature in the period 1880 to 2016 or 135-year time series. Global surface temperature refers to average over land and ocean. The 135-year record is expressed as an anomaly or difference relative to the 1951–1980 average temperature. Using data downloaded from the website we can visualize graphs as an annual average and as a 5-year average (Figure 2.7). We see a clear increasing trend and positive anomaly after 1980, that is in the last 36 years, rising to 0.99°C above

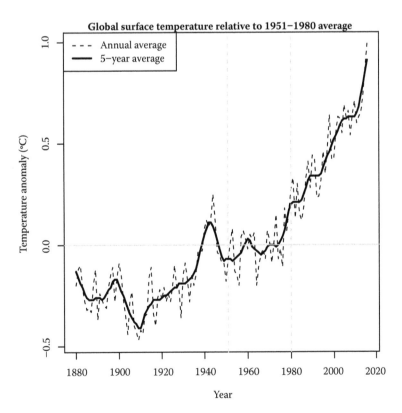

FIGURE 2.7   Global average temperature of Earth 1880–2016 as anomaly with respect to the 1951–1980 average. Produced using data downloaded from [9].

the 1951–1980 average. The 10 warmest years in the examined record have occurred in the last 16 years or since 2000 (except 1998).

It is interesting to do your own analysis of the data. Examine the data that has been downloaded and saved as a text file extdata/global-temp-anomaly.txt. The first few lines are as shown here.

```
1880    -0.20    -0.13
1881    -0.12    -0.16
1882    -0.10    -0.19
1883    -0.21    -0.21
1884    -0.28    -0.24
1885    -0.32    -0.26
1886    -0.31    -0.27
```

The first column is the year, the second column is the annual average, and the third column is the 5-year average.

You can use the following lines of code in R to read the data file GlobTempAnom.txt, create a data set, convert to time series, and plot as shown in Figure 2.7:

```
gt <- read.table("extdata/GlobTempAnom.txt",col.names=c("Year","Avg","Avg5yr"))
gt.ts <- ts(gt[,2:3],start=1880, deltat=1)
ts.plot(gt.ts,type="l",lty=2:1,lwd=1:2,xlab="Year",ylab="Temperature
Anomaly (°C)")
abline(h=0, col='gray')
abline(v=1951,lty=2,col='gray'); abline(v=1980,lty=2,col='gray')
legend("topleft",lty=2:1,,c("Annual average","5-year average"), lwd=c(1,2))
title("Global Surface Temperature Relative to 1951-1980 Average",cex.main=0.9)
```

Recall that alternatively you could have used system.file to access the file or simply used the renpow dataset and assign gt <- GlobTempAnom.

Example 2.6

Does the global temperature data show an exponential increase? We can proceed as we did for CO_2 in Example 2.4. We add an arbitrary positive value (+1.0) to the anomaly to make all anomaly values positive:

```
t <- gt[,1]-gt[1,1]
gtp <- gt[,3]+1
y <- log(gtp/gtp[1])
summary(lm(y ~ 0+t))
```

The results indicate an estimate of $k = 0.003$ per year, with poor R^2 (0.57) but significant (negligible p-value).

```
> summary(lm(y ~ 0+t))
Call:
lm(formula = y ~ 0 + t)
Residuals:
     Min      1Q   Median      3Q     Max
-0.47859 -0.19951 -0.13312 0.03278 0.37738
Coefficients:
   Estimate Std. Error t value Pr(>|t|)
t 0.0030072 0.0002222  13.54  <2e-16 ***
—
Signif. codes: 0 '***' 0.001 '**' 0.01 '*' 0.05 '.' 0.1 ' ' 1
Residual standard error: 0.2046 on 136 degrees of freedom
Multiple R-squared: 0.574,   Adjusted R-squared: 0.5708
F-statistic: 183.2 on 1 and 136 DF, p-value: < 2.2e-16
```

This means a 0.3% per year rate coefficient, which translates to a doubling time of $0.693/0.003 \approx$ 231 years. However, we know the R^2 is not very good. Moreover, by plotting this log of ratios,

```
plot(t,y,type="l",ylab="ln(ratio)",xlab="Time (Years)",lwd=c(2,1))
abline(lm(y ~ 0+ t),lty=2)
legend('topleft',leg=c("Data", "Model"),lty=1:2,lwd=c(2,1))
```

we realize that a straight line is not at all a good estimate (Figure 2.8)! Therefore, the rate may be a nonlinear function of temperature. This implies that prediction of doubling time would be off and the doubling time could be shorter. The trajectory after $t \sim 100$ years (~ 1980) indicates a faster increase.

2.2.5 DOUBLY EXPONENTIAL INCREASE

We just saw that simple exponential growth $X(t) = X(0)\exp(kt)$ cannot account for data on a CO_2 increase. There are various ways of modeling a faster increase than the one predicted by an exponential. One of these is by a doubly exponential

$$X(t) = X(0)\exp(\exp(kt) - 1) \tag{2.8}$$

where you take the exponential of an exponential, meaning the rate coefficient is itself increasing exponentially. More flexibility in fitting data is to use two rate coefficients

$$X(t) = X(0)(\exp(k_1 \exp(k_2 t)) - \exp(k_1) + 1) \tag{2.9}$$

where we add the term $-\exp(k_1) + 1$ to force the function $X(t)$ through $X(0)$ for $t = 0$.

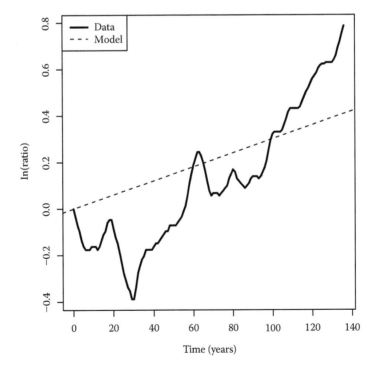

FIGURE 2.8 Determining possible exponential increase of global temperature.

How do we calculate doubling time in a doubly exponential model? Make $X(t_d) = 2X(0)$ in Equation 2.9 and group terms $2 + \exp(k_1) - 1 = \exp(k_1 \exp(k_2 t_d))$. Now take the natural log of both sides $\ln(2 + \exp(k_1) - 1) = k_1 \exp(k_2 t_d)$ divide by k_1 and take the logarithm again to get $\ln\left(\dfrac{\ln(2 + \exp(k_1) - 1)}{k_1}\right) = k_2 t_d$ and solving for t_d we get

$$t_d = \ln\left(\frac{\ln(2 + \exp(k_1) - 1)}{k_1}\right)/k_2 \qquad (2.10)$$

We use this opportunity to introduce nonlinear regression, which consists of a numerical procedure to minimize the error in the fit of a function to the data. A convenient way of doing this is to use function nls of R.

Example 2.7

Use the CO_2 annual data and apply the doubly exponential model. Update the estimate of doubling time. First, use quick trial-and-error to see that k_1 must be around 0.1 and k_2 around 0.01 so that we have initial estimates of the rate coefficients:

```
CO2.yr <- read.table("extdata/CO2annual.txt",skip=57)[,1:2]
names(CO2.mo) <- c("Year","Annual Avg")
t <- CO2.yr[,1]- CO2.yr[1,1]
y <- CO2.yr[,2]; y0 <- y[1]
dexp <- nls(y~ y0*(exp(k1*exp(k2*t))-exp(k1)+1),start=list(k1=0.1,k2=0.01))
summary(dexp)
```

Of course, alternatively you could have used system.file to access the file or simply used the renpow dataset and assign CO2.yr <- CO2annual.

We can look at the nonlinear regression summary:

```
>> summary(dexp)
Formula: y ~ y0 * (exp(k1 * exp(k2 * t)) - exp(k1) + 1)
Parameters:
   Estimate Std. Error t value Pr(>|t|)
k1 0.1700861 0.0056174   30.28  <2e-16 ***
k2 0.0141638 0.0003972   35.66  <2e-16 ***
-
Signif. codes: 0 '***' 0.001 '**' 0.01 '*' 0.05 '.' 0.1 ' ' 1
Residual standard error: 0.9003 on 56 degrees of freedom
Number of iterations to convergence: 6
Achieved convergence tolerance: 5.68e-06
>
```

Note that the coefficients have values $k_1 = 0.170$ and $k_2 = 0.014$. Now we predict the values of CO_2 using these coefficient and plot for comparison:

```
yest <- predict(dexp)
plot(t,y,   type="l",ylab="CO2   (ppm)",xlab="Time   (Years)",lwd=c(2,1),ylim=c
(300,400))
lines(t,yest, lty=2)
legend('topleft',leg=c("Data", "Model"),lty=1:2,lwd=c(2,1))
```

The result shown in Figure 2.9 illustrates a good fit to the data. Then calculate doubling time using Equation 2.10:

```
> k1=0.17; k2=0.014
> log(log(2+exp(k1)-1)/k1)/k2
[1] 108.9817>
```

This means our estimate is now ~109 years instead of 177 years as obtained by the simple exponential.

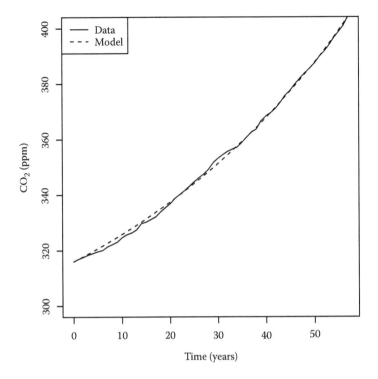

FIGURE 2.9 Modeling CO_2 increase using the doubly exponential.

At the time of this writing, CO_2 just passed the 400 ppm mark. Looking back in time, we may guess that 109 years ago (first decade of last century) the concentration would have been half of this value or about 200 ppm. Looking ahead in time, double CO_2 concentration with respect to 1959 ($\sim 2 \times 315 = 730$ ppm) would be obtained in year 2068. These are predictions based only on the trend and do not take into account modifications due to changing emissions.

Example 2.8

Apply the doubly exponential to the global temperature data:

```
# global temp data
gt <- read.table("extdata/GlobTempAnom.txt",col.names=c("Year","Avg","Avg5yr")
# time in years
t <- gt[,1]-gt[1,1]
# shift up to make positive
gtp <- gt[,3]+1
y <- gtp; y0 <- y[1]
# non linear doubly exp
dexp <- nls(y~ y0*(exp(k1*exp(k2*t))-exp(k1)+1),start=list(k1=0.1,k2=0.01))
```

As before, you could have used system.file to access the file or simply used the renpow dataset and assign gt <- GlobTempAnom.

In this case the coefficient values are $k_1 = 0.013$, $k_2 = 0.030$ as we can see from the summary:

```
> summary(dexp)
Formula: y ~ y0 * (exp(k1 * exp(k2 * t)) - exp(k1) + 1)
```

```
Parameters:
  Estimate Std. Error t value Pr(>|t|)
k1 0.013354  0.003474  3.844 0.000185 ***
k2 0.030248  0.002038 14.840 < 2e-16 ***
—
Signif. codes: 0 '***' 0.001 '**' 0.01 '*' 0.05 '.' 0.1 ' ' 1
Residual standard error: 0.1135 on 135 degrees of freedom
Number of iterations to convergence: 16
Achieved convergence tolerance: 5.258e-06
```

Estimate and plot

```
yest <- predict(dexp)
# shift back down by 1
ya <- y-1; yaest <- yest-1
plot(t,ya, type="l",ylab="Anomaly",xlab="Time (Years)",lwd=1,ylim=c(-1,1))
lines(t,yaest, lty=2,lwd=1)
legend('topleft',leg=c("Data", "Model"),lty=1:2,lwd=c(1,1))
abline(h=0, col="gray")
```

to get Figure 2.10. Then calculate doubling time

```
> k1=0.013; k2=0.030
> log(log(2+exp(k1)-1)/k1)/k2
[1] 132.8552
```

which means ~133 years instead of the ~231 years we obtained when using the simple exponential. To put this in perspective we would double the current anomaly (~+0.9°C) of 2016 as we enter the next century.

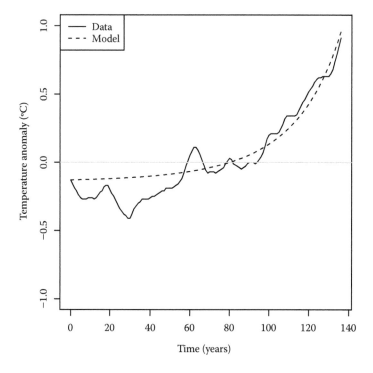

FIGURE 2.10 Modeling temperature increase using the doubly exponential.

2.3 GEOLOGIC HISTORY AND AGE OF FOSSIL FUELS

2.3.1 GEOLOGIC HISTORY

Studying fundamental aspects of Earth's geologic history would help us understand our fossil fuel resource, particularly the long time it takes to form this resource, which makes their use non-renewable. At the same time, this study helps us to obtain a perspective on long-term climatic changes. Succinctly, Earth is about 4.54 billion years old and its geologic history comprises successively smaller frames of time: *eons*, *eras*, and *periods*. A large proportion (4 billion) of Earth's geologic history was formerly known as the Precambrian and is now divided in individual eons (Hadean, Archean, and Proterozoic), which we will not discuss here.

For our purposes, we are more interested in the most recent eon, the *Phanerozoic*, that goes back to about 0.54 billion years ago, which corresponds to the appearance of macroscopic fossils. This eon comprises three eras: *Paleozoic*, *Mesozoic*, and *Cenozoic*. We will focus on some *periods* of each one of these eras: the *Carboniferous* and *Permian* periods of the Paleozoic, the Jurassic period of the Mesozoic, and the *Quaternary* of the Cenozoic (Figure 2.11).

The Paleozoic was the era in between 540 and 252 million years ago. The word *Paleo* means "old or ancient" and *zoic* refers to "animal." Keep in mind that we will only give approximate numbers here. We will abbreviate million years ago or before the present by *Mya*. In the Paleozoic, we find the Carboniferous period in between 360 and 280 Mya, which corresponds to large accumulation of remains of algae and zooplankton on the seafloor, as well as deposits of remains of large plants. Recall that algae and zooplankton are in the food chain of aquatic ecosystems; these remains under heat and pressure will become oil and gas. Remains of large plants did not decay after dying; these remains became coal under pressure and heat. You can see that the word *Carboniferous* contains *carbon* for the element making up large amounts of oil and gas (hydrocarbons) and coal formed in this period. Toward the onset of the Carboniferous, continental plates started coming together to form a supercontinent named *Pangaea*. Naturally, oil and gas formed at different times in different parts of the Earth. For example, in Alaska, oil started forming toward the end of Paleozoic and continued during the Mesozoic. Let us next describe more details about the Mesozoic.

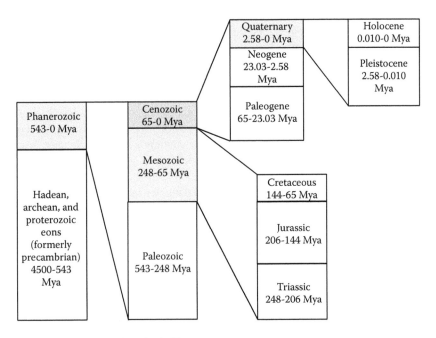

FIGURE 2.11 Brief summary of geologic history.

The Mesozoic is the era in between 252 and 65 Mya, known for dominance of reptiles and conifers. The word *Meso* means "middle" and it applies here in regard to life. The Mesozoic comprises three periods: *Triassic* (252–206 Mya), *Jurassic* (206–144 Mya), and *Cretaceous* (144–65 Mya). Toward the end of the Triassic, dinosaurs started to become dominant in the terrestrial environment and were prevalent during the Jurassic. Pangaea started to break up during the Jurassic. Oil in Alaska formed by the same process described earlier during the Triassic and Jurassic.

The end of the Cretaceous, or last period of the Mesozoic, corresponds to a very large mass extinction of plant and animals on Earth, including dinosaurs, and is marked by the *Cretaceous–Paleogene* (K–Pg) boundary (formerly called the Cretaceous–Tertiary K–T boundary) (65 Ma). This is the beginning of the Cenozoic, or more recent era, culminating in the Quaternary (2.58 Ma to the present). We will have more to say about the Cenozoic and Quaternary in connection with climate. For now, let us turn our attention to understanding the age of deposits that are used for fossil fuels.

Oil and natural gas are contained in "reservoir rocks" where they were trapped in between grains of the sediments (that gave origin to the rock) together with other substances (e.g., water, salt). In fact, *petroleum* is a term also used for oil and gas and this word means "rock oil." Natural gas contains simple hydrocarbons and "crude" oil contains complex hydrocarbons. Natural gas is a fuel to generate electricity, whereas crude oil is refined to make fuels, for example, gasoline and diesel, used mostly for transportation but also for electricity generation.

As tectonic plates moved and collided, large forces pushed reservoir rocks up and down, lifting or burying them, thus yielding petroleum reservoirs at different depths below the surface of the Earth, sometimes even showing at the surface. Similarly, coal deposits moved with tectonic forces. Now that we have a basic overview of fossil fuel formation and ages, let us look briefly at the methods to date rocks.

2.3.2 Radiometric Dating

Much of our knowledge of the age of fossils used for fossil fuels comes from understanding Earth's geologic history, building inferences on what species of plants and animals were present in different periods. The distribution of fossils along a sequence of layers of sedimentary rocks together with the age of these layers help reconstruct this history. Using measurements based on *radioactive decay* it is possible to date the age of rocks and fossils.

Let us briefly review radioactive decay. Recall from Chapter 1 that many elements have several forms or isotopes according to the atomic mass [8]. Radioactive decay is the emission of particles or electromagnetic radiation by unstable isotopes and due to disintegration of nuclei. One form of decay is emission of an alpha particle α (a nucleus with two protons and two neutrons), thereby reducing the atomic number by two and the atomic mass by four yielding a different element. Another decay process is to emit an electron (or beta particle β) yielding a proton from a neutron, therefore increasing the atomic number by one but leaving the atomic mass about the same. Those elements that have isotopes, parent or unstable (e.g., ^{235}U) and daughter (e.g., ^{207}Pb), can be used for radiometric dating by measuring the ratio of parent to daughter isotopes. For this, we use the half-life of an isotope, that is, how long it takes half of the atoms to decay to the daughter form. Dating rocks by isotopes helps confirm some of the hypothesized relations among fossils and rocks. Let us look at how this is calculated.

Consider the variable $X(t)$ to be the number of atoms of an isotope at time t. The simplest assumption is linearity or first-order decay stating that the rate of decay is proportional to the number of atoms

$$\frac{dX(t)}{dt} = -\lambda X(t) \qquad (2.11)$$

where λ is a decay rate constant. We encountered a similar equation in our first attempt to describe increase of CO_2 in the atmosphere. From that discussion, recall that Equation 2.11 is a linear ODE. The same equation applies to first-order degradation of compounds due to various chemical, physical, or biological mechanisms and agents to drive the reaction, for instance, oxidation by chemicals, photolysis by light, and biodegradation by bacteria.

As we demonstrated earlier in this chapter, the linear model 2.11 has an exponential solution

$$X(t) = X_0 \exp(-\lambda t) \tag{2.12}$$

where $X_0 = X(0)$ is the initial number of atoms. Use t_h to denote the half-life (time it takes for the concentration to decay to one-half of the initial concentration) and substitute it in the solution 2.12 with $X(t) = (1/2)X_0$. Thus we obtain $1/2 = exp(-\lambda t_h)$. Now take the natural log of both sides, $\ln(1) - \ln(2) = \ln(\exp(-\lambda t_h))$. Then $-\ln(2) = -\lambda t_h$, the negative sign cancels on both sides, and finally solve for the half-life

$$t_h = \frac{\ln(2)}{\lambda} = \frac{0.693}{\lambda} \tag{2.13}$$

as a function of the decay rate coefficient λ.

Example 2.9

Assume the parent/daughter $^{235}U/^{207}Pb$ pair with a half-life of 7.1×10^8 years. What is the decay rate constant? Answer: $\lambda = \frac{\ln(2)}{t_h} = \frac{0.693}{7.1 \times 10^8 \text{ yr}} = 9.76 \times 10^{-10} \text{ yr}^{-1}$.

So far, we have used R for functions and programs. However, please note that it can also be used for quick calculations.

```
> format(0.693/(7.1*10^8), scientific=TRUE)
[1] "9.760563e-10"
>
```

This half-life range is useful to date rocks from about 10 Mya to the age of the Earth was formed. In its simplest form, we can date a rock in the following manner. Denote by $X(t_a)$ the measured concentration of the parent isotope at a time t_a representing the age of the rock. Assume that when the rock was formed all isotopes were of the parent form with a concentration X_0; therefore, we can use Equation 2.12 to write $X(t_a) = X_0 \exp(-\lambda t_a)$, move X_0 to the left-hand side, take log $\ln(X/X_0) = -\lambda t_a$ and solve for t_a to get

$$t_a = -\frac{\ln(X/X_0)}{\lambda} \tag{2.14}$$

Example 2.10

Suppose the parent/daughter $^{235}U/^{207}Pb$ pair and that we measure a ratio of 0.9 of parent to total (parent plus daughter) isotope concentration. What is the age of the rock? Answer: Use Equation 2.14 and the rate coefficient calculated in Example 2.9, $t_a = -\frac{\ln(0.9)}{9.76 \times 10^{-10}} = 1.086 \times 10^8 \text{ yr} \sim 109 \text{ Myr}$ or a rock formed during the Cretaceous. This calculation can easily be performed in R.

```
> format(-log(0.9)/(9.7*10^-10), sci=TRUE)
[1] "1.086191e+08"
>
```

However, the assumption that there were no daughter isotopes when the rock was formed is not necessarily correct. To address this problem, we use a stable isotope (nonradiogenic and thus constant concentration through time) of the element and calculate the ratio of the parent and daughter to the stable isotope.

Another application of isotopes is to use the ratio of two isotopes in a sample to a standard. This technique is the basis of carbon-14 (^{14}C) or radiocarbon dating, which can be used to date remains of living material. Most carbon atoms are ^{12}C but the isotope ^{14}C is formed in the atmosphere from nitrogen by cosmic ray action. It has a half-life of ~5700 years, and therefore it is too short lived to date fossil fuels [10]. The delta ^{14}C is $\delta^{14}C = \left(\dfrac{\left(^{14}C/^{12}C\right)_{sample}}{\left(^{14}C/^{12}C\right)_{standard}} - 1 \right) \times 1000$, which is expressed in per thousand. Carbon 14 oxidizes to $^{14}CO_2$, which is taken by plants the same way as ^{12}C. Therefore, the isotopic ratio can be used to date remains of living material. This isotope is also used as a tracer to understand C cycle pathways and rates.

2.3.3 THE QUATERNARY PERIOD AND CLIMATE

There is a large amount of iridium in clays deposited at the time of the K–Pg boundary (~65 Mya), or transition between the Cretaceous and Tertiary, which coincides with large species extinction including dinosaurs. Iridium is a material not commonly found on the Earth's surface and more common in asteroids and molten rock deep in the Earth. It has been hypothesized that the unusual large quantity of iridium came from the impact of a large meteorite on Earth or alternatively by intense volcanic activity. Either one of these causes would have led to high concentration of particulate matter in the atmosphere, thus blocking solar radiation and reducing primary productivity, which would have shut down the food chains and eventually led to mass extinction, including dinosaurs. Current consensus favors the meteorite impact as the cause of mass extinction.

Starting at the K-Pg boundary, Earth went into its most recent Era, the Cenozoic, comprised by the *Paleogene* (65–23.03 Mya), the *Neogene* (23.03–2.58 Mya), and *Quaternary* (2.58 Ma–present) periods. The Cenozoic was formerly divided into the Tertiary and Quaternary. During the Paleogene and Neogene there were substantial rearrangements of land masses to roughly today's continent configuration and a number of climatic events that could shed light on the climate of the past [8]. However, there is more information on what may have been the climate during the Quaternary; thus, we turn our attention to this period of the Cenozoic. The Quaternary is comprised by the Pleistocene (2.58 Mya–10 Kya), characterized by periodic glaciations and large mammals, and the Holocene (10 Kya–present), the most recent, characterized by a warming trend since the end of the last glacial period.

Periodic glaciations occurred with cycles of periods of 100 kyr, 41 kyr, and 23 kyr, determined by dynamics of Earth's orbital parameters: eccentricity, precession, and tilt [8]. In Chapter 14, we will encounter these cycles again when we study solar radiation. Data from air bubbles trapped in Antarctic ice cores dating back 160 kyr indicated a strong relationship between CO_2, CH_4, and temperature. The latter is estimated by the ratio of stable isotopes of oxygen and hydrogen. Of interest are ^{18}O and ^{2}H or deuterium. These isotopes are less abundant in cold periods compared to warm periods. Further drilling for ice cores in Antarctica have now extended the available data to 800 kya from Dome C, and 400 kya from Vostok [11].

Such analysis indicates that for the last 800 kyr, atmospheric CO_2 has fluctuated between 170 ppm (~660 kya) and 300 ppm (~330 kya) according with glacial and interglacial periods. These values are given as ppm by volume [12]. As we studied earlier in this chapter, atmospheric CO_2 has increased considerably in recent times, reaching today's values of 400 ppm and increasing at about 2 ppm/year. Such a look into the geologic past allows an assessment that the high values of CO_2 observed today are above the range of natural variation observed during the Pleistocene and could be due to human activities including combustion of fossil fuels.

2.3.4   OTHER FUELS IN SEDIMENTARY DEPOSITS

The fossil fuels used for electricity production, coal, natural gas, and oil (Chapter 1) are the ones most understood in terms of their reserves, extraction, utilization, and conversion technologies. A large amount of methane is stored in the ocean floor in the form of *methane hydrates* and has come into the discussion of a potential yet untapped source of natural gas. *Clathrates* are solids composed of one molecule providing a "cage" structure that encloses another type of molecule. When the cage molecule is water, we have a *hydrate*; when the trapped molecule is a gas we have a *gas hydrate*. The resulting solid is similar to water ice. Methane hydrates are gas hydrates where the gas is methane. They are also called methane ice.

Methane hydrates form under specific conditions of high-pressure from the water column (greater depths) and cold temperatures and predominantly on the continental seafloor. They have been found in the ice-cores drilled in Vostok and Dome C providing data on methane concentration looking back 800 kya. Researchers are investigating technologies to liberate the gas trapped in the hydrate so that it can be used for power production. It is estimated that methane hydrate deposits in the ocean floor exceed the known deposits of natural gas. Potential exploitation of methane hydrates is controversial since inadvertent destabilization of the deposits would liberate the methane to the atmosphere, not only wasting the potential resource, but also leading to accelerated greenhouse effect and global warming. Recall that methane is a much more powerful greenhouse gas than CO_2. We will discuss methane hydrates further in Chapter 9 when we talk about natural gas.

2.4   SHORTENING THE CYCLE AND SEQUESTERING CARBON

We have just studied that fossil fuels were formed from geologic forces operating for a long time on remains of energy captured by ecosystems on land (coal formed from plants) and water (petroleum formed by algae and zooplankton). In addition, we have also seen how combustion of fossil fuels lead to increased CO_2 concentration in the atmosphere. Several approaches exist to mitigate these two issues.

Regarding the possibility of shortening the time to store energy chemically in matter that can be used as fuel, two major approaches are to use *biomass* produced by nature and to devise *solar fuels* (Figure 2.12). We will talk more about these approaches in Chapter 9. The tenant of the biomass approach is that primary producers are constantly taking CO_2 from the atmosphere, as we saw at the beginning of this chapter. The *solar fuel* approach is to emulate nature and use sunlight to derive fuels but in a shorter time scale, for example, to produce a synthetic chemical *energy carrier* (fuel) from sunlight through photochemical reactions. This differs from converting sunlight to electricity

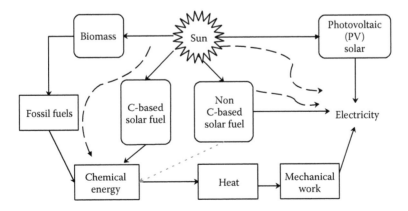

FIGURE 2.12   C-based would still exchange CO_2 with the atmosphere (uptake and release). Non-C-based could be used as carrier (fuel) and offer an opportunity for direct conversion to electricity.

using photovoltaics (PV) (Chapter 14) because the carrier can be stored and transported, very much like fossil fuels.

Carbon sequestration attempts to capture the CO_2 and avoid emission to the atmosphere. Methods include indirect methods, such as forest growth, geologic storage, and conversion to stable materials. We will discuss this topic further in Chapter 7.

EXERCISES

2.1. Consider the C cycle given in Figure 2.2. Assume $f_{12} = 100$, $f_{21} = 102$, $f_{13} = 90$, $f_{31} = 92$, $u_d = 2$, and $u_f = 5.3$ Gt/yr. Assume that the flow from rocks to atmosphere is negligible compared to fluxes from the biota and fossil fuel combustion. What is the total flow of C into the atmosphere and what is the total flow of C out of the atmosphere? Assuming these fluxes remain constant, what would be the C content in the atmosphere 50 years from now?

2.2. In the previous problem, assume that we reduce C emission from fossil use by 1% with respect to 5.3 Gt/yr. What would be the C content in the atmosphere in 50 years? What is the percent difference between this value and the one calculated in the previous problem?

2.3. Assume a simple energy budget of planet Earth (Figure 2.6). Assume the albedo is $\alpha = 0.28$ (incoming reflected loss is 28%) and $f = 0.39$ (outgoing trapped loss by GH effect is 39%). What is the Earth's temperature in °C?

2.4. Assume we reduce GH due to CO_2 and other active gases by 1% (relative) to the value given above (that is to say reduce by 1% of 0.39). What would be the new Earth's temperature? What is the percent change with respect to the one calculated in the previous problem? Give answer for all cases in °C.

2.5. Use annual data from the NOAA's Earth System Research Laboratory, Global Monitoring Division website for CO_2 measured at Mauna Loa [6]. Calculate the rate coefficient for the 1960–2005 interval and for the 2005–2017 interval. Compare.

2.6. Research about origin, formation process, and age of oil. Summarize what you learned taking into account these questions: What type of biota contributed to oil formation? What range of temperatures is required for oil formation? Are there distinctive geologic ages of oil formation? What is the duration of these ages in million years? What biological compound from plants has been proposed to date oil deposits?

2.7. Research basic facts about coal deposits in the United States and U.S. coal mine production. Approximately, what was the U.S. annual production of coal from mines in recent years? What was the coal production of the Appalachian or Eastern region? Focusing on this region, in which segment of geologic history is this deposit located? How old is this coal deposit? How does the sulfur content vary as a general trend? What is the primary use of this coal?

2.8. Consider the parent/daughter $^{238}U/^{206}Pb$ pair with a half-life of 4.5×10^9 years or 4.5 Gyr [8]. What is the decay rate constant? Suppose there is 60% parent atoms to total (parent plus daughter). What is the age of the rock? In what geologic period was the rock formed?

REFERENCES

1. Christopherson, R.W., and G.E. Birkeland, *Geosystems: An Introduction to Physical Geography*. 10th edition. 2018: Pearson. 610 pp.
2. Gates, D.M., *Climate Change and Its Biological Consequences*. 1993: Sinauer. 280 pp.
3. Acevedo, M.F., *Simulation of Ecological and Environmental Models*. 2012, Boca Raton, Florida: CRC Press, Taylor & Francis Group. 464 pp.
4. Vaughan, H. et al., Monitoring long-term ecological changes through the Ecological Monitoring and Assessment Network: Science-based and policy relevant. *Environmental Monitoring and Assessment*, 2001. 67: 3–28.
5. Lovett, G.M. et al., Who needs environmental monitoring? *Frontiers in Ecology and the Environment*, 2007. 5(5): 253–260.

6. NOAA. Trends in Atmospheric Carbon Dioxide. 2017. NOAA, Earth System Research Laboratory, Global Monitoring Division. Accessed June 2017. Available from: http://www.esrl.noaa.gov/gmd/ccgg/trends/.

7. US EPA. Landfill Methane Outreach Program (LMOP). 2017. Accessed August 2017. Available from: https://www.epa.gov/lmop.

8. Graedel, T.E., and P.J. Crutzen, *Atmospheric Change: An Earth System Perspective.* 1993, Freeman: New York. 446 pp.

9. NASA. Global Temperature. 2017. Accessed June 2017. Available from: http://climate.nasa.gov/vital-signs/global-temperature/.

10. Harte, J., *Consider a Spherical Cow: A Course in Environmental Problem Solving.* 1988: University Science Books. 283 pp.

11. CDIAC. 800,000-Year Ice-Core Records of Atmospheric Carbon Dioxide (CO2). 2017. Accessed July 2017. Available from: http://cdiac.ornl.gov/trends/co2/ice_core_co2.html.

12. Lüthi, D. et al. *EPICA Dome C Ice Core 800KYr Carbon Dioxide Data.* IGBP PAGES/World Data Center for Paleoclimatology Data Contribution Series # 2008-055. NOAA/NCDC Paleoclimatology Program, Boulder CO, USA. 2008. Available from: http://cdiac.ornl.gov/trends/co2/ice_core_co2.html.

3 Fundamentals of Direct Current Electric Circuits

This chapter is one of several providing a very basic review of those major concepts of electrical circuits that we need in order to understand electrical power systems. In this chapter, we review basic electrical quantities and circuits, introducing Ohm's law and the fundamentals of circuit analysis methods including Kirchhoff's voltage and current laws, nodal and mesh analysis, and Thévenin and Norton theorems. In preparation for understanding power transfer, we cover modeling of voltage and current sources and the basics of maximum power transfer. In this chapter, we will focus on direct current (DC) circuits, and later, in Chapter 5, we introduce alternating current (AC) circuits. This topic is expanded in Chapter 8 discussing AC circuit analysis and power calculation for AC circuits, and in Chapter 10 when we introduce transformers and three-phase circuits. Basic DC circuit analysis is covered in many textbooks used in introductory circuit analysis courses and can serve as supplementary material [1–3]. Reviews are also available in textbooks devoted to renewable energy [4,5].

3.1 BASICS OF ELECTRIC CIRCUITS

3.1.1 PRINCIPLES OF ELECTRICAL QUANTITIES

Electrical *charge* is a fundamental property of matter that can both generate and interact with electromagnetic fields. Charge can be positive or negative; at the subatomic level, protons represent positive charge, whereas electrons have negative charge. The unit of charge is the coulomb or C, where 1 C is the equivalent charge of 6.2×10^{18} electrons. In a conductor, free electrons can flow and represent a movement of negative charge.

Voltage is the potential energy difference between two points in an electric field, measured per unit charge. Being potential energy means it is available to perform the work of moving a unit charge against an electric field. Intuitively, voltage is the energy available to cause electrons to flow through a conductor. Its unit is volt or V, which is defined as joule/coulomb (J/C), and is named volt in honor of Alessandro Volta. In general, work and charge vary with time. Denoting charge by q, voltage by v, and work by w, the definition of voltage is

$$v(t) = \frac{dw}{dq} \tag{3.1}$$

Current is charge (electrons) flow rate through a material (e.g., a conductor) or in other words, the rate of electric charge motion through a conductor. Its unit is the ampere (A), or "amp" for short, in honor of André M. Ampère. One amp is equivalent to is one coulomb of electrons passing by a given point in a circuit in 1 second of time. That is to say 1 A is 1 C/s. Denoting current by $i(t)$, we can use the derivative with respect to time:

$$i(t) = \frac{dq}{dt} = \dot{q} \tag{3.2}$$

As we can see, voltage and current are related by the fundamental notion of flow of charge and work required to move it. Multiplying Equations 3.1 and 3.2 we obtain a very interesting result:

$$v(t)i(t) = \frac{dw}{dq}\frac{dq}{dt} = \frac{dw}{dt} = p(t) \tag{3.3}$$

In short, multiplying voltage and current we obtain power! Make sure you relate the last step in this derivation to the very first equation of Chapter 1. Dimensionally, V × A or $\dfrac{J}{C} \times \dfrac{C}{s}$ = W. In other words, electric power is the rate of change of work required to move charge at a given rate. This relation is a fundamental equation that we use throughout this book:

$$p(t) = v(t)i(t) \tag{3.4}$$

Thus, high power can occur due to high voltage and low current, or to low voltage and high current. Neither voltage nor current alone constitute power. We need both. If $v(t) = 0$, then the power will be zero (no potential to do work), and if $i(t) = 0$, then the power will be zero (no charge movement).

Example 3.1

Suppose voltage is 12 V constant for 1-day (24 h) interval during which current drawn is 1 A for the first half-day and 0 A later. What is the power drawn (consumed) during the day? What is the energy consumed at the end of the day? Answer: During the first half-day, power is 12 × 1V × A = 12 W. During the second half-day, the power is 12 × 0 V × A = 0 W. Therefore

$$p(t) = \begin{cases} 12 \text{ W in } 0 \le t \le 12 \text{ h} \\ 0 \text{ W in } 12 \text{ h} \le t \le 24 \text{ h} \end{cases}$$

The energy consumed or work done by the voltage source is

$$e(t) = w(t) = \int_0^{24 \text{ h}} p(x)dx = 12 \int_0^{12 \text{ h}} dx + 0 \int_{12h}^{24 \text{ h}} dx = 12 \times 12 \text{ Wh} = 144 \text{ Wh}.$$

Relating these electrical concepts to mechanical energy and power described in Chapter 1, voltage is analogous to the work done in lifting a weight against the pull of gravity; current is analogous to the movement of that weight. Similarly, voltage is analogous to water stored in an elevated tank; current is analogous to water flow in a pipe draining the tank.

3.1.2 ELECTRIC CIRCUITS

The most common form of constant voltage, current, or power is *direct current* (DC), which refers to the case where these quantities remain the same during a period in consideration. There is no change of polarity. This is the type of electricity provided by a battery. We use uppercase V, I, and P to signify constant voltage, current, and power values, respectively, in DC. This type of electrical quantities is in contrast to a time-varying voltage, current, or power defined as *alternating current* (AC), which refers to a periodic variation of these quantities. The most common form is that of a sinusoidal variation. Polarity changes in a cyclical fashion. This is the type of power supplied by an AC generator. AC is introduced in Chapters 7 and 8.

Various devices are made to generate, store, and consume electrical power according to specific relationships of voltage and current. The most basic devices are *circuit elements* that dissipate power (resistors, explained in this chapter) or store energy (capacitors and inductors, explained in Chapter 5). An *electric circuit* is a combination of devices (e.g., circuit elements) connected by electrical conductors (e.g., wires). Voltage is measured or defined across a pair of *nodes* or points of equal potential energy of a circuit, whereas current is defined through a *loop* of a circuit. We will define these concepts more precisely later. However, for now, Figure 3.1 illustrates the simplest case of two nodes and one loop. In this simple circuit, there is one power *source* and one *load* or power consumption element. By convention, polarity of current is opposite to electron flow. Polarity of

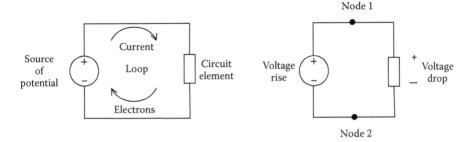

FIGURE 3.1 Conventions. Current flow direction is opposite to electron flow. Potential drops from positive to negative.

voltage is considered a *drop* as it goes from + to − (the load side), whereas it is a *rise* going from − to + (the source) (Figure 3.1).

The diagram of Figure 3.1 shows the typical situation where power is consumed by a load when the current is drawn from the source or supply at a rated voltage V_s, which remains nearly constant as we change the load. This is a type of source is called a *voltage source* (Figure 3.2). We can distinguish another type of source, a *current source*, which maintains a nearly constant current I_s, regardless of voltage (Figure 3.2). Ideally, a voltage source would plot out as a vertical line on a graph of current versus voltage (the *I-V* plane), whereas a current source would plot out as a horizontal line on the *I-V* plane. In reality, voltage provided from a voltage source may drop as current increases, and current provided by a current source may diminish as voltage increases (Figure 3.2). These effects are due to internal losses (e.g., voltage drop in the voltage source) occurring at higher power demand from the load or higher value of the product *VI*.

We provide DC power using a variety of processes and devices. Two important ones are to use chemical reactions, e.g., batteries and fuel cells, and to use radiant energy, e.g., photovoltaic (PV) cells. Determining the nature of the source, whether voltage and current source, may be complicated. A battery normally behaves as a voltage source, but a fuel cell and a PV cell may exhibit current or voltage source behavior depending on the voltage and current range. For example, a PV cell behaves as a nonideal current source for lower values of voltage and can behave as a nonideal voltage source for higher values of voltage. Keep in mind that the current and voltage source concepts are just models that we can use to analyze real sources.

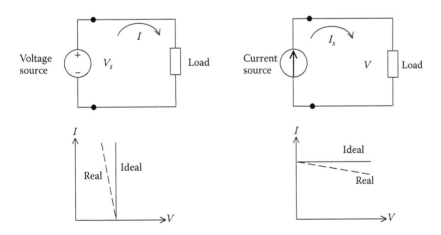

FIGURE 3.2 Voltage and current sources. The bottom graphs in the voltage-current (I-V) plane illustrate the departure from constant voltage (left-hand side) and departure from constant current (right-hand side).

A load in general is a circuit performing a function, consuming power supplied by the source. The function performed by the load varies. For example, resistors convert current to heat dissipation and capacitors store charge as potential energy for later use. A circuit may have more than one source and more than one load. The power supplied by all sources must equal the power consumed by all loads. This is known as Tellegen's theorem, which is just the law of conservation of energy.

Example 3.2

Consider the simple example of Figure 3.2 of one voltage source and one load. Assume no power losses, that is, ideal voltage source and no power loss in the wire. The ideal source provides 12 V and the load consumes 60 W. What is the current? Then assume the voltage source is nonideal and voltage diminishes 0.025 V per W supplied. What is the current if the power remains constant? Answer: Power supplied must equal the power consumed. For an ideal source $I = \dfrac{P}{V_s} = \dfrac{60\ W}{12} = 5$ A. For a nonideal source the voltage supplied at 60 W is $V = V_s - 0.025 \times 60\ V = 12 - 1.5 = 10.5$ V and the current is $I = \dfrac{P}{V} = \dfrac{60\ W}{10.5} = 5.71$ A. The current increases to keep the same power when voltage decreases.

3.2 RELATIONSHIP BETWEEN CURRENT AND VOLTAGE

3.2.1 OHM'S LAW

Ohm's law is an important relationship between current and voltage for a circuit element and based on the concept of *resistance (R)* to current flow:

$$R = \frac{v(t)}{i(t)} \tag{3.5}$$

Such an element is called a *resistor* and it has a resistance value of R. For DC, we can simply write $R = \dfrac{V}{I}$ where we use uppercase V and I to signify constant values (Figure 3.3, left-hand side). Resistance's unit is the ohm (Ω) defined as V/A. When R remains constant for all values of V, the relationship is *linear* (Figure 3.3, right-hand side), that is, voltage is proportional to current, and we say that the element follows Ohm's law or that the element is "ohmic." In such a case, the trace in the I-V plane is a straight line, with slope equal to the inverse of resistance. This slope is *conductance* $G = 1/R$ and therefore an equivalent statement of Ohm's law is that conductance G is the ratio of current to voltage $G = i(t)/v(t)$. Conductance has unit of siemens, abbreviated S, where 1 S = 1 A/V.

Example 3.3

Consider the DC circuits in Figure 3.4. Calculate current, voltage, and resistance as required. Answer: Left-panel current is $I = \dfrac{9\ V}{10\ \Omega} = 0.9$ A. Middle-panel voltage is $V = IR = 1mA \times 10\ k\Omega = 10$ V, and right-panel resistance is $R = \dfrac{V}{I} = \dfrac{9\ V}{1mA} = 9\ k\Omega$.

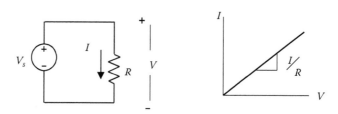

FIGURE 3.3 Circuit with a resistor and linear relationship in the I-V plane.

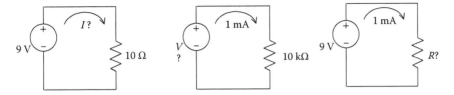

FIGURE 3.4 Examples of Ohm's law calculations.

Combining Equations 3.4 and 3.5 we can derive specific power expressions for an ohmic element: $p(t) = v(t)i(t) = (i(t) \times R) \times i(t) = i(t)^2 R$ or $p(t) = v(t)i(t) = v(t) \times \dfrac{v(t)}{R} = \dfrac{v(t)^2}{R}$. In summary, power can be calculated in three ways for ohmic elements:

$$p(t) = v(t)i(t) = i(t)^2 R = \frac{v(t)^2}{R} \tag{3.6}$$

For DC we can simply write $P = VI = I^2 R = \dfrac{V^2}{R}$.

The *Joule heating* effect, also called Joule's first law, implies that power absorbed by a resistor or a conductor of resistance R produces heat flow. This heat is proportional to power, and thus proportional to $i(t)^2$ or the square of the current; also, this heat flow is independent of the direction of the current. This effect is the basis of using a resistor for an intended purpose such as space heating, cooking, and heating water. It could also represent an unintended consequence of using a load, for example, undesired heating of wires and electric equipment.

Example 3.4

Consider the circuit in Figure 3.3 with $R = 1\ \Omega$ and DC voltage varies between –0.2 V and 1 V. What is the current in this range? What is the power consumed by the resistor in this range? Answer: By Ohm's law $I = \dfrac{V}{R}$. The lower and upper values of current are $I = \dfrac{-0.2}{1} = -0.2$ A and $I = \dfrac{1}{1} = 1$ A. The lower and upper values of power are $P = \dfrac{(-0.2)^2}{1} = 0.04$ W and $P = \dfrac{1^2}{1} = 1$ W.

Note that power is still positive when reversing the direction of current. There is electron flow, but in the other direction, this requires work to move the charge. Figure 3.5 shows the current and

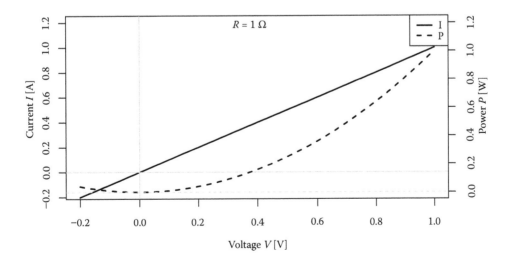

FIGURE 3.5 Resistor characteristics on the *I-V* plane. Current is linear and power is nonlinear.

power on the *I-V* plane. Observe from this figure that power is nonlinear with voltage even though the current is linear with voltage. This figure was produced with functions resistor and ivplane of renpow.

```
V <- seq(-0.2,1,0.01) # volts
x <- resistor(V,R=1)
ivplane(x)
```

Object *x* contains all the calculated values and ivplane plots the graphs. Note the parabolic behavior of power.

In AC circuits, we employ the concept of *impedance* to extend the concept of resistance. The impedance concept includes capacitors and inductors; it is defined similarly to DC using the ratio of voltage to current as in Ohm's law. We will introduce impedance with detail in Chapters 5 and 8.

3.2.2 Nonohmic

When the relationship between *V* and *I* is nonlinear, the element is *nonohmic* and Equation 3.5 does not hold. Instead, current is a nonlinear function of voltage, say, for DC

$$I = f(V) \tag{3.7}$$

In this case, the proportion between *I* and *V* changes with *V*. An example of a nonlinear element is a *diode*, which is an element made from semiconductor material. Considering the circuit in Figure 3.6 the voltage source produces a voltage drop across the diode and a current into the diode that follows a nonlinear relation as the one given in Equation 3.7.

The basic diode model relates current *I* and voltage *V* by an exponential function that depends on temperature *T* (in K)

$$I = I_0 \left[\exp\left(\frac{qV}{kT} \right) - 1 \right] \tag{3.8}$$

where I_0 is the reverse current or the current for negative values of *V*, *q* is the charge of an electron 1.6×10^{-19} C, and *k* is the *Boltzmann constant* with value $k = 1.380 \times 10^{-23}$ J/K relating average kinetic energy to temperature. At reference ambient *T* of 25°C or 298 K we have $\frac{qV}{kT} = 38.9 \times V$. For example, for $I_0 = 1$ nA we would have an I-V relationship of

$$I = [\exp(38.9 \times V) - 1]\text{nA} \tag{3.9}$$

which is illustrated in Figure 3.7. For small voltages (left-hand side) we can see the small reverse current, whereas for larger voltages (right-hand side) we see a large increase of forward current as we

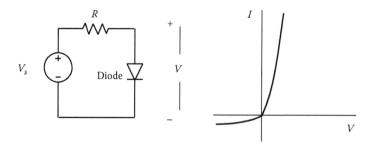

FIGURE 3.6 Circuit with a diode, which is the nonlinear element as shown on *I-V* plane response.

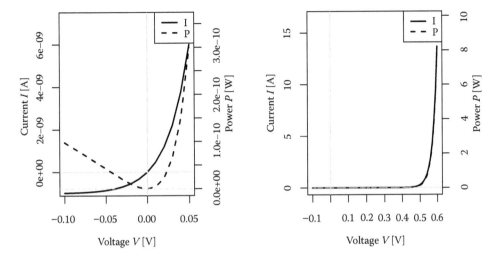

FIGURE 3.7 Diode characteristics on the *I-V* plane. Both current and power are nonlinear.

approach the 0.5 V value. In other words, a diode shows a sharp increase of current with voltage once voltage overcomes a threshold. Power increases in the same proportion. This graph was obtained using code in the same manner as for the resistor but using the function diode twice: first for small voltages and the second for larger voltages:

```
# diode
V <- seq(-0.1,0.05,0.01) # volts
x <- diode(V)
ivplane(x)
V <- seq(-0.1,0.6,0.01) # volts
x <- diode(V)
ivplane(x)
```

3.3 CIRCUIT ANALYSIS METHODS

3.3.1 KIRCHHOFF'S VOLTAGE AND CURRENT LAWS

We form circuits by connecting circuit elements in a network. We already mentioned that two very useful concepts to understand circuits are those of nodes and loops (Figure 3.1). In this section we give more details and indicate how these concepts support circuit analysis. Recall that a node is a point at a distinct potential. Refer to Figure 3.8; the circuit has three nodes (1, 2, and G). The node denoted by G is *ground* or common, and it is a very important node defined to have zero potential or

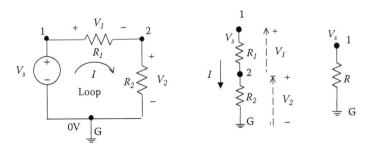

FIGURE 3.8 KVL, resistances in series and voltage divider.

zero volts. It serves as reference to measure the potential of all other nodes; for example, V_2 is voltage of node 2 with respect to ground. Points directly connected to ground have the same voltage (0 V).

Because ground is a reference, there are only two independent nodes (and voltages) in the circuit. We have a loop in this circuit by tracing the circuit starting at a given node (say, G) and returning to it. A simplified view, easier to draw, is offered in the middle part of Figure 3.8 where the loop is implicit in the fact that the current returns to ground.

Kirchhoff's voltage law (KVL) states that the sum of voltage around a circuit loop adds to zero. In other words, the sum of all voltage rises is equal to the sum of all voltage drops. Thus, in the figure we have $V_s = V_1 + V_2$.

The equivalent resistance of resistors connected in *series* is calculated as the sum of all resistances. To see this, apply KVL to the circuit in Figure 3.8. The two resistors R_1 and R_2 are connected in series, the current I is the same for both resistors, and it is also the same that would flow through their equivalent R (see the right-hand side of the figure). By KVL and Ohm's law $V_1 + V_2 = IR_1 + IR_2 = I\underbrace{(R_1 + R_2)}= I\underbrace{R} = V_s$, therefore equating terms grouped by the curly brackets

$$R = R_1 + R_2 \tag{3.10}$$

We form a *voltage divider* using two resistances in series. The voltage across one resistance is proportional to the input voltage multiplied by the fraction of that resistance to the total equivalent resistance. Consider V_2; since the current is the same we have $I = \dfrac{V_2}{R_2} = \dfrac{V_s}{R}$ and rearrange to obtain

$$V_2 = V_s \times \frac{R_2}{R_1 + R_2} \tag{3.11}$$

Similarly for V_1, $I = \dfrac{V_1}{R_1} = \dfrac{V_s}{R}$ yielding

$$V_1 = V_s \times \frac{R_1}{R_1 + R_2} \tag{3.12}$$

The voltage divider is an important concept often employed for faster calculations. A variable resistor or *potentiometer* is in essence a voltage divider. The position of the cursor determines the proportions R_1 and $R_2 = R - R_1$ of the total potentiometer resistance R. Therefore, the voltage across R_1 is given by Equation 3.12 at any cursor position.

Example 3.5

The circuit of Figure 3.8 is built with $V_s = 12$ V, $R_1 = 12$ kΩ, and $R_2 = 24$ kΩ. What is the current? What are the voltages? Answer: Using resistance in series, current is $I = \dfrac{V_s}{R_1 + R_2} = \dfrac{12 \text{ V}}{36 \text{ k}\Omega} = 0.33$ mA. Calculating voltages is easy using a voltage divider; use Equation 3.12 $V_1 = 12 \times \dfrac{12}{36} V = 4$ V and then V_2 is just the difference $V_2 = 12 - 4 = 8$ V by KVL.

Next, let us look at Figure 3.9. The circuit has only two nodes (source and G) but three loops, one of these is the outer path and two other are internal. An independent loop is called a *mesh*. The internal loops in this circuit are independent loops and therefore we have two meshes. The currents I_1 and I_2 are currents through each resistor. The simplified view of the middle part of this figure is easy to draw and it does include the loops because the currents return to ground. *Kirchhoff's current law* (KCL) states that the sum of currents in and out of a circuit node adds to zero. In other words, all currents leaving a node must equal the sum of all currents entering a node. If you take the node at the top of the circuit (Figure 3.9) KCL dictates that $I = I_1 + I_2$, where I is the current flowing out of the source.

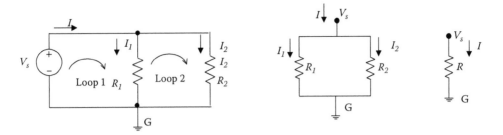

FIGURE 3.9 KCL, resistances in parallel and current divider.

The inverse of the equivalent resistance of resistances connected in *parallel* is calculated as the sum of inverses of all resistances. To see this, apply KCL in Figure 3.9, the two resistances R_1 and R_2 are connected in parallel, and the voltage V_s is the same for both resistors and their equivalent R (see the right-hand side of the figure). By KCL and Ohm's law $I_1 + I_2 = V_s/R_1 + V_s/R_2 = \underbrace{V_s(1/R_1 + 1/R_2)} = V_s \times \underbrace{(1/R)} = I$, therefore equating terms grouped by the curly brackets, we conclude that two resistances R_1 and R_2 connected in parallel (Figure 3.9) yield an equivalent resistance R given by

$$\frac{1}{R} = \frac{1}{R_1} + \frac{1}{R_2} \tag{3.13}$$

Or manipulating algebraically, we can simply use

$$R = \frac{R_1 R_2}{R_1 + R_2} \tag{3.14}$$

When the two resistances in parallel are equal, Equation 3.14 reduces to simply $R = \dfrac{R_1 R_1}{R_1 + R_1} = \dfrac{R_1 R_1}{2R_1} = R_1/2$, which says that the equivalent is just half of the value of each.

Recall that conductance G is the inverse of resistance $G = 1/R$ and therefore a more practical interpretation of Equation 3.13 is that the equivalent conductance of a parallel combination G and is the sum of individual conductance values G_1 and G_2.

We form a *current divider* using two resistances in parallel. The current through one resistance is proportional to the input voltage multiplied by the fraction of the other resistance to the total equivalent resistance. The current I_2 through resistance R_2 in Figure 3.9 is

$$I_2 = I \times \frac{R_1}{R_1 + R_2} \tag{3.15}$$

Whereas the other current I_1 through resistance R_1 is

$$I_1 = I \times \frac{R_2}{R_1 + R_2} \tag{3.16}$$

Example 3.6

The circuit of Figure 3.9 is built with $V_s = 12$ V, $R_1 = 12$ kΩ, and $R_2 = 24$ kΩ. What is the total current? What are the branch currents? Answer: Use resistance in parallel, $R = \dfrac{R_1 R_2}{R_1 + R_2} = \dfrac{12 \times 24}{36}$ kΩ = 8 kΩ. Note that the equivalent resistance of a parallel combination is smaller than the smallest resistance. Current is $I = \dfrac{V_s}{R} = \dfrac{12\ V}{8\ k\Omega} = 1.5$ mA V. Currents are easy by current divider; use Equation 3.16 $I_1 = 1.5 \times \dfrac{24}{36}$ mA = 1 mA and then I_2 is just the difference $I_2 = 1.5 - 1 = 0.5$ mA by KCL. Note that the higher currents flow through the smallest resistance.

3.3.2 COMBINING CIRCUIT ANALYSIS METHODS

For complicated circuits, we calculate voltages and currents by combining various methods, such as Ohm's law, KCL and KVL, series and parallel combinations, and voltage and current dividers, as well as other tools to be discussed later. It is common practice to analyze circuits starting from the load side (typically drawn on the rightmost side of circuit diagrams) and moving leftward toward the source (typically on the leftmost side). The easiest way to explain this process is by following an example.

Example 3.7

Please look at Figure 3.10 and suppose we have the following values: $V_s = 12V$, $R_1 = 12\ k\Omega$, $R_2 = 24\ k\Omega$, $R_3 = 12\ k\Omega$, $R_4 = 12\ k\Omega$. Assume R_4 is the load. Calculate load current, voltage, and power. Answer: Start at the rightmost branch and use resistance in series $R_3 + R_4 = (12 + 12)k\Omega = 24\ k\Omega$. Now use parallel of this equivalent resistance with R_2 to get $R_2\|(R_3 + R_4) = 24\ k\Omega\|24\ k\Omega = 12\ k\Omega$ and this in series with R_1 gives the total resistance seen from the source $R_1 + ((R_3 + R_4)\ \|\ R_2) = 12 + 12 = 24k\Omega$. Now we can calculate the total current from the source using Ohm's law $I_s = \dfrac{12V}{24\ k\Omega} = \dfrac{1}{2}$mA. We have used two vertical lines to denote parallel of resistances. Now we move toward the right; this current sees a current divider of $R_2\ \|\ (R_3 + R_4) = 24\ \|\ 24$. Because the divider has equal resistances, the current is split in half. The load current is then $I_{load} = I_s \times \dfrac{1}{2} = \dfrac{1}{2} \times \dfrac{1}{2}$mA $= \dfrac{1}{4}$mA. Use Ohm's law again to calculate the voltage drop across the load $V_{load} = I_{load} \times R_4 = \dfrac{1}{4}$mA $\times 12\ k\Omega = \dfrac{1}{3}$V. Then power is $P_{load} = V_{load}I_{load} = \dfrac{1}{4}$mA $\times \dfrac{1}{3}$V $= \dfrac{1}{12}$mW.

3.3.3 NODAL ANALYSIS

Two systematic methods to analyze a circuit that leads to an easily implementable algorithm are *nodal* analysis and *mesh* analysis. These methods are based on KCL and KVL, respectively, and both consist of setting up a system of linear equations solvable by matrix algebra. In nodal analysis the unknowns in the system of equations are the independent node voltages, whereas in mesh analysis the unknowns are the independent loop (mesh) currents. These methods are practical for complicated circuits and an excellent option when the circuits are very large. We will expand these methods to AC in Chapter 8 and then applicable to an electric power system or grid with thousands of nodes (Chapter 11).

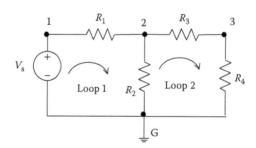

FIGURE 3.10 Example to explain analysis strategy.

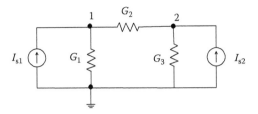

FIGURE 3.11 A simple circuit to explain nodal analysis.

To explain nodal analysis look at the circuit in Figure 3.11, where we have written the elements as conductance. Assign unknowns V_1 and V_2 to the independent nodes (nodes 1 and 2), and assume you know the sources I_{s1}, I_{s2}. Apply KCL and Ohm's law to each of these nodes: $V_1G_1 + (V_1 - V_2)G_2 = I_{s1}$ and $V_2G_3 + (V_2 - V_1)G_2 = I_{s2}$. Now rearrange as a system of equations

$$V_1(G_1 + G_2) + V_2(-G_2) = I_{s1}$$
$$V_1(-G_2) + V_2(G_3 + G_2) = I_{s2}$$

(3.17)

or in matrix form

$$\begin{bmatrix} G_1 + G_2 & -G_2 \\ -G_2 & G_3 + G_2 \end{bmatrix} \begin{bmatrix} V_1 \\ V_2 \end{bmatrix} = \begin{bmatrix} I_{s1} \\ I_{s2} \end{bmatrix}$$

(3.18)

Using matrix notation we write succinctly

$$\mathbf{GV} = \mathbf{I}_s$$

(3.19)

where $\mathbf{G}$ is the *conductance matrix*, vector $\mathbf{V} = \begin{bmatrix} V_1 \\ V_2 \end{bmatrix}$ is a vector of unknown node voltages, and $\mathbf{I}_s = \begin{bmatrix} I_{s1} \\ I_{s2} \end{bmatrix}$ is a vector of current sources or injected currents into the nodes.

An easy way to set up the matrix equation is by main diagonal G_{ii} and off-diagonal G_{ij} entries. For instance, in this case

$$\mathbf{G} = \begin{bmatrix} G_{11} & G_{12} \\ G_{21} & G_{22} \end{bmatrix}$$

(3.20)

The main diagonal entries are the sum of all conductance connected to a node, whereas an off-diagonal entry corresponds to the negative of shared conductance between a pair of nodes. In this simple case $G_{11} = G_1 + G_2$, $G_{22} = G_2 + G_3$ and $G_{12} = G_{21} - G_2$, resulting in matrix $\mathbf{G}$ being symmetric. This is not necessarily true always, but when it is the matrix setup is simpler.

The solution for the unknown $\mathbf{V}$ is found by matrix algebra taking the inverse of $\mathbf{G}$:

$$\mathbf{V} = \mathbf{G}^{-1}\mathbf{I}_s$$

(3.21)

Example 3.8

Take the circuit in Figure 3.11, with $R_1 = R_2 = 5\ \Omega$, $R_3 = 10\ \Omega$, and sources I_{s1}, I_{s2} are 1 A and –0.25 A respectively. Set up and solve the matrix equation. Answer: The conductance of each element is

$G_1 = G_2 = 0.2S$, $G_3 = 0.1S$. The matrix equation is $\begin{bmatrix} 0.2+0.2 & -0.2 \\ -0.2 & 0.2+0.1 \end{bmatrix} \begin{bmatrix} V_1 \\ V_2 \end{bmatrix} = \begin{bmatrix} 1 \\ -0.5 \end{bmatrix}$ or $\begin{bmatrix} 0.4 & -0.2 \\ -0.2 & 0.3 \end{bmatrix} \begin{bmatrix} V_1 \\ V_2 \end{bmatrix} = \begin{bmatrix} 1 \\ -0.25 \end{bmatrix}$. The easiest way in practice is to use a computer to solve this matrix equation. Using R we will code it as

```
G=matrix(c(0.4,-0.2,-0.2,0.3),ncol=2,byrow=TRUE)
Is=c(1,-0.25)
Vn <- solve(G,Is)
```

Note the matrix, current, and voltage vectors are

```
> print(list(G=G,Is=Is,Vn=Vn))
$G
     [,1] [,2]
[1,]  0.4 -0.2
[2,] -0.2  0.3
$Is
[1]  1.00 -0.25
$Vn
[1] 3.125 1.250

>
```

The solution is $\begin{bmatrix} V_1 \\ V_2 \end{bmatrix} = \begin{bmatrix} 3.125 \\ 1.250 \end{bmatrix}$ V.

Observe that the current sources can be thought of current *injection* at node 1 and current *export* (or negative injection) at node 2. These values have forced the voltages V_1 and V_2 to acquire the values shown. We can think of power injection at node 1 $P_1 = V_1 I_{s1}$ and power injection (negative or export) at node 2 $P_2 = V_2 I_{s2}$. This terminology becomes useful later when we talk about power flow in the grid.

Example 3.9

Calculate power injected at node 1 as $P_1 = V_1 I_{s1} = 3.125 \times 1 = 3.125$ W and power injected (negative or exported) at node 2 as $P_2 = V_2 I_{s2} = 1.25 \times (-0.25) = -0.3125$ W. The difference of 2.8125 W must be the power dissipated in the circuit. If we calculate the sum of power dissipated in all three resistors, we get $P_{diss} = V_1^2 G_1 + V_2^2 G_3 + (V_2 - V_1)^2 G_2 = 2.8125$ W.

3.3.4   MESH ANALYSIS

Mesh analysis is the dual of nodal analysis when applied to independent loops (i.e., meshes). Think of the circuit in Figure 3.12 and setting up a system of equations for the unknown currents using KVL

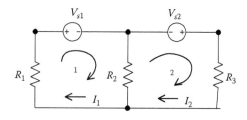

FIGURE 3.12   A simple circuit to explain mesh analysis.

and Ohm's law to each one of these meshes $I_1 R_1 + (I_1 - I_2)R_2 = V_{s1}$ and $I_2 R_3 + (I_2-I_1)R_2 = V_{s2}$. Now rearrange as a system of equations

$$I_1 (R_1 + R_2) + I_2(-R_2) = V_{s1}$$
$$I_1 (-R_2) + V_2 (R_3 + R_2) = V_{s2}$$

(3.22)

or in matrix form

$$\begin{bmatrix} R_1 + R_2 & -R_2 \\ -R_2 & R_3 + R_2 \end{bmatrix} \begin{bmatrix} I_1 \\ I_2 \end{bmatrix} = \begin{bmatrix} V_{s1} \\ V_{s2} \end{bmatrix}$$

(3.23)

The main diagonal has sum of all resistance encountered traversing a mesh, whereas an off-diagonal entry corresponds to the negative of shared resistance by a pair of meshes.

Now use matrix notation for Equation 3.23: $\mathbf{RI} = \mathbf{V}_s$, where the matrix $\mathbf{R}$ and vectors $\mathbf{I}$ and $\mathbf{V}_s$ have obvious correspondence to Equation 3.23 terms. The solution for the unknown $\mathbf{I}$ is

$$\mathbf{I} = \mathbf{R}^{-1}\mathbf{V}$$

(3.24)

In this simple case, the resulting matrix $\mathbf{R}$ is symmetric. This is not necessarily true always, but when it is, the matrix setup is simpler.

You should realize now that Equations 3.21 and 3.24 are matrix generalizations of Ohm's law. To see this, rewrite Ohm's law as $V = RI = \dfrac{1}{G}I = G^{-1}I$ and $I = \dfrac{V}{R} = \dfrac{1}{R}V = R^{-1}V$. Now match these scalar equations to Equations 3.21 and 3.24 recognizing the matrix generalization!

Example 3.10

Take the circuit in Figure 3.12 with $R_1 = R_2 = 5\ \Omega$, $R_3 = 10\ \Omega$, and sources V_{s1}, V_{s2} 12 V and 0 V, respectively. Set up and solve the matrix equation. Answer: Directly use the resistance of each element to write the matrix equation $\begin{bmatrix} 5+5 & -5 \\ -5 & 5+10 \end{bmatrix}\begin{bmatrix} I_1 \\ I_2 \end{bmatrix} = \begin{bmatrix} 12 \\ 0 \end{bmatrix}$ or $\begin{bmatrix} 10 & -5 \\ -5 & 15 \end{bmatrix}\begin{bmatrix} I_1 \\ I_2 \end{bmatrix} = \begin{bmatrix} 12 \\ 0 \end{bmatrix}$. Using R we will code it as

```
Vs=c(12,0)
R=matrix(c(10,-5,-5,15),ncol=2,byrow=TRUE)
Im <- solve(R,Vs)
```

Note the matrix, current, and voltage vectors are

```
> print(list(R=R,Vs=Vs,Im=Im))
$R
     [,1] [,2]
[1,]   10   -5
[2,]   -5   15

$Vs
[1] 12  0

$Im
[1] 1.44 0.48
>
```

The solution is $\begin{bmatrix} I_1 \\ I_2 \end{bmatrix} = \begin{bmatrix} 1.44 \\ 0.48 \end{bmatrix}$ A.

These are extremely simple circuits, but the process is the same for complex circuits. Once computerized, the system can easily be calculated repeatedly when changing the parameters (matrices $\mathbf{G}$ and $\mathbf{R}$).

3.3.5 SUPERPOSITION

Circuits consisting of interconnected resistances and powered by ideal voltage and current sources are *linear*, that is to say the effect of a voltage or current source is a linear relationship of the values of those sources. For instance, when we apply a voltage source V to a series combination of resistances, we obtain a current I that depends linearly on the voltage. When you double the voltage, you will double the current. In practice, this means that we do not have to analyze a circuit when just changing the value of the source, instead we can just scale the result up or down accordingly.

Another consequence of linearity is that we can add the effect of each source as if each one was acting alone. For instance, if we have two sources, we can calculate the effect of one source while making the other one null; then calculate the effect of the second source while nullifying the first. Finally, we can just add both effects to obtain the combined effect. This is the *principle of super-position*. To nullify a voltage source we make its contribution equal to zero voltage, and to nullify a current source we make it contribute zero current. Observe that a voltage source contributes zero voltage when replaced by a short circuit, whereas a current source injects zero current when replaced by an open circuit.

Example 3.11

Use the principle of superposition to calculate current delivered to the load in the circuit shown on the left-hand side of Figure 3.13. Answer: The circuit has two sources. First, nullify the current source by replacing it by an open circuit (Figure 3.13, top right-hand side). The resulting current drawn from the voltage source is $\dfrac{6}{6 + 2 \parallel (3 + 3)}$ mA $= \dfrac{6}{7.5}$ mA = 0.8 mA, then use a current divider to get the current injected to the load $0.8 \times \dfrac{2}{2 + 3 + 3}$ mA = 0.2 mA. Now replace the voltage source by a short circuit (Figure 3.13, bottom right-hand side). The current injected to the load is a divider $2 \times \dfrac{6 \parallel 2 + 3}{6 \parallel 2 + 3 + 3}$ mA $= 2 \times \dfrac{4.5}{7.5}$ mA = 1.2 mA. Finally add the two contributions to get $(0.2 + 1.2)$ mA = 1.4 mA.

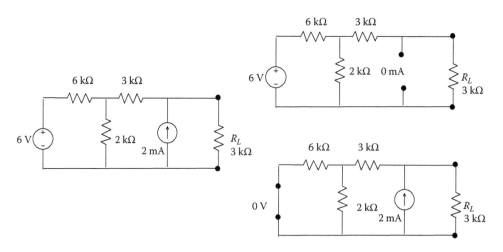

FIGURE 3.13 A simple circuit with two sources may be analyzed by superposition.

3.3.6 THÉVENIN AND NORTON THEOREMS

As we have seen in this chapter, it is interesting to conceptualize a circuit as a power source and a load. For complicated circuits, with more than one voltage or current source, and multiple resistive interconnects, it is often possible to identify a pair of nodes as the connection to the load. Thévenin's theorem allows modeling the entire network seen from the load as a single voltage source and an equivalent resistance in series. It is then possible to study the effect of varying the load in this circuit using the equivalent model (Figure 3.14, right-hand side).

The equivalent voltage source has value equal to the open-circuit voltage V_{oc} across the load terminals when the load is disconnected. In finding this voltage, we may use any combination of circuit analysis tools that result convenient to simplify the calculation. The top-left drawing of Figure 3.14 illustrates this concept.

To find the equivalent resistance, called Thévenin resistance, we determine the resistance seen from the load terminals when all sources are "nullified," that is, voltage sources contribute zero voltage and current sources inject zero current. See the bottom-left drawing of Figure 3.14. Observe that a voltage source contributes zero voltage when replaced by a short circuit, whereas a current source injects zero current when replaced by an open circuit.

Example 3.12

Calculate the Thévenin equivalent of the circuit shown on the left-hand side of Figure 3.15. Calculate load voltage and current. Answer: First, determine V_{oc}; see top right-hand side of Figure 3.15. The circuit has two sources, therefore, use superposition to determine V_{oc}. With the

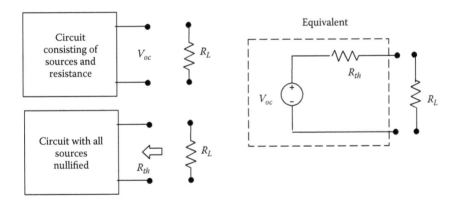

FIGURE 3.14   Determining the Thévenin equivalent circuit.

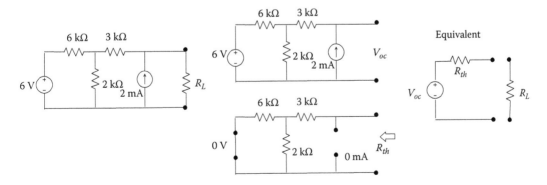

FIGURE 3.15   Example of determining the Thévenin equivalent.

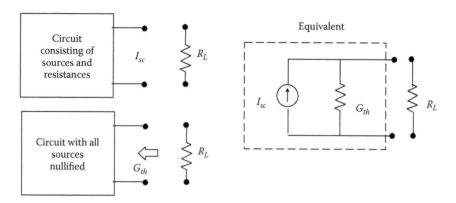

FIGURE 3.16 Determining the Norton equivalent circuit.

current source replaced by an open circuit, V_{oc} due to the voltage source is a voltage divider $6 \times \dfrac{2}{2+6} = 1.5$ V. With the voltage source replaced by a short circuit, the total drop due to the current source is $2 \times (3 + 6\|2) = 9$ V. Thus, total V_{oc} by superposition is $V_{oc} = 9 + 1.5 = 10.5$V. To calculate R_{th}, refer to the bottom right-hand side of Figure 3.15. The resistance is $R_{th} = 3 + 6\|2 = 4.5$ kΩ. Finally, reconnect the load; we can calculate current $I_L = \dfrac{V_{oc}}{R_{th} + R_L} = \dfrac{10.5}{7.5\,\mathrm{k}} = 1.4$ mA, voltage $V_L = V_{oc} \times \dfrac{R_L}{R_{th} + R_L} = 10.5 \times \dfrac{3}{7.5} = 4.2$ V.

The Norton equivalent circuit is the corresponding model in current; it is composed of a short-circuit current source I_{sc} and a conductance G_{th}, which is the reciprocal of the Thévenin resistance. The equivalent current source has value equal to the current across the load terminals when the load is disconnected and replaced by a short circuit. In finding this current, we may use any combination of circuit analysis tools that result convenient to simplify the calculation. The top-left drawing of Figure 3.16 illustrates this concept.

To find the equivalent conductance, called Norton conductance, we determine the reciprocal of Thévenin resistance, that is, resistance seen from the load terminals when all sources are null (i.e., voltage sources contribute zero voltage and current sources inject zero current). See the bottom-left drawing of Figure 3.16.

3.4 MODELING VOLTAGE AND CURRENT SOURCES

3.4.1 I-V Characteristics

We can use Thévenin and Norton equivalent circuits to model voltage and current sources, respectively. Start with a voltage source modeled by a Thévenin equivalent with an open circuit voltage V_{oc} and a series resistance R_s (Figure 3.17, left-hand side). The source current is $I = \dfrac{V_{oc} - V}{R_s}$.

When $V = 0$ (short circuit) the current is $I = \dfrac{V_{oc}}{R_s}$; conversely, when $V = V_{oc}$ (open circuit) the current is $I = 0$. These are the intersection points with the vertical and horizontal axes that you see in Figure 3.18 (left-hand side). The power that can be supplied by the source is $P = IV$ and as illustrated in that figure showing a maximum when $V = 0.5V_{oc}$.

A similar result is seen for a current source using a Norton equivalent with short-circuit current I_{sc} and parallel resistance R_p (Figure 3.17, right-hand side). The current is $I = I_{sc} - \dfrac{V}{R_p}$.

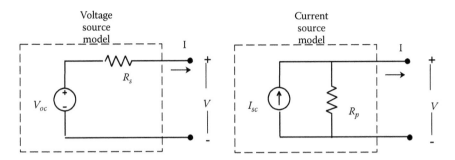

FIGURE 3.17 Modeling voltage and current sources.

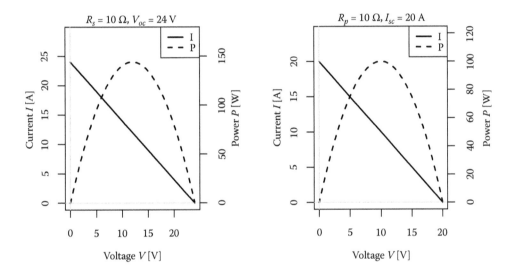

FIGURE 3.18 Voltage and current source models on the I-V plane.

When $V = 0$ (short circuit), the current is $I = I_{sc}$; conversely, when $V = I_{sc}R_p$ (open circuit), the current is $I = 0$. These are the intersection points with the vertical and horizontal axes that you see in Figure 3.18 (right-hand side). The power is $P = IV$ and is illustrated in that figure showing a maximum when $V = 0.5I_{sc}R_p$.

The plots of Figure 3.18 were obtained using functions vsource and isource of renpow.

```
x <- vsource(Voc=24,Rs=1)
ivplane(x)
x <- isource(Isc=20,Rp=1)
ivplane(x)
```

Example 3.13

Consider a voltage source with $V_{oc} = 24$ V and $R_s = 1\ \Omega$. Calculate the current for a voltage of 0.5 Voc. What is the power? Answer: $I = \dfrac{V_{oc} - V}{R_s} = \dfrac{24 \times (1 - 0.5)}{1} = 12$A. The power is $P = VI = 12 \times 12$ W $= 144$ W.

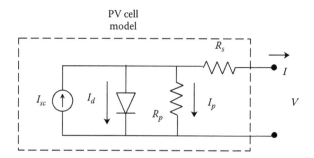

FIGURE 3.19 PV cell model. Combination of resistors and nonlinear element.

3.4.2 A More Complicated Source Model: A PV Cell

A PV cell (we will study solar cells with detail in Chapter 14) is modeled using a current source with short-circuit current I_{sc}, a series resistance R_s, a resistance in parallel R_p, and a diode [5] (Figure 3.19). The diode is included because the PV cell is manufactured as a junction of p and n semiconductor material. Short-circuit current is proportional to intensity of solar radiation.

While the voltage is smaller than the diode elbow or threshold (~0.6 V), the diode is "turned off," that is, allows very little current I_d, and therefore most of the I_{sc} current is delivered to the cell terminals. Part of the current is leaked via the parallel resistor R_p; the remaining is limited by the series resistor R_s. As the voltage increases and the diode "turns on," more current is diverted through the diode, and the current arriving to the terminals decreases abruptly.

Applying KCL, the current is $I = I_{sc} - I_d - I_p$ and using the diode equation (with V_d the voltage across the diode) and Ohm's law we can write

$$I = I_{sc} - I_0[\exp(38.9 \times V_d) - 1] - \frac{V_d}{R_p} \tag{3.25}$$

$$V = V_d - I \times R_s \tag{3.26}$$

To calculate current we apply Equation 3.25 given V_d, then to calculate voltage we use Equation 3.26. These equations are implemented in function PVcell of renpow package for a range of values of V and I.

This response is demonstrated in Figure 3.20 showing a slowly decreasing current with voltage while the voltage is low, and a sharp decrease past an elbow as the voltage approaches the diode threshold. This graph was obtained using function PVcell of renpow:

```
x <- list(I0.A=1, Isc.A=40, Area=100, Rs=0.05, Rp=1, Light=1)
# units: I0.A pA/cm2 Is c.A mA/cm2 Area cm2 Rs ohm Rp ohm
X <- PVcell(x)
ivplane(X,x0=T,y0=T)
```

In this case, we see that power increases with V, reaches a maximum (~1.25 W) when V is ~0.42 V, but then declines as the current decreases sharply. As solar radiation changes, I_{sc} changes, and this in turn changes the voltage at which maximum power occurs.

Example 3.14

Use function PV cell to show the I-V characteristics of a PV cell when solar radiation is reduced to 50%. What is the maximum power? Answer: Use Light = 0.5 in the argument list for PVcell:

```
x <- list(I0.A=1, Isc.A=40, Area=100, Rs=0.05, Rp=1, Light=0.5)
# units: I0.A pA/cm2 Isc.A mA/cm2 Area cm2 Rs ohm Rp ohm
```

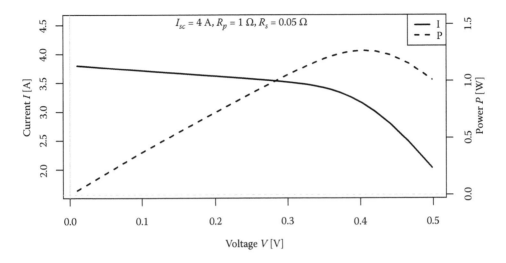

FIGURE 3.20 PV cell characteristics on the *I-V* plane.

```
X <- PVcell(x)
ivplane(X,x0=T,y0=T)
```

From the resulting curve the maximum power is slightly above ~0.6 W and it occurs for ~0.5 V.

3.4.3 POWER TRANSFER AND EFFICIENCY OF POWER TRANSFER

In electric power systems, we are interested in how much power is transferred to a load (or output) for a given power injected or provided by the source (or input). We define *efficiency of power transfer* as the ratio $\eta_p = \dfrac{P_{out}}{P_{in}}$. This relationship occurs at certain input voltages and output voltages. We can generalize this relationship for any circuit based on its Thévenin equivalent. The load voltage V_L is given by a voltage divider $V_L = V_{oc} \times \dfrac{R_L}{R_{th} + R_L}$ and can be interpreted as if V_{oc} is the input voltage V_{in} and V_L is the output voltage V_{out}. Renaming and rearranging, we obtain $\dfrac{V_{out}}{V_{in}} = \dfrac{R_L}{R_{th} + R_L}$. The power consumed by the load is $P_{out} = IV_{out} = IV_L = I^2 R_L$, whereas the power supplied by the circuit is $P_{in} = IV_{in} = IV_{oc} = I^2 (R_{th} + R_L)$. Therefore,

$$\eta_p = \frac{P_{out}}{P_{in}} = \frac{R_L}{R_{th} + R_L} \tag{3.27}$$

Assuming that R_{th} is fixed, efficiency η_p is near zero when $R_L \ll R_{th}$. Then it increases with R_L, reaching 0.5 when $R_{th} = R_L$. Finally, it has an asymptote of 1 as $R_L \gg R_{th}$. We can see this behavior in Figure 3.21 where efficiency is plotted versus the ratio R_L/R_{th}.

In the same graph, we plotted input power and output power. Observe that output power shows a maximum at $R_L/R_{th} = 1$, which occurs when the load and Thévenin resistances match, that is, $R_L = R_{th}$. In order to see this, let us look at P_{out} with more detail:

$$P_{out} = IV_L = \frac{V_{oc}}{R_L + R_{th}} \times \frac{V_{oc} \times R_L}{R_L + R_{th}} = \left(\frac{V_{oc}}{R_L + R_{th}}\right)^2 R_L \tag{3.28}$$

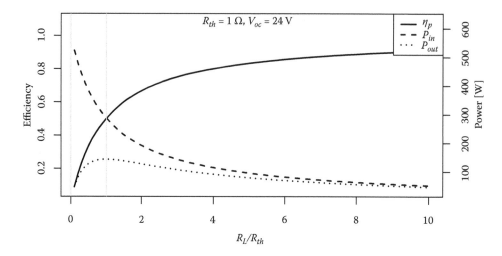

FIGURE 3.21 Power transfer and efficiency.

Take the derivative of output power with respect to R_L and make it equal to zero: $\dfrac{dP_{out}}{dR_L} =$

$$\frac{d}{dR_L}\left(\left(\frac{V_{oc}}{R_L + R_{th}}\right)^2 R_L\right) = \frac{V_{oc}^2(R_L + R_{th})((R_L + R_{th}) - 2R_L)}{(R_L + R_{th})^4} = 0.$$

For this to be zero, $(R_L + R_{th}) - 2R_L = 0$, which yields $R_L = R_{th}$. Now note some important facts for

this condition: The efficiency of power transfer is 0.5, the maximum output power is $\max(P_{out}) = \left(\dfrac{V_{oc}}{R_{th} + R_{th}}\right)^2 R_{th} = \dfrac{V_{oc}^2}{4R_{th}}$, the input power is $P_{in} = 2 \times \max(P_{out}) = \dfrac{V_{oc}^2}{2R_{th}}$, and half of this power is dissipated at R_{th}.

Example 3.15

Suppose $R_{th} = 10\ \Omega$, $V_{oc} = 24\ V$. What is the maximum power transfer to the load? What is the value of input power when we attain maximum output power? Visualize variations of power and efficiency versus R_L using renpow's function eff.pow. Answer: $\max(P_{out}) = \dfrac{24^2}{4 \times 10} = 14.4\ W$. The input power is double or 28.8 W. We can visualize these relations using

```
x <- list(Rth=10,Voc=24)
eff.pow(x)
```

which produces the plots in Figure 3.21.

Before we leave this section, please realize from Figure 3.21 that greater values of efficiency are achieved when $R_L \gg R_{th}$, that is, toward the right end of the horizontal axis, and in that range both input and output power are much lower than the maximum output power and the corresponding input power.

3.5 RESISTIVITY, WIRES, AND POWER LOSS IN THE WIRE

Resistivity is a property of a conductive material and determines the resistance of an element with certain geometry. Resistivity is denoted by ρ and its units are ohms-m (Ω-m). For instance, take a wire of length l and cross-sectional area A. We calculate resistance by

$$R = \rho \frac{l}{A} \tag{3.29}$$

Resistance increases for longer and thinner wires, whereas it decreases for shorter and thicker wires.

Example 3.16

Take a copper wire 10 m long and 2 mm diameter. The resistivity of copper is 1.72×10^{-8} Ω-m. Therefore, using Equation 3.29, we obtain a resistance of

$$R = 1.72 \times 10^{-8} \frac{10}{(\pi/4)(2 \times 10^{-3})^2} = \frac{1.72 \times 0.1}{\pi} = 0.055\Omega$$

There are various standards for wire size, for example, the American Wire Gauge (AWG) standard. A partial table for wire size in the 4 to 12 gauge is in Table 3.1. Number 4 is the heaviest duty wire in this set; it can carry 95 A and has resistance of 0.8152 Ω/km. Number 12 is the smallest in this set with 5.211 Ω/km and can carry 20 A.

The smaller the wire (higher AWG number), the larger the resistance for the same length and current, and therefore the voltage drop and power loss in the wire increases. Suppose you want to deliver DC power P_{load} to a load that draws a current I, and that the source is x m away. The wire resistance is calculated from the resistance per unit length, call it r_{wire}, times double the distance x, $R_{wire} = r_{wire} \times 2x$. It is double because the current needs to return, so it is actually two conductors. The voltage drop in the wire is $V_{wire} = IR_{wire}$. The power loss in the wire is $P_{wire} = I^2 R_{wire}$.

Is this amount significant? In absolute terms, it depends on load usage so that we can compute total energy per month and its cost. In relative terms, we can answer that question comparing to the power consumed by the load $\dfrac{P_{wire}}{P_{load}} = \dfrac{I^2 R_{wire}}{I^2 R_{load}} = \dfrac{R_{wire}}{R_{load}}$ or to the total power $\dfrac{P_{wire}}{P_{load} + P_{wire}} = \dfrac{R_{wire}}{R_{load} + R_{wire}}$.

In order to reduce the loss $P_{wire} = I^2 R_{wire}$ in absolute terms without increasing the wire size, we must reduce the current. This can be done for the same power P by increasing the voltage V, so that $I = \dfrac{P}{V}$ decreases. This is why power is transmitted at higher voltages.

Example 3.17

Suppose we have a 0.36 kW load that can run at 12 V or 24 V DC and is located 20 m from the source. For each voltage what size of wire would you use and what is the power loss compared

---

TABLE 3.1

Partial Information on AWG Wire Sizes from Gauge 4 to 12

AWG	Area (mm²)	Resistance per Unit Length (Ω/km)	Ampacity (A),[a] with 90°C Insulation
4	21.2	0.8152	95
6	13.3	1.296	75
8	8.37	2.061	55
10	5.26	3.277	40
12	3.31	5.211	20

[a] Ampacity is allowable current. This is solid wire only and has to be corrected for stranded wire. The ampacity depends on insulation. Given here is the highest (corresponding to 90°C).

to the load? Assume the best wire insulation. Answer: Start with 12 VDC, and calculate the current as $I = \frac{P}{V} = \frac{360}{12} = 30$ A. From the wire size table select wire #10 that has ampacity of 40 A, which is above the demand of 30 A. This gauge has resistance 3.277 Ω/1 km. Therefore its resistance is $R_{wire} = r_{wire} \times 2x = 3.277$ Ω/km $\times 2 \times 20$ m $= 0.131$ Ω. The power loss is $P_{wire} = I^2 R_{wire} = 900 \times 0.131$ W $= 117.9$ W. The load has a resistance of $R_{load} = \frac{P_{load}}{I^2} = \frac{0.36 \text{ kW}}{30^2} = 0.4$ Ω. Thus, $\frac{P_{wire}}{P_{load}} = \frac{R_{wire}}{R_{load}} = \frac{0.131}{0.4} \simeq 0.32$. This is an undesirable power loss since it is 32% of the load! To reduce loss, use larger wire, say #8 with resistance 2.061 Ω/1 km, which would give us $R_{wire} = 2.061$ Ω/km $\times 2 \times 20$ m $= 0.08244$ Ω. The loss is lower $\frac{P_{wire}}{P_{load}} = \frac{0.0824}{0.4} \simeq 20\%$, but still undesirable.

The extra expenditure on heavy-duty wire may or not compensate for energy savings; it depends on load usage. What happens at 24 VDC? The current is half, 15 A; this means we could use smaller wire say #12 with ampacity of 20 A, but use #10 as before for the sake of comparison. The power loss is $P_{wire} = I^2 R_{wire} = 225 \times 0.131$ W $\simeq 29.5$ W. Therefore, the loss has been reduced to a quarter by doubling the voltage. In relative terms $R_{load} = \frac{P_{load}}{I^2} = \frac{0.36 \text{ kW}}{15^2} = 1.6$ Ω or four times larger. The relative loss in the wire would decrease by a factor of 4, or to 8% for #10 wire. This example illustrates how supplying power by long wires can cause high power loss and that increasing the voltage allows for decreased power loss.

3.6 MEASURING VOLTAGES, CURRENTS, AND RESISTANCES

A *voltmeter* measures voltage (AC or DC), an *ammeter* measures current (AC or DC), and an *ohmmeter* measures resistance. A *multimeter* combines all these functions in one instrument. Voltmeters and ammeters derive their power from the circuit under measurements, whereas an ohmeter requires a battery. An ohmeter measures current and indirectly measures resistance by using Ohm's law. Therefore, do not use an ohmmeter while a circuit is hot or powered on. Many multimeters denote the resistance function by the symbol "Ω" or the word "ohms." The ohmmeter is also useful for continuity test, that is, detecting whether there is a continuous electrical connection from one point to another.

Digital multimeters (DMM) convert the voltage, current, or ohms to a digital number, and then displays it on a numerical readout. Polarity is given by the sign on the readout. The number of digits is an important specification of the display. For example, a 3½-digit digital display would show $\pm 1XXX$, where X denotes a full digit (0–9). Seven segments make up a digit. The left-most digit is the leading digit or most significant digit. It is the "1/2 digit" and can indicate "1" as a maximum. The range is 0 to 1999, positive or negative. This is why we also call these meters a "2000-count" meter. Similarly, a 4½ digit would show $\pm 1XXXX$ and therefore has a range of 0 to 19999, positive or negative. In resistance mode these meters indicates noncontinuity by a nonnumerical code on the display (e.g., "OL" for Open-Loop, or "---").

Voltmeters select the measuring range by switching resistances in a voltage divider circuit (resistances in series), whereas ammeters select range by changing resistance in a current divider circuit (resistances in parallel). A DMM may select the range automatically (autoranging) or manually (you would have to switch it to find the best range). For measuring voltage, select AC or DC and use correct polarity; typically using the red color test lead to positive (+) and the black test lead to negative (–). Select the range: Start with the highest range. If the reading is small, set to the next lower range. The reading should be larger now; you would iterate if the reading continues to be small. Use the lowest-range setting that does not "overrange" the meter. An overranged digital meter displays the letters "OL," or a series of dashed lines, or some other symbol (this indication is manufacturer specific).

To measure resistance, start with a simple test of continuity: Set the meter to its highest resistance range, and touch the two test probes to check for 0 ohms (short circuit). To measure a resistance value: Connect the test probes across the resistor, and obtain a reading. If the reading is close to zero, select a lower resistance range on the meter, and iterate until you use the appropriate range.

For measuring current, open the circuit and insert an "ammeter" in series with the circuit so that all current flowing through the circuit also have to go through the meter. Measuring current in this manner makes the meter part of the circuit. Ideally, it should not cause a voltage drop, assuming it has very little internal resistance. Therefore, the ammeter will act as a short circuit if placed in parallel to a source of voltage, causing a high current and potentially damaging the meter.

3.7 BATTERIES AND ELECTROCHEMICAL CELLS

Electrochemical cells generate electricity from chemical reactions or conversely promote chemical reactions by providing them with electrical power. An electrochemical cell is composed of two parts called *half-cells*, where each part has an electrode and an electrolyte (Figure 3.22). The electrodes are conductors and the electrolyte can dissociate into ions. The two half-cells can share the electrolyte or have separate electrolytes. *Oxidation* occurs at one of the half-cells (loses electrons), whereas *reduction* occurs at the other half-cell (gains electrons). Thus, there is flow of electrons from the oxidation to the reduction half-cell. The oxidation side becomes positive as it loses electrons and is named the *anode*; whereas, the reduction side becomes negative and is the *cathode* (Figure 3.22). Conventional current is opposite to electron flow.

Each half-cell develops an electrical potential in volts that can be predicted using the Nernst equation,

$$E = E_0 + \frac{RT}{zF} \ln(a) = E_0 + 2.303 \frac{RT}{zF} \log(a) \tag{3.30}$$

where R is the gas constant (8.31 J/mol K), T is temperature in K, F is Faraday constant (96,500 C), z is the valence of the ion of interest, and a is the activity. For low concentrations, activity is approximately the same as concentration. E_0 is a constant (the standard electrode potential). We are using E for electric potential (in V), which is the common notation in electrochemistry; please do not confuse it with energy E in J.

The difference in half-cell potential between anode E_a and cathode E_c is the cell voltage or electromotive force (emf):

$$V_{cell} = E_a - E_c \tag{3.31}$$

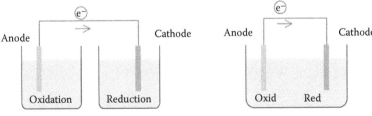

Anode   Cathode   Anode   Cathode

Oxidation   Reduction   Oxid   Red

Separate electrolyte   Same electrolyte

FIGURE 3.22 Electrochemical cells.

Cell voltages vary over a wide range from 0 to several volts. A ~1.5 V cell is common. For electrolyte based on liquid solutions, the voltage may be as high as 2.5 V. Other electrode-electrolyte technologies yield higher voltages ~3 V.

When the reduction-oxidation is spontaneous, it releases energy that can be converted to do useful work. We then have a power source and this cell can form a *battery* by combining it with other cells in series to increase the voltage or parallel to increase the current.

Example 3.18

How many 1.5 cells would form a 9 V battery? How many 2.25 V cells are needed to make a 12 V battery? Answer: Cells are connected in series, then their voltages add up, therefore 9V/1.5 V = 6 and 12 V/2.25 V = 6.

Conversely, providing electric current to a cell can promote chemical reactions. For example, a current can produce electrolysis, or splitting water into hydrogen and oxygen. Hydrogen can be used in an electrochemical cell, for example, a fuel cell, to generate electric power. We will discuss more on fuel cells in Chapter 6.

EXERCISES

3.1. A single conductor in a transmission line dissipates 6 MWh of energy over one day (24 h), while the current was maintained constant at 100 A. Calculate power dissipated and resistance of the conductor.

3.2. Take the circuit in Figure 3.23, with $R_1 = 5$ kΩ, $R_2 = 10$ kΩ, $R_3 = 15$ kΩ, $R_4 = 5$ kΩ, and sources I_{s1}, I_{s2}, and I_{s3} are 1 mA, 2 mA, and 3 mA, respectively. Set up and solve the matrix equation for nodal analysis.

3.3. Take the circuit in Figure 3.24, with $R_1 = 5$ Ω, $R_2 = 10$ Ω, $R_3 = 15$ Ω, $R_4 = 5$ Ω, $R_5 = 20$ Ω, $R_3 = 15$ Ω and sources V_{s1} and V_{s2} are 6 V and 9 V, respectively. Set up and solve the matrix equation for mesh analysis.

3.4. Calculate the Norton equivalent circuit for the network shown on the left-hand side of Figure 3.15.

3.5. Consider a current source with $I_{sc} = 20$ A and $R_s = 1$ Ω. Calculate the voltage for a current of $0.5 I_{sc}$. What is the power?

3.6. Suppose $R_{th} = 1$ Ω, $V_{oc} = 36$ V. What is the maximum power transfer to the load? What is the value of input power when we attain max output power? Visualize variations of power and efficiency versus R_L using renpow's function eff.pow.

3.7. Based on Problem 2.9 from Masters [5], one 12 V battery is used to power lights in a cabin (two 60 W @12 V light bulbs in parallel). What gauge wire should be used (as a minimum) to carry the current drawn by the bulbs? Suppose the battery is 20 m away from the bulbs.

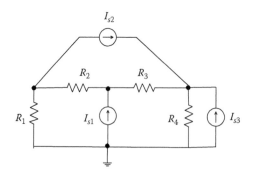

FIGURE 3.23   Circuit to solve by nodal analysis.

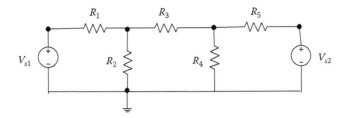

FIGURE 3.24   Circuit to solve by mesh analysis.

What is the power lost in the wires and power delivered to the bulbs? Discuss the results. How could you reduce wire power loss?

3.8. Based on Problem 2.11 from Masters [5], in a home, lights draw 20 A and are on 10 hr every day, and are located about 30 m from the wiring panel. You know that 12-gauge wire can carry this current but you are considering whether to buy 10-gauge wire to save money as potential lower wire loss. Assume "Romex" wire (2 wire + ground) costs $1.50 per m for 12 gauge and $2.00 per ft for 10 gauge, and utility rates are $0.10 per kWh. How many years (simple payback period) would it take to pay the extra cost of heavier duty wire?

REFERENCES

1. Irwin, J.D., and R.M. Nelms, *Basic Engineering Circuit Analysis*. 11th edition. 2011: Wiley. 688 pp.
2. Hayt, W., J. Kemmerly, and S. Durbin, *Engineering Circuit Analysis*. Eighth edition. 2012: McGraw Hill. 880 pp.
3. Alexander, C.K., and M.N.O. Sadiku, *Fundamentals of Electric Circuits*. Third edition. 2007: McGraw Hill. 1056 pp.
4. Hemami, A., *Electricity and Electronics for Renewable Energy Technology: An Introduction*. Power Electronics and Applications Series, M.H. Rashid. 2016, Boca Raton, Florida: CRC Press/Taylor & Francis. 796 pp.
5. Masters, G.M., *Renewable and Efficient Electric Power Systems*. Second edition. 2013: Wiley-IEEE Press. 690 pp.

4 Thermodynamics

The central concerns of thermodynamics are heat and temperature and how they relate to energy and work. Because of this basic focus, thermodynamics is very important for power generation, particularly those conversion processes using fossil-fuel sources. We will employ thermodynamics in this book to understand basic functioning of various forms of generating power when heat and chemical reactions are involved.

The *laws of thermodynamics* enunciate the constraints imposed by nature on any process that involves energy, and its conversion from one form to another. Therefore, these laws allow estimating the *efficiency* of various energy conversion processes and involve devices such as a boiler, steam turbine, condenser, internal combustion engine, and gas turbine. Knowledge of thermodynamics is essential for the design and operation of all these devices as well as for predicting their performance.

In this chapter, we introduce the important difference between state and path functions, present the first law of thermodynamics and reversible processes, the ideal gas law, and basic thermodynamic paths as one variable (temperature, volume, pressure) is kept constant. We conclude the chapter introducing the very important model of heat engine and its idealization that leads to the Carnot limit. Further reading is available in fundamentals of thermodynamics textbooks [1,2]. In addition, several textbooks on energy systems contain reviews of thermodynamics that can supplement the presentation given here [3–7]. Thermodynamics have applications to environmental, ecological, and biological systems; we will not discuss these topics in this book, but the reader is referred to other texts [8,9].

4.1 FIRST LAW OF THERMODYNAMICS

4.1.1 STATE AND PATH FUNCTIONS

Thermodynamics is typically concerned with *changes* in state and other properties of a system; therefore, we will often use a differential notation to write fundamental equations. Such differential quantities are then integrated to obtain total amounts. However, there is an additional subtlety in considering differentials for thermodynamic computations. Our typical dx involved in the definition of derivative is called an *exact* differential and we integrate as usual; its evaluation depends only on final and initial values.

For example, take $\int_{x_1}^{x_2} dx = x_2 - x_1$ and we see that the result is just the difference of initial and final values.

A function that is independent of the path of change is called a *state function*.

Energy is a state function; an object has so many joules of energy regardless of the path taken to change energy. However, heat (thermal energy transfer) and work (mechanical energy transfer) are energy transfer processes and therefore dependent on the path of change. When integrating to obtain total heat and work, we need to consider the path, or in other words, the conditions under which the change is made. These are called *path functions*. Work and heat can be transferred at different conditions of temperature, pressure, and volume. For path functions, instead of using dx we use δx to make it clear that the change depends on the conditions under which it occurs. The differential δx is called an *inexact differential*.

An intuitive way of understanding path functions and inexact differentials is to calculate distance traveled along a line (one dimensional) for two cases (Figure 4.1): (a) from point 1 to point 2 and (b) distance traveled from point 1 to point 2 and back to point 1 (a loop or cycle). In the

first case using exact differentials $\int_{x_1}^{x_2} dx = x_2 - x_1$ we obtain that distance is just the difference in

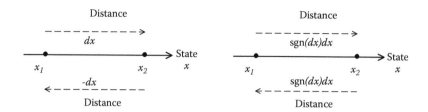

FIGURE 4.1 State, path, and inexact differentials. Left: Integrating with exact differential calculates zero distance. Right: Inexact differential using $\delta x = \text{sgn}(dx)dx$.

one-dimensional coordinates x (the "state"). Now in the second case use the exact differential $\int_{x_1}^{x_1} dx = x_1 - x_1 = 0$. This means that there is no change of state but we know that we traveled twice as much as in the first case. We can clarify the situation by integrating separately for each path, first in the positive x direction and second in the negative x direction: $\int_{x1}^{x2} dx + \int_{x2}^{x1} (-dx) = (x_2 - x_1) + (-1)(x_1 - x_2) = 2(x_2 - x_1)$. We traveled twice as much and this makes more sense. We can generalize this process with an inexact differential defined as $\delta x = \text{sgn}(dx)dx$, where sgn is the

sign or signum function. The sign function is defined as $\text{sgn}(x) = \begin{cases} -1 \text{ if } x < 0 \\ 0 \ \text{ if } x = 0 \\ +1 \text{ if } x > 0 \end{cases}$

and it is the derivative of the absolute value of x. Then, $\int_{x1}^{x2} \delta x + \int_{x2}^{x1} \delta x = \int_{x1}^{x2} \text{sgn}(dx)dx + \int_{x2}^{x1} \text{sgn}(dx)dx =$

$\int_{x1}^{x2} dx + \int_{x2}^{x1} -dx = 2(x_2 - x_1)$. In essence, we have used an *integration factor* to convert an inexact differential to an exact one.

Example 4.1

Suppose you move in a straight line from point $x_1 = 10$ m to point $x_2 = 20$ m and then return halfway between these two points. What is the change of state? What is the distance traveled? The answer is straightforward; the change of state is 5 m and the distance is 15 m, but use inexact differentials and the sgn function for practice. Answer: The change of state is

$$\int_{x1}^{(x_2+x_1)/2} dx = ((x_2 + x_1)/2 - x_1) = (x_2 - x_1)/2 = (20 - 10)/2 = 5\,\text{m}$$

The distance is found by integration with inexact differential using the sign function

$$\int_{x1}^{x2} \delta x + \int_{x2}^{(x_2+x_1)/2} \delta x = \int_{x1}^{x2} \text{sgn}(dx)dx + \int_{x2}^{(x_2+x_1)/2} \text{sgn}(dx)dx$$

Integrate

$$\int_{x1}^{x2} dx + \int_{x2}^{(x_2+x_1)/2} -dx = x_2 - x_1 - ((x_2 + x_1)/2 - x_2) = 1.5 \times (x_2 - x_1)$$

Now substitute $1.5 \times (x_2 - x_1) = 1.5 \times (20 - 10) = 15$m.

4.1.2 WORK AND HEAT ARE PATH FUNCTIONS

Let us go back to Chapter 1 when we defined work done to move an object as product of force and distance, and we simply assumed that force F was constant. Actually, being more rigorous, we integrate force along a path from point x_1 to point x_2, that is to say $W = \int_{x1}^{x2} F(x)dx$, where force varies with x. When force $F(x)$ is a constant F we obtain the simplest result $W = \int_{x1}^{x2} Fdx = F \times (x_2 - x_1)$.

Think now what happens if you apply the same force to bring the object back to x_1. Obviously, you perform twice the amount of work. However, if you just use an exact differential the integration result is zero. To avoid this inconsistency, we must use an inexact differential and an integration factor as in the previous section

$$F\int_{x1}^{x2} \delta x + F\int_{x2}^{x1} \delta x = F\left(\int_{x1}^{x2} \text{sgn}(dx)dx + \int_{x2}^{x1} \text{sgn}(dx)dx\right) = F \times 2(x_2 - x_1)$$

Now the result indicates that we performed double the amount of work.

We can see that that work is a path function. Therefore, in general we could write that

$$W = \int_{1}^{2} \delta W \tag{4.1}$$

to make clear that the integration is path dependent.

Example 4.2

Suppose you displace a mass a distance of 1 m with a constant 10 N force and then displace it back applying a constant force of 5 N (Figure 4.2). What is the work done? Use inexact differential.

Answer: $10\int_{0}^{10} \delta x + 5\int_{10}^{0} \delta x = 10N \times 10m + 5N \times (-1) \times (-10)m = (100 + 50)J = 150J.$

We will see shortly that in thermodynamics we will be concerned with work done to a gas to compress it and work done by a gas as it expands.

A similar argument applies to heat. To make the story shorter, assume you *transfer heat in* (add) to a mass of water to raise its temperature from $T_1°$C to $T_2°$C (from state 1 to state 2) and then *transfer heat out* (remove) of this water to lower its temperature from $T_2°$C back to $T_1°$C (from state 2 back to state 1). The change of energy from state 1 and back to 1 is zero because the water ends up at the

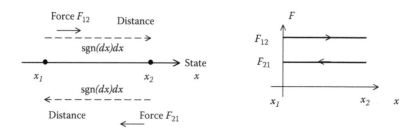

FIGURE 4.2 Work and inexact differentials. Force has different value in positive x direction compared to negative x direction.

same temperature; however, the heat exchanged was not zero. If we integrate with exact differential, we would get an incorrect result of zero energy transfer. Using an inexact differential resolves the issue. Therefore we could write $Q = \int_{1}^{2} \delta Q$ as in Equation 4.1. For a gas, we can transfer heat at constant volume and therefore the pressure will vary, or we can transfer heat at constant pressure and thus the volume will change. Specific heat capacity differs whether the process is at constant volume or constant pressure, therefore the heat transfer depends on the path.

4.1.3 Statement of the First Law: Conservation of Energy

The first law of thermodynamics is the principle of *conservation of energy*: In order to "produce" energy in one form we need to spend energy in another form. In other words, energy can only be transformed, not created. Said another way, we cannot devise a machine or process that continuously produces energy without any corresponding input. A formal statement of the first law requires considering the *internal energy*, U, of a *closed* system and its relationship to possible external inflow and outflow of energy. Thermodynamically, a closed system is one that *does not exchange matter* with its surroundings; it only exchanges heat Q and work W (Figure 4.3).

Internal energy is a state function. When the system is closed and there is no exchange of energy in the form of work or heat, its internal energy is constant and therefore its change is zero:

$$dU = 0 \qquad (4.2)$$

As the system exchanges heat and work with its surroundings, there is a change in internal energy:

$$dU = \delta W + \delta Q \qquad (4.3)$$

This is the *first law of thermodynamics*. Note that work and heat are given with inexact differentials to acknowledge the fact that they are path functions, as explained in the previous section. However, internal energy is a state function and given by an exact differential.

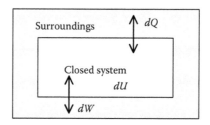

FIGURE 4.3 Conceptualization of a closed system.

We will adopt the convention that heat transferred from the surroundings to the system is positive (a gain from the system point of view) and, conversely, heat transferred from the system to the surroundings is negative (a loss from the system point of view). The same convention applies to work. Work done by the system is negative, whereas work done on the system is positive. Therefore, the balance of change of internal energy depends on the sign of each term. For example, if the system gains heat that performs work, the internal energy balance would be $\Delta U = Q - W$.

Example 4.3

Suppose a closed system does 100 Nm of work against the surroundings while it gains 200 J of heat from the surroundings. What is the change of internal energy? Answer: $\Delta U = Q - W = 200J - 100J = 100J$.

4.1.4 JOULE'S EXPERIMENT

An experiment performed by James Prescott Joule in the mid-1800s illustrates the conversion between mechanical and thermal energy. Joule's experiment apparatus consists of a vessel with water, paddles, a thermometer, a string, a pulley, a mass, and a yardstick (Figure 4.4). The mass changes its potential energy after it drops, pulling on the string and spinning the paddles that cause an input of heat, which causes the water temperature to rise. Joule's calculations consisted of comparing the mechanical energy "disappearing" as the weight falls with the heat energy "manifested." The mechanical energy is the work $W = mg\Delta h$, where m is the mass of the falling body, g is acceleration of gravity, and Δh is the change in height. The heat manifesting is calculated by $Q = m_w c(T_f - T_i)$, where m_w is the mass of water, c is specific heat capacity of water, and T_f and T_i are the final and initial temperature of water, respectively.

4.2 *PV* PATHS AND STATES

When analyzing changes of energy in conversion systems, we want to calculate changes in *states* of a system based on pressure P, volume V, temperature T, and composition (moles X_i of each species i). In this section, we will describe a simple *equation of state* (the ideal gas law) and its application to energy conversion. As already discussed, work and heat are path functions. Heat can be transferred at constant volume (an *isochoric* path) or at constant pressure and doing work (an *isobaric* path).

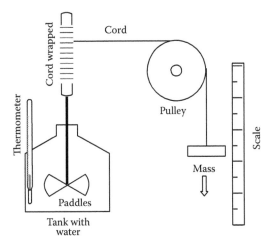

FIGURE 4.4 Joule's experiment.

Work can be done at constant temperature (an *isothermal* path), or without adding heat (an *adiabatic* path). In this section, we learn how to calculate heat and work for all of these processes.

4.2.1 IDEAL GAS LAW

A simple *equation of state* is the ideal gas law

$$PV = nRT \qquad (4.4)$$

where P is pressure (in pascals, Pa), V is volume (in m^3), T is absolute temperature in K, n is number of moles of the gas, and R is the *universal gas constant* 8.314 J/(mol K). Both sides of this equation have units of energy. When R is multiplied by moles and K, we get J. The left-hand side has units of work; take 1 Pa $= 1\dfrac{\text{N}}{\text{m}^2}$ and multiply by m^3 to get $\dfrac{\text{N}}{\text{m}^2}$m^3 = Nm = J. Both sides of the Equation 4.4 represent energy of the gas irrespective of the work or heat transferred to get there.

As discussed in Chapter 1, temperature T is an intensive property because it does not depend on the amount of fluid under consideration. Pressure P is also an intensive property. However, volume V is an extensive property because it is proportional to the amount of fluid; gas in this case. To make V independent of the amount, we divide by the mass m, which is $m = nM$, where M is the molar mass, g/mol, of this gas. Thus, we have *specific volume* denoted by lowercase letter $v = \dfrac{V}{m}$ in m^3/kg. Note that specific volume is the reciprocal of density $v = 1/\rho$.

An alternative way of expressing the ideal gas law is to use the specific volume or the density. Substitute moles n using $m = nM$ in Equation 4.4 to get $PV = \dfrac{m}{M}RT$ and divide by mass $Pv = \dfrac{R}{M}T$ or $P = \rho\dfrac{R}{M}T$. The term R/M depends on the gas and it is called the *specific gas constant*. For brevity, R/M is many times called the gas constant R, requiring a different symbol for the universal gas constant. You should avoid confusion when reading tables listing R for various gases; these would be values of specific gas constant for those gases, not the universal gas constant.

In many cases, it is useful to consider standard conditions of pressure and temperature. Two important ones are Standard Ambient Temperature and Pressure (SATP) conditions 25°C and 1 atm (101.325 kPa), and Standard Temperature and Pressure (STP) conditions 0°C and 100 kPa. It is common to express pressure in bar (1 bar $= 10^5$ Pa $= 100$ kPa), thus STP is 0°C and 1 bar, and SATP is 25°C and about 1 bar (slightly larger). When considering standard conditions, the conversion from Celsius to K uses the more exact 273.15 instead of the commonly used 273. For SATP 1 mol of gas occupies 24.46 liters as we verify using the gas law $V = \dfrac{nRT}{P} = \dfrac{1 \times 8.314 \times (273.15 + 25)}{101.325} = 24.46\,\text{l}$. Throughout this book, we will use 273.

Example 4.4

What is the volume V_1 of 1 mol of air at 25°C and 1 bar? What is the specific volume? What is the pressure P_2 of one mole of air at 30°C if it occupies $V_2 = 10$ l? Assume the molecular mass M of air is 28.97 g/mol (dry air, which is close to that of N_2, the most abundant gas in the atmosphere).

Answer: The volume is $V_1 = \dfrac{nRT}{P} = \dfrac{1 \times 8.314 \times (273 + 25)}{100} = 24.78\,\text{l}$. The specific volume is $v = \dfrac{V_1}{m} = \dfrac{0.02478\text{m}^3}{1\text{mol} \times 28.97\text{g/mol}} \approx 0.86\,\dfrac{\text{m}^3}{\text{kg}}$. The pressure $P_2 = \dfrac{nRT}{V_2} = \dfrac{1 \times 8.314 \times (273 + 30)}{0.01} = 251.9\,\text{kPa}$ or $P_1 = 251.9$ kPa $\times (1/100) = 2.52$ bar.

4.2.2 Internal Energy of a Gas

The internal energy of a gas is given by the average kinetic energy of the molecules. The *Boltzmann constant* k relates the average kinetic energy of molecules in a gas with its temperature; it is given by $k = \dfrac{R}{N_A}$, where R is the universal gas constant and N_A is Avogadro's number. Because R is in J/(mol K), the units of k are J/K or energy divided by temperature. The value is $k = 1.380 \times 10^{-23}$ J/K. The average kinetic energy of translational movement in three dimensions is $\dfrac{3}{2}kT$. Using this average, the internal energy of n moles of an ideal gas is $U = nN_A \dfrac{3}{2}kT$ and, therefore, $U = \dfrac{3}{2}nRT$. This expression can be generalized to $U = \dfrac{f}{2}nRT$, where f is the number of degrees of freedom of molecular motion. Monoatomic (single atom) gases (e.g., He) have $f = 3$, the basic 3-D translational degrees of freedom. Diatomic gases (e.g., O_2) have two additional degrees of freedom due to rotational freedom, therefore $f = 5$. Polyatomic gases have three additional rotational degrees of freedom for $f = 6$. Actually, f depends on temperature. These values correspond to standard temperature of 300 K (used often instead of 25°C, which is 298 K).

Internal energy U is an extensive property since it depends on the amount of fluid. Dividing by mass $u = \dfrac{U}{m}$ we have specific internal energy denoted by lowercase letter u.

Example 4.5

What is the internal energy of one mole of gas as in Example 4.4? Assume diatomic gas. Answer: Assume $f = 5$, $U = \dfrac{f}{2}nRT = \dfrac{5}{2} \times 1 \times 8.314 \times 303\,K = 6.298\,kJ$. Specific internal energy would be $u = \dfrac{U}{m} = \dfrac{U}{nM} = \dfrac{6.298\,kJ}{1\,mol \times 28.97\,g/mol} = 0.217\dfrac{kJ}{g} = 217\dfrac{kJ}{kg}$.

4.2.3 Pressure-Volume Work

We already know from the gas law that the product PV has units of work. Pressure-volume work is defined as $\delta W = PdV$. An *expanding* gas would do work $\delta W = -PdV$ against its surroundings, or conversely the surroundings can perform work on the gas by *compressing* it $\delta W = PdV$. The total work when changing from state P_1V_1 to state P_2V_2 is found by integration:

$$W = \int_1^2 \delta W = \int_1^2 PdV \tag{4.5}$$

Keep in mind that the integral depends on the path; for example, if volume is kept constant while the pressure changes (an isochoric path) $dV = 0$, the integral will be zero, meaning no work performed $W = 0$. Conversely, if pressure is kept constant while the volume changes (an isobaric path), the integral is

$$W = \int_1^2 PdV = P\int_1^2 dV = P(V_2 - V_1) \tag{4.6}$$

because P is constant and taken out of the integral. Since V is extensive, W is also extensive; we can express this equation in terms of *specific work* $w = \dfrac{W}{m}$ dividing both sides by mass $w = P\int_1^2 dv = P(v_2 - v_1)$.

Example 4.6

Calculate work and specific work for an isobaric expansion of one mole of air from 25 to 35 liters at 1 bar. Answer: Apply Equation 4.6 and use the negative sign because the gas is doing work $W = -P$ $(V_2 - V_1) = -100\,\text{kPa} \times 0.010\,\text{m}^3 = -1.0\,\text{kJ}$. Dividing by mass the specific work is $w = \dfrac{W}{nM} = \dfrac{1.0\,\text{kJ}}{1\,\text{mol} \times 28.97\,\text{g/mol}} = 34.5\,\dfrac{\text{kJ}}{\text{kg}}$.

4.2.4   REVERSIBLE AND IRREVERSIBLE PATHS

The simplest model to keep in mind when analyzing gas expansion and compression is of gas in a *cylinder* at a certain pressure P and subject to an external pressure P_{ext} (surroundings) by a *piston*. When the gas pressure equals the external pressure, the gas is in *equilibrium* with the surroundings. If P_{ext} is adjusted gradually to correspond with P while the gas expands or compresses from state 1 to state 2, then the gas was in equilibrium during the path. Thermodynamically, this is a *reversible* process. We can reverse it at any time by adjusting P_{ext} by a minute amount in the opposite sign. Therefore, we can go back from state 2 to state 1 following the same path. On the other hand, if P differs from P_{ext} during the path, we have an *irreversible* path. The gas was not in equilibrium with the surroundings during the path. An example would be to suddenly let a gas at pressure P_1 expand freely to $P_{ext} \ll P_1$. We can compress it back to P_1 but not following the same path. We will have more to consider about reversible and irreversible paths in the remainder of the chapter and in Chapter 6 when we discuss the second law of thermodynamics.

4.2.5   ISOCHORIC AND ISOBARIC PATHS: HEAT CAPACITIES

We can relate U to heat and heat capacity since internal energy is directly related to temperature. For this, we start by assuming an isochoric path that performs zero work $\delta W = PdV = 0$. The first law $dU = \delta Q + \delta W$ reduces to $dU = \delta Q$, that is, the change of energy is solely due to heat. Note that since U is extensive, heat Q must be extensive, so we can define *specific heat* as $q = \dfrac{Q}{m}$, in units of kJ/kg.

Using the definition of *molar heat capacity*, heat is written as $\delta Q = c_v ndT$, where c_v is the constant-volume molar heat capacity, which has the same units as R, that is, J/(mol K). The subscript v specifies constant volume. Equating to $dU = \delta Q$ we have

$$dU = c_v ndT \tag{4.7}$$

Now differentiate the kinetic energy formula $U = \dfrac{f}{2}nRT$ to get $dU = \dfrac{f}{2}nRdT$ and equating to $dU = c_v ndT$ we can rewrite $c_v ndT = \dfrac{f}{2}nRdT$ and conclude that

$$c_v = \dfrac{f}{2}R \tag{4.8}$$

Since f increases with temperature, so does c_v.

Next, assume pressure is held constant (isobaric process). An expanding gas would do work $\delta W = -PdV$ against its surroundings, which can be rewritten as a function of dT using the ideal gas law as $\delta W = -PdV = -nRdT$. The heat would be $\delta Q = c_p ndT$, where c_p is the constant-pressure molar heat capacity. The first law requires these changes of heat and work to equal changes in internal energy

$dU = \delta Q + \delta W$. Using Equation 4.7 and substituting δQ and δW, we have $c_v n dT = c_p n dT - nRdT$. Dividing by ndT we conclude that

$$c_p = c_v + R \qquad (4.9)$$

Note that the heat capacity at constant pressure is larger than at constant volume, since the added heat goes not only into increasing the temperature but also into performing work.

The ratio $\gamma = \dfrac{c_p}{c_v}$ is an important coefficient in heat engine models for energy conversion and in determining speed of sound in gases. We can see that $\gamma = \dfrac{c_p}{c_v} = \dfrac{c_v + R}{c_v} = 1 + \dfrac{R}{c_v} = 1 + \dfrac{2}{f}$ or

$$\gamma = \frac{c_p}{c_v} = 1 + \frac{2}{f} \qquad (4.10)$$

Example 4.7

What are the values of molar heat capacity for an ideal diatomic gas at 300 K? Assume $f = 5$. What is the ratio of these heat capacities? Answer: $c_v = \dfrac{5}{2} R = \dfrac{5}{2} \times 8.314$ J/(mol K) = 20.785 J/(mol K) and $c_p = 20.785 + 8.314 = 29.099$ J/(mol K). The ratio is $\gamma = \dfrac{c_p}{c_v} \simeq 1.4$. Equivalently, $\gamma = \dfrac{c_p}{c_v} = 1 + \dfrac{2}{f} = 1 + \dfrac{2}{5} = 1.4$.

A polyatomic gas with $f = 6$ has $\gamma = 1 + \dfrac{2}{6} = 1.333$. Vibrational motion contributes more degrees of freedom in polyatomic gases making γ smaller. Air is mostly N_2 and O_2 and thus it is common to use $\gamma = 1.4$ as an approximation for dry air at 300 K.

Heat capacity can also be expressed as *specific heat capacity* on a per mass unit basis, dividing the molar heat capacity by M, the molar mass. Thus c_v/M and c_p/M are the specific heat capacities expressed in kJ/(kg K). For notational simplicity, throughout the book we will use the same symbols c_p and c_v for molar and specific heat capacities; its interpretation will be clear from the context. The ratio of specific heat capacities on a per mass basis is the same as the ratio of molar heat capacities since the molar mass would cancel out in the ratio.

Example 4.8

Using the molar heat capacities from Example 4.7, calculate specific heat capacities for air at 300 K. Assume molecular mass M of air is 28.9 g/mol. Answer: $c_v = 20.785/28.97$ J/(mol K) (g/mol) = 0.717 kJ/(kgK) and $c_p = 29.099/28.97$ kJ/(kgK) = 1.004 kJ/(kgK). The ratio is $\gamma = \dfrac{c_p}{c_v} \simeq 1.4$.

Because of their dependence on f, both c_v and c_p increase with temperature T, and their ratio γ decreases with temperature. These changes $c_v(T), c_p(T), \gamma(T)$ are modeled using polynomials $\sum_{i=0}^{np} a_i T^i$ with coefficients estimated from empirical data and given in tables.

Functions cpcv.cal and cp.cv of package renpow facilitate calculation of heat capacity and γ as a function of T. For example, third-order polynomials for air can be calibrated from a data file using your local extdata folder or the system's by employing system file

```
> ref.coef.air <- cpcv.cal(datafile="extdata/AirCvCpTK.csv")
> ref.coef.air
$TK.ref
[1] 300
```

```
$cv.ref
[1] 20.80046

$cpcv.ref
[1] 1.4

$cv.coef
       TK    I(TK^2)     I(TK^3)
 2.992416e-03 6.344972e-06 -3.890969e-09

$cpcv.coef
       TK    I(TK^2)     I(TK^3)
-6.587279e-05 -7.156517e-08 5.381396e-11

>
```

The result of cpcv.cal can be used as value to argument ref of function cp.cv. This argument defaults to RefCoefAir which corresponds to the ref.coef.air calculated above. Function cp.cv calculates heat capacities at a given temperature, say, 100°C:

```
> cp.cv(TC=100)
$cv
[1] 21.051
$cp
[1] 29.363
$cp.cv
[1] 1.395
$cv.kg
[1] 0.727
$cp.kg
[1] 1.014
>
```

Here the last two values cv.kg and cp.kg are specific heat capacities (per kg), whereas cv and cp are molar heat capacities (per mole).

For quick calculations, a reasonable approximation is to use the value of specific heat at the middle of the range of temperature being considered. For instance, if considering the range 30°C to 100°C, we could use $(30 + 100)/2 = 65°C$ at which the specific heat values are

```
> cp.cv(TC=(30+100)/2)
$cv
[1] 20.923
$cp
[1] 29.238
$cp.cv
[1] 1.397
$cv.kg
[1] 0.722
$cp.kg
[1] 1.009
>
```

Example 4.9

Consider one mole of air at 1 bar and 25°C (nearly SATP). Recall from Example 4.4 that it should occupy 24.78 liters. Assume we pressurize it to 2 bar keeping the volume constant. What are the specific work, heat, and internal energy changes? Answer: For an isochoric process $W = 0$, and to calculate heat use $Q = \int_1^2 c_v(T)ndT$. To simplify, $c_v(T)$ is assumed constant and equal to its value at middle of the temperature range, which is from 25°C to $T_2 = \dfrac{P_2V_1}{nR} = \dfrac{200\,k \times 24.78 \times 10^{-3}}{1 \times 8.314} =$ 596.1K or 323.1°C. The middle of the range is approximately $(323 + 25)/2 \simeq 174°C$ and at that temperature c_v the specific heat can be found quickly from

```
> cp.cv(TC=(323+25)/2)$cv
[1] 21.365
>
```

Note that we have used $cv to get the molar cv component directly. Then use 21.365 kJ/(molK) to calculate $Q = c_v n\Delta T = 21.365 \times (323.1 - 25) \simeq 6.37$ kJ. This result can be expressed as specific heat divided by mass $q = Q/(nM) = 6.37/(1 \times 28.97/10^3) \simeq 219.8$kJ/kg. The specific internal energy is $\Delta u = q - w \simeq 219.8$ kJ/kg.

Here we have calculated total heat using molar heat capacity and then divided by mass to obtain specific heat and specific internal energy. Alternatively, we can first calculate mass, use specific heat capacity on a per kg basis, and calculate specific heat, work, and internal energy directly. The following example illustrates this alternative method.

Example 4.10

Consider one mole of air at 1 bar and 25°C (nearly SATP). Recall from Example 4.4 that it should occupy 24.78 liters. We let it expand to 34.78 l keeping pressure constant at 1 bar. What are the specific work, heat, and internal energy changes? Answer: The mass in kg is $m = nM = 1$mol $\times 28.97$g/mol $\times 10^{-3}$kg/g $\simeq 0.02897$ kg. For this isobaric path, specific work is $w = -P\Delta V / m = -100\,k \times 0.01/0.02897 = -34.52$kJ/kg. The final temperature is $T_2 = \dfrac{P_1V_2}{nR} =$ $\dfrac{100\,k \times 34.78 \times 10^{-3}}{1 \times 8.314} = 418.33$K or ~145.33°C. The c_p value for the middle of the range is found from

```
> cp.cv(TC=(145+25)/2)$cp.kg
[1] 1.012
>
```

using $c_p = 1.012$ kJ/(kgK) constant, $q = c_p\Delta T = 1.012 \times (145.33 - 25) = 121.74$ kJ/kg. The specific internal energy change is $\Delta u = q - w = (121.77 - 34.52)$ kJ/kg $= 87.25$kJ/kg.

These calculations are time consuming and not exact if there is wide variation in T since we assumed the heat capacities to be constant. We can expedite calculations using function path.calc of the renpow package, which includes temperature dependency of heat capacities and the particular relationship for each type of thermodynamic path. For instance, using values of Example 4.9:

```
# arguments: V(l), P(bar), T(°C), n(mol), M(g/mol)
# default n=1,M=28.9

# example specify V and P
x <- list(V=c(24.78,NA),P=c(1,2),path='isochor',lab=c("1","2"))
y <- path.calc(x)
```

Among the function arguments used here, V is an array with values of V1 and V2 (V1 is specified at 24.78, but V2 is left unspecified as NA), and P is array of corresponding pressure values. In this case, T is not given and will be calculated from P and V. Argument path declares the type of process, and lab is labels for the initial and final states of the path. Default values are one mole of dry air with $M = 28.97$. Other calculation modes such as providing T instead of P, to calculate temperature from pressure and volume, are explained in the function's help. The call y < - path.calc(x) produces y that contains values of specific volume (m3/kg), volume (liters), pressure (bar), temperature (°C), specific heat W and work Q in kJ/kg, heat capacities in kJ/kgK, and gamma γ. By default there are 1001 points calculated. Object y resulting from path.calc is used by other functions for plotting a variety of graphs to be described later. The last statement, path.summary(y), provides a quick way to view the results:

```
> path.summary(y)
$start.end
 v(m3/kg) V(l) P(b) T(C) s(J/Kkg) S(J/K) W(kJ/kg) Q(kJ/kg) cv(kJ/Kkg)
1  0.855 24.8  1 25.1  0.035   1.0     0    0.0    0.718
2  0.855 24.8  2 323.1 0.545  15.8     0  220.2    0.764
 cp(kJ/Kkg) cp/cv
1    1.005 1.400
2    1.051 1.376

$WQtot
 Wtot(kJ/kg) Qtot(kJ/kg)
1      0      220.2

$pts
[1] 1001

$call
   V1    V2   P1   P2  path  lab1  lab2
 24.78  <NA>   1   2 isochor   1    2

$nM
 n(mol) M(g/mol)  m(kg)
1   1   28.97 0.02897

>
```

Component $start.end provides the values at the start and end of the path; $s and $S correspond to entropy and will be explained in Chapter 6. The component $WQtot are the final or total values for work and heat, and $call reminds us of the values used in the call to the function. Note that heat Qtot 220.2 kJ/g is approximately the ones calculated in Example 4.9, where we assumed c_v is constant at the middle of the temperature range.

Similarly, the isobaric path of Example 4.10 can be calculated by

```
x <- list(V=c(24.78,34.78),P=c(1,NA),path='isobar',lab=c("3","4"))
y <- path.calc(x)
path.summary(y)
```

Examining the results in path.summary(y) we note the work and heat are –34.5 and 121.7 kJ/kg, respectively, as calculated in Example 4.9 by assuming c_p is constant in the middle of the range.

```
> path.summary(y)
$start.end
 v(m3/kg) V(l) P(b) T(C) s(J/Kkg) S(J/K) W(kJ/kg) Q(kJ/kg) cv(kJ/Kkg)
1   0.855 24.8   1  25.1   0.035    1.0     0.0      0.0       0.718
2   1.201 34.8   1 145.3   0.377   10.9   -34.5    121.7       0.733
 cp(kJ/Kkg) cp/cv
1    1.005 1.400
2    1.020 1.391

$WQtot
 Wtot(kJ/kg) Qtot(kJ/kg)
1    -34.5      121.7

$pts
[1] 1001

$call
   V1    V2   P1   P2  path  lab1  lab2
24.78 34.78    1 <NA> isobar    3     4

$nM
 n(mol) M(g/mol)  m(kg)
1    1    28.97 0.02897

>
```

These paths are visualized in the *P-v* plane using function path.lines of renpow,

```
x1 <- list(V=c(24.78,NA),P=c(1,2),path='isochor',lab=c("1","2"))
x2 <- list(V=c(24.78,34.78),P=c(1,NA),path='isobar',lab=c("3","4"))
path.lines(list(x1,x2))
```

which produces the left-hand side of Figure 4.5. The legend provides quick listing of calculated temperature, work, and heat. We can shade the area of the curve for the isobar as shown on the right-hand side of Figure 4.5 using shade.under=TRUE for the isobar list as follows:

```
x1 <- list(V=c(24.78,NA),P=c(1,2),path='isochor',lab=c("1","2"))
x2<-list(V=c(24.78,34.78),P=c(1,NA),path='isobar',lab=c("3","4"),
shade.under=TRUE)
path.lines(list(x1,x2))
```

4.2.6   Isothermal Paths

A consequence of the ideal gas equation of state is that for a fixed number of moles, changing from state 1 to state 2 at constant temperature, that is, *isothermally*, the product *PV* should be the same for

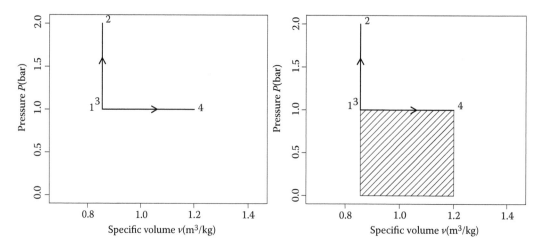

FIGURE 4.5 *P-v* graph of isochoric and isobaric paths. Shading under the curve illustrates work as integral of the curve. Left legend: 1–2: isochor 25.1°C–> 323.1°C, $W = 0$ kJ/kg, $Q = 220.2$ kJ/kg; 3–4: isobar, expansion 25.1°C–> 145.3°C, $W = -34.5$ kJ/kg, $Q = 121.7$ kJ/kg. Right: 1–2: isochor 25.1°C–> 323.1°C, $W = 0$ kJ/kg, $Q = 220.2$ kJ/kg; 3–4: isobar, expansion 25.1°C–> 145.3°C, $W = -34.5$ kJ/kg, $Q = 121.7$ kJ/kg.

the two states, implying no net change of energy. For state 1, $P_1V_1 = nRT$ and for state 2, $P_2V_2 = nRT$. Dividing we obtain $\dfrac{P_2}{P_1} = \dfrac{V_1}{V_2}$ or $P_1V_1 = P_2V_2$ or, generalizing, an isotherm is given by the hyperbola

$$PV = K \qquad (4.11)$$

where K is a constant.

Example 4.11

Suppose the temperature of one mole of gas is 30°C and is kept constant while the gas expands from state 1 at 0.01 m^3 to state 2 at 0.03 m^3. What is the pressure at each state? Plot pressure versus specific volume; assume the molecular mass M of this gas is 28.97 g/mol (dry air, which is close to that of N$_2$, the most abundant gas in air). Answer: $P_1 = \dfrac{nRT}{V_1} = \dfrac{1 \times 8.314 \times (273 + 30)}{0.01} = 251.9\text{kPa} = 2.52\,\text{bar}$. Because the process is isothermal, we have $\dfrac{P_2}{P_1} = \dfrac{V_1}{V_2}$ and therefore, $P_2 = P_1 \dfrac{V_1}{V_2} = 2.52 \times \dfrac{0.01}{0.03} = 0.84\,\text{bar}$. The plot is shown in Figure 4.6 (left-hand side) and it is obtained using function path.lines of the renpow package:

```
# example specify V and T
x <- list(V=c(10,30),T=c(30,NA),path='isotherm',lab=c("1","2"))
path.lines(x)
```

P are corresponding pressure values, but not used in this example because P1 will be calculated from *T*.

For an isotherm we can use $P = \dfrac{nRT}{V}$ with *T* constant. When substituted in the integral to compute work when going from state 1 to state 2, $W = \displaystyle\int_1^2 P dV$, we can integrate $W = \displaystyle\int_1^2 nRT\,\dfrac{dV}{V} = nRT\displaystyle\int_1^2 \dfrac{dV}{V} =$

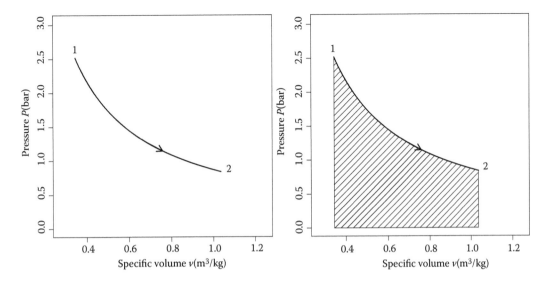

FIGURE 4.6 *P-v* graph of isotherm expansion. Work is the integral of *P* or the shaded area under the curve. Left: 1–2: isotherm, expansion 30°C–> 30°C, $W = -95.5$ kJ/kg, $Q = 95.5$ kJ/kg. Right: 1–2: isotherm, expansion 30°C–> 30°C, $W = -95.5$ kJ/kg, $Q = 95.5$ kJ/kg.

$nRT \ln \left(\dfrac{V_2}{V_1} \right)$. Here nRT is taken out of the integral because the path is isothermal. You must realize that to keep T constant while the gas expands we must add heat to avoid cooling. The first law then predicts that the heat added must equal the work done to avoid a change in U, that is to say, $\Delta U = 0 = Q - W$ and therefore $Q = W$. The isotherm just obtained assumes the path is reversible; it is done gradually to keep the gas in equilibrium with the surroundings at all times. In other words, the gas pressure is the same as the external pressure P_{ext} while traversing the path.

If the process were to be irreversible, the work is reduced. For example, expanding the gas freely against the external pressure $P_2 = P_{ext}$, the work is $W = P_{ext} \displaystyle\int_1^2 dV = P_2(V_2 - V_1) = nRT \dfrac{V_2 - V_1}{V_2}$. This work is less than the one for a reversible process since $\ln \left(\dfrac{V_2}{V_1} \right) > \dfrac{V_2 - V_1}{V_2}$. We will further interpret this difference when we discuss entropy in Chapter 6.

Using Equation 4.11 we can express work indistinctly using the ratio of pressures or the ratio of volumes

$$W = nRT \ln \left(\frac{V_2}{V_1} \right) = nRT \ln \left(\frac{P_1}{P_2} \right) \tag{4.12}$$

This work is the area under an isotherm curve like the one shown in Figure 4.6 (right-hand side), which is obtained using argument shade.under=TRUE in the list of x:

```
x<-list(V=c(10,30),T=c(30,NA),path='isotherm',lab=c("1","2"),shade.under
=TRUE)
path.lines(x)
```

Example 4.12

Consider reversible isothermal expansion of Example 4.10, where one mole of air previously compressed three times with respect to final volume and held at 30°C. What is the specific work done? Answer: We use Equation 4.12 $W = nRT \ln\left(\dfrac{V_2}{V_1}\right) = 1 \times 8.314 \times 303 \times \ln\left(\dfrac{3}{1}\right) J = 2.767$ kJ. Then divide by mass 0.02897 kg to get $w = -95.51$ kJ/kg. Note that this value of work is printed inside the legend box in Figure 4.6. It is equal and opposite sign to the heat because we assume it to be a reversible process.

Figure 4.7 shows two isotherm paths at different temperatures and the corresponding work as shaded areas. Work is larger for the higher-temperature isotherm (left-hand side) and includes the work done by the lower-temperature one. The difference of work between the two corresponds to the shaded area in between (right-hand side). This figure is obtained using PV.lines as follows:

```
x1 <- list(V=c(10,30),T=c(200,200),path='isotherm',lab=c("3","4"))
x2 <- list(V=c(10,30),T=c(30,30),path='isotherm',lab=c("1","2"))
x <- list(x1,x2)
path.lines(x,shade.between=TRUE)
```

4.2.7 ADIABATIC PATHS

Consider PV work to be done by a closed system with an expanding gas $\delta W = -PdV$ but without adding heat $\delta Q = 0$. This is an *adiabatic* process; the first law reduces to $dU = \delta W + \delta Q = -PdV$. Equate this to Equation 4.7 so that we can write $ndT = -\dfrac{PdV}{c_v}$. Now use the gas law $ndT = -\dfrac{nRT}{V}\dfrac{dV}{c_v}$ and rearrange to get ready for integration of each side $\dfrac{dT}{T} = -\dfrac{R}{c_v}\dfrac{dV}{V}$. If we assume c_v to be constant, we can integrate between state 1 and 2 $\ln\left(\dfrac{T_2}{T_1}\right) = -\dfrac{R}{c_v}\ln\left(\dfrac{V_2}{V_1}\right)$ or $\dfrac{T_2}{T_1} = \left(\dfrac{V_1}{V_2}\right)^{\frac{R}{c_v}}$. Now because

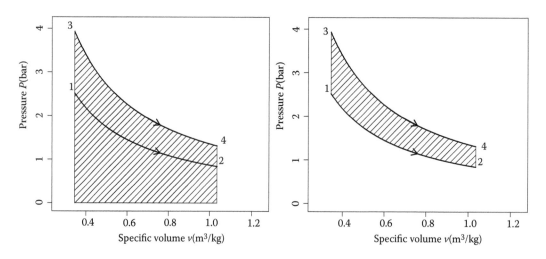

FIGURE 4.7 Two isotherms. Difference in work done by each expansion is the shaded area between the curves. Left: 1–2: isotherm, expansion 30°C–> 30°C, $W = -95.5$ kJ/kg, $Q = 95.5$ kJ/kg; 3–4: isotherm 200°C–> 200°C, $W = -149.1$ kJ/kg, $Q = 149.1$ kJ/kg. Right: 1–2: isotherm expan 30°C–> 30°C, $W = -95.5$ kJ/kg, $Q = 95.5$ kJ/kg; 3–4: isotherm, expansion 200°C–> 200°C, $W = -149.1$ kJ/kg, $Q = 149.1$ kJ/kg.

$\gamma = \dfrac{c_p}{c_v} = \dfrac{c_v + R}{c_v} = 1 + \dfrac{R}{c_v}$ we have $\dfrac{T_2}{T_1} = \left(\dfrac{V_1}{V_2}\right)^{\gamma-1}$ or, generalizing, $TV^{\gamma-1} = K$, where K is a constant. Moreover, using the ideal gas law $\dfrac{P_2 V_2}{P_1 V_1} = \left(\dfrac{V_1}{V_2}\right)^{\gamma-1}$ or $\dfrac{P_2}{P_1} = \left(\dfrac{V_1}{V_2}\right)^{\gamma}$ or, generalizing,

$$PV^\gamma = K_A \tag{4.13}$$

where K_A is a constant. It is important to realize that Equation 4.13 is only valid within an interval such that γ can be assumed constant. For wider variation of temperature, we need to use a function $\gamma(T)$ as we discussed earlier in this chapter. The adiabatic path is an important equation in energy conversion as well as in atmospheric dynamics.

Example 4.13

Suppose one mole of air at 30°C occupying 0.03 m³ (state 3) is compressed adiabatically to 0.01 m³ (state 4) to reach 100°C. What is the pressure at each state? Plot pressure versus specific volume. Answer: The pressure P_3 can be calculated from the gas law $P_3 = \dfrac{nRT}{V_3} = \dfrac{1 \times 8.314 \times (273 + 30)}{0.03} = 839.71\,\text{kPa} = 0.84\,\text{bar}$. Before using Equation 4.13 estimate γ at the middle of the temperature range,

```
> cp.cv(TC=(30+100)/2)$cp.cv
[1] 1.397
>
```

Then, we can calculate K_A from Equation 4.13 as $P_3 V_3^\gamma = K_A = 83971 \times 0.03^{1.397} = 626.1346$ Pa × m$^{3 \times 1.397}$. Assuming the specific heat ratio constant with the value at midtemperature range, we can calculate the pressure at state 4 as $P_4 = \dfrac{K_A}{V_4^\gamma} = \dfrac{626.1346}{0.01^{1.397}} = 3.896\,\text{bar}$. The plot is shown in Figure 4.8 (left-hand side, path 3–4). On the same plot, we see an isotherm reaching a lower pressure because the final temperature is lower (path 1–2).

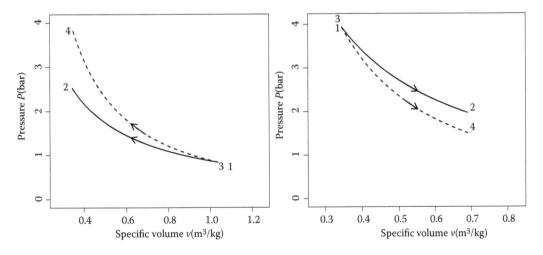

FIGURE 4.8 Comparing isothermal and adiabatic paths for compression and expansion. Left: 1–2: isotherm, expansion 200°C–> 200°C, $W = -94.1$ kJ/kg, $Q = 94.1$ kJ/kg; 3–4: adiabat expan 200°C–> 87.6°C, $W = -82.4$ kJ/kg, $Q = 0$ kJ/kg. Right: 1–2: isotherm, expansion 200°C–> 200°C, $W = -94.1$ kJ/kg, $Q = 94.1$ kJ/kg; 3–4: adiabat, expansion 200°C–> 87.6°C, $W = -82.4$ kJ/kg, $Q = 0$ kJ/kg.

Both lines are obtained using function path.lines of the renpow package:

```
x1 <- list(V=c(30,10),T=c(30,NA),path='isotherm',lab=c("1","2"))
x2 <- list(V=c(30,10),T=c(30,NA),path='adiabat',lab=c("3","4"))
path.lines(list(x1,x2))
```

We specify the initial and final volume, and only the initial temperature (second entry of T is NA); for the adiabat, the calculation yields the final temperature once it reaches the specified final volume. The work, shown in the legend, is positive for both paths because it is done on the gas. The adiabat shows an increase in T; this thermal energy increase is not due to heat ($\delta Q = 0$), but to an increase in internal energy or higher average molecular kinetic energy. We study work next.

The work done by adiabatic expansion or compression is calculated from $W = \int_1^2 P dV$ upon

substituting P for $P = \dfrac{K_A}{V^\gamma}$ and integrating $W = \int_1^2 \dfrac{K_A}{V^\gamma} dV = K_A \int_1^2 \dfrac{dV}{V^\gamma}$. Use $\int x^a dx = \dfrac{x^{1+a}}{1+a}$ to obtain

$$W = \frac{K_A}{1-\gamma}\left(V_2^{1-\gamma} - V_1^{1-\gamma}\right) \tag{4.14}$$

This is valid as long as the temperature range is small enough to have nearly constant γ.

Example 4.14

Calculate work done to one mole of air given in Example 4.12, that is, starting at 30°C and 0.03 m³ (state 1) air is compressed adiabatically to 0.01 m³ (state 2). Answer: We already calculated K_A, therefore $W = \dfrac{K_A}{1-\gamma}\left(V_2^{1-\gamma} - V_1^{1-\gamma}\right) = \dfrac{626.1346}{-0.397}\left(0.01^{-0.397} - 0.03^{-0.397}\right) P_a m^3 = -3.469\,kJ$.

The sign is negative because work is done on the gas. Then, when writing $dU = -W$ the internal energy change results positive. Calculate specific work dividing by mass of 0.02897 kg to get $w \simeq -119.7\,kJ/kg$. This work corresponds to curve 3–4 in Figure 4.8 (left-hand side), where this value is obtained by numerical evaluation instead of Equation 4.14.

Isothermal and adiabatic expansion is specified by the two values of V for the paths as shown here:

```
x1 <- list(V=c(10,20),T=c(200,NA),path='isotherm',lab=c("1","2"))
x2 <- list(V=c(10,20),T=c(200,NA),path='adiabat',lab=c("3","4"))
x <- list(x1,x2)
path.lines(x)
```

Expansion (from 10 to 20 liters, starting at 200°C). The resulting graphs are in Figure 4.8 (right-hand side). Work, heat, and temperature are interpreted as we did for compression.

Isothermal ($PV = K$) compression and expansion paths are different from adiabatic ($PV^\gamma = K$) compression and expansion paths in the P-v plane (Figure 4.9). Adiabatic lines have steeper slopes because of exponent γ. As we will see in the next section, isothermal and adiabatic (compression and expansion) are combined to obtain cyclic thermodynamic processes, which are the basics of thermal energy conversion for power generation.

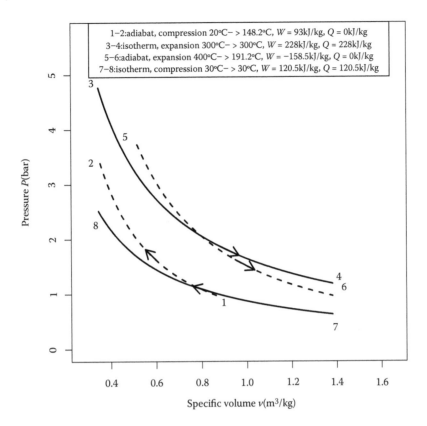

1–2:adiabat, compression 20°C– > 148.2°C, W = 93kJ/kg, Q = 0kJ/kg
3–4:isotherm, expansion 300°C– > 300°C, W = 228kJ/kg, Q = 228kJ/kg
5–6:adiabat, expansion 400°C– > 191.2°C, W = −158.5kJ/kg, Q = 0kJ/kg
7–8:isotherm, compression 30°C– > 30°C, W = 120.5kJ/kg, Q = 120.5kJ/kg

FIGURE 4.9   Isotherm and adiabatic curves for various conditions.

The lines of Figure 4.9 are obtained using renpow by specifying a path for each line, binding them in a list, and calling path.lines as follows:

```
# four lines
x1 <- list(V=c(25,10),T=c(20,NA),path='adiabat',lab=c("1","2"))
x2 <- list(V=c(10,40),T=c(300,NA),path='isotherm',lab=c("3","4"))
x3 <- list(V=c(15,40),T=c(400,NA),path='adiabat',lab=c("5","6"))
x4 <- list(V=c(40,10),T=c(30,NA),path='isotherm',lab=c("7","8"))
x <- list(x1,x2,x3,x4)
path.lines(x)
```

Besides application in energy conversion, another example of adiabatic compression related to the themes of this book is the warming or "heating" of a descending mass of air in Earth's atmosphere as it increases its pressure when pushed down, say, down a mountain. The opposite example would be the "cooling" of an air mass when lifted adiabatically experiencing a drop in pressure.

4.3   HEAT ENGINE, CYCLES, AND CARNOT LIMIT

4.3.1   Carnot Limit

In this section, we describe the basics of maximum efficiency of a *heat engine*, a contribution of Nicolas Sadi Carnot in 1824. First, let us recall that the steam engine was one of the major inventions that triggered the industrial revolution in the 1700s. Condensing steam in a cylinder drove a piston that connected to a rod in order to do mechanical work. This process is repeated indefinitely as long as steam is provided by boiling water. The steam becomes a high temperature reservoir (HTR) or *heat source*; the steam performs mechanical work releasing its heat to the low temperature reservoir (LTR) or *heat sink*. The importance of steam engines drew attention to improvements of its performance or efficiency. Carnot's contribution was a theoretical limit to this efficiency, that is, any technical attempt to increase efficiency will hit this theoretical upper limit. This concept is related to the development of the second law of thermodynamics, which we will discuss in Chapter 6.

The repetitive process of sequential thermodynamic paths, that is, compression and expansion, is described as a *thermodynamic cycle*. Thermodynamic cycles apply to a variety of processes, such as for a steam engine, used extensively in the past for transportation on rails (and nowadays for electrical power generation), and internal combustion engines (e.g., based on gasoline, diesel, and gas), used mostly for transportation but still used some in power generation. We briefly describe how the thermodynamic cycles relate to power generation in the next section.

Figure 4.10 shows a conceptual diagram of a heat engine illustrating what happens in each cycle. Heat engines attempt to perform work from heat flow, they lose heat to the environment, and therefore only a part of the input energy is converted to output energy. Thermal efficiency is the ratio of work output to the heat input

$$\eta = \frac{W}{Q_H} = \frac{Q_H - Q_L}{Q_H} = 1 - \frac{Q_L}{Q_H} \qquad (4.15)$$

The Carnot limit is a maximum $\eta_{\max}$ of this efficiency and is given by the temperature difference of the two reservoirs

$$\eta_{\max} = \frac{T_H - T_L}{T_H} = 1 - \frac{T_L}{T_H} \qquad (4.16)$$

where T_H and T_L are the *absolute* temperatures of the HTR or heat source and LTR or heat sink, respectively, and given in K. We will demonstrate how to derive this limit in the next section after we calculate heat and work for each path of the cycle.

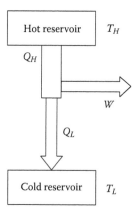

FIGURE 4.10 Conceptual diagram of a heat engine.

Example 4.15

What is the maximum efficiency (Carnot limit) for a heat engine with a heat source at 600°C and a heat sink at 25°C? Answer: First convert to absolute temperature $T_H = 600 + 273 = 873$ K and $T_C = 25 + 273 = 298$ K. Now calculate η_{max} from Equation 4.16 $\eta_{max} = 1 - \dfrac{298}{873} = 0.658 \simeq 0.66$. The Carnot limit is ~66%.

In general, high and low temperature reservoirs can occur at different parts of the system. The HTR corresponds to the products of combustion, for example, in the firebox of a boiler, inside the combustion chamber of gasoline and diesel engines, and inside the combustion chamber of a gas turbine. The LTR or heat sink is typically the atmosphere, which receives the "waste" heat.

A Carnot cycle is an ideal cycle comprising all ideal processes and used to demonstrate that the thermal efficiency is given by Equation 4.16. Carnot proved that the value predicted for this ideal cycle is the highest value allowed by nature. All real engines have a value lower than this maximum because of other losses such as friction.

4.3.2 The Ideal Carnot Cycle in the *P-v* Plane

A common representation of a thermodynamic cycle is to portray the changes of state of a working fluid, say, a gas, on the *P-v* plane. We will use the Carnot cycle of Figure 4.11 as an example to focus the description while demonstrating the Carnot limit. In a cycle, the system returns repetitively to state 1 by four sequential paths. From state 1 to 2, then from state 2 to 3, then from 3 to 4, and finally returning from 4 to 1.

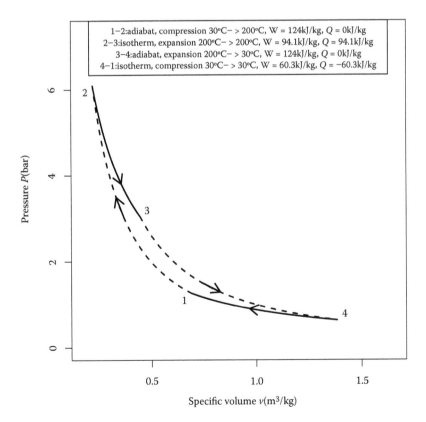

1–2:adiabat, compression 30°C– > 200°C, W = 124kJ/kg, Q = 0kJ/kg
2–3:isotherm, expansion 200°C– > 200°C, W = 94.1kJ/kg, Q = 94.1kJ/kg
3–4:adiabat, expansion 200°C– > 30°C, W = 124kJ/kg, Q = 0kJ/kg
4–1:isotherm, compression 30°C– > 30°C, W = 60.3kJ/kg, Q = −60.3kJ/kg

FIGURE 4.11 Carnot thermodynamic cycle on *P-v* plane.

The system goes from state 1 to 2 in an adiabatic reversible path, thus the heat transferred is zero; the only contribution to internal energy is work done on the gas δW to compress it. Using the first law, Equation 4.3, $U_2 - U_1 = \int_1^2 \delta W = W_{12}$. The temperature increases from that one of the LTR $T_L = \dfrac{P_1 V_1}{nR}$ to the HTR $T_H = \dfrac{P_2 V_2}{nR}$.

Then, from state 2 to 3 the gas performs work while it expands at constant temperature T_H (isothermal reversible expansion at the HTR) requiring heat input Q_H. Since $dT = 0$, there is zero change in internal energy $\Delta U = U_3 - U_2 = 0 = Q_H - W_{32}$; therefore, $Q_H = \int_2^3 \delta W$. The integral is

$W_{32} = \int_2^3 PdV$ and it is the area under the curve of the upper trace of the cycle. Using Equation 4.12,

$$Q_H = W_{32} = nRT_H \ln\left(\frac{V_3}{V_2}\right).$$

From state 3 to 4, the gas expands along a reversible adiabatic path, which means that $U_4 - U_3 = -\int_3^4 W = -W_{34}$, dropping from the HTR to the LTR. Finally, the working fluid will return to state 1 by reversible isothermal (at T_C) compression, losing heat Q_C, thus the change in internal energy is zero $U_1 - U_4 = 0 = Q_L - W_{41}$. During this part of the cycle, work is done on the fluid. The integral $W_{41} = \int_4^1 PdV$ is the area under the curve of the lower trace $Q_L = W_{41} = nRT_L \ln\left(\frac{V_1}{V_4}\right)$.

Combining the changes of internal energy for all the paths of the cycle, we have $(U_2 - U_1) + (U_3 - U_2) + (U_4 - U_3) + (U_1 - U_4) = U_1 - U_1 = 0$ because we have returned to a state with the same energy. However, this does not mean that heat or work are zero because they are path functions integrated along the cycle.

Integration of a path function along a cyclical path is represented by $X = \oint \delta x$. Succinctly, we can write $\oint \delta Q = \oint \delta W$ to say that in a cycle, net exchanges of work and heat are equal so that the internal energy does not change. Consider heat and integrate along individual paths that make up the cycle

$$Q = \oint \delta Q = \int_1^2 \delta Q + \int_2^3 \delta Q - \int_3^4 \delta Q - \int_4^1 \delta Q = 0 + Q_H - 0 - Q_L \text{ or simply } Q = Q_H - Q_L. \text{ Similarly for}$$

work, $W = \oint \delta W = W = \oint PdV = \int_1^2 \delta W + \int_2^3 \delta W - \int_3^4 \delta W - \int_4^1 \delta W = W_{12} - W_{23} - W_{34} + W_{41}$, which is the area of the polygon enclosed by the cycle (Figure 4.12). We conclude that $Q_H - Q_L = W$, which is the heat engine equation.

The net difference in heat is

$$Q_H - Q_L = nRT_H \ln\left(\frac{V_3}{V_2}\right) - nRT_L \ln\left(\frac{V_1}{V_4}\right) \tag{4.17}$$

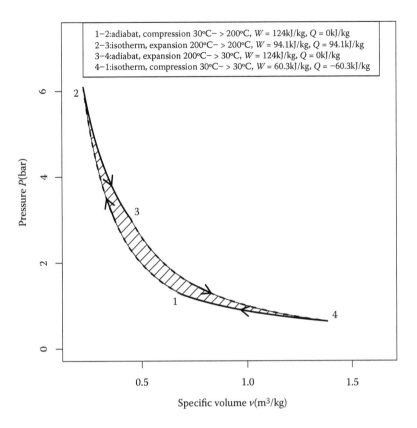

1–2:adiabat, compression 30°C– > 200°C, $W = 124$kJ/kg, $Q = 0$kJ/kg
2–3:isotherm, expansion 200°C– > 200°C, $W = 94.1$kJ/kg, $Q = 94.1$kJ/kg
3–4:adiabat, expansion 200°C– > 30°C, $W = 124$kJ/kg, $Q = 0$kJ/kg
4–1:isotherm, compression 30°C– > 30°C, $W = 60.3$kJ/kg, $Q = -60.3$kJ/kg

FIGURE 4.12 Carnot thermodynamic cycle on P-v plane. Net work is shaded area inside the cycle.

At the LTR, the temperature $T_L = \dfrac{P_1 V_1}{nR} = \dfrac{P_4 V_4}{nR}$ and at the HTR $T_H = \dfrac{P_2 V_2}{nR} = \dfrac{P_3 V_3}{nR}$. The ratio of these temperatures is $\dfrac{T_H}{T_L} = \dfrac{P_2 V_2}{P_1 V_1} = \dfrac{P_3 V_3}{P_4 V_4}$. Following the adiabat from 1 to 2, we must have $P_1 V_1^\gamma = P_2 V_2^\gamma$ or rearranging $\dfrac{V_1^\gamma}{V_2^\gamma} = \dfrac{P_2}{P_1}$, which when substituted in $\dfrac{T_H}{T_L} = \dfrac{P_2 V_2}{P_1 V_1}$ yields $\dfrac{T_H}{T_L} = \dfrac{V_1^{\gamma-1}}{V_2^{\gamma-1}}$. In a similar way for states 3 and 4, $\dfrac{T_H}{T_L} = \dfrac{V_3^{\gamma-1}}{V_4^{\gamma-1}}$. Combining these last two equations, we have $\dfrac{V_2^{\gamma-1}}{V_1^{\gamma-1}} = \dfrac{V_3^{\gamma-1}}{V_4^{\gamma-1}}$, which means $\dfrac{V_1}{V_4} = \dfrac{V_3}{V_2}$. Using this last expression in Equation 4.17 we conclude that $Q_H - Q_L =$

$$nRT_H \ln\left(\frac{V_3}{V_2}\right) - nRT_L \ln\left(\frac{V_1}{V_4}\right) = nR \ln\left(\frac{V_3}{V_2}\right)(T_H - T_L).$$

Now, let us demonstrate the Carnot limit from this cycle. Efficiency is $\eta = \dfrac{W}{Q_H} = 1 - \dfrac{Q_L}{Q_H}$ and

using the isothermal formulas for Q_H and Q_L we have $\eta = 1 - \dfrac{nRT_L \ln\left(\dfrac{V_1}{V_4}\right)}{nRT_H \ln\left(\dfrac{V_3}{V_2}\right)}$ and because $\dfrac{V_1}{V_4} = \dfrac{V_3}{V_2}$,

the efficiency is simply $\eta = 1 - \dfrac{T_L}{T_H}$, which is the Carnot limit presented in the previous section.

Example 4.16

Assume a Carnot heat engine with a heat source at 200°C and a heat sink at 30°C. The isotherm compression in the LTR is from $V_4 = 40$ l to $V_1 = 20$ l. Use one mole of air. This cycle corresponds to Figure 4.12. What are the heat Q_H and Q_L from HTR and LTR? Express these also as specific heat values. Draw the cycle. What is the maximum efficiency? Answer: We know $\frac{V_1}{V_4} = \frac{20\,l}{40\,l} = \frac{1}{2}$, then

using $\frac{V_3}{V_2} = \frac{V_1}{V_4} = \frac{1}{2}$. Next, $Q_H = -nRT_H \ln\left(\frac{V_3}{V_2}\right) = -1 \times 8.314 \times (273 + 200) \ln\left(\frac{1}{2}\right) = 2.726$ kJ

and $Q_L = nRT_L \ln\left(\frac{V_1}{V_4}\right) = 1 \times 8.314 \times (273 + 30) \ln\left(\frac{1}{2}\right) = -1.746$ kJ. Dividing by mass 0.02897 kg, the specific values are $q_H = 94.10$ kJ/kg and $q_L = 60.27$ kJ/kg. These correspond near the values displayed in the legend of Figure 4.12. The difference is $q_H - q_L \simeq 33.83$ kJ/kg. The max efficiency $\eta_{max} = 1 - \frac{Q_L}{Q_H} = 1 - \frac{1.75}{2.73} \simeq 0.36$. Note that this is the same as $\eta_{max} = 1 - \frac{q_L}{q_H} \simeq 0.36$, because the mass cancels out when taking the ratio. It is important to note that this efficiency value is the same as obtained directly from the temperature $\eta_{max} = 1 - \frac{273 + 30}{273 + 200} \simeq 0.36$.

The efficiency is low because the temperature difference is not large. The work values are printed on the legend of Figure 4.12. This figure is obtained using four paths

```
TH=200;TL=30;V1=20;V4=40
x1 <- list(V=c(V1,NA),T=c(TL,TH), path='adiabat',lab=c("1","2"))
y1 <- path.calc(x1); V2 <- y1$V[length(y1$V)]; V3=V2*V4/V1
x2 <- list(V=c(V2,V3),T=c(TH,NA), path='isotherm',lab=c("2","3"))
x3 <- list(V=c(V3,NA),T=c(TH,TL), path='adiabat',lab=c("3","4"))
x4 <- list(V=c(V4,V1),T=c(TL,NA), path='isotherm',lab=c("4","1"))
z <- list(x1,x2,x3,x4)
path.lines(z,lab.cycle=TRUE,shade.cycle=TRUE)
```

where the second path x_2 is forced to V_2 and V_3 calculated from the output y_1 so that their ratio matches V_1 and V_4 as required by an ideal Carnot cycle. Asserting the argument lab.cycle =TRUE skips printing the label of the ending state for each path to avoid duplicate labels on the graph; this is just for display simplicity. Finally, making the argument shade.cycle equals to TRUE shades the area within the cycle; the default is FALSE.

These paths are combined in a single function path.cycles of renpow and can be employed simply by one call

```
x <- list(TH=200,TL=30,V1=20,V4=40,cty='carnot')
y <- path.cycles(x,shade.cycle=TRUE)
```

Then, we can query a summary

```
> path.cycles.summary(y)
$WQnet
  Qin(kJ/kg) Qout(kJ/kg)  Eff
1     94.1      -60.3 0.36

$end.state
   v(m3/kg) V(l) P(b) T(C) s(J/Kkg) S(J/K) W(kJ/kg) Q(kJ/kg) cv(kJ/Kkg)
1-2  0.223  6.5  6.1  200   0.035    1.0    124.0    0.0      0.742
2-3  0.446 12.9  3.0  200   0.233    6.8    -94.1   94.1      0.742
3-4  1.381 40.0  0.6   30   0.233    6.8   -124.0    0.0      0.718
4-1  0.690 20.0  1.3   30   0.035    1.0     60.3  -60.3      0.718
   cp(kJ/Kkg) cp/cv
```

```
1-2    1.029 1.387
2-3    1.029 1.387
3-4    1.005 1.400
4-1    1.005 1.400

$pts
[1] 1001

$nM
 n(mol) M(g/mol)  m(kg)
1    1   28.97 0.02897
```

or more succinctly when we just want to have the efficiency

```
> path.cycles.summary(y)$WQnet
 Qin(kJ/kg) Qout(kJ/kg) Eff
1     94 .1     -60.3 0.36
>
```

We will further discuss other heat engine models based on cyclic processes in the *P-v* plane with more details in forthcoming chapters. The various devices used for conversion of heat energy into mechanical work have common features. All of them receive heat energy from a source of heat; convert a fraction of this heat into work and deliver the remainder to a heat sink. Their performance can be predicted by modeling their working in terms of a thermodynamic cycle.

4.3.3 HEAT ENGINE AND ELECTRIC POWER GENERATION

Heat engines are used widely for electric power generation. In the most general sense, thermal electrical generation proceeds as follows (Figure 4.13). First, use some energy source to produce heat; for example, combustion of fossil fuels, nuclear reactions, and solar radiation. This heat is used to convert water into high-pressure steam in a boiler. The high-pressure steam then expands going through a turbine, causing it to spin at high speed. This turbine is mechanically coupled to an electric generator, and thus the turbine drives a generator, which rotates and produces electrical power. We will describe this entire process with more detail in Chapters 7 and 9.

Please realize that the efficiency we have calculated in this chapter is the efficiency of the thermal process alone; that is to say, of converting heat transfer to available mechanical work. It does not include the efficiency of producing heat from burning fuel, the efficiency of the turbine to produce torque from the expanding gas, and the efficiency of the generator to convert mechanical work to electrical power. We will discuss these other efficiencies in later chapters.

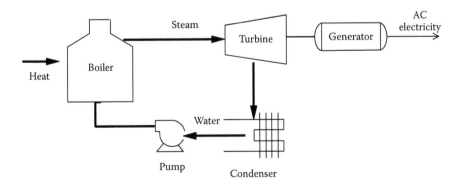

FIGURE 4.13 Use of heat engines for electric power generation.

We should emphasize that a thermal conversion process, that is, converting heat transfer (from a HTR to a LTR) to work, does not necessarily imply burning carbon-based fuels to generate the heat needed. We can use the existing renewable temperature gradient created by solar radiation, such as thermal gradients in ocean water, or existing gradients from Earth's internal processes (geothermal). The maximum thermal efficiency of each conversion process is predictable from the Carnot limit. Let us consider one case of renewable temperature gradient to illustrate how to apply these concepts.

4.3.4 OTEC (Ocean Thermal Energy Conversion) Power Plant

The Carnot's limit given in Equation 4.16) can be used to predict the efficiency of an OTEC plant (Ocean Thermal Energy Conversion scheme) from the knowledge of temperature of the surface water (used as a heat source) and deep waters (which serve as a heat sink) (Figure 4.14). The heat engine operates by exploiting the temperature difference between the surface of the ocean and the deep layers of water.

Example 4.17

Calculate the maximum thermal efficiency that can be obtained when the surface waters at a temperature of 22°C are used as the heat source and the ocean waters at a depth of 1000 m having a temperature of 4°C are used as the heat sink. Answer: The highest possible value of thermal efficiency is the Carnot efficiency given by $\eta_{max} = \dfrac{T_H - T_L}{T_H} = 1 - \dfrac{T_L}{T_H}$, where T_H and T_L are the absolute values of the heat source (HTR, hot surface waters at 22°C) and heat sink temperatures (LTR, cold water at 4°C for a depth ~1000 m), respectively. Use absolute temperatures K $T_H = 273 + 22 = 295$ K, $T_L = 273 + 4 = 277$ K. Therefore $\eta_{max} = 1 - \dfrac{277}{295} = 0.061$ or 6.1%.

As we can see from this example, the Carnot efficiency applies here as the upper limit for a scheme used for converting heat into work. In practice, the efficiency of the actual power plant will be lower than the value predicted above due to the various losses. In addition, there is a 20% to 40% reduction of power, which is used to pump water from deep layers. As we progress through this book, we will see that 6.1% maximum efficiency is very low compared with other heat engines. However, note that in OTEC, the input heat is solar energy and not derived from burning fuels.

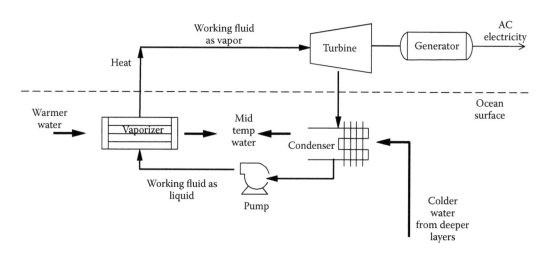

FIGURE 4.14 Ocean Thermal Energy Conversion (OTEC).

4.3.5 STEADY FLOW

Many thermodynamic processes operate in *steady flow* conditions in which flow or rates of change are constant with time. For example, consider heat Q in J, and heat flow $\dot{Q}$ in J/s as a rate. Recall from Chapter 1 that it is convenient to use a dot on top of a symbol to denote its derivative with respect to time. In the same manner, we can write rate of work $\dot{W}$, which is mechanical power $P = \dot{W}$.

Heat engine calculations can be expressed using rates. By invoking the first law, $\dot{Q}_H = \dot{Q}_C + \dot{W}$. Now divide both sides by $\dot{Q}_H$ to get $1 = \dfrac{\dot{Q}_C}{\dot{Q}_H} + \dfrac{\dot{W}}{\dot{Q}_H}$ and rearranging we get $1 - \dfrac{\dot{Q}_C}{\dot{Q}_H} = \dfrac{\dot{W}}{\dot{Q}_H}$, which gives us an expression for efficiency based on rates or flows $\eta = \dfrac{\dot{W}}{\dot{Q}_H}$. We can rewrite it as mechanical power $P = \dot{W} = \eta\dot{Q}_H$. Note that $\dot{Q}_C = \dot{Q}_H - \dot{W}$. All these terms can be written in specific terms dividing by mass.

Example 4.18

Consider a heat engine with $\dot{Q}_H = 1$ kJ/s, $\eta = 40\%$. What is the power from work? What is the heat flow to the LTR? Answer: We can use $P = \dot{W} = \eta\dot{Q}_H$ to obtain $P = 0.4 \times 1$ kW = 400 W. Note that $\dot{Q}_L = \dot{Q}_H - \dot{W}$ and then $\dot{Q}_L = 1$ kW $- 400$ W $= 600$ W.

EXERCISES

4.1. Suppose a closed system transfers 100 Nm of heat to the surroundings while the surrounding performs 300 J of work on the system. What is the change of internal energy?

4.2. How much mechanical energy (in J) would it take to change the temperature of 10 kg of water from 10°C to 20°C? Assume heat capacity of water is $C = 4.187$ kJ/(kg K).

4.3. Assume one mole of dry air at 100°C s pressurized at 2 bar. What is the volume? What is the specific volume?

4.4. What is the internal energy of two moles of dry air at 100°C?

4.5. Consider one mole of air and an isochoric process from 25 l to go from 2 bar to 1 bar and then followed by an isobaric process at 1 bar from 25 l to 30 l. What are the specific work, heat, and internal energy changes for each path? What is the specific work, heat, and internal energy changes for the entire path?

4.6. Assume a reversible isothermal compression of one mole of dry air at 40°C from state 1 at 0.03 m³ to state 2 at 0.02 m³. What is the pressure at each state? Plot pressure versus specific volume using renpow functions.

4.7. Calculate specific work done by adiabatic expansion of one mole of air (starting at 300°C and 10 l) to 30 l. Draw this path in the P-v plane. Use renpow functions.

4.8. Suppose we could concentrate solar radiation by parabolic surfaces to focus on an HTR. The LTR is atmospheric ambient conditions at 25°C. What temperature is needed at the HTR to achieve a Carnot efficiency of 50%?

4.9. Consider a heat engine and suppose we know the flow to the LTR, its temperature, and the efficiency. $\dot{Q}_C = 0.55 \times 10^6$ J/s, $\eta = 45\%$. What is the heat flow from the HTR? What is the power from work?

REFERENCES

1. Borgnakke, C., and R.E. Sonntag, *Fundamentals of Thermodynamics*. Eighth edition. 2012: Wiley. 912 pp.
2. Tester, J.W., *Thermodynamics and Its Applications*. Third edition. 1996: Prentice Hall. 960 pp.

3. Fay, J.A., and D.S. Golomb, *Energy and the Environment. Scientific and Technological Principles.* Second edition. MIT-Pappalardo Series in Mechanical Engineering, M.C. Boyce and G.E. Kendall. 2012, New York: Oxford University Press. 366 pp.
4. Tester, J.W. et al., *Sustainable Energy: Choosing Among Options.* Second edition. 2012: MIT Press. 1021 pp.
5. Vanek, F., L. Albright, and L. Angenent, *Energy Systems Engineering: Evaluation and Implementation.* Third edition. 2016: McGraw-Hill. 704 pp.
6. da Rosa, A., *Fundamentals of Renewable Energy Processes.* Third edition. 2013: Elsevier, Academic Press. 884 pp.
7. Demirel, Y., *Energy: Production, Conversion, Storage, Conservation, and Coupling.* Second edition. Green Energy and Technology. 2016: Springer. 616 pp.
8. Jorgensen, S.E., and Y.M. Svirezhev, *Towards a Thermodynamic Theory for Ecological Systems.* 2004: Elsevier. 380 pp.
9. Morowitz, H., *Foundations of Bioenergetics.* 1978: Academic Press. 344 pp.

5 Electrical Storage Elements, Basics of AC Circuits, and AC-DC Conversion

In this chapter, we expand our knowledge of electrical circuits acquired in Chapter 3 by looking at energy storage elements, that is, capacitors (storage of electrical energy) and inductors (storage of magnetic energy). When they are used in circuits, the response includes a transitory component, which leads to the concept of time constant. The magnetic field concept is used to study the basics of two electromechanical devices, that is, electrical generators and motors. We start our study of AC circuits using both sinusoidal functions and phasor diagrams. The chapter concludes with circuits to convert AC to DC and between various levels of DC voltage. Fundamentals are given in basic circuit analysis textbooks [1–3] and reviews geared toward renewable power systems are given in Masters [4] and Hemami [5].

5.1 PRINCIPLES OF CIRCUITS WITH ENERGY STORAGE ELEMENTS

This section provides a very basic and quick review of some concepts of energy storage in circuit elements in the form of electrical and magnetic energy. These elements play an important role in electrical power systems.

5.1.1 ELECTRIC FIELD AND CAPACITORS

An electric field emanates from an electric charge Q and can exert a force (Coulomb force) on a test charge q. The field is a vector field and has direction and magnitude. The field direction is defined as the one corresponding to the force applied to a positive test charge q, whereas the magnitude is given as force per unit charge $E_f = F/q$ in units of N/C. This vector field is visualized by lines of the electric field (Figure 5.1).

The Coulomb force is proportional to the product of the source charge and the test charge, and inversely proportional to the square of the distance d between them $F = \dfrac{kQq}{d^2}$. The constant k is a function of the medium in which the charges are immersed. Therefore, the electric field magnitude is $E_f = \dfrac{kQq}{qd^2} = \dfrac{kQ}{d^2}$. The electric field resultant from several source charges can be obtained by super-position of the effects of these charges.

A common way of accounting for the effect of the medium is the concept of *permittivity* ε of the medium in which the charges are located and which has complicated units, but we will use a simpler alternative unit once we define capacitance. An interesting configuration of charges is one of parallel charged plates (one positive and the other negative) separated by a *dielectric* material (noncon-ductive) and impeding movement of these charges across (Figure 5.1). *Capacitance* is defined as the charge in the plates per unit of voltage or potential difference between the plates, that is to say,

$$C = \frac{q}{v} \ \text{ or } \ Cv = q \tag{5.1}$$

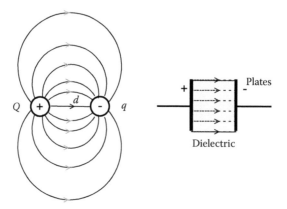

FIGURE 5.1 Electric field and capacitor.

in units of farad (F), which is coulomb/volt (C/V). This configuration is defined as a *capacitor*. Taking into account the surface area A of the plates, the distance d between them, and the permittivity ε of the dielectric, we can calculate capacitance of such an element as

$$C = \varepsilon \frac{A}{d} \tag{5.2}$$

We can see that permittivity should have units of F/m and this provides a convenient operational unit for permittivity. Permittivity of free space is $\varepsilon_0 = 8.85 \times 10^{-12}$ F/m and permittivity of dielectric materials is given relative to this value using $\varepsilon = \varepsilon_r \varepsilon_0$, where ε_r is relative permittivity and can vary according to several factors. For example, some values of ε_r for common materials are paper 3.85, mica 5.6–8, and porcelain 4.5–6.7.

Example 5.1

How much plate surface area would it take to build a capacitor of 100 μF using paper as dielectric? Assume a distance of 0.1 mm (about the same as paper thickness). Answer: Solve for A in Equation 5.2 to get $A = \dfrac{Cd}{\varepsilon}$. Substitute values $A = \dfrac{Cd}{\varepsilon} = \dfrac{100 \times 10^{-6}\text{F} \times 10^{-4}\text{m}}{3.85 \times 8.85 \times 10^{-12}\text{F/m}} = 293.4\,\text{m}^2$.

This is an area equivalent to the square footage of a large home.

This example illustrates that it requires a lot of area to obtain a moderate value of C with a moderate dielectric and thickness. Capacitance for the same area can be increased by using dielectric with higher permittivity or by decreasing the thickness. We will later describe supercapacitors, which are devices that achieve higher capacitance.

From a circuit point of view, we can define the relationship between current and voltage in a capacitor. Recall from Chapter 1 that current is rate of change of charge or $i(t) = dq(t)/dt$ and take the derivative of Equation 5.1. Therefore, $C\dfrac{dv(t)}{dt} = \dfrac{dq(t)}{dt} = i(t)$ or simply

$$i(t) = C\frac{dv(t)}{dt} \tag{5.3}$$

Please realize that this current does not flow through the dielectric between the plates. In reality, this current is a displacement current, meaning that charges flow around the circuit in which the capacitor is connected to separate themselves; this charge separation manifests as the voltage across the plates.

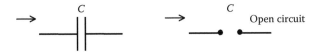

FIGURE 5.2　Circuit symbol for capacitor. The capacitor behaves as an open circuit when the voltage is steady.

The circuit symbol is a couple of parallel lines (Figure 5.2). Voltage can be determined from Equation 5.3 using integrals to obtain $v(t) = \frac{1}{C} \int_{t_0}^{t} i(x)dx + v(t_0)$, which requires the initial conditions for voltage. A consequence of Equation 5.3 is that a capacitor behaves as an *open circuit* (i.e., no current flows through it) when the voltage is steady, such that $\frac{dv(t)}{dt} = 0$ (Figure 5.2). Thus, in DC circuits once a capacitor acquires a constant value, it therefore can be considered as an open circuit.

Another consequence of Equation 5.3 is that a capacitor would resist changing its voltage instantaneously. Mathematically, the derivative would have infinite value for zero time; physically, it would require the charges to rearrange in zero time and this requires infinite power. To see this, calculate power as product of current and voltage (Chapter 1) $p(t) = v(t)i(t) = Cv(t)\frac{dv(t)}{dt}$, and reason that an instantaneous change of voltage means an infinite derivative on the right-hand side, and, therefore, power $p(t)$ would have to be infinite.

5.1.2 Magnetic Field and Inductors

The magnetic field is the result of moving electric charges (electric currents) and moments of particles of magnetic materials. The field can be represented at any point by the direction and magnitude (in units of webers, or Wb for short), thus a vector field. This vector field is then visualized by lines of magnetic flux typically denoted by ϕ (Figure 5.3). The two basic magnetic effects interact in the three basic observations of the 1800s by Oersted, Ampère, and Faraday that influence the field of electromagnetism. These were that (1) current through a wire moves a nearby magnet, (2) current in a wire effects a force on another wire carrying current in the opposite direction, and (3) moving a magnet near a coil of wire induces a current in the coil.

For our purposes in this chapter, these three effects are subsumed in the statements that electric current through a wire creates a magnetic field and that moving a magnetic field creates a current in a coil. A coil that creates a magnetic field is named an *electromagnet* to distinguish this magnetic field from the one created by a permanent magnet. Direction of the magnetic field created by a current is given by the "right-hand rule," which states that if you imagine wrapping your right hand around a wire and the thumb pointed in the direction of the current, then the magnetic field direction is given by the direction of the fingers (Figure 5.4). Using this rule, we can predict the direction of the magnetic field created by current flowing through a coil (Figure 5.4). In this figure, we visualize the magnetic field created by the coil in two ways: one as a longitudinal view and another as a cross-sectional view where dots represent flowing current coming out of the page and the cross is current flowing into the page.

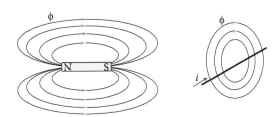

FIGURE 5.3　Magnetic field from a magnet and from current flowing through a wire.

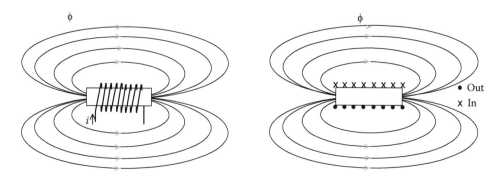

FIGURE 5.4   Right-hand rule and magnetic field created by a coil carrying current.

An additional property of the magnetic field is its magnitude, which is a function not only of the strength of the current that creates it but also relates to the medium that sustains the field via a property of the medium named magnetic *permeability* and denoted by μ. We will define the units as newton per ampere square (N/A^2) but will discuss alternative units a little bit later. Permeability of free space μ_0 is low ($\mu_0 = 4\pi \times 10^{-7}$ N/A^2). The higher the permeability, the greater the magnetic flux density. Permeability of materials is given with respect to air using μ_r relative permeability and the expression $\mu = \mu_r \mu_0$. Relative permeability is higher for ferromagnetic materials, for example, iron ($\mu_r = 5000$), cobalt-iron (18000), and nickel (100–600).

So now, if we wrap the wire (*N* turns) carrying current around a core of ferromagnetic material as shown in Figure 5.5 (left-hand side), the magnetic flux will prefer to remain within the core because it has a much higher permeability than air. The magnetic flux ϕ created by the coil, interacts "back" with the coil *inducing* a voltage across the coil that is proportional to the rate of change of magnetic flux. The constant of proportionality is the number of *turns* (*N*) of the wire making the coil. This voltage is named an *electromotive force* or *emf* for short. It is often denoted by *e* but we will reserve *e* for energy.

$$emf = N\frac{d\phi(t)}{dt} \tag{5.4}$$

This important relationship is *Faraday's law of electromagnetic induction*. Significantly, this emf has a polarity opposing the current that created the magnetic field (*Lenz's law*).

Our next step is to introduce concepts analogous to electric circuits that help us to analyze magnetic circuits. A more detailed explanation will be given in Chapter 3. First, the *magnetomotive force*, or *mmf* for short, denoted by $\mathcal{J}$ can be thought of as the driving force, and is a function of strength *i*(*t*) of the current and the number *N* of turns of the coil (Figure 5.5, left-hand side). Therefore, we can write $\mathcal{J} = Ni(t)$ in units of A-turns.

The response to this driving force is the magnetic flux ϕ via a quantity called *reluctance*, denoted by $\mathcal{R}$ and can be thought of as a property describing opposition to flux of a magnetic field. We can then write $\mathcal{J} = \mathcal{R}\phi$, in analogy to Ohm's law. Reluctance units are A-turns/Wb and is a function of

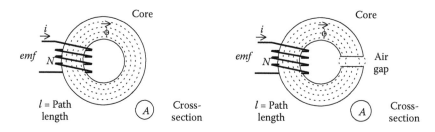

FIGURE 5.5   Coil with ferromagnetic core.

permeability and the geometry of the core. Assume l is the path length of magnetic field inside core, μ is permeability of material, and A the cross-sectional area (Figure 5.5, left-hand side). We can write an expression for reluctance $\mathfrak{R} = \dfrac{l}{\mu A}$. This last equation allows for defining alternative units for permeability as Wb/(A-turns m).

However, this is not as simple because permeability (and thus reluctance) varies nonlinearly with flux; the core may saturate if the flux is too high. To achieve a stable reluctance, the core may include an air gap or powdered ferromagnetic material, which stabilizes permeability and makes reluctance depend mainly on geometry (factors l and A). A simplified model of an air gap is shown in Figure 5.5 (right-hand side). This is further discussed in Chapter 10.

Consider a coil (N turns) of wire wrapped around a ferromagnetic core. Let us link Equation 5.4 with reluctance $emf = N\dfrac{d\phi}{dt} = N\dfrac{d(\mathfrak{I}/\mathfrak{R})}{dt} = N\dfrac{d(Ni/\mathfrak{R})}{dt} = \dfrac{N^2}{\mathfrak{R}}\dfrac{di}{dt}$ and ignoring variation of reluctance with time $emf = \dfrac{N^2}{\mathfrak{R}}\dfrac{di}{dt}$. Then, we define a new quantity called *inductance* $L = \dfrac{N^2}{\mathfrak{R}}$ to rewrite the last equation as $emf = L\dfrac{di}{dt}$. Inductance is given in units called henries (H for short) and is proportional to the square number of turns N and inverse to reluctance $\mathfrak{R}$ of the core.

Example 5.2

Assume we make an inductor with an iron core of 2 mm diameter and path length of 1 cm. The core is air-gapped to stabilize the relative permeability to 1000. What is the reluctance of this core? How many turns would be required to make a 10 mH inductor with this core? Answer: Substitute values to calculate reluctance

$$\mathfrak{R} = \frac{l}{\mu A} = \frac{l}{(\mu_r\mu_0)\pi r^2} = \frac{1\times 10^{-2}}{1000\times 4\pi\times 10^{-7}\times\pi\times(1\times10^{-3})^2} = 2.53\,\text{MA-turns/Wb}$$

Now solve for n in $L = \dfrac{N^2}{\mathfrak{R}}$ to obtain $N = \sqrt{\mathfrak{R}L}$ and substitute $N = \sqrt{2.53\times10^6\times10\times10^{-3}} \simeq$ 159 turns.

A couple of functions in renpow can assist with these calculations. Function reluctance uses as arguments μ_r relative permeability, path length, and cross-area. Its output is in MA-turn/Wb. Function inductor uses as arguments N and reluctance. Its output is in mH.

```
> mu=1000; l=1*10^-2; A=pi*(1*10^-3)^2
> reluc <- reluctance(x=list(c(mu,l,A)))
> reluc$prnt
[1] "Rel= 2.53 MA-turn/Wb"
> ind <- inductor(x=list(N=159,rel=reluc$rel))
> ind$prnt
[1] "L= 9.99 mH"
>
```

Assuming no losses, the voltage v applied to the coil is the same as the *emf* and therefore we obtain a very important relationship

$$v = L\frac{di}{dt} \qquad (5.5)$$

That is, the voltage applied to the coil is proportional to rate of change of current. Schematically we represent an inductor or coil as shown in Figure 5.6. This last equation can be integrated on both sides

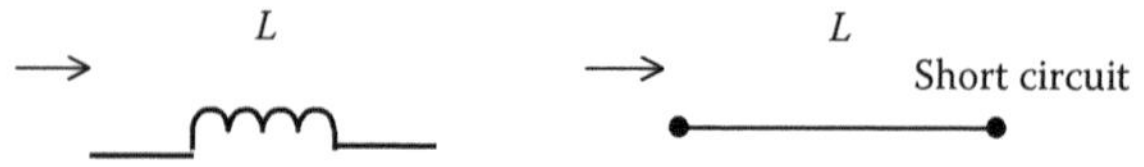

$$\xrightarrow{} \quad \overset{L}{\underset{\text{\textmu}}{}} \qquad\qquad \xrightarrow{} \quad \overset{L}{\bullet\!\!-\!\!\!-\!\!\!-\!\!\bullet} \quad \text{Short circuit}$$

FIGURE 5.6   Circuit symbol for inductor or coil. The coil behaves as a short circuit when the current is steady.

to solve for current $i(t) = \frac{1}{L}\int_{t_0}^{t} v(x)dx + i(t_0)$, which requires the initial conditions for current.

A consequence of Equation 5.5 is that an inductor behaves as a *short circuit* (i.e., no voltage drop across it) when the current is steady such that $\frac{di(t)}{dt} = 0$ (Figure 5.6). Thus, in DC circuits once an inductor achieves a constant value, it can be considered as short circuit. Another consequence of Equation 5.5 is that an inductor would resist changing its current instantaneously. Mathematically, the derivative would have infinite vale for zero time; physically, it would require the magnetic field to change in zero time and this requires infinite power. Similarly to our reasoning for a capacitor, calculate power as product of current and voltage $p(t) = v(t)i(t) = Li(t)\frac{di(t)}{dt}$, and see that an instantaneous change of current is an infinite derivative on the right-hand side and therefore power $p(t)$ would have to be infinite.

5.1.3   ENERGY RELATIONSHIPS IN CAPACITORS AND INDUCTORS

Capacitors and inductors store energy. Capacitors store it in the form of electrical energy related to the voltage across the plates, and inductors store it as magnetic energy related to the current flowing through the coil. We will now calculate the energy stored in a capacitor as a function of voltage and in an inductor as a function of current. Using $p(t) = v(t)i(t)$, we can calculate energy as an integral of power $e(t) = \int_{0}^{t} v(x)i(x)dx + e(0)$.

For a capacitor we have $i(t) = C\frac{dv(t)}{dt}$ and let us assume for simplicity that initially (i.e., at time zero $t = 0$) the voltage is zero. We calculate energy stored at time t by

$$e(t) = \int_{0}^{t} v(x)i(x)dx = \int_{0}^{t} v(x)C\frac{dv(x)}{dx}dx = C\int_{0}^{t} v(x)dv = \frac{1}{2}Cv^{2}(t)$$

In summary,

$$e(t) = \frac{1}{2}Cv^{2}(t) \tag{5.6}$$

For an inductor $v(t) = L\frac{di(t)}{dt}$ and assume that the current at $t = 0$ is zero:

$$e(t) = \int_{0}^{t} v(x)i(x)dx = \int_{0}^{t} i(x)L\frac{di(x)}{dx}dx = L\int_{0}^{t} i(x)dv = \frac{1}{2}Li^{2}(t)$$

In summary,

$$e(t) = \frac{1}{2}Li^{2}(t) \tag{5.7}$$

Equations 5.6 and 5.7 emphasize that energy storage is proportional to the square of the quantity (voltage or current) related to the storage. Mathematically, in both cases this happens because the integral of a linear function is a parabola. In one case, energy is proportional to capacitance C and in the other to inductance L. The factor ½ comes about in both cases as part of the integration of a linear function.

A capacitor is "charged" to a given voltage v from a source. Informally by analogy, we will refer to sustaining a current i through an inductor as "charging" it. Using the energy standpoint, we can further understand that a capacitor resists instantaneous changes of voltage and an inductor resists instantaneous changes of current. Take for example a capacitor and reason that an instantaneous change of voltage is sudden change of energy according to Equation 5.6 or, in other words, a finite change of energy in zero time. The same reasoning applies to an inductor for current when you consider Equation 5.7.

Example 5.3

Consider a 100 µF capacitor and a 10 mH inductor. The capacitor has been discharged to zero volts and the inductor to zero current. Now we charge the capacitor to 10 V and the inductor to 0.1 A. What is the energy stored in each element after they have been charged to those values?

Answer: Use Equations 5.6 and 5.7. For the capacitor $E = \frac{1}{2} 100 \times 10^{-6} \times 10^2 = 5 \times 10^{-3} \text{ J} = 5 \text{ mJ}$

and for the inductor $E = \frac{1}{2} 10 \times 10^{-3} \times 1^2 = 5 \times 10^{-3} \text{ J} = 5 \text{ mJ}$.

5.1.4 TRANSIENT RESPONSE OF CAPACITORS AND INDUCTORS

Let us see what happens to the voltage $v(t)$ across a capacitor as it charges. Refer to the circuit in Figure 5.7. You can see that the voltage source V_s produces a current through the resistor and capacitor, and that the voltage drop across each one of these elements should add up to V_s by Kirchhoff's voltage law (KVL); thus, we can write $Ri(t) + v(t) = V_s$. Now use $i(t) = C\dfrac{dv(t)}{dt}$ and substitute it in this equation to obtain $RC\dfrac{dv(t)}{dt} + v(t) = V_s$, which is a first-order ordinary differential equation (ODE). Define the quantity RC as a *time constant* $\tau = RC$. It has units of s when C is in F and R in Ω. Rewrite the ODE as

$$\frac{dv(t)}{dt} + av(t) = aV_s \tag{5.8}$$

where a is the inverse of the time constant $a = 1/\tau$. The solution to this ODE is the sum of a constant (since the right-hand side is constant) and an exponential, say, $v(t) = K_1 + K_2 \exp(-at)$ where the constants K_1, K_2 are unknown. We can easily check this by taking the derivative of $v(t)$ and substituting in the ODE.

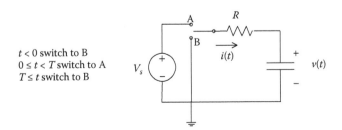

$t < 0$ switch to B
$0 \le t < T$ switch to A
$T \le t$ switch to B

FIGURE 5.7 RC capacitor charging and discharging.

To determine the value of these constants we use initial and final conditions. Evaluate the solution at $t = 0$ to get $v(0) = K_1 + K_2$ because the exponential becomes 1. For simplicity assume the capacitor is discharged at $t = 0$, then $K_1 = -K_2$. Next, evaluate the solution at $t = \infty$ (meaning a very long time after the capacitor is charged), thus $v(\infty) = K_1$ because the exponential becomes negligible. After a long time the capacitor charges to V_s and the current ceases, that is, the capacitor behaves as an *open circuit*; therefore, $K_1 = V_s$ and consequently $K_2 = -V_s$. So we can write $v(t) = V_s - V_s \exp(-at) = V_s (1-\exp(-at))$. Because $a = 1/\tau$, we can rewrite the solution as

$$v(t) = V_s(1 - \exp(-t/\tau)) \tag{5.9}$$

When $t = \tau$ the exponential is $\exp(-1) \approx 0.37$ and therefore the capacitor has charged to ~67% of the source V_s. After a sufficient time elapses (long compared to the time constant), the capacitor is almost fully charged since the exponential term goes to zero for t/τ large. Actually, this approximately happens by $t = 5\tau$ since $\exp(-5) \approx 0.007$.

Let us see what happens to voltage $v(t)$ as a capacitor is discharging. Refer back to Figure 5.7 and assume T is long enough for the capacitor to fully charge to V_s. Now the switch will go to position B and the capacitor will discharge. In this case, the capacitor is charged initially to $v(0) = V_s$ and then discharges completely to 0 V. The ODE of Equation 5.8 holds, but the right-hand side is 0. Therefore, the solution is simply $v(t) = K_2 \exp(-at)$ because there is no forcing term on the right-hand side. Evaluating at $t = 0$ we get $K_2 = V_s$ and therefore

$$v(t) = V_s \exp(-t/\tau) \tag{5.10}$$

Example 5.4

Consider the circuit in Figure 5.7 with $V_s = 12$ V, $R = 200$ kΩ, $C = 1$ μF. Calculate charging and discharge dynamics. Answer: The time constant is $\tau = RC = 0.2M\Omega \times 1\mu F = 0.2s$. The charging dynamics is given by $v(t) = 12(1 - \exp(-t/0.2))$V. The discharge is $v(t) = 12 \exp(-t/0.2)$V.

These trajectories can be visualized using function transient of renpow whose arguments are the source, the time constant, and labels:

```
R=0.2;C=1 # Mohm and uF
transient(ys=12,tau=R*C,ylabel="Vc(t) [V]",yslabel="Vs [V]")
```

That is how the graphs of Figure 5.8 were produced. Note that as expected, after 1 s (5 times the time constant), the capacitor is close to fully charged or fully discharged.

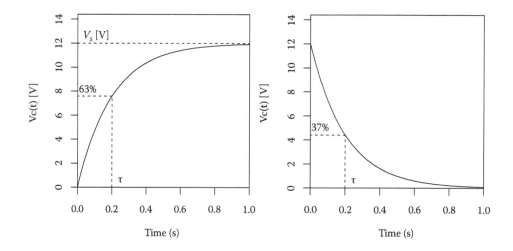

FIGURE 5.8   RC circuit transient response.

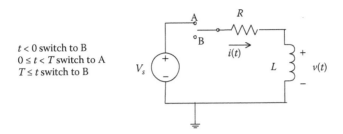

FIGURE 5.9 RL circuit charging and discharging.

Our next task is to see what happens to current $i(t)$ through an inductor as it charges. Refer to the circuit in Figure 5.9. You can see that the voltage source V_s produces a current through the resistor and inductor, and that the voltage drop across each one of these elements should add up to V_s by KVL. Denote the current by $i(t)$, and $v(t)$ the voltage across the coil. We can write $Ri(t) + v(t) = V_s$. Now use $v(t) = L\dfrac{di(t)}{dt}$ to obtain $L\dfrac{di(t)}{dt} + Ri(t) = V_s$ or equivalently $\dfrac{di(t)}{dt} + \dfrac{R}{L}i(t) = \dfrac{1}{L}V_s$, which is a first-order ODE. Define the quantity L/R as a *time constant* $\tau = L/R$. It has units of s when L is in H and R in Ω. Rewrite the ODE as

$$\frac{di(t)}{dt} + ai(t) = \frac{a}{R}V_s \tag{5.11}$$

where a is the inverse of the time constant $a = 1/\tau$.

We already know by analogy to the capacitor how to solve this ODE and find the unknown constants. The solution to this ODE is $i(t) = K_1 + K_2 \exp(-at)$ with unknowns K_1, K_2. Evaluate the solution at $t = 0$ to get $i(0) = K_1 + K_2$. For simplicity assume the inductor is discharged at $t = 0$, then $K_1 = -K_2$; evaluate the solution at $t = \infty$, that is, $v(\infty) = K_1$. After a long time the inductor charges to V_s/R and the voltage of the inductor becomes zero, that is, the inductor behaves as a short circuit, therefore $K_1 = V_s/R$ and consequently $K_2 = -V_s/R$. So we can write, $i(t) = \dfrac{V_s}{R}(1 - \exp(-at))$. Because $a = 1/\tau$, we can rewrite the solution as

$$i(t) = \frac{V_s}{R}(1 - \exp(-t/\tau)) \tag{5.12}$$

Finally, let us see what happens to current $i(t)$ as the inductor is discharging. Refer to the circuit in Figure 5.9. The inductor has been charged to V_s/R and then discharges completely to 0 A. The ODE of Equation 5.11 still holds but the right-hand side is 0. Therefore, the solution is simply $i(t) = K_2 \exp(-at)$. Evaluating at $t = 0$ we get $K_2 = V_s/R$, and therefore

$$i(t) = \frac{V_s}{R}\exp(-t/\tau) \tag{5.13}$$

Example 5.5

Consider the circuit in Figure 5.9 with $V_s = 12$ V, $R = 20$ kΩ, $L = 1$ mH. Calculate charging and discharge dynamics. Answer: The time constant is $\tau = \dfrac{L}{R} = \dfrac{1\,\text{mH}}{20\,\text{k}\Omega} = 0.05$ s. The charging dynamics is given by $v(t) = \dfrac{12}{20}(1 - \exp(-t/0.05))$ V. The discharge is $v(t) = \dfrac{12}{20}\exp(-t/0.2)$V.

Similarly, as for the RC circuit, we can visualize these trajectories using function transient() of renpow:

```
R=20;L=1; # kohm and mH
transient(ys=12/R,tau=L/R,ylabel="iL(t) [A]",yslabel="Vs/R [V]")
```

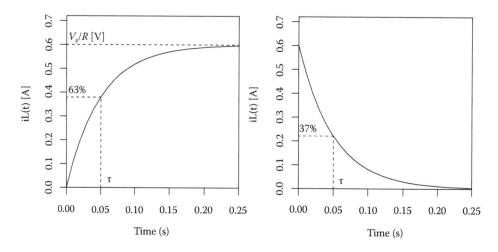

FIGURE 5.10   RL circuit transient response.

See the graphs of Figure 5.10. Note that as expected, after 1 s (5 times the time constant), the capacitor is close to fully charged or fully discharged.

5.1.5   COMBINING CAPACITORS AND INDUCTORS

We can use Kirchhoff's current law (KCL) to calculate the equivalent capacitance of capacitors in parallel. Refer to Figure 5.11, the voltage $v(t)$ is common to both capacitors and the currents through each one, calculated from $i(t) = C\dfrac{dv(t)}{dt}$, add up to the total current; that is to say,

$$i(t) = i_1(t) + i_2(t) = C_1\frac{dv(t)}{dt} + C_2\frac{dv(t)}{dt} = (C_1 + C_2)\frac{dv(t)}{dt} = C\frac{dv(t)}{dt}$$

Therefore, the equivalent capacitance C is the sum of capacitances $C = C_1 + C_2$. This can easily be generalized to write the equivalent C of many, say N, capacitors as

$$C = C_1 + C_2 + \ldots + C_N \tag{5.14}$$

Now, consider the capacitors in series (Figure 5.11). We can use KVL to calculate the equivalent capacitance of capacitors in series. This time, the current $i(t)$ is common to both capacitors and the

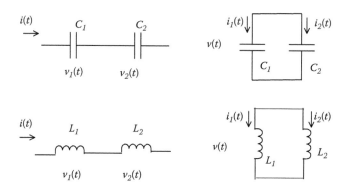

FIGURE 5.11   Capacitors and inductors in parallel and in series.

voltages across each one, calculated from $v(t) = \dfrac{1}{C}\displaystyle\int i(x)dx$, add up to the total voltage; that is to say,

$$v(t) = v_1(t) + v_2(t) = \frac{1}{C_1}\int i(x)dx + \frac{1}{C_2}\int i(x)dx = (1/C_1 + 1/C_2)\int i(x)dx = \frac{1}{C}\int i(x)dx$$

Therefore, the inverse of equivalent capacitance C is the sum of the inverse of the capacitance of each capacitor:

$$\frac{1}{C} = \frac{1}{C_1} + \frac{1}{C_2} \tag{5.15}$$

For this case of two capacitors, simply write $C = \dfrac{C_1 C_2}{C_1 + C_2}$. We can easily generalize Equation 5.15 to write the equivalent C of many, say N, capacitors as

$$\frac{1}{C} = \frac{1}{C_1} + \frac{1}{C_2} + \ldots + \frac{1}{C_N} \tag{5.16}$$

Now let us do similar deductions for inductors in series and parallel. Starting with series, we can use KVL (Figure 5.11), where the current $i(t)$ is common to both inductors and the voltages across each one, calculated from $v(t) = L\dfrac{di(t)}{dt}$, add up to the total voltage $v(t)$; that is to say,

$$v(t) = v_1(t) + v_2(t) = L_1\frac{di(t)}{dt} + L_2\frac{di(t)}{dt} = (L_1 + L_2)\frac{di(t)}{dt} = L\frac{di(t)}{dt}$$

Therefore, the equivalent inductance L is the sum of inductances:

$$L = L_1 + L_2 \tag{5.17}$$

Equation 5.17 is easily generalized to write the equivalent L of many, say N, inductors as

$$L = L_1 + L_2 + \ldots + L_N \tag{5.18}$$

Now, consider the inductors in parallel (Figure 5.11). We can use KCL, with a common voltage $v(t)$ to both inductors and the currents through each one, calculated from $i(t) = \dfrac{1}{L}\displaystyle\int v(x)dx$, add up to the total current

$$i(t) = i_1(t) + i_2(t) = \frac{1}{L_1}\int v(x)dx + \frac{1}{L_2}\int v(x)dx = (1/L_1 + 1/L_2)\int v(x)dx = \frac{1}{L}\int v(x)dx$$

Therefore, the inverse of equivalent inductance L is the sum of inverse of the inductance of each inductor:

$$\frac{1}{L} = \frac{1}{L_1} + \frac{1}{L_2} \tag{5.19}$$

For this case of two inductors, simply write $L = \dfrac{L_1 L_2}{L_1 + L_2}$. We can generalize to write the equivalent L of many, say N, inductors as

$$\frac{1}{L} = \frac{1}{L_1} + \frac{1}{L_2} + \ldots + \frac{1}{L_N} \tag{5.20}$$

In summary, we see that capacitance and inductance combine additively (increases) when the variable related to energy storage is common, that is, voltage for capacitors (in parallel) and current for inductors (in series). In other words, energy storage adds up. Conversely, the capacitance and inductance decrease when the variable related to energy storage is divided, that is, voltage for capacitors (in series) or current for inductors (in parallel).

Example 5.6

How much more energy would you store in two 100 μF capacitors in parallel compared to a single 100 μF? How much more energy would you store in two 0.1 mH inductors in series compared to a single one 0.1 mH? Answer: A simple capacitor of 100 μF would store $E = \frac{1}{2}Cv^2$; capacitances of two capacitors in parallel add up, therefore $E = \frac{2}{2}Cv^2 = Cv^2$, or double the energy. The same argument applies to inductors in series since the inductance would double.

5.1.6 SUPERCAPACITORS OR ULTRACAPACITORS

Supercapacitors or ultracapacitors are capacitors primarily intended for energy storage. They are similar to a battery because they use an electrolyte, electrodes, and cells. A battery relies on electrochemistry to store energy, whereas the supercapacitor uses electrostatic principles to accomplish storage. It is similar to a capacitor inasmuch as energy storage occurs because of charge separation. In the supercapacitor, this occurs at the electrode and electrolyte interface.

A supercapacitor is considered a double-layer capacitor. For each cell, there is a voltage applied across the electrodes or *collectors*. The negative-side electrode or collector attracts the positive ions of the electrolyte; the other electrode or collector (positive) attracts the negative ions of the electrolyte. A dielectric is used to impede flow of charge from one electrode to another, and therefore the device produces separation of charge as in a capacitor.

5.2 ELECTROMECHANICAL DEVICES

Now that we have some understanding of magnetic fields and induction, let us study how these concepts are applied to build electromechanical devices in which electric quantities (voltages and currents) and magnetic fields interact with mechanical quantities (rotational speed and torque).

First, let us consider the fundamentals of an *electric generator*. As explained already in this chapter, many thermal-based power plants boil water into steam, run it through a turbine to perform mechanical work, which turns the shaft of a generator to obtain electrical power. In this case, the turbine is the generator's *prime mover*. Several other processes that we will cover in later chapters use the same principle of electric power generation. The prime mover may be a turbine driven by water, as in hydroelectrical generation, or a wind turbine, as in wind power generation.

5.2.1 ELECTRICAL GENERATOR

A generator is based on relative motion between a magnetic field and a set of coils where the output is generated. The latter are called *armature* coils, whereas we call *field* coils those used to generate the magnetic field (when the magnetic field is generated by an electromagnet instead of permanent magnets). The magnetic field can be fixed and the armature rotates, requiring slip rings to pick up the generated voltage.

However, we will focus our explanation on the opposite configuration, which has fixed armature and called a *stator*, surrounding a *rotor* that generates a magnetic field, by permanent magnets or electromagnets (Figure 5.12). This structure is typically in a cylindrical case that holds the stator coils and therefore does not require slip rings to pick up the output. Figure 5.12 shows a cross-section of

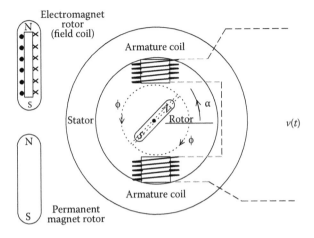

FIGURE 5.12 Simplified view of an electrical generator.

this cylinder. A prime mover exerts input torque and moves the rotor shaft to rotate the magnetic field element (electromagnet or permanent magnet) of the rotor, then a voltage is induced at the armature coil and this becomes the output voltage.

The rotational movement of the magnetic field translates into a sinusoidal wave of magnetic flux $\phi(t)$ as seen at the stator coil, which results in a sinusoidal wave of induced voltage according to $v(t) = nd\phi(t)/dt$. To grasp this intuitively, think of an arrow representing the flux emanating from the N pole and making an angle α with respect to the right-hand side of the horizontal axes. When the arrow points at 0°, the flux does not link with the stator coil and we have zero flux since $\sin \alpha = 0$. As the arrow moves counterclockwise, the linkage increases in the same proportion as the sine of the angle, $\sin\alpha$. When the arrow reaches 90° the flux fully links with the stator coil and we have $\sin \alpha = 1$. As rotation continues, the linkage decreases in magnitude but in the opposite direction, until the arrow points at 180°, then again we have no flux ($\sin \alpha = 0$). Linkage increases in magnitude again but negative until it reaches 270° when $\sin \alpha = -1$. Finally, after one full revolution the magnetic flux is again zero at 0°. In other words, the flux variation in time is proportional to $\sin \alpha$, which can be represented as a function of time, as shown in Figure 5.13.

The rotor mechanical speed ω_{mech} in revolutions per minute (rpm) determines the frequency f of the sinusoidal wave, in cycles per sec, or Hz, in the following manner. In the simplified drawing of

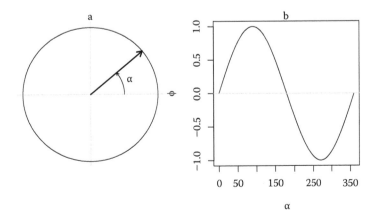

FIGURE 5.13 Rotation and magnetic flux time variation.

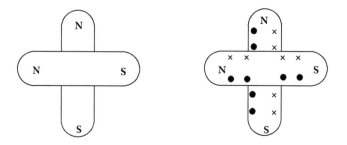

FIGURE 5.14 Multiple poles.

Figure 5.12, the rotor has two magnetic poles (N and S), that is to say, $p = 2$, where p is the number of poles. During one revolution the stator coil sees one pole twice, or the N pole once. So, the number of cycles in a revolution is $p/2$ and therefore the number of cycles per minute is $\frac{p}{2}\omega_{mech}$. To convert this to Hz we need to scale by 60 s/min, that is to say, $f = \frac{\omega_{mech} p/2}{60}$ or equivalently $\omega_{mech} = \frac{60f}{p/2}$. Increasing the number of poles allows reducing rpm to achieve the same frequency because the stator coil will experience a maximum flux more times during a revolution (Figure 5.14). Doubling the number of poles cuts the required rpm for a given f by half.

Example 5.7

What is the mechanical rotational speed for $f = 60$ Hz and $p = 2$. What would the rotational speed be for $p = 4$? Answer: the rotational speed for $p = 2$ must be $\omega_{mech} = \dfrac{60\,\text{s}/\min \times 60\,\text{s}^{-1}}{2/2} = 3600$ rpm.

For $p = 4$, $\omega_{mech} = \dfrac{60\,\text{s}/\min \times 60\,\text{s}^{-1}}{4/2} = 1800$ rpm.

Recall from Chapter 1 that power is torque multiplied by mechanical angular speed

$$P_{mech} = \tau\omega_{mech} \times \frac{\pi}{30}\ \text{W} \tag{5.21}$$

where torque τ is in Nm, and angular speed ω_{mech} is in revolutions per minute (rpm).

Example 5.8

Consider a prime mover rotating at 1800 rpm to give the required 3.6 kW mechanical power to a 4-pole electrical generator. What is the torque in Nm? Answer: Use Equation 5.21 and solve for torque $\tau = \dfrac{P_{mech}}{\omega_{mech} \times \dfrac{\pi}{30}}$; evaluate $\tau = \dfrac{3600}{1800 \times \dfrac{\pi}{30}}$ Nm $= 19.09$ Nm.

We can now calculate electrical angular frequency ω in radians per sec (rad/sec) from the frequency f in Hz using 2π radians $\omega = 2\pi f$. The product of angular frequency and time ωt would be the angle of rotation α described earlier and then the flux varies as $\sin(\omega t)$ and would have a magnitude ϕ_m that depends on the strength of the field produced by the rotor.

Induced voltage is proportional to the rate of change of magnetic flux $v(t) = n\dfrac{d\phi(t)}{dt}$. Substituting $\phi(t) = \phi_m \sin(\omega t)$ and taking its derivative, we have $v(t) = n\dfrac{d(\phi_m \sin(\omega t))}{dt} = n\omega\phi_m \cos(\omega t)$. We see that the voltage is a sinusoidal with magnitude $v_m = n\omega\phi_m$, which would be constant for a given rotational speed.

Our discussion of generators in this section is only introductory. Later, in forthcoming chapters we continue the description of generators when we discuss three-phase generators, as well as synchronous and asynchronous generators.

5.2.2 MOTOR

An electric motor has the same structure as an electrical generator but has the opposite functionality. Generators produce voltage and require a prime mover, which provides shaft rotation. However, a motor produces torque and shaft rotation from voltage. The input voltage applied to the stator coils produce a rotating magnetic field that moves the rotor, which then exerts an output torque on the shaft.

5.3 BASICS OF AC SYSTEMS

As we just saw in the previous section, electrical generators produce voltages that vary as sinusoidal waves with time. This is an AC voltage. In this section, we describe the fundamentals of methods of AC circuit analysis using trigonometric functions (sine and cosine) and phasors (complex numbers).

5.3.1 SINUSOIDAL WAVES AND PHASORS

AC voltages and currents are given in sinusoidal form using the angular frequency ω in radians per sec (rad/sec), which is calculated from the frequency f in cycles per sec or Hz using 2π radians $\omega = 2\pi f$. Therefore, we can write a voltage $v(t)$ as $v(t) = V_m \cos(\omega t)$, where V_m is magnitude and t is time. We refer to this as the *time-domain* representation of AC because time t appears explicitly. For example, $V_m = 170$ V and $f = 60$ Hz, which is the frequency of the grid in the United States, would give $\omega = 2\pi$ rad $\times 60$ s$^{-1} = 377$ rad/s, rounding up to the nearest 1 rad/sec. Then the AC voltage is $v(t) = 170 \cos(377t)$. See Figure 5.15. You may be wondering why the consumer side of the grid is said to be 120 V. We will study in Chapter 8 the concept of root mean square and that 170 V corresponds to 120 V in root mean square value, which is the value typically given when referring to the grid.

An *oscilloscope* allows us to see time-varying voltages by sweeping horizontally at a speed such that the horizontal deviation is proportional to real time. The proportionality is a time scale. On the scope screen, we then see waveforms or AC voltages such that vertical deviations correspond to voltage values and horizontal deviations correspond to time. When measuring DC, the vertical deviation is constant and we just see a horizontal line. When measuring transient DC voltages, we may see a horizontal line moving down or up (if the sweep is too fast compared to the transient), or a

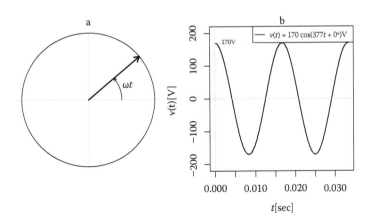

FIGURE 5.15 Rotation and AC voltage in time domain.

trace describing the transient (if the sweep speed is commensurate with the transient). Adjust the volts/division knob on the oscilloscope until the voltage appears on the screen. Estimate the voltage by counting the divisions and multiplying by the number of volts per division.

When the sinusoidal wave is out of phase by an angle θ_v, the time-domain equation of the voltage wave with V_m and phase angle θ_v is

$$v(t) = V_m \cos(\omega t + \theta_v) \text{ V}$$

For example, take $V_M = 170$ V and phase angle of $\theta_v = \pi/3 = 60°$:

$$v(t) = 170 \cos(377t + \pi/3) \text{ V} = 170 \cos(377t + 60°) \text{ V}$$

Please note that although technically ωt is in radians, it is customary to write the phase angle in degrees as in $v(t) = 170 \cos(377t + 60°)$V. Of course, this is technically incorrect because it mixes units but facilitates identifying the angle. We will use this convention in this book.

To visualize AC plots in the time domain we provide function waves and ac.plot of package renpow. Amplitude and phase are arguments to function waves, together with a default frequency 60 Hz, and two cycles to plot. Function ac.plot takes the output of waves as argument.

Example 5.9

Plot two cycles of an AC voltage with magnitude of 170 V with phase 0, and a voltage of 170 V with phase 30°. Use default frequency 60 Hz, and two cycles to plot. Answer:

```
x <- list(c(170,0)); v.t <- waves(x); ac.plot(v.t)
```

which is the graph already shown in Figure 5.15b.

```
x <- list(c(170,30)); v.t <- waves(x); ac.plot(v.t)
```

which produces the graph shown in Figure 5.16.

To illustrate, how these functions work for more than one voltage, take as an example voltage $v_1(t)$ with magnitude of 170 V with phase 0°, and $v_2(t)$ with magnitude 160V at 30° (60 Hz). We can use waves and ac.plot to show both waves as in Figure 5.17.

```
x <- list(c(170,0),c(160,30)); v.t <- waves(x);
v.lab <- c("v1(t)","v2(t)"); v.units <- rep("V",2)
ac.plot(v.t,v.lab,v.units)
```

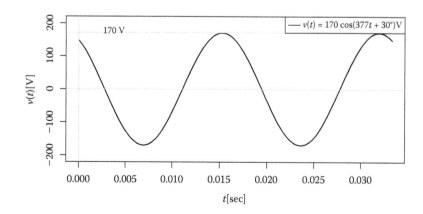

FIGURE 5.16 AC voltage in time domain basics of phase angle.

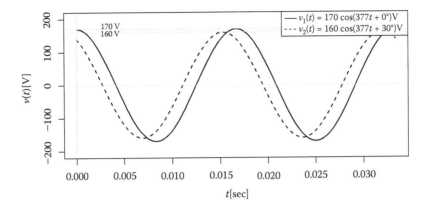

FIGURE 5.17 Two voltages in time domain with different amplitude and angle.

A convenient manner of working with AC is to write $v(t) = V_m \cos(\omega t + \theta_v)$V as a *phasor*, which is a complex number of magnitude V_m and angle θ_v expressed in polar coordinates:

$$\mathbf{V} = V_m \angle \theta_v \; V$$

We use boldface for a phasor in this book to clarify that the quantity is a complex number. For example, $v(t) = 170 \cos(377t + 10°)$ V would be $\mathbf{V} = 170\angle 10°$ V. See Figure 5.18. We refer to this manner of representing AC as the *frequency domain* representation. Please note that in the frequency domain, you do not write the time and frequency explicitly. In other words, you do not write symbols t and ω or the values of frequency ω. The frequency is specified separately, for example, that $f = 60$ Hz or ω is 377 rad/s.

We can facilitate plotting phasors by using the function phasor.plot of the renpow package. For example, to plot only one phasor of magnitude 170 and phase 10°, with label V for voltage (default)

```
v.p <- list(c(170,10)); phasor.plot(v.p)
```

which gives us the plot shown already in Figure 5.18. We will see examples of plotting more than one phasor in the next few sections.

Example 5.10

Assume $f = 60$ Hz, write the time domain of voltage wave with $V_m = 170$V and consider two phase angles, 0° and 90°. Write these voltages as phasors (frequency domain). Draw phasor diagrams. Answer: For 0° $v(t) = 170 \cos(377t + 0°)$ V and for 90° $v(t) = 170 \cos(377t + 90°)$ V. Then, in the frequency domain, the phasors are $\mathbf{V} = 170\angle 0°$ V and $\mathbf{V} = 170\angle 90°$ V. The phasor diagrams are shown in Figure 5.19 and is drawn using the phasor.plot function:

```
# one phasor Re
v.p <- list(c(170,0)); phasor.plot(v.p)
# one phasor Im
v.p <- list(c(170,90)); phasor.plot(v.p)
```

Note that for 0° the phasor is purely real and for 90° is purely imaginary.

The use of phasors are better understood using Euler's identity that gives the real and imaginary parts of a complex number $\exp(j\omega t)$ using sin and cos functions as $\exp(j\omega t) = \cos(\omega t) + j \sin(\omega t)$. It is convenient to formulate the response of circuits containing capacitors and inductors assuming that AC sources are given by $v(t) = V_m \exp(j\omega t + \theta_v)$, instead of just $v(t) = V_m \cos(\omega t + \theta_v)$. Although it seems awkward at first, it turns out that it easier to simplify after taking derivatives of voltages and currents to calculate the response of a circuit, because the term $\exp(j\omega t)$ is present in all terms and can be canceled.

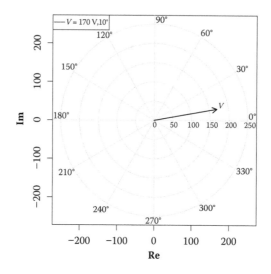

FIGURE 5.18   AC voltage as a phasor in frequency domain.

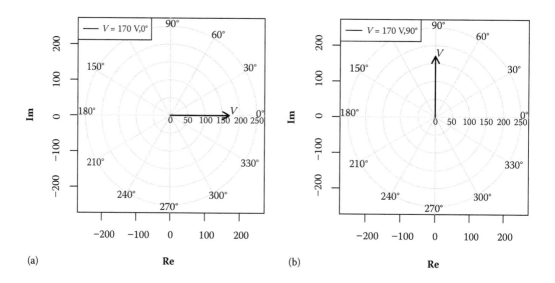

FIGURE 5.19   Example: Phasor diagram for (a) purely real and (b) purely imaginary.

We can see that our time-domain representation is just the real part of this more general complex number, that is to say, $\text{Re}(\exp(j\omega t + \theta_v)) = \cos(\omega t + \theta_v)$. Therefore, we can simplify the algebra and then assume that whatever we calculated is the real part. In actuality then, when we write a phasor as a complex number, it is implicit that it is multiplied by $\exp(j\omega t)$ and that the actual response corresponds to the real part or $\cos(\omega t + \theta_v)$.

5.3.2 PHASORS IN POLAR AND RECTANGULAR COORDINATES

A phasor given as $\mathbf{V} = V_m \angle \theta_v$ is in polar coordinates (magnitude and angle) and we can convert it to rectangular coordinates using $\mathbf{V} = V_m \cos\theta_v + jV_m \sin\theta_v$. It is easy to add or subtract phasors in rectangular coordinates because we just add or subtract the real and imaginary parts separately.

However, it is more difficult to multiply or divide phasors in rectangular coordinates. These operations are facilitated in polar. To convert a phasor in rectangular coordinates, say, $x + jy$, to polar, we can use $\sqrt{x^2 + y^2} \angle \tan^{-1}\dfrac{y}{x}$. To multiply (or divide) two phasors use polar, and the resulting magnitude will be the product (or ratio) of magnitudes and the resulting phase will be obtained by addition (or subtraction of phase angles).

Example 5.11

Consider two voltages $\mathbf{V}_1 = 20\angle 45°$ and $\mathbf{V}_2 = 5\angle 30°$. What is the ratio $\mathbf{V} = \dfrac{\mathbf{V}_1}{\mathbf{V}_1 + \mathbf{V}_2}$?

Answer: $\mathbf{V} = \dfrac{20\angle 45°}{20\angle 45° + 5\angle 30°}$. To calculate the denominator we convert each voltage to rect-

angular $\mathbf{V} = \dfrac{20\angle 45°}{20(\cos 45° + j\sin 45°) + 5(\cos 30° + j\sin 30°)}$ or $\mathbf{V} = \dfrac{20\angle 45°}{14.142(1 + j1) + 4.330 + j2.5}$

and then perform addition $\mathbf{V} = \dfrac{20\angle 45°}{18.472 + j16.642}$. Now convert the denominator back to

polar so that we can divide $\mathbf{V} = \dfrac{20\angle 45°}{\sqrt{18.472^2 + 16.642^2}\;\angle \tan^{-1}\dfrac{16.642}{18.472}} = \dfrac{20\angle 45°}{24.863\angle 42.017°}$. Finally

$\mathbf{V} = \dfrac{20\angle 45°}{24.863\angle 42.017°} = \dfrac{20}{24.863}\angle(45° - 42.017°) = 0.804\angle 2.983°$.

This example illustrates that multiple operations on complex numbers are time consuming. Many low-cost scientific calculators include functions to convert from polar to rectangular and vice versa, making it possible to expedite these calculations. It is also convenient to program these conversions.

R has functions to manipulate complex numbers. We can declare a number to be complex using function complex. The function arguments are real and imaginary parts or are magnitude and angle. The latter are named "modulus" and "argument," respectively. Argument is in radians; we can multiply by $\pi/180°$ to input an angle in degrees. R uses lowercase i for $\sqrt{-1}$. This is very common in mathematics and many disciplines. We use j for $\sqrt{-1}$ in this book because it is the most common notation in circuit analysis. Let us see the numbers from Example 5.11: $\mathbf{V}_2 = 5\angle 30°$ and $\mathbf{V}_1 = 20\angle 45°$. We can input both as arrays to complex:

```
> V1.V2 <- complex (modulus=c(20,5), argument=c(45,30)*pi/180)
> round(V1.V2,3)
[1] 14.142+14.142i 4.330+ 2.500i
>
```

The result is $\mathbf{V}_1 = 14.142(1 + j1)$ and $\mathbf{V}_2 = 4.330 + j2.5$.

Once an object is complex, we can perform normal arithmetic operations and other operations specific of complex numbers. For example, we can calculate $\mathbf{V} = \dfrac{\mathbf{V}_1}{\mathbf{V}_1 + \mathbf{V}_2}$

```
> V <- V1.V2 [1]/sum(V1.V2)
> round(V,3)
[1] 0.803+0.042i
>
```

convert to polar by using Mod and Arg

```
> V.p <- c(Mod(V),Arg(V)*180/pi)
> round(V.p,3)
[1] 0.804 2.983
>
```

and recover rectangular coordinates by using Re and Im functions

```
> V.r <- c(Re(V),Im(V))
> round(V.r,3)
[1] 0.803 0.042
>
```

Alternatively, you can start by making the complex number from rectangular coordinates. For example, $V = 2 + j1$

```
> V <- complex(real=2,imaginary=1)
> V
[1] 2+1i
```

In the renpow package, I have also written functions to work with complex numbers. Two of these, polar and recta, facilitate coordinate conversion by treating the phasors as arrays. To use this take, for example, converting $V = 2 + j$ to polar:

```
> polar(c(2,1))
[1]  2.236 26.565
>
```

That is, $V = 2.236\angle26.565°$. Conversely, convert $V = 2\angle45°$ to rectangular:

```
> recta(c(2,45))
[1] 1.414 1.414
>
```

That is, $V = 1.414 + j1.414$.

When cast this way, adding or subtracting in rectangular form is simply addition of arrays in R. We can also program functions for division and multiplication of phasors. See functions mult.polar and div.polar of renpow.

For example, dividing two phasors in polar $V_1 = 20\angle45°$ and $V_2 = 5\angle30°$:

```
> v1.p <- c(20,45); v2.p <- c(5,30)
> div.polar(v1.p,v2.p)
[1]  4 15
>
```

Or $V = \dfrac{V_1}{V_2} = 4\angle15°$.

Example 5.12

Solve Example 5.11 using the renpow complex number functions. That is, calculate $V = \dfrac{V_1}{V_1 + V_2}$ given voltages $V_1 = 20\angle45°$ and $V_2 = 5\angle30°$. Answer:

```
> v1.p <- c(20,45); v2.p <- c(5,30)
> v.p <- div.polar(v1.p,polar(recta(v1.p)+recta(v2.p)))
> v.p
[1] 0.804 2.983
>
```

Or $V = 0.804\angle2.983°$, which is equal to the result obtained earlier.

5.3.3 PHASE DIFFERENCE

We can determine the phase angle difference between two voltages (or two currents, or a voltage and a current). For example, consider two voltages, $v_1(t) = V_1 \cos(\omega t + \theta_1)$ and $v_2(t) = V_2 \cos(\omega t + \theta_2)$, that have to have the same frequency. Then we use $\theta = \theta_1 - \theta_2$ to calculate the phase difference. A positive phase difference θ means $v_1(t)$ *leads* $v_2(t)$, whereas a negative phase difference θ means that $v_1(t)$ *lags* $v_2(t)$.

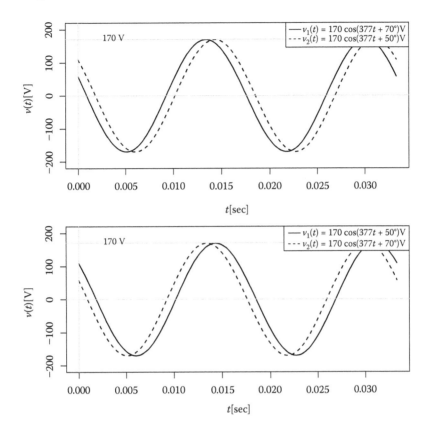

FIGURE 5.20 Phase difference illustrated in the time domain.

Figure 5.20 illustrates this concept in the time domain, using voltages with the same frequency (60 Hz) and the same magnitude, say, 170 V. The top panel graph is for $\theta_1 = 70°$ and $\theta_2 = 50°$ ($v_1(t)$ leads $v_2(t)$ by 20°) and the bottom panel graph is for $\theta_1 = 50°$ and $\theta_2 = 70°$. The phase angle is $\theta = 50° - 70° = -20°$ and the voltage $v_1(t)$ lags the voltage $v_2(t)$. Figure 5.20 was obtained using the function ac.plot in the following manner:

```
v.units <- rep("V",2)
v.lab <- c("v1(t)","v2(t)")
# V1 leads V2
x <- list(c(170,70),c(170,50)); v.t <- waves(x)
ac.plot(v.t,v.lab,v.units)
# V1 lags V2
x <- list(c(170,50),c(170,70)); v.t <- waves(x)
ac.plot(v.t,v.lab,v.units)
```

The lead or lag angle in terms of time is (20° × π/180°)/366 = 0.00092s.

In the frequency domain, write them as phasors, $\mathbf{V}_1 = V_1 \angle \theta_1$ and $\mathbf{V}_2 = V_2 \angle \theta_2$. For example, $\theta_1 = 70°$ and $\theta_2 = 50°$, the phase angle is $\theta = 70° - 50° = 20°$. The voltage $v_1(t)$ leads the voltage $v_2(t)$. But if, for example, $\theta_1 = 50°$ and $\theta_2 = 70°$, the phase angle is $\theta = 50° - 70° = -20°$. The voltage $v_1(t)$ lags the voltage $v_2(t)$. See Figure 5.21. These plots are obtained using

```
# two phasors illustrate phase difference
v.lab <- c("V1","V2");v.units <- rep("V",2)

# V1 leads V2
```

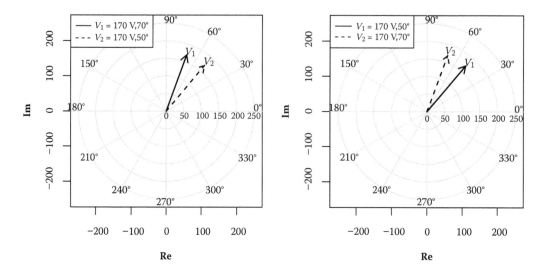

FIGURE 5.21   Phase difference illustrated by phasors.

```
v.p <- list(c(170,70),c(170,50))
phasor.plot(v.p,v.lab,v.units)

# V1 lags V2
v.p <- list(c(170,50),c(170,70))
phasor.plot(v.p,v.lab,v.units)
```

Example 5.13

Determine the phase angle between a voltage $v(t) = 170 \cos(377t + 60°)$ V and a current $i(t) = 60 \cos(377t + 30°)$ A. Which leads which? Write them as phasors. Draw a time-domain plot and a phasor plot. Answer: Phase angle is $\theta = 60° - 30° = 30°$. The voltage leads the current $\mathbf{V} = 170\angle60°$ V and $\mathbf{I} = 60\angle30°$ A. See plots in Figure 5.22. These plots are obtained using

```
# illustrate V and I
v.lab <- c("v(t)","i(t)")
v.units <- c("V","A")
y.lab <- "v(t) [V] or i(t) [A]"
x <- list(c(170,60),c(60,30)); v.t <- waves(x)
ac.plot(v.t,v.lab,v.units,y.lab)

v.lab <- c("V","I")
v.units <- c("V","A")
v.p <- list(c(170,60),c(60,30))
#phasors
phasor.plot(v.p,v.lab,v.units)
```

5.3.4   AC Voltage and Current Relations for Circuit Elements

For each circuit element (R, L, or C), we assume that we apply a voltage $v(t) = V_m \cos(\omega t + \theta_v)$ and produce a current $i(t) = I_m \cos(\omega t + \theta_i)$ with unknown I_m and θ_i. Then for each element, we use the voltage–current relationship for the element to determine these unknowns.

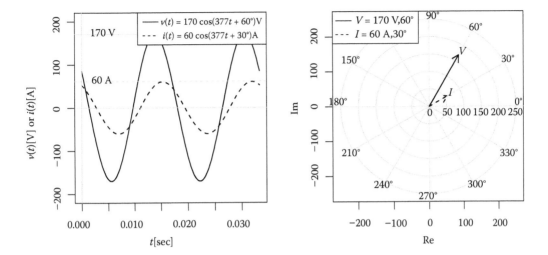

FIGURE 5.22 Example: Phase difference between voltage and current.

For a resistor R we know that $i(t) = \dfrac{v(t)}{R}$ and, therefore, $i(t) = \dfrac{V_m}{R} \cos(\omega t + \theta_v)$. In the frequency domain the current phasor **I** would be the voltage phasor **V** divided by a real number R (which is the same as a complex number with magnitude R and phase $0°$):

$$\mathbf{I} = \frac{\mathbf{V}}{R} = \frac{V_M \angle \theta_v}{R \angle 0°} = \frac{V_M}{R} \angle \theta_v \tag{5.22}$$

In other words, voltage and current are in phase but have different magnitude, with current being lower. Figure 5.23 shows this in the time-domain and the complex plane (current phasor is shorter but the angle remains the same). This illustration is obtained using the following code:

```
w <- 377 # ang frq
v.s <- c(170,60) # voltage source
R = 2 # ohm
v.units <- c("V","A")
# current response
i.res <- c(v.s[1]/R,v.s[2])

# time domain
v.lab <- c("v(t)","i(t)")
y.lab <- "v(t) [V] or i(t) [A]"
ac.plot(waves(list(v.s,i.res)),v.lab,v.units,y.lab)
# frequency domain
v.lab <- c("V","I")
phasor.plot(list(v.s,i.res),v.lab,v.units)
```

Let us consider a capacitor, which has a voltage–current relationship $i(t) = C\dfrac{dv(t)}{dt}$. In the time domain, take derivative of $v(t)$ as the cosine function. The derivative of cos is $-\sin$; but sin is an odd function, then $-\sin$ is equal to sin phased by $180°$; but sin is cos phased by $-90°$

$$i(t) = C\frac{dv(t)}{dt} = C\frac{d}{dt}\left(V_m \cos(\omega t + \theta_v)\right) = -\omega C V_m \sin(\omega t + \theta_v) = \omega C V_m \cos(\omega t + \theta_v + 90°)$$

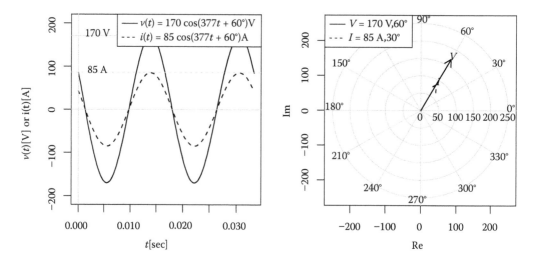

FIGURE 5.23 Time domain and phasor relations for resistor.

Make it equal to current $i(t) = I_m \cos(\omega t + \theta_i) = \omega C V_m \cos(\omega t + \theta_v + 90°)$, and equate magnitude and phase to get $I_m = \omega C V_m$ and $\theta_i = \theta_v + 90°$. In the frequency domain the voltage is $\mathbf{V} = V_m \angle \theta_v$ and $\mathbf{I} = j\omega C\mathbf{V} = \omega C V_m \angle \theta_v \angle 90° = \omega C V_m \angle(\theta_v + 90°)$. Therefore, current leads voltage by 90°. Figure 5.24 shows this in the time-domain and the complex plane (current phasor is shorter and 90° ahead). This illustration is obtained using the following code:

```
w <- 377 # ang frq
v.s <- c(170,60) # voltage source
v.units <- c("V","A")
C = 1000*10^-6 # uF
# current response
i.res <- c(v.s[1]*(w*C),v.s[2]+90)

# time domain
v.lab <- c("v(t)","i(t)")
```

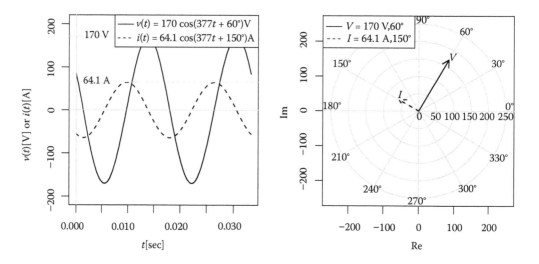

FIGURE 5.24 Time domain and phasor relations for capacitor.

```
y.lab <- "v(t) [V] or i(t) [A]"
ac.plot(waves(list(v.s,i.res)),v.lab,v.units,y.lab)

# frequency domain
v.lab <- c("V","I")
phasor.plot(list(v.s,i.res),v.lab,v.units)
```

Now, consider an inductor L and use $v(t) = L\dfrac{di(t)}{dt}$ or conversely $i(t) = \dfrac{1}{L}\displaystyle\int v(x)dx$. Substitute $v(t) = V_M \cos(\omega t + \theta_v)$ V inside the integral

$$i(t) = \frac{1}{L}\int v(x)dx = \frac{V_m}{L}\int \cos(\omega x + \theta_v)dx = \frac{V_m}{\omega L}\sin(\omega t + \theta_v) = \frac{V_m}{\omega L}\cos(\omega t + \theta_v - 90°)$$

and solve for unknowns to get $I_m = \dfrac{V_m}{\omega L}$ and $\theta_i = \theta_v - 90°$. In the frequency domain the voltage is $\mathbf{V} = V_m \angle \theta_v$ and $\mathbf{I} = \dfrac{\mathbf{V}}{j\omega L} = \dfrac{V_m}{\omega L}\dfrac{\angle \theta_v}{\angle 90°} = \dfrac{V_m}{\omega L} \angle (\theta_v - 90°)$. Therefore, the current lags the voltage by 90°.

Figure 5.25 shows this in the time-domain and the complex plane (current phasor is shorter and 90° behind). This illustration is obtained using the following code:

```
w <- 377 # ang frq
v.s <- c(170,60) # voltage source
v.units <- c("V","A")
L = 10*10^-3 # mH
# current response
i.res <- c(v.s[1]/(w*L),v.s[2]-90)
# time domain
v.lab <- c("v(t)","i(t)")
y.lab <- "v(t) [V] or i(t) [A]"
ac.plot(waves(list(v.s,i.res)),v.lab,v.units,y.lab)

# frequency domain
v.lab <- c("V","I")
phasor.plot(list(v.s,i.res),v.lab,v.units)
```

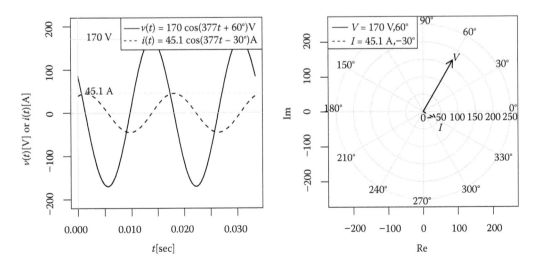

FIGURE 5.25 Time domain and phasor relations for inductor.

The expressions $\mathbf{I} = \dfrac{\mathbf{V}}{R}$, $\mathbf{I} = j\omega C\mathbf{V}$, and $\mathbf{I} = \dfrac{\mathbf{V}}{j\omega L}$ can be rearranged as $\dfrac{\mathbf{V}}{\mathbf{I}} = R$, $\dfrac{\mathbf{V}}{\mathbf{I}} = j\omega L$ and $\dfrac{\mathbf{V}}{\mathbf{I}} = \dfrac{1}{j\omega C}$ as a generalized Ohm's law to define the concept of impedance $\mathbf{Z} = \dfrac{\mathbf{V}}{\mathbf{I}}$ as an AC analog to resistance. We will discuss impedance further in Chapter 8. For now, note that impedance $\mathbf{Z} = \dfrac{\mathbf{V}}{\mathbf{I}}$ is given in units of Ω; it is a complex number. For storage elements, C and L depends on frequency.

Example 5.14

Suppose R = 1 kΩ, C = 10 μF, and L = 10 mH. What is the impedance of each element at f = 60Hz?
Answer: At 60 Hz we have 377 rad/s.
For R, $\mathbf{Z}_R = R = 1k\Omega$.
For L, $\mathbf{Z}_L = j\omega L = j377 \times 10 \times 10^{-3}\Omega = j3.77\Omega$.
For C, $\mathbf{Z}_C = 1/j\omega C = 1/(j377 \times 10 \times 10^{-6})\ \Omega = -j265.252\Omega$.
Recall that $1/j = 1/\sqrt{-1} = -\sqrt{-1} = -j$.

5.4 AC TO DC AND DC TO DC CONVERSION

Now that we have studied DC and AC, we will discuss how to convert from one form to the other. A converter from AC to DC is called a *rectifier*, and one from DC to AC is called an *inverter*. These devices play an important role in modern electric power systems since some renewable power harvesters, such as photovoltaic (PV) cells (Chapter 14), are DC. In addition, there are needs to connect AC systems operating at different frequencies, and these can be matched by converting to DC and then back to AC at another frequency. In this chapter, we study the rectifier and wait to talk about the inverter until Chapter 8.

Another important form of conversion is DC-DC where the DC voltage level is converted to another level. We will see how this is useful to match output of a PV module or panel and a load (Chapter 14). We will start our study of DC-DC conversion in this chapter and continue in later chapters.

5.4.1 AC TO DC: RECTIFIER

Recall the solid-state device called a *diode* discussed in Chapter 3, which has a sharp increase of current with voltage once the latter passes a threshold. For negative voltage, the current is very small, whereas for positive voltage (above the threshold ~0.5 V) the current is very large. In essence, the diode will behave as an open circuit for negative voltage and short circuit for positive voltage (see Figure 5.26). When we connect an AC source to a load using a diode, the positive half of the wave will appear at the load since the diode will act as a short circuit. However, the negative

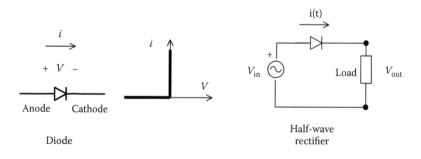

FIGURE 5.26 Diode symbol, idealized i-v curve, and its use in a half-wave rectifier.

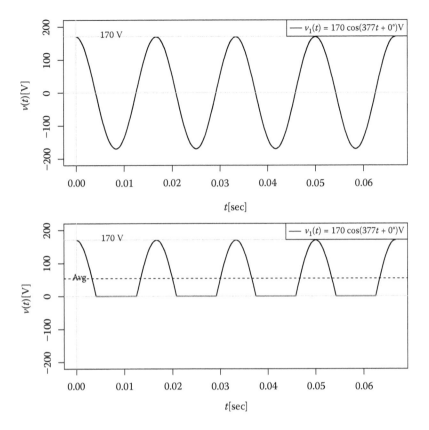

FIGURE 5.27 Input and output of a half wave rectifier.

half will be blocked by the diode acting as an open circuit. This is a half-wave rectifier (Figure 5.26). The function rectifier of renpow will compute the output of the rectifier as shown in Figure 5.27.

```
v.AC <- list(c(170,0))
v.t <- waves(v.AC,nc=4)
V.t <- rectifier(v.t)
ac.plot(v.t)
ac.plot.rect(V.t)
```

This voltage is now all positive, with an average value of $V_{avg} = V_m/\pi$, which would be the equivalent to the DC value. But this is not a good DC since it is a pulsating rather than a continuous voltage. To improve this, we use a capacitor in parallel with the load, so that it stores the voltage as it rises during a half-wave and will try to hold it as the voltage decreases (Figure 5.28, left-hand side). This capacitor together with the load act as a *filter*, meaning that the pulsating nature of the wave has been reduced to a relatively constant value but with smaller amplitude oscillations, called *ripple*. Its amplitude peak-to-peak depends on the ratio of the input period T to the time constant τ formed by the capacitor and the load

$$V_{ripple} = V_m \frac{T}{\tau} = V_m \frac{T}{RC} = \frac{I_{load}}{f \times C} \tag{5.23}$$

We can see that the ripple is made as small as possible by using a large capacitance. The result is shown in Figure 5.29. Once we add the filter, the new DC or average value would be approximately equal to the amplitude of the wave.

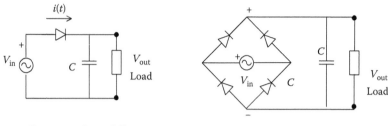

Half-wave rectifier and filter   Full-wave rectifier and filter

FIGURE 5.28 Left: Half-wave rectifier and capacitor filter. Right: bridge rectifier together with filter making a full-wave rectifier.

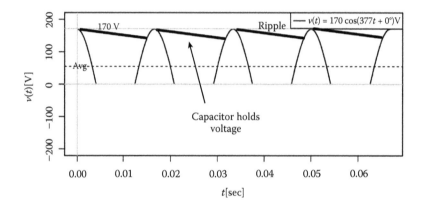

FIGURE 5.29 Filtered half-wave rectifier output showing ripple.

A half-wave rectifier is not common in practice since it can be much improved by using a *full wave rectifier*, which employs a *diode bridge* as shown in Figure 5.28 (right-hand side). The bridge allows for two paths to the load, one for the positive half of the wave and another for the negative half. It can be visualized as taking the absolute value of the wave as illustrated in Figure 5.30. This figure was obtained using the rectifier function with argument full=TRUE:

```
v.AC <- list(c(170,0))
v.t <- waves(v.AC,nc=4)
V.t <- rectifier(v.t,full=TRUE)
ac.plot(v.t)
ac.plot.rect(V.t)
```

The rectified wave still has a pulsating nature, but the gap is reduced because the frequency of the rectified wave is now double to that of the input voltage frequency. The average value is now double with respect to a half-wave and equal to $V_{avg} = 2V_m/\pi$. Once we add a capacitor, the filtered wave is more of a continuous voltage since the capacitor has shorter discharge time; consequently, the ripple is reduced and we have a better implementation of a DC voltage (Figure 5.31). The peak-to-peak amplitude of the ripple voltage is

$$V_{ripple} = V_m \frac{T/2}{RC} = \frac{I_{load}}{2f \times C} \tag{5.24}$$

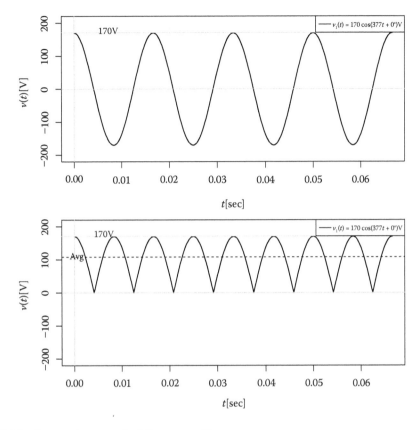

FIGURE 5.30   Input and output of a full-wave rectifier.

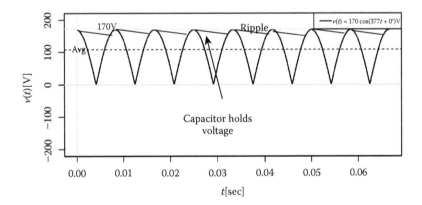

FIGURE 5.31   Filtered full-wave rectifier output showing ripple.

Example 5.15

Suppose the input voltage to a full bridge rectifier is 170 V amplitude at 60 Hz. The filter capacitor is 100 μF and the load is 10 kΩ. What is the average value of the rectified voltage without the filter? What is the average value once we add the capacitor (approximately)? What is the amplitude of the ripple once we add the capacitor? What is the ripple as a percent of the output voltage? Answer: The average value without a filter would be $V_{avg} = 2V_m/\pi = \dfrac{2 \times 170}{\pi} = 108.23\text{V}$.

With a filter, it should be approximately 170 V. The peak-to-peak ripple is $V_{ripple} = \dfrac{I_{load}}{f_{ripple} \times C} \simeq$ $\dfrac{170/(10 \times 10^3)}{60 \times 2 \times 100 \times 10^{-6}} = 1.42\text{V}$. As a percent the output is $1.42/170 = 0.83\%$.

5.4.2 DC-DC Converters

The simplest DC-DC converters are the *buck* or step-down type of converter shown in Figure 5.32 (left-hand side) and the *boost* or step-up type of converter (Figure 5.32, right-hand side). The transistor switch utilized in both circuits represents a device made from an *insulated gate bipolar transistor (IGBT)* and driven by a pulse waveform produced by *pulse width modulation (PWM)* (Figure 5.32, bottom).

The IGBT is a hybrid between a field effect and a bipolar transistor; the gate G is controlled by the voltage from the PWM. When the gate voltage is high enough to turn the transistor on, there is flow of large current between the collector C and the emitter E. Otherwise, the switch is off and there is negligible current from C to E. The periodic pulse waveform or signal is composed of a pulse in the on state that lasts a fraction D of the period T. For the remainder time $(1 - D)T$, the signal is in the off state. The switch receives this signal at its gate G. The switch closes when the signal is on, and it closes when the signal is off. The fraction D is called the *duty cycle* of the pulse wave.

Referring to the buck converter, when the switch is closed, the diode is an open circuit, and the input voltage drives a current through the load that obeys the differential equation $\dfrac{di(t)}{dt} + \dfrac{R}{L} i(t) = \dfrac{1}{L} V_{in}$ with solution $i(t) = \dfrac{V_s}{R}(1 - \exp(-t/\tau))$. Recall from Section 5.1.4 that $\tau = L/R$. When the switch opens, the current from the source is zero, but the current through the load will decrease according to $i(t) = \dfrac{V_s}{R} \exp(-(t - DT)/\tau)$. For a time constant much larger than the width DT, the current through the load will approximate a triangular wave, whereas the current from the source is triangular when the pulse is on and zero when the pulse is off.

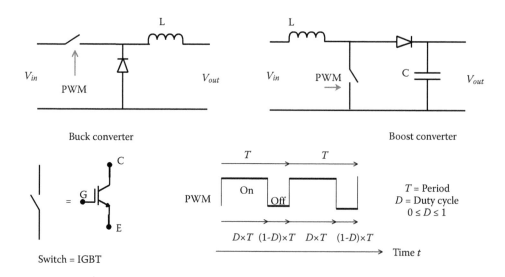

FIGURE 5.32 DC-DC converters: buck converter and boost converter. PWM signal.

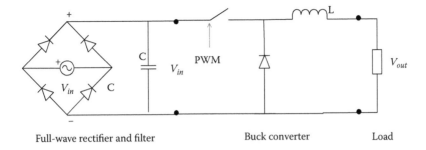

Full-wave rectifier and filter   Buck converter   Load

FIGURE 5.33 Switching power supply. V_{in} AC and V_{out} DC with value controlled by the duty cycle.

The average input current is $\langle I_{out} \rangle = \dfrac{1}{D} \langle I_{in} \rangle$ where the angular braces denote time average and therefore we must have $\langle V_{out} \rangle = D V_{in}$ in order for the output power to be approximately the same as input power (minus some losses) $V_{in} \langle I_{in} \rangle \simeq \langle I_{out} \rangle \langle V_{out} \rangle$. In other words, the output voltage is a fraction D of the input voltage and the output current increases by the reciprocal $\approx 1/D$.

For a boost converter, when the switch is closed the input source charges the coil. When the switch opens, the coil injects its current to the capacitor charging it. Then when the switch closes again, the capacitor holds the charge, which keeps building up, thus achieving a higher voltage at the output. In this case $\langle V_{out} \rangle = \dfrac{1}{1-D} V_{in}$ and the current is $\langle I_{out} \rangle = (1-D)\langle I_{in} \rangle$.

Example 5.16

What would be the duty cycle for a buck converter to step down the voltage from 15 to 10 V? What would be the duty cycle for a boost converter to step up from 12 to 24 V? Answer: For a buck converter $\langle V_{out} \rangle = D V_{in}$, thus $D = \langle V_{out} \rangle / V_{in} = 10/15 = 0.66$ or the pulse will be on 66.66% of the period. For a boost converter $\langle V_{out} \rangle = \dfrac{1}{1-D} V_{in}$, then $D = 1 - \dfrac{V_{in}}{\langle V_{out} \rangle} = 1 - \dfrac{12}{24} = 0.5$ or the pulse should be on 50% of the period.

5.4.3 SWITCHING POWER SUPPLY

A full-wave rectifier and a buck converter are used together to make a power supply (Figure 5.33). For instance, we may take AC from the consumer side of the grid (170 V amplitude and 60 Hz), rectify and filter to DC 170 V with little ripple, and then convert to 24 V using the buck converter set with appropriate value of D. The switching power supply is very common nowadays in many electronic devices; for instance, it is the basic component of AC-DC adapters or of a battery charger.

EXERCISES

5.1. What is the required path length to make an air-gapped inductor with reluctance 4 MA-turns/ Wb using an iron core of 2 mm diameter and relative permeability to 1000? How many turns would be required to make a 10 mH inductor with this core?

5.2. What is the capacitance required to store 10 mJ of energy when a capacitor is fully charged at 10 V?

5.3. Consider the circuit in Figure 5.7 with $V_s = 12$ V, $R = 200$ kΩ. What value of capacitance yields a discharge of $v(t) = 12 \exp(-t/0.1)$V?

5.4. Consider the circuit in Figure 5.9 with $V_s = 12$ V, $R = 10$ kΩ. What value of inductance yields charging dynamics $v(t) = 1.2(1 - \exp(-t/0.1))$V?

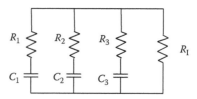

FIGURE 5.34 Supercapacitor bank.

5.5. Figure 5.34 illustrates a model for one cell of an ultracapacitor or supercapacitor. A module of the "ultracap" is made of N identical cells. For energy storage applications, it is of interest to know how quickly the supercap charges and discharges. $Rl = 1\ K\Omega$ is leakage. Assume $R1 = 500\ \mu\Omega$, $C1 = 3\ kF$, $R2 = 1\ \Omega$, $C2 = 0.3\ kF$, $R3 = 3\ \Omega$, $C3 = 0.6\ kF$. Calculate the time constant τ of each one of the three RC branches. Which one is the slow τ branch? Which one is the medium τ branch? Which one is the fast τ branch? Assuming the cell is fully charged to 12 V, what amount of energy is stored in one cell? Assuming $N = 10$ cells in a module, how much energy is stored in a module if all cells are fully charged?

5.6. Assume $f = 60$ Hz, write time domain of voltage wave with $V_m = 170$ V and consider two phase angles, $0°$ and $90°$. Write these voltages as phasors (frequency domain). Draw phasor diagrams.

5.7. An inductor is driven by a 60 Hz AC current modeled by a sine wave $i(t) = I_m \cos(\omega t)$. Determine the voltage. What is the phase difference between voltage and current? Draw the time functions for voltage and current. Why do we say that voltage "leads" the current?

5.8. Consider a 4-pole generator. How fast (in rpm) do you have to rotate the shaft to get a 60 Hz wave? Use renpow to plot one cycle of single phase 60 Hz sine wave voltage.

5.9. Using the renpow complex number functions calculate $I = \dfrac{I_1}{I_1 + I_2}$ given voltages $I_1 = 15\angle 45°$ and $I_2 = 10\angle 30°$.

5.10. Determine the phase angle between a voltage $v(t) = 170 \cos(377t + 45°)$ V and a current $i(t) = 60 \cos(377t + 25°)$ A. Which leads which? Write them as phasors. Draw a time-domain plot and a phasor plot.

5.11. Suppose $R = 1\ k\Omega$, $C = 100\ \mu F$, and $L = 1\ mH$. What is the impedance of each element at $f = 60$ Hz?

5.12. Suppose the input voltage to a full bridge rectifier is 170 V amplitude at 60 Hz and the load is $1\ k\Omega$. How large should the filter capacitor be to have less than 1% ripple?

5.13. What would be the duty cycle for a buck converter to step down the voltage from 24 to 12 V? What would be the duty cycle for a boost converter to step up from 10 to 15 V?

REFERENCES

1. Irwin, J.D., and R.M. Nelms, *Basic Engineering Circuit Analysis*. 11th edition. 2011: Wiley. 688 pp.
2. Hayt, W., J. Kemmerly, and S. Durbin, *Engineering Circuit Analysis*. Eighth edition. 2012: McGraw Hill. 880 pp.
3. Alexander, C.K., and M.N.O. Sadiku, *Fundamentals of Electric Circuits*. Third edition. 2007: McGraw Hill. 901 pp.
4. Masters, G.M., *Renewable and Efficient Electric Power Systems*. Second edition. 2013: Wiley-IEEE Press. 690 pp.
5. Hemami, A., *Electricity and Electronics for Renewable Energy Technology: An Introduction*. Power Electronics and Applications Series, M.H. Rashid. 2016, Boca Raton, Florida: CRC Press/Taylor & Francis. 796 pp.

6 More Thermodynamics State Functions
Entropy, Enthalpy, and Free Energy

In this chapter, we continue our study of thermodynamics. We expand from Chapter 4 by looking at the *second law of thermodynamics* and the related concept of *entropy*. In short, the second law establishes the bounds for what can happen in the real world. To formalize the second law, we expand the review of heat engines and reversible processes initiated in Chapter 4. An important tool to understand thermodynamic cycles is the temperature and entropy (*T-s*) plane, which we employ to discuss efficiency and the Carnot limit. We further examine the concept of internal energy by including enthalpy and *Gibbs free energy*. Armed with these concepts we study thermochemical processes, or those that involve chemical change. We end the chapter applying this knowledge to fuel cells, using hydrogen fuel cells as an example. Compared to a heat engine, in a fuel cell conversion goes directly from chemical to electrical energy. As in Chapter 4, further reading is available in fundamentals of thermodynamics textbooks [1,2]. In addition, several textbooks on energy systems [3–5] contain reviews of thermodynamics that can supplement the presentation given here. Thermodynamics have applications to environmental, ecological, and biological systems, which will not be discussed in this book. However, the reader is referred to Jorgensen and Svirezhev [6] and Morowitz [7]. A good source for further reading on the thermochemical section including fuel cells is Masters [8].

6.1 ENTROPY AND THE SECOND LAW OF THERMODYNAMICS

6.1.1 CARNOT CYCLE: Q AND T

As we discussed in Chapter 4, a reversible process allows returning to an initial state along the same path. We also saw that a heat engine cannot convert all heat input to work; some of the heat is rejected to a heat sink or low temperature reservoir (LTR). For a Carnot cycle, consisting of reversible paths, the ratio of heat transfer at the high temperature reservoir (HTR) to the heat transfer at the LTR is the same as the ratio of temperature at the HTR to the one at the LTR. That is, $\dfrac{Q_L}{Q_H} = \dfrac{T_L}{T_H}$. Rearranging this ratio as

$$\frac{Q_H}{T_H} = \frac{Q_L}{T_L} \tag{6.1}$$

we see that the idealized cycle efficiency is possible only when the ratio of heat transfer to temperature at the sink is the same as the ratio at the source.

In 1865, Rudolf Clausius observed that the equality of these ratios must not hold for irreversible paths. In the simplest scenario of transferring Q from a source to its surroundings irreversibly, we would have $\dfrac{Q}{T_H} < \dfrac{Q}{T_L}$, because $T_H > T_L$. This means that the ratio Q to T at the sink must be larger than the one at the source, $\dfrac{Q_H}{T_H} \leq \dfrac{Q_L}{T_L}$ being equal only for reversible paths. Equivalently,

$$\frac{Q_H}{T_H} - \frac{Q_L}{T_L} \leq 0 \tag{6.2}$$

In other words, the Q/T ratio increases in the direction of heat flow.

Example 6.1

Suppose a heat engine with T_H = 1000K, T_L = 300K, Q_H = 1000 kJ, η = 40%. What is the difference in Q/T from HTR to LTR? Answer: First, solve for Q_L from efficiency definition $\eta = 1 - \dfrac{Q_L}{Q_H}$ to get $Q_L = Q_H(1-\eta)$. Calculate LTR heat transfer Q_L = 600 kJ and now calculate the ratio for both reservoirs $\dfrac{Q_H}{T_H} = \dfrac{10^3 kJ}{10^3 K} = 1\ kJ/K$ and $\dfrac{Q_L}{T_L} = \dfrac{0.6 \times 10^3 kJ}{300K} = 2\ kJ/K$. The difference is $\dfrac{Q_H}{T_H} - \dfrac{Q_L}{T_L} = -1\ kJ/K$. We note that this is negative as predicted by Equation 6.2.

6.1.2 ENTROPY: A STATE FUNCTION

Generalizing Equation 6.2 for any cycle consisting of a sequence of paths with heat transfer Q_i to the surroundings, which have temperature T_i, we should have $\sum\limits_{i} \dfrac{Q_i}{T_i} \leq 0$ where the equality holds only for reversible paths. Using a differential amount of heat, we can convert the sum to an integral

$$\oint \frac{\delta Q}{T} \leq 0 \tag{6.3}$$

The equality holds only for reversible paths

$$\oint \frac{\delta Q_{rev}}{T} = 0 \tag{6.4}$$

where δQ_{rev} emphasizes that heat is transferred reversibly.

Equation 6.4 implies that $\dfrac{\delta Q_{rev}}{T}$ is a differential of a state function because there is no net change at the end of a cycle. This state function is defined as *entropy* (*S*). Thus, we write its exact differential as

$$dS = \frac{\delta Q_{rev}}{T} \tag{6.5}$$

The units of entropy are therefore J/K or a ratio of energy to absolute temperature. In Equation 6.5, we convert an inexact differential of heat δQ_{rev} to an exact differential of entropy dS using $1/T$ as an integration factor. Now using Equation 6.3, we could write $\dfrac{\delta Q}{T} \leq dS$ or

$$dS \geq \frac{\delta Q}{T} \tag{6.6}$$

which is the *Clausius inequality*. This inequality implies that if $\delta Q = 0$, which is an adiabatic process, then $dS \geq 0$, meaning that entropy change is always positive or zero.

Define a system and its surroundings as an isolated system with no heat exchange across its boundary ($\delta Q = 0$). Using the Clausius inequality, the second law of thermodynamics can be stated saying that the entropy of this isolated system always increases or remains the same (for a reversible process), but never decreases. Therefore, an adiabatic reversible process is *isentropic* (entropy remains the same). The second law has many consequences as we study processes to generate work from heat.

Example 6.2

Consider a Carnot heat engine with $T_H = 1000$ K, $T_L = 300$ K, $Q_H = 1000$ kJ. What is the entropy of the HTR and the LTR? Answer: Recall that a Carnot heat engine is based on reversible processes. Then calculate HTR entropy change from reversible heat transfer $\Delta S_H = \dfrac{Q_H}{T_H} = \dfrac{1000\,\text{kJ}}{1000\,\text{K}} = 1\,\text{kJ/K}$.

Next since Equation 6.1 holds, solve for Q_L from this equation $Q_L = Q_H \dfrac{T_L}{T_H} = 1000\dfrac{300}{1000} = 300\,\text{kJ}$.

Calculate LTR entropy change $\Delta S_L = \dfrac{Q_L}{T_L} = \dfrac{300\,\text{kJ}}{300\,\text{K}} = 1\,\text{kJ/K}$. We see the entropy changes are the same. They have opposite signs and there is no net entropy change.

6.1.3 ENTROPY CHANGE OF ISOTHERMAL EXPANSION

To further illustrate entropy and the second law, consider isothermal reversible expansion of a gas from state 1 (P_1, V_1) to state 2 (P_2, V_2). Recall from Chapter 4 that this reversible expansion performs mechanical work W that must equal heat transfer Q into the gas to keep the temperature constant at T:

$$Q_{rev} = W_{rev} = nRT \ln\left(\frac{V_2}{V_1}\right) = -nRT \ln\left(\frac{P_2}{P_1}\right) \tag{6.7}$$

We use subscript *rev* to emphasize this is reversible. The change of entropy is $\Delta S = Q_{rev}/T$ and therefore

$$\Delta S = nR \ln\left(\frac{V_2}{V_1}\right) = -nR \ln\left(\frac{P_2}{P_1}\right) \tag{6.8}$$

This represents the entropy change and it is the same amount of entropy change in the surroundings.

Now, consider irreversible isothermal expansion between the same states (P_1, V_1) and (P_2, V_2) against the pressure P_{ext} of the surroundings. The heat Q_{irr} exchanged is not the same as in the reversible path. To see this, assuming $P_2 = P_{ext}$ we can calculate the PV work involved to get $W_{irr} = P_2(V_2 - V_1)$, which is less than the work done reversibly. A heat flow Q_{irr} equal to this work is transferred from the surroundings to keep the temperature constant at T. Now using the ideal gas law $Q_{irr} = W_{irr} = nRT\left(1 - \dfrac{P_2}{P_1}\right) = nRT\dfrac{V_2 - V_1}{V_2}$. Therefore, the ratio $\dfrac{Q_{irr}}{T}$ due to this heat flow is $\dfrac{Q_{irr}}{T} = nR\left(1 - \dfrac{P_2}{P_1}\right) = nR\dfrac{V_2 - V_1}{V_2}$. Comparing to Equation 6.8 we see that $\Delta S > \dfrac{Q_{irr}}{T}$ as predicted by the Clausius inequality. The difference $\Delta S - \dfrac{Q_{irr}}{T}$ is entropy generated in the surroundings.

Example 6.3

Compare reversible and irreversible isothermal expansion of 40 moles of a gas previous compressed 10 times respect to atmospheric pressure and held at standard room temperature (25°C or 298 K). What is the change of entropy of the reversible expansion? What is the ratio of heat transfer to temperature of the irreversible expansion? Answer: We will use Equation 6.7 to calculate work performed by reversible expansion,

$$Q_{rev} = W_{rev} = -nRT \ln\left(\frac{P_2}{P_1}\right) = -40 \times 8.314 \times 298 \times \ln\left(\frac{1}{10}\right) J$$

$$= 40 \times 8.314 \times 298 \times 2.303 \, J = 228.235 \, kJ$$

and then the change of entropy is $\Delta S = \frac{Q_{rev}}{T} = \frac{228.235 \, kJ}{298 \, K} = 765.89 \, J/K$. Now we calculate irreversible work done against the atmospheric pressure $Q_{irr} = W_{irr} = nRT\left(1 - \frac{P_2}{P_1}\right) = nRT(1 - 1/10)$

and substituting values $Q_{irr} = 40 \times 8.314 \times 298 \times \left(\frac{9}{10}\right) J = 89.192 \, kJ$. Then, $\frac{Q_{irr}}{T} = \frac{89.192 \, kJ}{298 \, K} =$

$299.3 \, J/K$. Compare to $\Delta S = 765.89 \, J/K > 299.3 \, J/K = \frac{Q_{irr}}{T}$.

6.1.4 CARNOT LIMIT AND ENTROPY

We will now examine the Carnot limit using the concept of entropy and the second law. Since entropy increases, the entropy change at the LTR heat engine should be larger than that at the HTR, $\Delta S_L \geq \Delta S_H$. Using the definition of entropy on both sides we get $\frac{Q_L}{T_L} \geq \frac{Q_H}{T_H}$, which can be rewritten as $\frac{Q_L}{Q_H} \geq \frac{T_L}{T_H}$ and recalling that thermal efficiency is $\eta = 1 - \frac{Q_L}{Q_H}$, we can change the sign of the inequality and add 1 to both sides to obtain $1 - \frac{Q_L}{Q_H} \leq 1 - \frac{T_L}{T_H}$ or $\eta \leq 1 - \frac{T_L}{T_H}$. We can see that the Carnot limit $\eta_{max} = 1 - \frac{T_L}{T_H}$ is the right-hand side of the inequality and this means that efficiency is always less or equal to the Carnot limit $\eta \leq 1 - \frac{T_L}{T_H} = \eta_{max}$. In other words, ideal efficiency corresponds to the case of equal entropy change $\Delta S_L = \Delta S_H$. It should be noted that a reversible process is an idealized model, which is not achieved in practice due to other losses such as friction or wasted power.

We can divide both sides of Equation 6.5 by dt to obtain the rate of change of entropy $\frac{dS}{dt} = \frac{\delta Q_{rev}/dt}{T}$ as a function of the rate of change of heat or heat flow for a reversible process. Using dot notation for derivative, we can write for brevity $\dot{S} = \frac{\dot{Q}}{T}$, but note that this is not rigorous because δQ is not an exact differential.

Example 6.4

Consider a heat engine with $T_H = 1000 \, K$, $T_L = 300 \, K$, $\dot{Q}_H = 10^6 \, J/s$, $\eta = 40\%$. What is the difference in entropy rates? Answer: First calculate change of entropy of the HTR $\dot{S}_H = \frac{\dot{Q}_H}{T_H} = \frac{10^6 \, J/s}{10^3 \, K} = 10^3 \, J/s \, K^{-1}$. Then, the LTR $\dot{S}_L = \frac{\dot{Q}_L}{T_L} = \frac{0.6 \times 10^6 \, J/s}{300 \, K} = 2 \times 10^3 \, J/s \, K^{-1}$. Subtract to obtain $\Delta \dot{S} = -\dot{S}_H + \dot{S}_L = -1000 + 2000 = 1000 \, J/s \, K^{-1}$.

6.2 THE *T-s* PLANE

In addition to the representation of a thermodynamic cycle on a *P-v* plane as we studied in Chapter 4, it is common to use the *T-s* plane or temperature versus entropy plane. Because Q depends on mass, entropy S is an extensive property. Dividing by mass $s = \dfrac{S}{m}$ we have specific entropy given in (kJ/K)/kg. The horizontal axis of a *T-s* diagram is commonly assigned to specific entropy.

6.2.1 PATHS IN THE *T-s* PLANE

Reversible isotherms and adiabats that plot as curves on the *P-v* plane will display as straight lines on the *T-s* plane; isotherms as straight horizontal lines and adiabats (isentropic) as straight vertical lines (Figure 6.1).

Conversely, reversible isobaric and isochoric processes, which display as straight lines on the *P-v* plane will display as curves on the *T-s* plane (Figure 6.2). To see why these paths show as curves in the *T-s* plane, take, for example, an isobaric process. Using the constant-pressure specific heat capacity of a fluid c_p, we can write $\delta Q = c_p m dT$. By the definition of entropy, when this process is reversible, we can write $\delta Q = TdS$. Then equating these two expressions, we get $c_p m dT = TdS$, and finally dividing by mass, we have $c_p dT = Tds$ using specific entropy. We can rearrange terms and integrate both sides to get $\displaystyle\int_1^2 \frac{1}{T} dT = \int_1^2 \frac{1}{c_p} ds$, where the 1 and 2 denote initial and final states. The integration is $\ln(T_2/T_1) = \dfrac{s_2 - s_1}{c_p}$, assuming c_p is constant; therefore

$$T = T_1 \exp\left(\frac{s - s_1}{c_p}\right) \tag{6.9}$$

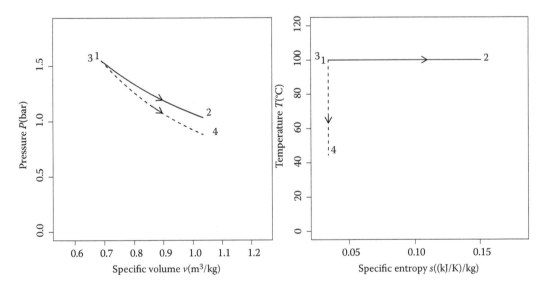

FIGURE 6.1 Reversible isothermal and adiabatic paths in the *P-V* and *T-s* plane. Left: 1–2: isotherm, expansion 100°C–> 100°C, $W = 43.4$ kJ/kg, $Q = 43.4$ kJ/kg; 3–4: isobar, expansion 100°C–> 44.6°C, $W = -40.1$ kJ/kg, $Q = 0$ kJ/kg. Right: 1–2: isotherm, expansion 100°C–> 100°C, $W = -43.4$ kJ/kg, $Q = 43.4$ kJ/kg; 3–4: adiabat, expansion 100°C–> 44.6°C, $W = -40.1$ kJ/kg, $Q = 0$ kJ/kg.

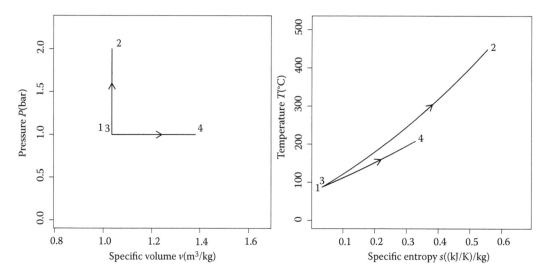

FIGURE 6.2 Reversible isochoric and isobaric paths in the *P-V* and *T-s* plane. Left: 1–2: isochor, 87.8°C–> 448.7°C, *W* = 0 kJ/kg, *Q* = 272.5 kJ/kg; 3–4: isobar, expansion 87.8°C–> 208.1°C, *W* = –34.5 kJ/kg, *Q* = 122.7 kJ/kg. Right: 1–2: isochor, 87.8°C–> 448.7°C, *W* = 0 kJ/kg, *Q* = 272.5 kJ/kg; 3–4: isobar, expansion 87.8°C–> 208.1°C, *W* = –34.5 kJ/kg, *Q* = 122.7 kJ/kg.

an isobaric process, will be represented by an exponential on the *T-s* plane. A similar equation applies for an isochoric process when replacing c_p for c_v. As we know from Chapter 4, specific heat capacity varies with temperature, and therefore these equations will change for a broader temperature range.

To plot paths in the *T-s* plane we can use the same function path.lines of the renpow package, which we introduced in Chapter 4. But we use argument plane='Ts' when making the call, that is to say, path.lines(x, plane='Ts'). The following code produces Figure 6.1. We use plane='Pv' just to be explicit but it is not needed because argument plane defaults to 'Pv'.

```
x1 <- list(V=c(20,30),T=c(100,NA),S=c(1,NA),path='isotherm',lab=c("1","2"))
x2 <- list(V=c(20,30),T=c(100,NA),S=c(1,NA),path='adiabat',lab=c("3","4"))
x <- list(x1,x2)
path.lines(x,plane='Pv')
path.lines(x,plane='Ts')
```

Note that we have used an additional argument value to specify the initial value of entropy. The function calculates entropy starting from this value.

Example 6.5

Use path.lines to produce Figure 6.2. Answer:

```
x1 <- list(V=c(30,NA),P=c(1,2),S=c(1,NA),path='isochor',lab=c("1","2"))
x2 <- list(V=c(30,40),P=c(1,NA),S=c(1,NA),path='isobar',lab=c("3","4"))
x <- list(x1,x2)
path.lines(x,plane='Pv')
path.lines(x,plane='Ts')
```

which indeed produces Figure 6.2.

Function path.lines uses the values calculated by path.calc. For instance,

```
x <- list(V=c(25,NA),P=c(1,2),S=c(1,NA),path='isochor',lab=c("1","2"))
y <- path.calc(x)
path.summary(y)
```

As explained in Chapter 4, the output y contains values of volume, pressure, and temperature; these can be summarized using path.summary. It also produces values of entropy; we just did not employ them in Chapter 4. Specific entropy is component $s and extensive entropy is $S. We can see that $s starts at 0.035 and ends at 0.545 (kJ/K)/kg.

```
> path.summary(y)
$start.end
 v(m3/kg) V(l) P(b) T(C) s(J/Kkg) S(J/K) W(kJ/kg) Q(kJ/kg) cv(kJ/Kkg)
1   0.863   25   1  27.7   0.035    1.0       0      0.0      0.718
2   0.863   25   2 328.4   0.545   15.8       0    222.3      0.765
 cp(kJ/Kkg) cp/cv
1    1.005 1.400
2    1.052 1.375
$WQtot
 Wtot(kJ/kg) Qtot(kJ/kg)
1     0          222.3
$pts
[1] 1001
$call
   V1    V2    P1    P2    S1    S2  path  lab1  lab2
   25  <NA>    1     2     1   <NA> isochor   1     2
$nM
 n(mol) M(g/mol)  m(kg)
1    1   28.97 0.02897
>
```

Entropy is calculated by numerical integration of dQ/T. Each one of these two terms, dQ and T, are calculated according to the path equations.

6.2.2 AREA UNDER THE CURVE

Using the definition of entropy, for a reversible process we can write $\delta Q = TdS$ and the net change of heat as $Q = \int TdS$ or area under the curve of a line on the T-s plane. When plotting temperature in °C instead of K, the area is adjusted accordingly. For illustration, Figure 6.3 shows the area under the curve for an isotherm at 200°C (left-hand side), which includes the area for another isotherm at 30°C. This area corresponds to the heat Q reported in the caption. Subtracting the area under the curve for these two isotherms shows as the area between the curves (right-hand side); this will be the difference in heat of the two isotherms.

The left-hand side plot of Figure 6.3 can be produced using path.lines when applied to paths that include argument shade.under = TRUE. In the following, area under path x2 is shaded but not the one for path x1. It produces the left-hand side panel of Figure 6.3.

```
x1 <- list(V=c(10,30),T=c(30,NA),S=c(1,NA),path='isotherm',lab=c("1","2"))
x2 <- list(V=c(10,30),T=c(200,NA),S=c(1,NA),path='isotherm',lab=c("3","4"),
shade.under=TRUE)
x <- list(x1,x2)
path.lines(x,plane='Ts')
```

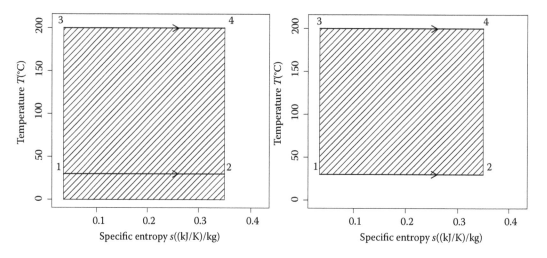

FIGURE 6.3   Integration or area under the curve in the *T*-*s* plane for a reversible isotherm and difference in area under the curve for two isotherms. Left: 1–2: isotherm, expansion 30°C–> 30°C, *W* = –95.5 kJ/kg, *Q* = 95.5 kJ/kg; 3–4: isotherm, expansion 200°C–> 200°C, *W* = –149.1 kJ/kg, *Q* = 149.1 kJ/kg. Right: 1–2: isotherm, expansion 30°C–> 30°C, *W* = –95.5 kJ/kg, *Q* = 95.5 kJ/kg; 3–4: isotherm, expansion 200°C–> 200°C, *W* = –149.1 kJ/kg, *Q* = 149.1 kJ/kg.

For an isotherm, the area under the curve $Q = \int TdS$ simplifies to $Q = T\Delta S$ because T is constant and can be taken out of the integral. We can quickly check that the heat reported in the caption corresponds to the area under the curve. The start and end values of entropy can be queried from path.summary, and then calculate the area of the rectangle by $q = (200 + 273)K \times (0.350 - 0.035)J/kgK \simeq 149$ kJ/kg

```
> path.summary(path.calc(x1))$start.end[5]
  s(J/Kkg)
1   0.035
2   0.350
> (200+273)*(0.350-0.035)
[1] 148.995
>
```

To obtain the area in between the curves use argument shade.between=TRUE for function path.lines

```
x1 <- list(V=c(10,30),T=c(30,NA),S=c(1,NA),path='isotherm',lab=c("1","2"))
x2 <- list(V=c(10,30),T=c(200,NA),S=c(1,NA),path='isotherm',lab=c("3","4"))
x <- list(x1,x2)
path.lines(x,plane='Ts',shade.between=TRUE)
```

which produces the right-hand side of Figure 6.3. The difference in heat would be (149.1 – 95.5) kJ/g = 53.6 kJ/kg.

Example 6.6

Using path.lines, illustrate an isobaric expansion of one mole of air from 20 l to 40 l at 2 bar assuming that the initial entropy is 1 J/K. Plot this path in the *P*-*v* plane and the *T*-*s* plane. Include area under the curve, and compare heat and work calculated. Answer:

```
x<- list(V=c(20,40),P=c(2,NA),S=c(1,NA),path='isobar',lab=c("1","2"),shade.under=TRUE)
path.lines(x,plane='Pv')
path.lines(x,plane='Ts')
```

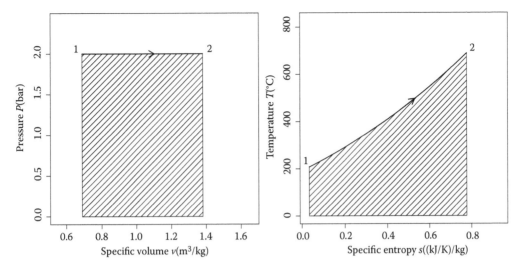

FIGURE 6.4 Isobaric expansion in the *P-v* and *T-s* planes. Left: 1–2: isobar, expansion 208.1°C–> 689.2°C, W = −138.1 kJ/kg, Q = 518.9 kJ/kg. Right: 1–2: isobar, expansion 208.1°C–> 689.2°C, W = 138.1 kJ/kg, Q = 518.9 kJ/kg.

The plots are shown in Figure 6.4. The work is 138.1 kJ/kg, which can be easily verified as area in the rectangle shown in the *P-v* plane. The heat is 518.9 kJ/kg, which could be checked from the temperature difference assuming c_p constant of about

```
> cp.cv(TC=(208+689)/2)$cp.kg
[1] 1.078
>
```

1.078 kJ/kgK in the middle of that temperature range. We get $q = c_p \Delta T = 1.078 \times (689.2 - 208.1) \simeq$ 518.5 kJ/kg, which is close to the heat reported in the graph and obtained by integration using the variation of c_p with temperature.

6.2.3 CARNOT CYCLE IN THE *T-S* PLANE

For a Carnot cycle, the change in entropy is the same for both LTR and HTR. The upper path is isothermal expansion from 2 to 3 with addition of heat δQ_H from the heat source (LTR) to the engine. We can write

$$\int_2^3 \delta Q_H = \int_2^3 T_H ds = T_H \int_2^3 ds = T_H(s_2 - s_1) = T_H \Delta s_H \tag{6.10}$$

The lower path is isothermal compression from 4 to 1 where the engine loses heat δQ_L to the LTR

$$\int_4^1 -\delta Q_L = -T_L \int_4^1 ds = T_L(s_4 - s_1) = T_L \Delta s_L \tag{6.11}$$

Paths 1 to 2 (adiabatic compression) and 3 to 4 (adiabatic expansion) are isentropic (same entropy), whereby the fluid changes from low to high temperature and high to low, respectively. For these paths, $ds = 0$ and therefore $\int_1^2 \delta Q = \int_3^4 \delta Q = 0.$

As illustrated in Figure 6.5 (right-hand side), the difference of input and output heat is the work $W = Q_H - Q_L$. This heat difference is illustrated in Figure 6.6. On the left-hand side we see that the

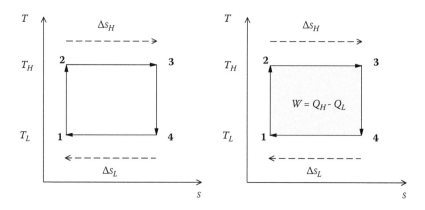

FIGURE 6.5 Carnot cycle on the *T-s* plane illustrating equality of entropy change at LTR and HTR.

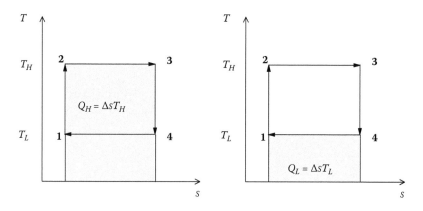

FIGURE 6.6 Graphical illustration of Carnot heat engine calculations as area under the curve on the *T-s* plane.

area under the curve for the upper path (2 to 3) is $Q_H = \int T_H ds = T_H \Delta s_H$, per Equation 6.10. On the right-hand side, the area under the curve for the lower path (4 to 1) is $Q_L = \int T_L ds = T_L \Delta s_L$, per Equation 6.11. The difference between both areas is precisely the area of the box in Figure 6.5 (right-hand side). In other words,

$$W = \oint T ds = Q_H - Q_L \tag{6.12}$$

The cyclic integral is the area of the box between the curves.

We can plot a Carnot cycle on the *T-s* plane using path.cycles function of renpow, as shown in the following example.

Example 6.7

Plot the Carnot cycle for one mole of air with $T_H = 200°C$, $T_L = 30°C$ and expanding from 20 l to 40 l at 200°C, assuming that the initial entropy is 1 J/K. Answer:

```
x <- list(TH=200,TL=30,V1=20,V4=40,S1=1,n=1,cty='carnot')
path.cycles(x,plane='Ts')
path.cycles(x,plane='Ts',shade.cycle=TRUE)
```

The results are shown in Figure 6.7; the area within the cycle is shown in Figure 6.8.

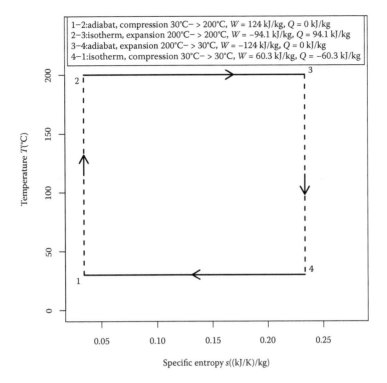

FIGURE 6.7 Example of Carnot cycle on *T-s* plane.

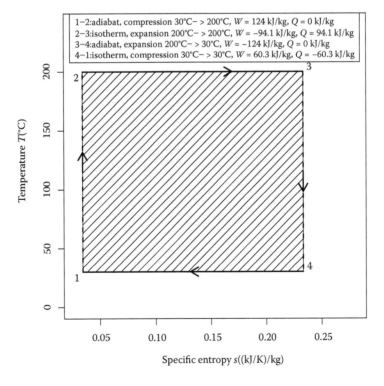

FIGURE 6.8 Example of Carnot cycle on *T-s* plane showing area inside the cycle.

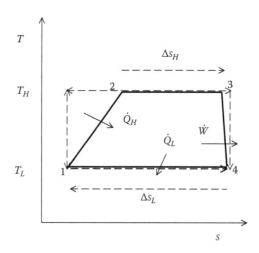

FIGURE 6.9 Nonideal cycle on the *T-s* plane. Cycle goes from state 1 to 2, to 3, to 4, and returns to 1. The change in entropy for the LTR is larger than the one for the HTR.

Example 6.8

Consider a Carnot heat engine with $T_H = 1000$ K, $T_L = 300$ K, $Q_H = 10^6$ J/s. What is the power? What are the entropy changes? Answer: Start by calculating the Carnot limit $\eta_{max} = 1 - \dfrac{T_L}{T_H} = 1 - \dfrac{300}{1000} =$ 0.7. Then, recall that power is $P = \dot{W} = \eta \dot{Q}_H$, then calculate $P = \dot{W} = 0.7 \times 10^6$ J/s = 700 × 10^3 J/s = 700 KW. Now, we calculate entropy HTR $\dot{S}_H = \dfrac{\dot{Q}_H}{T_H} = \dfrac{10^6 \text{ J/s}}{10^3 \text{ K}} = 10^3$ J/s K^{-1} and entropy LTR $\dot{S}_L = \dfrac{\dot{Q}_L}{T_L} = \dfrac{0.3 \times 10^6 \text{J/s}}{300 \text{ K}} = 10^3$ J/s K^{-1}. Finally, subtract $\Delta \dot{S} = -1000 + 1000 = 0$ J/s K^{-1}. We see that for this ideal cycle there is no change of entropy.

What makes the Carnot cycle ideal is that both changes of entropy ΔS_H and ΔS_L are the same because the processes are reversible. Just to understand the implication of the second law in this context, a less than ideal cycle would have smaller ΔS_H compared to ΔS_L, which is illustrated by the hypothetical cycle in Figure 6.9. In this case, the rectangular box is replaced by a trapezoid. The area of this trapezoid is obviously smaller than the area of the rectangle corresponding to the ideal cycle. The path from 3 to 4 is nearly isentropic, and thus the major reason for the smaller ΔS_H is that the path from state 1 to 2 departs greatly from an isentropic process.

In Chapters 7 and 9, we will analyze cycles as models for real processes such as the Rankine cycle for steam-based heat engines, Brayton cycle for gas turbines, and Otto and Diesel cycle for internal combustion engines. The trapezoid shown in Figure 6.9 for illustration is an extreme simplification of a Rankine cycle, which we will start discussing in the next section.

6.3   ENTHALPY AND FREE ENERGY

6.3.1   ENTHALPY

We now introduce another useful thermodynamic state function: *enthalpy* (*H*), which is defined as the sum of two state functions expressed in energy units:

$$H = U + PV \qquad\qquad (6.13)$$

Recall U is the internal energy of the system and PV is the product of pressure and volume. Enthalpy then is a measure of total energy accounting for both internal energy and the pressure volume state at which the system is in equilibrium with its surroundings; or, in other words, enthalpy accounts for the internal energy and the work done to make room for the system within the surroundings. Both U and V are extensive properties. Enthalpy is then an extensive property, and therefore it is convenient to work with specific enthalpy $h = \dfrac{H}{m}$ dividing it by mass.

In differential form, enthalpy is

$$dH = dU + d(PV) = dU + PdV + VdP \tag{6.14}$$

But it is often used to facilitate accounting for energy changes at constant pressure for which $d(PV) = PdV$ and therefore

$$dH = dU + PdV \tag{6.15}$$

We can relate change of enthalpy to heat as follows. When adding heat δQ to the fluid at constant pressure, the first law says that $\delta Q = dU + PdV$, and this is precisely the enthalpy change given by Equation 6.15. We conclude that enthalpy gain dH is the heat δQ added at constant pressure or $\delta Q = dH$. Therefore, the variation of enthalpy with respect to temperature is the same as constant-pressure heat capacity, in other words $\dfrac{\partial H}{\partial T} = \dfrac{\delta Q}{dT}$. Dividing by mass m, the variation of specific enthalpy $\dfrac{\partial h}{\partial T} = \dfrac{\delta Q}{mdT}$ is the same as constant-pressure specific heat capacity $\dfrac{\partial h}{\partial T} = \dfrac{\delta Q}{mdT} = c_p$.

Example 6.9

What is the change of enthalpy for an isobaric process that changes the temperature of one mole of air from 30°C to 40°C? Answer: The specific enthalpy change is $\Delta h = c_p \Delta T$. Estimate c_p of air at the middle of the range

```
> cp.cv(TC=(30+40)/2)$cp.kg
[1] 1.006
```

That is, 1.006 kJ/(kgK) and assume it is constant for this temperature range $\Delta h = 1.006$ kJ/(kg K) × 10 K = 10.06 kJ/kg.

This relationship between enthalpy and heat does not hold in general. At constant volume, the pdV work is null. The first law says $\delta Q = dU + 0 = dU$, that is to say, the heat added δQ becomes equal to the change of internal energy. The variation of specific internal energy with respect to temperature is the constant-volume specific heat c_v; in other words $\dfrac{\partial u}{\partial T} = \dfrac{\delta Q}{mdT} = c_v$. Differentiating Equation 6.13 with respect to temperature $\dfrac{\partial h}{\partial T} = \dfrac{\partial u}{\partial T} + \dfrac{\partial (Pv)}{\partial T}$. For an ideal gas, we could say $\dfrac{\partial (Pv)}{\partial T} = \dfrac{\partial (RT)}{\partial T} = R$ and thus $\dfrac{\partial h}{\partial T} = \dfrac{\partial u}{\partial T} + R$ or $c_p = c_v + R$, which is the relationship between c_p and c_v that we studied in Chapter 4.

One more important conclusion can be drawn when in addition to isobaric for which $\delta Q = dH$ the process is reversible, that is, $\delta Q = TdS$. Equating we have $dH = TdS$ or dividing by mass $dH = Tds$ in specific terms. Therefore, a total change of enthalpy along a reversible isobaric path in the T-s is the area under the T curve or $\Delta h = \displaystyle\int_a^b Tds$.

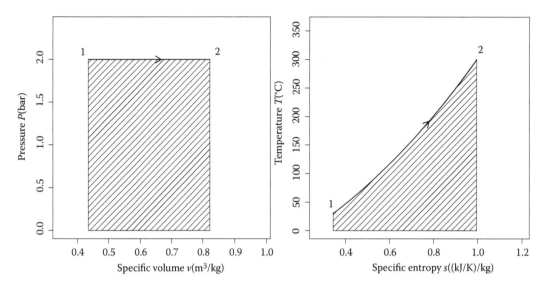

FIGURE 6.10   Isobaric process. Work is area under the curve on the *P-v* plot and enthalpy is equal to heat or area under the curve on the *T-s* plane. Left: 1–2: isobar, expansion 30°C–> 300°C, *W* = −77.5 kJ/kg, *Q* = 276.5 kJ/kg. Right: 1–2: isobar, expansion 30°C–> 300°C, *W* = −77.5 kJ/kg, *Q* = 276.5 kJ/kg.

Example 6.10

Use function path.lines of the renpow package to illustrate the *P-v* and *T-s* plot and calculate change of enthalpy for an isobaric expansion of one mole of air from 30°C to 300°C at 2 bar starting with entropy of 10 J/K. Answer: Use the following

```
# enthalpy
x<- list(P=c(2,2),T=c(30,300),S=c(10,NA),path='isobar',lab=c("1","2"),shade.under=TRUE)
path.lines(list(x),plane='Pv')
path.lines(list(x),plane='Ts')
```

to produce Figure 6.10, which shows the area under the curve on both planes. Work is −77.5 kJ/kg (area on *P-v* plane) and heat is 276.5 kJ/kg (area on *T-s* plane). The change of enthalpy is equal to heat for this reversible isobaric process, therefore Δ*h* = 276.5 kJ/kg.

6.3.2   GIBBS FREE ENERGY

The *Gibbs free energy* (or Gibbs' free energy or Gibbs energy or free enthalpy) is also a state function and combines enthalpy, entropy, and temperature, which are all state functions.

$$G = H - TS = U + PV - TS \qquad (6.16)$$

It is interpreted as that part of the total energy, accounted for enthalpy *H*, that is free or available to perform work; the rest will dissipate as heat, accounted for *TS*. This statement can be more precise for processes at constant pressure and temperature, as we see next. A full differential form of *G* is $dG = dU + d(PV) - d(TS) = dU + PdV + Vdp - Tds - SdT$. For processes at constant pressure and temperature, $dG = PdV - TdS$ or $dG + TdS = PdV$. Now recall Equation 6.6 saying that *dS* is at least $\delta Q/T$ and substitute to obtain $dG + \delta Q \geq PdV$. In other words, work is never larger than $dG + \delta Q$ achieving a maximum when equality holds, which occurs for a reversible process $dG + \delta Q_{rev}$. Thus, change in Gibbs energy establishes a limit on work *W* available after discounting for the heat transfer under reversible conditions.

Example 6.11

What is the change in G due to an isobaric and isothermal process at 300 K and 1 bar with a change of volume of 10 l if the change of entropy is 1 J/K? Answer: Work is $W = P\Delta V = 100$ kPa $\times 0.01$ m^3 = 1 kJ and TdS is $TdS = 300$ K $\times 1$ J/K $= 0.3$ kJ. The change in G is $\Delta G = P\Delta V - T\Delta S = (1 - 0.3)$ kJ $= 0.7$ kJ.

6.3.3 PHASE CHANGE

In addition to temperature, pressure, and volume, state functions entropy, enthalpy, and Gibbs energy have applications when the processes involve *phase change*, for example, converting liquid water to vapor by adding heat. In general, for three phases—liquid, gas, and solid—we have six processes (Figure 6.11): liquid to gas (vaporization) and vice versa (condensation), liquid to solid (freezing) and vice versa (fusion), gas to solid (deposition) and vice versa (sublimation).

For a substance, each one of these processes has temperature values at which the phase change occurs, for example, vaporization has a boiling point and fusion has a melting point. Also for a substance, each process has an associated enthalpy, given as specific enthalpy on a molar basis. We add heat for vaporization, fusion, and sublimation (endothermic or positive enthalpies, trending to the right in diagram of Figure 6.11), but remove heat for condensation, freezing, and deposition (exothermic or negative enthalpies, trending to the left in the diagram of Figure 6.11). At constant pressure, the heat is equal to the enthalpy. The various enthalpies are given as specific enthalpy on a molar basis in kJ/mol.

It takes more energy to go from liquid to gas than to go from solid to liquid, because the molecules will have to acquire more kinetic energy. Take water for example. Its melting point is 0°C and its boiling point is 100°C. The enthalpy of fusion is $\Delta H_{fus} = 6.01$ kJ/mol and the enthalpy of vaporization is $\Delta H_{vap} = 40.7$ kJ/mol. It takes more energy to vaporize than melt water. Using the same phase change temperature, the enthalpy of sublimation is the sum of the enthalpies of fusion and vaporization.

Example 6.12

How much heat is required per mole to bring water at 30°C to its boiling point of 100°C and vaporize it? Assume constant atmospheric pressure of 1 bar. Answer: First, calculate how much heat is required to bring water to its boiling point. Assume heat capacity of water is $c = 4.187$ kJ/(kg K) and molar mass is $M = 18.02$ g/mol. Calculate heat to get to boiling point

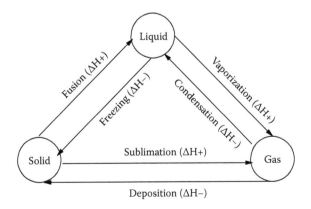

FIGURE 6.11 Phase changes and processes.

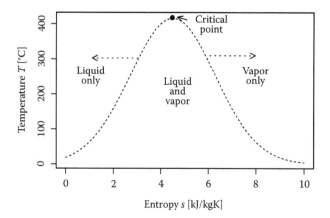

FIGURE 6.12 Approximate *T-s* diagram for water showing phase changes (shape of the curve and values are not exact).

$Q = Mc\Delta T = (18.02 \times 4.187 \times 70)$ J $= 5.28$ kJ/mol. Then add heat of vaporization $\Delta H_{vap} = 40.7$ kJ/mol for a total of 45.98 kJ/mol.

For a substance, phase changes can be represented in phase diagrams showing each phase in a region of the diagram separated by curves in a thermodynamic plane, for example, a *P-T* diagram, a *P-v* diagram, and a *T-s* diagram. For illustration, we show an approximate *T-s* diagram for water (Figure 6.12), which is used often in power generation processes involving steam. The left side of the curve is liquid only, whereas the right-hand side of the curve is vapor (steam) only. Inside the curve, vapor and liquid coexist. At the top of the dome, we have the *critical point* at which phase difference vanishes. It is the highest temperature of the liquid–gas boundary. Beyond this point, we have a supercritical fluid. In water, the critical point is 374°C and ~220 bar. In its supercritical state, CO_2 holds promise as a working fluid for solar thermal (Chapter 14).

6.3.4 STEAM-BASED POWER GENERATION

Thermodynamics of processes involving phase change have application in power generation. As an illustration, let us discuss the Rankine cycle (Figure 6.13). In this chapter, we describe a simplified version, and we will have more to say about this cycle in Chapter 7 when we discuss coal-fired power

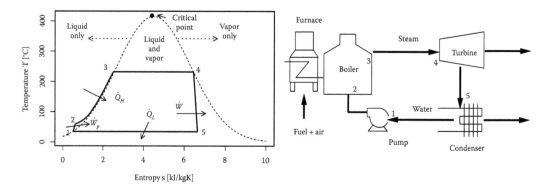

FIGURE 6.13 Left: Simplified Rankine cycle on the *T-S* plane superimposed on the saturation dome for reference (values are arbitrary and for illustration purposes only). Right: Simplified Rankine cycle heat engine.

plants. In Figure 6.13, the cycle is graphed on a T-s plane together with the curve for water undergoing phase changes, from liquid to vapor (steam).

The cycle proceeds as follows. From 2 to 4, liquid water is converted to steam in a boiler by adding heat $\dot{Q}_H$, resulting in increasing temperature and entropy from 2 to 3 (but mostly temperature along the phase curve), and increasing entropy from 3 to 4 (mostly inside the phase curve and isothermally). Most of this is isobaric, and then heat Q_H is nearly the same as enthalpy change from 2 to 4. $Q_H \simeq \Delta H_{42} = H_4 - H_2$. Then from 4 to 5 the steam goes through a turbine (which produces power $\dot{W}$), and lowering of its temperature following an almost vertical line (nearly isentropic as in the Carnot cycle). Once the steam has cooled, it is condensed isothermally back to water (from 5 to 1) in a condenser and rejecting heat $\dot{Q}_L$. Most of these changes are isobaric and then $Q_L \simeq \Delta H_{15} = H_1 - H_5$. At state 1, the liquid is pumped back to the boiler (this requires power to drive the pump and is performed isentropically). As the water reaches state 2, the cycle repeats.

Example 6.13

What is the thermal efficiency of this simplified Rankine cycle when modeled as a heat engine? Assume $\Delta H_{42} = 1000$ kJ and $\Delta H_{15} = 700$ kJ. Ignore the work required by the pump. Answer:

$$\eta = 1 - \frac{Q_L}{Q_H} \simeq 1 - \frac{0.7 \times 10^3}{10^3} = 0.3 \text{ or } 30\%.$$

6.4 THERMOCHEMICAL PROCESSES

In addition to state changes and phase changes, enthalpy, entropy, and Gibbs energy have application when the processes involve chemical change. In this case heat is liberated or absorbed by a chemical reaction, and there is work performed, but it is typically non-PV work, for example, electrical work. In this section, we focus on these thermochemical applications.

6.4.1 ENTHALPY OF FORMATION

Enthalpy is used to describe changes in chemical reactions as we discussed in Chapter 1. Recall the simple example of formation of liquid water H_2O (l) from hydrogen gas H_2 (g) and oxygen O_2 gas:

$$H_2 \text{ (g)} + \frac{1}{2}O_2(g) \rightarrow H_2O \text{ (l)} \tag{6.17}$$

This reaction is exothermic, the change of enthalpy per mole of hydrogen is $dH = -285.8$ kJ/mol and performs work, and the rest is heat. The reverse reaction

$$H_2O \text{ (l)} \rightarrow H_2 \text{ (g)} + \frac{1}{2}O_2(g) \tag{6.18}$$

is endothermic and has the same change of enthalpy but of opposite sign $dH = +285.8$ kJ/mol.

Enthalpy of formation ($H°$) of a compound is the change in enthalpy to form one mole of the compound at standard or reference conditions of 25°C and 1 bar. In other words, it is the energy contained in the compound because of its chemical composition. The stable form of an element is given an enthalpy value of zero. For example, $H_2(g)$ and $O_2(g)$ have $H° = 0$ kJ/mol. But H(g) and O(g) have $H° = 217.9$ kJ/mol and $H° = 247.5$ kJ/mol, respectively. Liquid mercury Hg(l) also has $H° = 0$ kJ/mol because it is the stable form. Values of enthalpy of formation are provided in tables. For example, see Haynes [9].

Liquid water has $H° = -285.8$ kJ/mol. The negative sign is used when the reaction forming that compound is exothermic and accounts for the fact that the product has less enthalpy. Enthalpy of formation is the difference between the enthalpy of the product and sum of the enthalpies of the reactants to form this product:

$$\Delta H = \sum_{products} H° - \sum_{reactants} H° \qquad (6.19)$$

Example 6.14

What is the change of enthalpy for formation of liquid water? Answer: For formation of liquid water (reaction 6.17) the sum of enthalpy of reactants is zero $\sum_{reactants} H° = (0 + 0)\,kJ/mol$ and of the products is $\sum_{products} H° = -285.8\,kJ/mol$. The difference is $\Delta H = \sum_{products} H° - \sum_{reactants} H° =$ $-285.8\,kJ/mol$.

6.4.2 STANDARD ENTROPY

Together with enthalpy of formation $H°$, a compound has a *standard entropy* ($S°$) (or absolute entropy or standard molar entropy). The standard entropy $S°$ is the entropy of one mole of the compound under standard conditions; it has units of entropy per mole, which is the same as J/(mol K). The reference value for this absolute scale is 0 as J/(mol K) for a pure crystal at 0 K. This is the *third law of thermodynamics*. For example, $H_2(g)$ and $O_2(g)$ have $S° = 0.130\,kJ/(mol\ K)$ and $S° = 0.205\,kJ/(mol\ K)$, respectively. Liquid water has $S° = 0.0699\,kJ/(mol\ K)$. Values of standard entropy are provided along with enthalpy of formation values in tables [9].

Changes of standard entropy for a chemical reaction are calculated similarly to enthalpy using the sum of entropies of reactants and products as in Equation 6.19, but we also need to account for the increase of entropy due to heat flow according to the second law

$$\frac{Q}{T} + \sum_{products} S° - \sum_{reactants} S° \geq 0 \qquad (6.20)$$

Equivalently, we can express the minimum heat as a function of change of standard entropies and temperature by moving terms across the inequality and changing signs:

$$Q \geq T\left(\sum_{reactants} S° - \sum_{products} S°\right) \qquad (6.21)$$

Example 6.15

Calculate minimum heat per mole for reaction 6.17 at 25°C and 1 bar. Answer: For formation of liquid water (reaction 6.17) the sum of entropy of reactants is $\sum_{reactants} S° = (0.130 + \frac{1}{2}0.205)$ kJ/(mol K) $= 0.2325\,kJ/(mol\ K)$ and of the products is $\sum_{products} S° = 0.0699\,kJ/(mol\ K)$. The difference $\sum_{reactants} S° - \sum_{products} S° = (0.2325 - 0.699)\,kJ/(mol\ K) = 0.1626\,kJ/(mol\ K)$. Now multiply by 298 K to obtain $Q \geq 48.45$ kJ/mol. Then $Q_{min} = 48.45$ kJ/mol of H_2 (g).

6.4.3 GIBBS FREE ENERGY

In addition to enthalpy of formation $H°$ and standard entropy $S°$, a compound has Gibbs free energy $G°$ given as free energy of one mole of the compound under standard conditions (25°C and 1 atm); it has units of J/mol. Similarly to enthalpy of formation, stable compounds such as H_2 (g) and O_2 (g) are given the value zero for $G°$, that is, $G° = 0$ kJ/mol 0 J/mol. Liquid water has $G° = -237.2$ kJ/mol. Values of Gibbs energy are available together with standard entropy and enthalpy of formation in tables [9].

The change of free energy of a reaction is calculated the same way as enthalpy of formation:

$$\Delta G = \sum_{products} G° - \sum_{reactants} G° \tag{6.22}$$

Recall from Equation 6.16 that $G = H - TS$ and particularly that $dG = dH - d(TS)$. We can see that the change ΔG for an exothermic reaction goes into two parts: a change of enthalpy ΔH and a heat loss represented in the term $\Delta(TS)$. The first part is free to be converted into work.

Example 6.16

What is the change of Gibbs free energy for formation of liquid water? Answer: For formation of liquid water (reaction 6.17) the sum of Gibbs energy of reactants is zero $\sum_{reactants} G° = (0 + 0)$ kJ/mol and of the products is $\sum_{products} G° = -237.2$ kJ/mol. The difference is $\Delta G = \sum_{products} G° - \sum_{reactants} G° = -237.2$ kJ/mol.

6.5 FUEL CELLS

We will apply the aforementioned thermochemical concepts to fuel cells, using hydrogen fuel cells as an example. The major concept to grasp when comparing this approach to a heat engine is that the fuel cell conversion consists of a single step (Figure 6.14); we go directly from chemical energy to electrical energy, thereby increasing efficiency and reducing emissions. Compare to the multi-step conversion of chemical energy to heat, then heat to mechanical, then mechanical to electrical (Figure 6.14); this process is constrained by thermal efficiency limits and has emissions from combustion of fuels.

This section follows the presentation in Masters [8], which is recommended for further reading. In a fuel cell, the reactants flow into the cell and the reaction products flow out, operating continuously as long as the flows are maintained. Electricity is generated through the reaction between a fuel and an oxidant, a reaction that is facilitated by an electrolyte catalyst. The fuel comes in on the

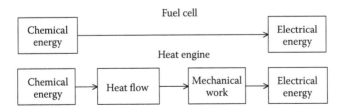

FIGURE 6.14 Comparing energy conversion using fuel cell to heat engine.

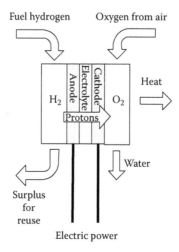

FIGURE 6.15   Schematic of hydrogen fuel cell.

anode side, whereas the oxidant comes in on the cathode side. Fuel cells have been known since the 19th century when Sir William Grove developed a gaseous voltaic battery. Toward the end of that century, a fuel cell delivering just over 1 W of power was developed. It was not until the middle of the 20th century that several kilowatts were produced from an alkaline fuel cell. More recently, fuel cells have been in development to power electric and hybrid vehicles and for space exploration.

A fuel cell consumes reactant from an external source, which must be replenished; thus, thermodynamically, it is an open system. Compared to batteries, these store electrical energy chemically, and therefore before recharging can be considered a thermodynamically closed system. Many combinations of fuels and oxidants are possible.

We will focus on hydrogen fuel cells, or proton exchange membrane (PEM) fuel cells, which use pure hydrogen as its fuel and oxygen (usually from air) as its oxidant (Figure 6.15). PEM fuel cells were used in the Gemini space program. They have 45% efficiency and a power range of 30 W to 250 kW. Typical specifications are 0.5 W/cm^2, 0.65 V, and 1 A/cm^2. These are given per unit area of membrane.

This fuel cell does not emit carbon and other air pollutants. Hydrogen is a simple atom: one proton, one electron, and is plentiful (in water, hydrocarbons, and biomass). The by-products are water and waste heat. Waste heat can be used for cogeneration or combined heat and power (CHP) (Chapter 9). Water can be used for drinking (this is an important process in space exploration).

The reaction in a hydrogen fuel cell is given by reaction 6.17, which is an exothermic reaction with $dH = -285.8$ kJ/mol and as we calculated in Example 6.15 $Q_{min} = 48.45$ kJ/mol of H_2 (g). Referring to Figure 6.16, we can calculate the maximum attainable efficiency of this fuel

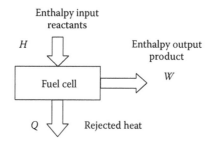

FIGURE 6.16   Balance of energy for a fuel cell.

cell using the concept that the maximum output is the enthalpy input minus the minimum heat $\eta_{max} = \dfrac{\Delta H - Q_{min}}{\Delta H}$. In other words,

$$\eta_{max} = 1 - \frac{Q_{min}}{\Delta H} \tag{6.23}$$

where the enthalpy is determined from Equation 6.19 as $\Delta H = \sum\limits_{products} H° - \sum\limits_{reactants} H°$ and the minimum heat is determined by Equation 6.21 as $Q_{min} = T \left(\sum\limits_{reactants} S° - \sum\limits_{products} S° \right)$.

An alternative calculation that yields the same result is obtained from Gibbs energy change of the reaction from Equation 6.22 as $\Delta G = \sum\limits_{products} G° - \sum\limits_{reactants} G°$ and using $\Delta G = \Delta H - \Delta(TS)$:

$$\eta_{max} = \frac{\Delta G}{\Delta H} \tag{6.24}$$

which states that maximum efficiency is simply the change of Gibbs energy divided by the input enthalpy.

Example 6.17

A fuel cell using hydrogen gas operates at 25°C and 1 atm forms liquid water according to reaction 6.17. Calculate maximum efficiency. Answer: Calculate minimum heat as in Example 6.15 to be $Q_{min} = 48.45$ kJ/mol. Then use change of enthalpy $\Delta H = -285.8$ kJ/mol of H_2 (g) to calculate maximum efficiency $\eta_{max} = 1 - \dfrac{Q_{min}}{\Delta H} = 1 - \dfrac{48.45}{285.8} = 0.83$ or 83%. Alternatively, this can be calculated from Gibbs free energy to yield the same result.

As we can see from this example, this is a very high theoretical maximum efficiency; higher than the ones typically obtained for a Carnot heat engine, which you may recall is 70% for a large temperature difference of 1000 to 300 K. However, 83% is not achieved in real fuel cells because of other losses such as the ones involved in conversion of work to electricity.

Activation loss occurs because the catalysts and electrodes require initial energy in order to activate the chemical reactions. *Ohmic losses* are due to internal resistance of the electrolyte membrane, the electrodes, as well as the various interconnections in the cell. *Fuel crossover loss* result from fuel (hydrogen gas) passing through the electrolyte without releasing its electrons to the external circuit. *Mass transport loss* is due to water build up at the cathode, impeding free flow of hydrogen and oxygen. After accounting for these losses, practical efficiency values are ~50% to 60%, which are still higher than the corresponding practical values for heat engines of about ~30% to 40%.

6.5.1 CURRENT, VOLTAGE, AND POWER

Suppose we supply fuel H_2 at a flow rate $\dot{n}$ in mol/s and connect the fuel cell to a load. The electric current is $I = 2\dot{n}Nq$, where N is Avogadro's number (6.022×10^{23} molecules/mol), q is charge of an electron (1.602×10^{-19} C/electron), and 2 comes from the number of electrons per H_2 molecule. Substituting we obtain $I = \dot{n}Nqe = \dot{n} \times 192{,}945$ A. Total current is proportional to the cross-sectional area of the cell, which is given in cm^2. The larger the area, the larger the current.

The free energy available for conversion into useful work is $\Delta G = 237.2$ kJ/mol and it can be used to calculate output power as $P = \dot{n} \times 237.2 \times 1000$ W $= \dot{n}\,237{,}200$ W. Once we have current

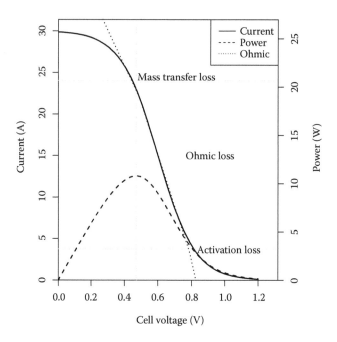

FIGURE 6.17 *I-V* characteristics and power of a fuel cell.

and power we can determine output voltage $V = P/I = (237,200\,\dot{n}/192,945\,\dot{n}) = 1.23$ V. Since the hydrogen flow rate cancels out, the output voltage is independent of fuel flow. Note that power P decreases with temperature T because ΔG decreases with T. Voltage is independent of the cell area.

The hydrogen consumption rate per kWh of electrical energy generated can be found as H_2 rate = $\dfrac{\dot{n} \times 2 \times 3600}{\dot{n} \times 237,200}\,\dfrac{\text{mol}}{\text{s}} \times \dfrac{\text{g}}{\text{mol}} \times \dfrac{\text{s}}{\text{h}} \times \dfrac{1}{\text{W}} = 30.35$ g/kWh.

The electrical *I-V* characteristic of a fuel cell with 15 cm^2 of area and nominal 0.6 V is illustrated in Figure 6.17. The horizontal axis is cell voltage V_{cell} in V and the vertical axis is I = current in A. Note that for low and high values of voltage, the current is nonlinear. In the middle range, the *I-V* curve can be approximated by a resistance Ro, that is to say, the *I-V* curve is fitted to an ohmic approximation. In this illustration, the ohmic approximation is $V_{cell} = 0.85 - I \times \dfrac{0.25\,\Omega\,\text{cm}^2}{15\,\text{cm}^2}$ and the resistance $Ro = \dfrac{0.25}{15}\,\Omega = 0.016\,\Omega$. We can see that the voltage of the cell decreases as the current supplied to the load increases.

Also shown in this plot is the product of voltage and current, or power in W. The value of voltage for maximum power is slightly below 0.5 V and corresponds to about 24 A (which is 1.6 A/cm^2), and this occurs at the beginning of the nearly linear segment that fits the ohmic approximation. The maximum power is nearly 11 W (which is about 0.73 W/cm^2).

When connected to a resistive load, the intersection of the load line and the *I-V* curve of the fuel cell will determine the operating voltage and current. Figure 6.18 illustrates this concept using a load resistance of 0.5 Ω. This plot is obtained using function fuel.cell of renpow using area of 15 cm^2 and load of 0.5 Ω. The intersection is at ~0.6 V and ~16 A, which intersects the power curve at ~10W.

```
x <- list(area.cm2=15,Rload.ohm=0.5)
fuel.cell(x)
```

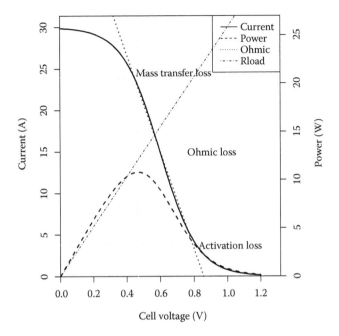

FIGURE 6.18 *I-V* characteristics of a fuel cell and a load line.

Example 6.18

Suppose we stack 80 cells in series operating at 0.6 V to generate 48 V DC. Assume the current is 16 A and it is the same for all cells when connected to a load. Each cell supplies 10 W. What is the power supplied to a load? Assume the cells operate in the ohmic approximation range. Answer: Total power is $10 \times 80 = 800$ W.

6.5.2 Types of Fuel Cell

There are other types of fuel cells based on different fuels and oxidants, for instance, solid oxide, direct methanol, phosphoric acid, alkaline, and molten-carbonate fuel cells. Solid oxide fuel cells (SOFCs) are based on O^{2-} instead of H^+, operate at high temperature (750°C–1000°C), and the rejected heat can be used to generate power using a thermodynamic cycle based on a turbine (70% efficiency) or can be used to reform hydrocarbons to make hydrogen. SOFCs can be used to build power stations. Similar application can use molten-carbonate fuel cells, which use CO_3^{2-} instead of H^+. Direct methanol fuel cells use the same electrolyte as PEM, and are applicable to electric vehicles and systems requiring liquid fueling.

6.5.3 Hydrogen Production

Steam reforming is a method to produce hydrogen from hydrocarbons, such as natural gas. This is achieved in a reformer, where methane and steam react at high temperature (~1000°C) to produce hydrogen

$$CH_4 + H_2O \rightarrow CO + 3H_2 \tag{6.25}$$

This reaction is endothermic ($dH = 206$ kJ/mol), so it requires energy to produce hydrogen. The CO can yield additional hydrogen by

$$CO + H_2O \rightarrow CO_2 + H_2 \tag{6.26}$$

which is exothermic. Note that this process yields CO_2. Thus, the reduction of CO_2 emission achieved by reforming natural gas and producing hydrogen for a fuel cell is limited.

Hydrogen is also produced by partial oxidation of natural gas

$$CH_4 + \frac{1}{2}O_2 \rightarrow CO + 2H_2 \tag{6.27}$$

which is exothermic and therefore does not require energy as reaction 6.25 does. Hydrogen is part of *syngas*, produced by gasification of biomass, coal, or wastes. We will discuss these processes in Chapters 7 and 9.

Hydrogen is also produced by electrolysis of water, or reaction 6.18, and endothermic reaction with $dH = +285.8$ kJ/mol, and therefore hydrolysis requires energy. One option is to power the hydrolysis of water using renewable power such as wind and solar PV, and using the hydrogen for a fuel cell. The water discharged by the fuel cell is recycled and sent back for hydrolysis.

EXERCISES

6.1. Suppose a heat engine with $T_H = 900$ K, $T_L = 300$ K, $Q_H = 900$ kJ, $\eta = 30\%$. What is the difference in Q/T from HTR to LTR? What is the difference in entropy rates?

6.2. Consider a Carnot heat engine with $T_H = 900$ K, $T_L = 300$ K, $Q_H = 900$ kJ. What is the entropy of the HTR and the LTR?

6.3. Compare reversible and irreversible isothermal expansion of 10 moles of a gas previously compressed 6 times of atmospheric pressure and held at standard room temperature (25°C or 298 K). What is the change of entropy of the reversible expansion? What is the ratio of heat transfer to temperature of the irreversible expansion?

6.4. Calculate the final value of entropy for an isothermal expansion of one mole of air from 25 l to 35 l at 100°C assuming that the initial entropy is 8 J/K. Plot this isotherm in the *T-s* plane and the area under the curve using renpow.

6.5. Plot the Carnot cycle for one mole of air with $T_H = 600$°C, $T_L = 30$°C and expanding from 25 l to 35 l at 600°C assuming that the initial entropy is 1 J/K.

6.6. What is the change of enthalpy for an isobaric process that changes the temperature of one mole of air from 40°C to 50°C?

6.7. Use function path.lines of the renpow package to illustrate the *T-s* plot and calculate change of enthalpy for an isobaric expansion of one mole of air from 35 to 45 l at 1 bar starting with entropy of 8 J/K.

6.8. What is the change in G due to an isobaric and isothermal process at 600 K and 1 bar with a change of volume of 20 l if the change of entropy is 1 J/K.

6.9. How much heat is required per mole to bring water at 60°C to its boiling point and vaporize it? Assume constant atmospheric pressure of 1 bar.

6.10. Calculate change of enthalpy and Gibbs free energy for hydrolysis of water.

6.11. A fuel cell using hydrogen gas operates at 25°C and 1 atm forming liquid water. Calculate maximum efficiency using Gibbs free energy. Compare to results obtained from minimum Q.

6.12. Using the ohmic approximation for a PME fuel cell of area 20 cm^2, calculate the current and power supplied to a load of 0.8 Ω. Use the fuel.cell function of renpow.

6.13. We stack 80 fuel cells at 0.6 V each to supply a load with 1 kW at 48 V. What is the load current? What cell area would be needed to attain this current? Assume the cells operate in the ohmic approximation range.

REFERENCES

1. Borgnakke, C., and R.E. Sonntag, *Fundamentals of Thermodynamics*. Eighth edition. 2012: Wiley. 912 pp.
2. Tester, J.W., *Thermodynamics and Its Applications*. Third edition. 1996: Prentice Hall. 960 pp.
3. Fay, J.A., and D.S. Golomb, *Energy and the Environment. Scientific and Technological Principles*. Second edition. MIT-Pappalardo Series in Mechanical Engineering, M.C. Boyce and G.E. Kendall. 2012, New York: Oxford University Press. 366 pp.
4. Tester, J.W. et al., *Sustainable Energy: Choosing Among Options*. Second edition. 2012: MIT Press. 1021 pp.
5. Vanek, F., L. Albright, and L. Angenent, *Energy Systems Engineering: Evaluation and Implementation*. Third edition. 2016: McGraw-Hill. 704 pp.
6. Jorgensen, S.E., and Y.M. Svirezhev, *Towards a Thermodynamic Theory for Ecological Systems*. 2004: Elsevier. 380 pp.
7. Morowitz, H., *Foundations of Bioenergetics*. 1978: Academic Press. 344 pp.
8. Masters, G.M., *Renewable and Efficient Electric Power Systems*. Second edition. 2013: Wiley-IEEE Press. 690 pp.
9. Haynes, W.M., editor, *CRC Handbook of Chemistry and Physics*. 97th edition. 2016: CRC Press. 2670 pp.

7 Coal- and Steam-Based Processes

This chapter is devoted to power generation using heat engines based on steam, which can be generated by various sources of heat, including those generated by burning coal, using nuclear energy, and geothermal processes. We will focus on burning coal to generate steam, which accounts for nearly 41% of the world's electric power generation (Chapter 1). Coal-fired power plants are large plants that require high capital investment but have lower operational costs due to lower cost of coal when compared to other fuels. These plants run continuously, and therefore are appropriate to use as baseload power plants. Before we examine coal-based power generation, a review of basic concepts of atmospheric processes is needed when examining environmental impacts of coal-fired power plants. The presentation is based on textbooks on introductory earth science as well as transport and fate of pollutants and are recommended for supplementary reading [1–3].We will study the basics of carbon sequestration, a mitigation approach to CO_2 emissions from coal-fired power plants; we will cover direct methods in this chapter following the presentation in Vanek et al. [4].

7.1 EARTH'S ATMOSPHERE

In this section, we expand our knowledge of Earth's atmosphere in preparation to understand more broadly the environmental effects of coal-fired power plants.

7.1.1 VERTICAL STRUCTURE

In Earth's atmosphere, air temperature and pressure change with increasing altitude. Pressure steadily declines as less air exerts less weight. However, temperature changes at varying rates and signs establishing an alternation of thermal layers starting from the planet's surface to the troposphere, stratosphere, mesosphere, thermosphere, and exosphere.

We will focus on the two layers closer to the planetary surface: the *troposphere* and the *stratosphere* (Figure 7.1). The troposphere contains most (~3/4) of the mass of the atmosphere and it extends from the surface to about 10 km, but is spatially and seasonally variable, being deepest in tropical areas and higher in summer. Air pressure is about 0.1 of the pressure at sea level. In this layer, temperature decreases with altitude because it is heated from below by longwave electromagnetic (EM) radiation emitted from the Earth's surface. This negative temperature gradient extends until the top of the troposphere or tropopause, at which point it becomes isothermal and then reverses showing an increasing temperature with elevation (Figure 7.1).

This positive gradient persists for about 40 km defining the stratosphere, and it occurs because of heating from above by incoming shortwave EM radiation from the sun, mostly absorption of ultraviolet (UV) radiation. The stratosphere has 10% to 20% of the mass of the atmosphere; thus the troposphere and stratosphere alone account for most of the mass of the atmosphere. The positive temperature gradient becomes null and switches to negative at the stratopause or boundary between stratosphere and the next layer up, the mesosphere. Air pressure drops to about 0.001 of pressure at sea level.

The lowest part of troposphere (~1 km) is the planetary *boundary layer* where thermal processes interact with details of the surface. Looking at tropospheric temperature gradients with more detail, the *lapse rate* is the slope of the temperature versus elevation profile. The lapse rate is driven by radiation and by heat from condensation of moist air (recall phase change from Chapter 6). The average lapse rate (or environmental lapse rate) determines the stability of rising air parcels.

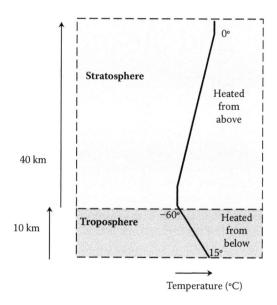

FIGURE 7.1 Atmosphere layers nearest to the surface: troposphere and stratosphere. Ozone processes in upper and lower atmosphere.

Recall that warm air rises in cooler surroundings and sinks in warmer surroundings. We will describe this process with more detail next.

7.1.2 CONVECTION

Convection transport is established by a density gradient in the fluid, for instance, in air and water. The density gradient is established by a difference in temperature; density decreases with increasing temperature. Take, for instance, air; recall from Chapter 4 that the ideal gas law can be written as $P = \rho \dfrac{R}{M} T$, where the term R/M depends on the gas and it is called the specific gas constant. Rearranging we see that $\rho = \dfrac{P}{(R/M)T}$ goes with the inverse of temperature. Therefore, warmer air has lower density, thus is more buoyant. Convective transport in the atmosphere is vertical; lighter parcels rise against gravity and denser parcels fall.

A parcel of air heated at the ground rises in the troposphere and cools as it ascends. When dry, it does not add heat to its surroundings and the air parcel cools adiabatically at ~–10°C/km. Wet or moist parcels of air release heat and warm the surrounding air as they condense, in turn cooling at a lower rate of ~–6°C/km (Figure 7.2). When the parcel cools slower than the surroundings (greater average lapse rate), it is warmer as it goes up and keeps rising, which are *unstable* conditions (Figure 7.2). However, when the parcels cool faster than the surroundings (lower average lapse rate), it becomes colder and it stops rising, creating *stable* conditions (Figure 7.2). As a relatively dry parcel rises at the adiabatic rate, the moisture content increases and the parcel will rise at the slower wet rate. At this point, we could form a cloud by *adsorption* of water droplets to condensation nuclei, such as small particulate matter. Adsorption is a process of adhering ions or molecules to the surface of a compound, the adsorbent. It differs from *absorption* in which ions or molecules are assimilated into the whole mass of the absorbent.

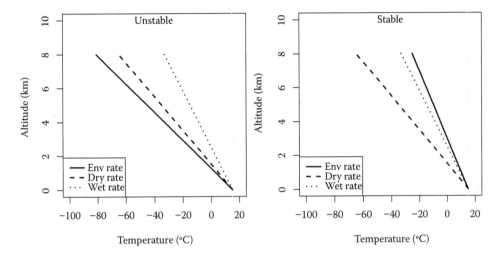

FIGURE 7.2 Lapse rate and parcel cooling rates: unstable and stable conditions.

Example 7.1

Consider two cases: (a) average lapse rate of 12°C/km, (b) average lapse rate is 5°C/km. Determine stability for each case. Answer: (a) 12°C/km > 10°C/km; the parcel remains warmer, and thus we have unstable conditions. (b) 5°C/km < 6°C/km; the parcel becomes cooler, and thus we have stable conditions.

The lapse rate may vary to zero (isothermal) or become positive in what is known as an inversion, which has consequences on how air pollutants may disperse in the lower troposphere. A parcel of air is unable to rise under inversion conditions, because the temperature is higher above, and the pollutants remain close to the ground.

7.1.3 COMPOSITION AND VERTICAL STRUCTURE

As we know from Chapter 2, almost all of the atmospheric gases are nitrogen N_2 and oxygen O_2; the rest (besides some argon) includes water vapor and trace gases such as CH_4, CO_2, and ozone O_3. The proportion or ratio of N_2 to O_2 remains relatively stable through the troposphere and stratosphere (and further up to about 90 km). However, water vapor and trace gas composition changes according to the atmosphere vertical structure.

Water vapor is present mostly in the troposphere and therefore most clouds are limited to the troposphere. CO_2 vertical distribution was thought to be relatively uniform but it is now known to vary with elevation.

Ozone is more abundant in the stratosphere, peaking at about 25 km above sea level, and is responsible for absorption of UV. Ozone's high abundance in the stratosphere is explained by photochemical principles. Molecular oxygen O_2 is photodissociated into two atoms of oxygen by a photon of UV, $h\nu + O_2 \rightarrow O + O$, then collision of atomic O and molecular oxygen O_2 produces ozone $O + O_2 \rightarrow O_3$. Ozone is destroyed by chlorine $Cl + O_3 \rightarrow ClO + O_2$. The ozone hole problem refers to depletion of stratospheric ozone, which causes increased UV at the planet surface, and represents a public health problem because excess UV produces skin cancer, cataracts, and other issues. In opposite manner, an increase of ozone in the lower troposphere or ground level leads to a local air pollution problem, causing respiratory health issues.

7.1.4 Air Quality

Ambient or ground-level tropospheric processes are of paramount importance because the resultant air quality relates primarily to human health, but also affects visibility, vegetation, crops, animals, and buildings. Pollutant concentrations are dependent on emissions and on meteorological conditions, which affect their movement and dynamics. For example, local thermal stability and wind conditions affect the dispersal of stack emissions, and regional wind patterns determines the distribution of pollutants over large distances.

Major pollutants of air include carbon monoxide (CO), lead (Pb), nitrogen dioxide (NO_2), ozone (O_3), particulate matter (PM), and sulfur dioxide (SO_2). Most of these are relevant to our study of coal-fired power plants. Some pollutants are emitted directly, whereas others are formed by precursors originating from reactions involving directly emitted pollutants. For example, PM and SO_2 are emitted directly, but O_3 is formed by precursors. In some cases, further precision in defining the pollutant is required according to its effect. For example, PM is divided into major types according to particle size; PM_{10} is for particles smaller than 10 μm and $PM_{2.5}$ is for particles smaller than 2.5 μm. Their effect on human health is different.

Atmospheric chemistry influences concentrations of all these compounds. For example, NO_2 has high concentration near the ground in the troposphere and at the stratosphere. Its main source is nitrous oxide (N_2O), which arises from biological processes. Nitrogen dioxide is the main natural sink of stratospheric ozone [5]. In the stratosphere, N_2O photodissociates into O and N_2, nitrous oxide reacts with O to form nitric oxide (NO), and then NO reacts with ozone to produce NO_2.

In the United States, the Clean Air Act [6,7] is the major legislation on air quality and requires the U.S. Environmental Protection Agency to set National Emissions Standards and National Ambient Air Quality Standards (NAAQS) [8]. The standards apply to six criteria pollutants: CO, Pb, NO_2, O_3, PM_{10}, $PM_{2.5}$, and SO_2. For example, NAAQS ozone is subject to both primary and secondary standards, by 8-hour averaged values, the annual fourth highest daily maximum 8-hour concentration, averaged over 3 years should not exceed 0.075 ppm. For $PM_{2.5}$, the annual standard is that the annual mean, averaged over 3 years, should not exceed 12 $\mu g.m^{-3}$ as the primary standard and 15 $\mu g.m^{-3}$ as the secondary standard. The 24-hour standard is that the 3-year average of the 98th percentile of a 24-hour average should not exceed 35.5 $\mu g\ m^{-3}$.

7.1.5 Advection

Transport of constituents as well as of energy (in the form of heat and enthalpy) in a fluid is characterized by several processes. *Advection* consists of transporting the constituents along with the bulk of the fluid moving with velocity v, which can be a vector (i.e., includes horizontal and vertical components). This occurs, for example, by wind or a current in the fluid with that velocity field. A simple example in liquids is transport within a pipe or a river. Advection typically describes horizontal transport in the atmosphere.

Given the concentration of a constituent X in mg/l (ppm) or mg/m^3 (ppb), and volume flow rate vA, where v is velocity (m/s) and A is the cross-section area (m^3/s), the flux density due to advection is $F = vX$ and given in $mg/(m^2s)$.

Example 7.2

Suppose air flows at a steady 3 m/s horizontally with 200 ppb concentration of SO_2. What is the SO_2 flux density? Answer: $F = vX = 3\,\dfrac{m}{s} \times 200 \times \dfrac{mg}{m^3} = 600\,\dfrac{mg}{m^2 s}$.

The same concept applies to energy, where instead of concentration we use heat or enthalpy in the fluid. Given the energy, say, heat q per unit mass in kJ/kg, the density ρ of the fluid in kg/m^3, and the volume flow rate vA, where v is velocity (m/s) and A is the cross-section area (m^3/s), we can calculate energy flux as $F = \rho q v$ in $kJ/(m^2s)$.

Example 7.3

Suppose water flows at 1 m/s carrying away heat from a source of 1 kJ/kg. Assume this process occurs at 30°C. What is the heat flux density? Answer: Actually, water density varies with temperature; we can look it up in tables or use polynomials. For simplicity, assume at 30°C it is $\sim$1000 $\frac{kg}{m^3}$. We can approximate by $F \simeq 1 \frac{kJ}{kg} \times 1000 \frac{kg}{m^3} \times 1 \frac{m}{s} = \frac{1\,kJ}{m^2 s} = \frac{1\,kW}{m^2}$.

7.2 COAL CHARACTERISTICS AND TYPES

As discussed in Chapter 2, coal was formed under pressure and heat from remains of vegetation that did not decay quickly, and large coal deposits date back to the Carboniferous period. Coal deposits are found in almost all continents of the Earth at various depths. The vegetation type that gave origin to the deposit and the physical and chemical conditions experimented during formation determine important characteristics of coal. These include carbon content, moisture content, sulfur content, minerals (ash) content, age, and heating value [9]. Several types of coal are distinguished based on these characteristics [10].

For the purposes of analyzing power generation and the environment, two important characteristics of coal are (1) its heat value (HV) or how much energy is produced by burning one unit of mass (in MJ/kg) and (2) the carbon (C) content by weight (in %). We studied the basics of heat value in Chapter 1 when we discussed energy released in combustion as the enthalpy of an exothermic reaction. This concept was further clarified in Chapter 6 when we discussed the thermodynamics of chemical reactions.

Also in Chapter 1, we studied that hydrocarbons formed by C and H bonds contain more C as the molecule increases; particularly alkanes contain more C atoms according to the formula C_nH_{2n+2}. In addition to C and H, coal includes O, N, and S atoms in their molecules. These came from the plant material that decayed and gave origin to coal. The molecular formula for various types of coal includes a different number of C atoms and this determines the C content by weight. For example, just for illustration, coal with the formula $C_{137}H_{97}O_9NS$ would have a molar mass of $137 \times 12 + 1 \times 97 + 9 \times 16 + 1 \times 14 + 1 \times 32 = 1931$ and C is $137 \times 12 = 1644$ or $1644/1931 = 0.85$ or 85%. This would be a high C content, and in reality coal has varying contents of water and other elements (e.g., Si, Ca, As, Pb, and Hg) reducing the percent of C content.

Sorting by increasing values of HV and C content, we can focus on four types of coal [9,10]. *Lignite*, a relatively young coal and used mostly for power generation, has a low C content (C $\sim$30%) and low HV (<20 MJ/kg). *Subbituminous* coal has slightly higher C content and HV (C $\sim$35%–45%, 20–24 MJ/kg). *Bituminous* coal has much higher C content and HV (C $\sim$45%–80%, >24 MJ/kg). *Anthracite*, a relatively old coal, has the highest carbon content and HV (C $\sim$80%–85%, $\sim$24–30 MJ/kg).

The C content is important in regard to contribution of atmospheric CO_2 and consequently climate change, as discussed in Chapter 1. The contents of minerals (ash), nitrogen, and sulfur are important in regard to other environmental impacts, such as air quality. Ash content varies from 3% to 20% by weight, being lowest for lignite and some bituminous, highest for anthracite, and intermediate ($\sim$10%–15%) for subbituminous. Sulfur content is variable by a tenfold difference (0.4%–4%); the lowest value (0.4%) corresponds to lignite and the highest occurs for some bituminous deposits.

Example 7.4

Assume we burn 1 kg of subbituminous coal with HV 20 MJ/kg and 40% C. What is the energy produced (in kWh) if only 33% of the released energy is converted to electricity? What is the mass of CO_2 produced? What is the mass of CO_2 per kWh of electricity? Answer: Electrical energy is $\frac{0.33 \times 20 \times 10^6}{3600} = 1.83$ kWh, then mass of C is 0.4 kg and by using the ratio 44:12 of molecular weight of CO_2 to atomic weight of C (Chapter 1) we have $0.4\,kgC \times \dfrac{44}{12} = 1.46\,kgCO_2$, then $1.46/1.83 \simeq 0.8\,kgCO_2/kWh$.

The efficiency used in this example (33%) is typical of many operating coal-fired power plants.

7.3   WORLD COAL CONSUMPTION AND RESERVES

A major theme of this book is how to provide electrical power in a sustainable manner. This is a complicated question particularly if we want electrical consumption to be equitable and accommodate for a growing population. In this section, we look into a partial answer when power provision is related to a nonrenewable resource, such as fossil fuels, and in particular coal. This partial question relates to the length of time it will take to deplete reserves of a particular resource.

Statistics on world energy supply and consumption are often given in *toe* units, which stands for *tonne of oil equivalent* (energy obtained by burning 1 ton of oil), and converts to 41.868 GJ. An estimate of annual world energy consumption, that is, total energy per year used by all countries from all sources, is by means of the TPES (total primary energy supply). A recent estimate (2014) of TPES [11] is ~13,700 Mtoe, which converts to 13700×10^6 toe $\times 41.868 \times 10^9$ J/toe $\simeq 5.74 \times 10^{20}$ J. For numbers this high, it is convenient to use EJ (E or exa is 10^{18}) and give the total annual consumption as 574 EJ/yr. Alternatively, we can express annual world energy consumption in the familiar units of Wh (using P or peta is 10^{15}) as 5.74×10^{20} J/yr/3600s/h $\simeq 159$ PWh/yr. Using the number of hours in a year 365d/yr $\times$ 24h/d = 8760h/yr, which is a number we will use often in this book, we can calculate an average world power consumption of (159 PWh/yr)/(8760h/yr) $\simeq 18$ TW where we have converted to TW (T or tera is 10^{12}).

Out of all this worldwide power consumption, in this book we want to focus on electrical power, and in this chapter on its production using coal-fired power plants. In 2014, electricity production was ~23.8 PWh/yr, which accounts for about 23.8/159 $\simeq 0.15$ or 15% of total energy world consumption. Worldwide percent contributions of electricity generation by coal was about 41% [12].

Example 7.5

What is the worldwide consumption of electrical power and energy? How much of the electrical power and energy is generated by coal? Answer: Electrical power worldwide 18 TW $\times$ 0.15 = 2.70 TW. Then focusing on coal, power is 2.70 TW $\times$ 0.41 = 1.11 TW and energy is 23.8 PWh/yr $\times$ 0.41 = 9.76 PWh/yr.

Estimates of world reserves of various fossil fuel resources vary with time as reserves are consumed and reservoirs are updated by exploration. We can calculate how long it will be before we run out of a particular resource under various scenarios of consumption. For example, one scenario is to continue consumption at current rates of this resource, and another plausible scenario is an increase in the consumption rates.

Let us start with the simplest scenario, which is to maintain current rate of consumption. Mathematically, assume the resource is $X(t)$, then we can assume a linear decrease rate $\dfrac{dX}{dt}$ given by current consumption rate $c(0)$; this is to say, $\dfrac{dX}{dt} = -c(0)$. Integrating, $X(t) = X(0) + \displaystyle\int_0^t c(0)d\tau = X(0) - c(0)t$.

We can calculate time t_f to depletion assuming $X(tf) = 0$ and then $X(tf) = 0 = X(0) - c(0)t_f$. Solving for t_f, we get $t_f = \dfrac{X(0)}{c(0)}$.

We will use the estimates of reserves and consumption of coal reported in Fay and Golomb [9]. The proven reserves are 25,000 EJ, consumption is 140 EJ/yr, and a scenario of consumption increase of 0.8%/yr. Let us first calculate time to deplete coal proven reserves at the current consumption rate.

Example 7.6

Calculate time to deplete coal reserves at the current consumption rate. Answer: Divide reserves by energy used in one year: $t_f = \dfrac{X(0)}{c(0)} = \dfrac{25,000EJ}{140EJ/yr} \simeq 178\,yr.$

This example has assumed that consumption does not increase. More realistically, we can ask how long it will take to exhaust the resource if consumption increases at a given rate. Assume consumption increases exponentially $c(t) = c(0)\exp(rt)$, where r is the rate coefficient in yr^{-1}. Write a new model for decreasing reserve $\dfrac{dX}{dt} = -c(t)$ and integrate to obtain an exponential decrease

$$X(t) = X(0) - \int_0^t c(\tau)d\tau = X(0) - (c(0)/r)(\exp(rt) - 1).$$

For depletion, set the resource to zero $X(t_f) = 0$, and using current reserves $X(0)$, we evaluate this equation to be $X(t_f) = 0 = X(0) - (c(0)/r)(\exp(rt_f) - 1)X(t_f) = 0 = X(0) - (c(0)/r)(\exp(rt_f) - 1)$. Rewrite $X(0) = (c(0)/r)(\exp(rt_f) - 1)$ and rearrange $\dfrac{X(0)}{c(0)/r} + 1 = \exp(rt_f)$, which can be solved for t_f using logarithms $t_f = \dfrac{1}{r}\ln\left(r\dfrac{X(0)}{c(0)} + 1\right).$

Example 7.7

Assume coal consumption increases exponentially $c(t) = c(0)\exp(rt)$ with a rate of $r = 0.8\%$ per year or $r = 0.008\,yr^{-1}$. Assume current consumption and reserves have the values used in Example 7.6. How long will it be before we run out of coal under this scenario? Answer: From Example 7.6, current rate is $c(0) = 140EJ/yr$ and reserves $X(0) = 25,000EJ$ then $t_f = \dfrac{1}{0.008yr^{-1}}\ln\left(0.008yr^{-1}\dfrac{25,000EJ}{140EJyr^{-1}} + 1\right) \simeq$ 111yr. Now the time t_f to depletion has been reduced substantially (by almost 40%) when we include exponential increase of consumption.

7.4   COAL-FIRED POWER PLANTS

7.4.1   HEAT ENGINE: RANKINE CYCLE

From an energy standpoint, a simple model of a coal-fired power plant is a heat engine (see Chapters 4 and 6) based on the Rankine cycle (see Chapter 6). A schematic of a coal-fired power plant is given in Figure 7.3. A Rankine cycle for a coal-fired power plant proceeds as explained in Chapter 6, but has an additional state 4a (Figure 7.4) to *superheat* steam in order to make dry steam and avoid water droplets reaching the turbine blades. From state 2 to 4, liquid water is converted to steam by adding heat $\dot{Q}_H$, resulting in increased temperature and pressure (~250°C, 40 bar) as well as increased entropy. At state 4, heat is still added to superheat the vapor to state 4a at about 500°C and 40 bar.

Then, from 4a to 5a and 5 the steam goes through a turbine (which produces power $\dot{W}$), and lowering its temperature to ambient values follows an almost vertical line (nearly isentropic). Once the steam has cooled down, it is condensed isothermally at relatively low pressure back to water (from 5 to 1) in a condenser and rejecting heat $\dot{Q}_L$. As we have discussed in Chapter 6, at state 1, the liquid is pumped back to the boiler and as the water reaches state 2 the cycle repeats.

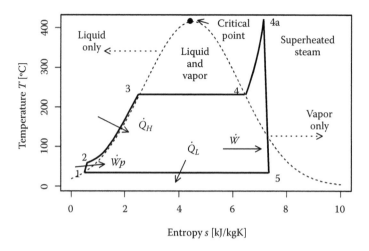

FIGURE 7.3   Schematic of a coal-fired power plant.

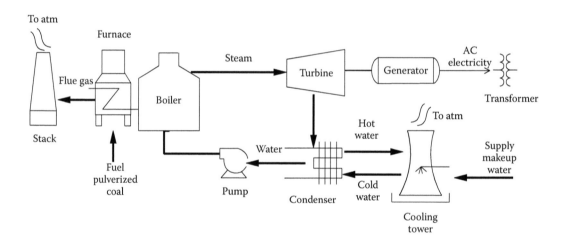

FIGURE 7.4   Rankine cycle for a coal-fired power plant.

We can calculate the thermal efficiency of an ideal Rankine cycle. Recall from Chapter 1 that the Carnot efficiency is given by $\eta_{max} = 1 - \dfrac{T_L}{T_H}$, where T_H and T_L are the absolute values in K of the heat source (HTR ~500°C) and heat sink temperatures (LTR ~30°C), respectively. In the present case $T_H = 273 + 500 = 573$ K, $T_L = 273 + 30 = 303$ K. Therefore, $\eta_{max} = 1 - \dfrac{298}{303} = 0.47$ or 47%.

Actual thermal efficiency is much lower than ideal, reaching ~35% to 40% as measured by the ratio of net mechanical power output to gross heat input. Several processes are implicated in making the actual efficiency lower than ideal. Efficiency is work output W_{out} divided by heat input Q_{in}, or $\eta = W_{out}/Q_{in}$. One type of loss relates to losing part of the input heat provided by the boiler (due to incomplete combustion, loss of heat up the stack in flue gas and other heat loss), and therefore the net input of heat Q_H to the thermal energy conversion is $Q_H = Q_{in}(1 - q_l)$, where q_l is a fraction of input heat lost. Another type of loss is due to employing mechanical work to operate pumps and other equipment, and turbine inefficiencies, thus reducing the net mechanical work, $W_{out} = W_t(1 - w_t)(1 - w_p)$, where W_t is the work done by the steam on the turbine rotor, w_p is a fraction of that work employed for pumps and other equipment, and w_t is turbine inefficiency. Therefore, we can write

$$\eta = \frac{W_{out}}{Q_{in}} = \frac{W_t(1-w_t)(1-w_p)}{Q_H/(1-q_l)} = \frac{W_t}{Q_H} \times (1-w_t)(1-w_p)(1-q_l)$$

and we see that the efficiency will be lowered from the ideal $\eta = W_t/Q_H$ by increased losses.

Example 7.8

Assuming a Carnot efficiency of 47%, what is the actual efficiency if heat and work losses were 8% and 2%, respectively, and the turbine is 90% efficient? Answer: The reduction of ideal efficiency is $0.9 \times (1-0.08) \times (1-0.02) \simeq 0.81$ and then $0.47 \times 0.81 = 0.38$. Therefore an ideal efficiency of 47% is lowered to 38%.

7.4.2 System Efficiency and Heat Rate

In addition to thermal efficiency, discussed in the previous section, we can use a system-level efficiency, which is that one of the entire power plant, meaning the ratio η_s of electrical energy output to the amount of thermal energy input. This efficiency is $\eta_s = \eta \times \eta_e$, where η_e is the electric generator efficiency. This efficiency η_e is typically greater than 90%, and therefore overall system efficiency is mostly limited by the thermal processes.

Example 7.9

Assume a 38% thermal efficiency as in the result of the previous example and a 93% efficient generator. What is the overall efficiency? Answer: $\eta_s = \eta \times \eta_e = 0.38 \times 0.93 = 0.35$ or 35%.

This example shows that overall efficiency may be 12% less than the ideal Carnot efficiency. Many power plants operating today have a ~33% system efficiency. Another popular measure of overall efficiency is *heat rate* (*HR*), which is defined as thermal energy input (in kJ) required to produce 1 kWh of electrical energy output. In other words, $HR = Q_{in}/W_E$, where W_E is electrical energy. Note that this definition is the inverse of unitless efficiency η_s, and that we are explicitly using two different units of energy, therefore after unit conversion *HR* is

$$HR = \frac{1\,(J/s)/W \times 3600\,s/h}{\eta_s} = \frac{3600}{\eta_s}J/Wh \qquad (7.1)$$

For example, for a system efficiency of 33%, $HR = \frac{3600\,J/Wh}{0.33} \approx 10,910\,J/Wh$. For practical values of power plant output, *HR* is often expressed as kJ per kWh. Thus, for example, this HR corresponding to 33% system efficiency would be written as 10,910 kJ/kWh.

New technologies, allowing for higher temperatures and pressures make coal-fired power plants more efficient, for example, supercritical plants and ultra-supercritical plants can have *HR* values of ~9500 and ~8500 kJ/kWh, respectively. Supercritical plants are based on *supercritical steam generators*, sometimes called a Benson boiler because of historical reasons, although it is not really a boiler. Water goes past the critical point (recall from Chapter 6 this is 474°C and 220 bar) directly into steam at high-pressure avoiding bubbles as in a regular boiler.

Example 7.10

Calculate the system efficiency of supercritical plants at 9500 kJ/kWh. Compare its CO_2 emission to a typical 33% efficiency using the same coal as in Example 7.4 (*HV* 20 MJ/kg and C content 40%). Answer: For supercritical plants we have $\eta_s = \frac{3600\,J/Wh}{HR} = \frac{3600\,J/Wh}{9500\,kJ/kWh} \approx 0.38$ or 38%.

Electricity produced increases to $\dfrac{0.38 \times 20 \times 10^6}{3600} = 2.11$ kWh, then using the same value calculated in Example 7.4 of 1.46 kgCO$_2$ we see that emissions decrease to $1.46/2.11 \simeq 0.69$ kgCO$_2$/kWh. This represents $(0.8 - 0.69)/0.8 \simeq 0.14$ or a 14% decrease in CO$_2$ emission.

7.4.3 COAL CONSUMPTION AND STORAGE AT THE PLANT

First, let us see how much coal is consumed by a typical power plant. We have already studied the concepts of heat rate HR for a plant and heat value HV for a given coal type. We can employ both to calculate a coal burn rate (CBR) as the mass of coal in kg required to produce 1 kWh of electricity

$$CBR = \frac{HR}{HV} \qquad (7.2)$$

The mass of coal required per day to feed a plant of installed capacity C (in W) operating at full capacity would be $CBR \times C \times 24$h.

Example 7.11

Take a typical 33% efficient plant and relatively low HV coal (say lignite) of 15 MJ/kg. What is the coal burn rate? What is the daily coal consumption? Answer: $CBR = \dfrac{HR}{HV} = \dfrac{10,910\,\text{kJ/kWh}}{15\,\text{MJ/kg}} = 0.727$ kg/kWh. Now consider, for illustration, a $C = 1$ GW plant which in one day would produce 24 GWh or 24×10^6 kWh. Multiply this value by the CBR to get 0.727 kg/kWh $\times$ 24h/d $\times 10^6$ kW = 17.45×10^3 t/d, which is about 17,450 tons of coal per day.

For continuous operation, coal is stored in mounds or piles that can supply a few weeks of operation [9]. The result of the previous example indicates that these mounds can have the equivalent of hundreds of thousands of tons of coal. For the sake of illustration, take a monthly supply 17.45×10^3t/d $\times$ 30d $\simeq 523.5 \times 10^3$t that is about half a million tons of coal. The coal arrives at the plant by train (or barge in coastal and riverine sites), typically prewashed, to remove part of the minerals and sulfur, and sized ready to be pulverized at the mill and carried by conveyor belt to a silo from which it is delivered at constant rate to the burner.

7.5 ENVIRONMENTAL IMPACTS OF COAL-FIRED POWER PLANTS

We will consider with detail the major environmental impacts of coal-fired power plants, looking at the process of power generation only, that is, we do not discuss details of environmental effects associated with coal mining and transportation to the plant. Deep-shaft mining presents considerable risks to the miners, and surface or strip mining affects the landscape and watersheds [9]. It should be noted that the environmental impacts of these plants on the environment causes not only effects on human health and human livelihood, but also on climate and ecosystem health.

7.5.1 CARBON EMISSION

As explained at the beginning of this chapter, two important characteristics of coal are its HV (in MJ/kg) and the C content by weight (in %). For example, lignite coal used often for power generation has C ~30% and $HV = 15$ MJ/kg. In Example 7.4 and in Example 7.10, we saw how to calculate CO$_2$ emissions. In this section, we formalize these concepts further.

We can sum up HV and C content, specifying a carbon content in coal per unit energy C_e, as kg of C in the equivalent mass of coal that would produce a kJ of energy. For example, if 1 kg coal with 30% C by weight produces 15 MJ, then $C_e = 0.3$ kgC/15 MJ = $2 \times 10 - 5$ kgC/kJ.

We can estimate carbon emission to the atmosphere a_C by simply multiplying the C content per unit energy C_e by heat rate HR, $a_C = C_e \times HR = 2 \times 10^{-5}$ kgC/kJ $\times$ 10,910 kJ/kWh $\simeq 0.22$ kgC/kWh.

An alternative way of calculating C emission is from the coal burn rate CBR and using C in % by weight $a_C = \%C \times CBR = 0.30$ kgC/kg $\times$ 0.727 kg/kWh $\simeq 0.22$ kgC/kWh.

This mass emission rate of carbon can be converted to CO_2 mass emission rate a_{CO2} by using the ratio 44:12 of molecular weight of CO_2 to atomic weight of C. Therefore,

$$a_{CO2} = a_C \times \frac{44}{12} = 0.22 \text{ kgC/kWh} \times 3.666 \text{ kgCO}_2/\text{kgC} = 0.80 \text{ kgCO}_2/\text{kWh}$$

We can see that for this coal and plant efficiency we emit almost one kg of CO_2 per kWh electrical energy generated. Note that this is the same result obtained in Example 7.4 by a coal with HV of 20 MJ/kg and C content of 40%.

Another useful indicator is the CO_2 emission per unit of HV, as $\dfrac{a_{CO2}}{HV} = \dfrac{0.80 \text{ kgCO}_2/\text{kWh}}{20 \text{ } MJ/kg}$.

Example 7.12

What would be C_e, and the C and CO_2 emission rate of a 33% efficient coal-fired power plant if we use bituminous coal with a higher heat value of 24 MJ/kg and higher carbon content of 50% by weight? Answer: The carbon per unit energy is $C_e = 0.5$ kgC/24MJ $\simeq 2.1 \times 10^{-5}$ kgC/kJ, which is only slightly larger than the one for lignite with 30% C and heat value of 15 MJ/kg: $a_C = C_e \times HR = 2.1 \times 10^{-5}$ kgC/kJ $\times$ 10,910 kJ/kWh $\simeq 0.23$ kgC/kWh and $a_{CO2} = a_C \times \frac{44}{12} = 0.23$ kgC/kWh $\times$ $\frac{44}{12}$ kgCO$_2$/kgC = 0.84 kgCO$_2$/kWh.

We can see that the potential decrease of C emission by higher HV is compensated by the higher carbon content and actually leading to higher C emissions.

7.5.2 Cooling Water

At the condenser, heat is transferred to *cooling water*, which may come from a lake, river, or a cooling tower. The warm water may return to the lake or river once used (once-through cooling) or to a cooling tower, where heat is released to the atmosphere and water is cooled down to circulate back to the condenser at a given flow rate.

A cooling tower is preferred over a receiving body of water because the release of warm water may have an impact of increased water temperature of the receiving water and therefore consequences on the aquatic ecosystem. However, a cooling tower implies evaporation loss and this has to be made up with water from the river, leading to reduced river flow or lake volume. This may be significant or not, depending on the water uptake and the total flow or volume. Optionally, treated reclaimed water from other processes could be used as cooling water makeup to reduce consumption of fresh water from the river or lake.

To protect the receiving water body, when using once-through cooling, the maximum temperature of discharge water is restricted, which imposes a need to increase cooling water flow. If heat to be rejected is Q_L and the maximum permissible increase in its temperature is ΔT, the mass of water required is $m_w = \dfrac{Q_L}{c_w \times \Delta T}$. This mass of water may be expressed per kWh by using reject heat on a per kWh basis. This can be done using HR in kJ/kWh, heat $Q_H = Q_{in}(1 - q_l)$ remaining after thermal losses, and assuming $Q_L \approx Q_H(1 - \eta)$ calculated by thermal efficiency:

$$m_w = \frac{Q_H(1-\eta)}{c_w \times \Delta T} = \frac{Q_{in}(1-q_l)(1-\eta)}{C_w \times \Delta T} = \frac{HR \times (1-q_l)(1-\eta)}{C_w \times \Delta T} \text{ kgWater/kWh} \qquad (7.3)$$

Example 7.13

Consider a coal-fired power plant with 33% system efficiency, thermal losses of 10%, and thermal efficiency of 40%. Find the minimum mass of cooling water per kWh if the maximum permissible increase in its temperature is 13°C. Answer: For 33% use HR 10,910 kJ/kWh, then use Equation 7.3 just given $m_w = \dfrac{(1 - 0.1) \times (1 - 0.4) \times 10910\,\text{kJ/kWh}}{4.184\,\text{kJ/(kg°C)} \times 13\,°\text{C}} = 108.31$ kg/kWh or about 108 l/kWh.

Alternatively using continuous flow from the cooling tower, assume the heat rejected per Kg of water evaporated is C_v, then the mass of water removed and lost by evaporation is $m_w = \dfrac{HR \times (1 - q_l)(1 - \eta)}{C_v}$ kgWater/kWh.

Example 7.14

Consider the same conditions as in the previous example but using a cooling tower. Find the mass of cooling water per kWh that is evaporated and removed from the river or lake. Answer: $m_w = \dfrac{(1 - 0.1) \times (1 - 0.4) \times 10910\,\text{kJ/kWh}}{136\,\text{kJ/(kg)}} = 43.32$ kg/kWh, which is about 43.32 liters per kWh.

7.5.3 OTHER EMISSIONS AND AIR QUALITY

Besides emitting carbon to the atmosphere, thus contributing to the greenhouse effect, and either consuming water or causing water thermal pollution, a coal-fired power plant has emissions to the atmosphere that present challenges to air quality. To mitigate air pollution, controls are imposed to reduce emissions of problematic contaminants. Harmful emissions that are controlled include PM (resulting from "fly ash"), sulfur oxides SO_x, nitrogen oxides NO_x, mercury, and other toxics. These emissions are controlled in the United States according to the National Emission Standards for Hazardous Air Pollutants (NESHAP). Depending on the technology, these controls may be expensive and represent a loss of energy conversion efficiency.

PM in the flue gas is due to the incombustible mineral matter contained in coal, and because PM constitutes a risk to human health and the environment, its emission is restricted. PM human health effects include decreased lung function, aggravated asthma, development of chronic bronchitis, and premature death. Environmental effects include visibility reduction and esthetic damage, atmospheric radiation balance, and global climate change.

Sulfur oxides (SO_x) occur because cells of living matter contain sulfur and it remains in the fossil matter. Sulfur oxides that would be emitted (mainly SO_2) from the stack can form sulfuric acid leading to acid precipitation and haze. Nitrogen oxides NO_x (mainly NO and NO_2) are produced not only by burning the organic nitrogen contained in coal, but also because of oxidation of the nitrogen in the air at high temperatures. NO_x are hazardous pollutants because of their effect on human respiratory system and because they are precursors of ozone, itself an important air pollutant. New U.S. 8h-average ozone standards then indirectly impose control on NO_x produced by power plants.

Controlling emissions include controls before combustion as well as combustion management. For example, mineral matter and sulfur content are reduced by washing the coal, and NO_x is reduced by controls of the combustion process. Major equipment and processes typically used to reduce air pollution after combustion from a coal-fired steam power plant are catalytic reduction (CR) equipment to convert the NO_x back to nitrogen, electrostatic precipitator (ESP) to remove fly ash, and a scrubber (flue gas desulfurization, FGD) to remove sulfur (Figure 7.5).

NO_x emission can be mitigated using catalytic reduction, for example, selective catalytic reduction (SCR) based on streaming ammonia together with the flue gas into a reactor to form

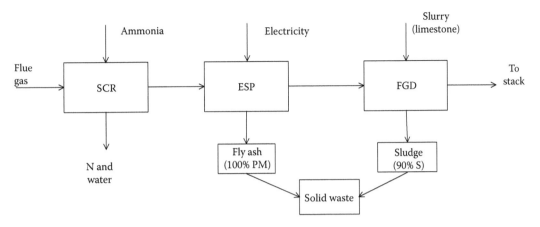

FIGURE 7.5 Air pollution control equipment (after combustion) for a coal-fired power plant. ESP, electrostatic precipitator; FGD, fluid gas desulfurization (scrubber); SCR, selective catalytic reduction.

elemental nitrogen and water. This removes ~80% to 90% of the NO_x. Because the SCR is poisoned by fly ash and SO_x, and it is upstream from the ESP and FGD, frequent replacement is needed making it expensive, representing 5% to 10% of power production.

Flue gas goes to an ESP, which electrically charges the particles so that they can be attracted by electrodes and collect them as fly ash. Particle collection efficiency of an ESP is almost 100% for particles greater than 2.5 μm [9]. In general, PM can be classified according to particle size. Coarse particles (PM_{10}) range in size from 2.5 to 10 μm in diameter and would be emitted from power plants but are almost completely removed by using an ESP. Coarse particles are prevalent near roadways and dusty industries. Fine particles ($PM_{2.5}$) have a diameter <2.5 μm, are found in smoke and haze, and are still emitted from power plants equipped with ESP. Even smaller, there are ultrafine particles of less than 100 nm in diameter, however, we will not discuss them here. U.S. standards existed for a long time to regulate PM_{10} emissions, but newer standards regulate $PM_{2.5}$ as well. The control of these fine particles require additional equipment based on fabric filters.

Example 7.15

Consider a 33% efficient coal-fired power plant that burns coal with a heat value of 15 MJ/kg and contains 10% mineral matter ("ash") by weight (assume that 70% of ash is PM as fly ash). What would be the uncontrolled PM emission (g per kWh)? What efficiency of PM removal equipment is needed to reduce PM emission to 0.013 g per MJ of heat input? Note: This is an old standard, and nowadays the standard is to use the Best Available Control Technology (which is the ESP). Answer: First, use HR as 10,910 kJ/kWh because efficiency is 33%; then calculate coal burn rate (CBR) for this coal:

$$CBR = \frac{HR}{HV} = \frac{10,910\,kJ/kWh}{15\,MJ/kg} = 0.727\,kg/kWh.$$ Now, the uncontrolled emission rate would be $a_{PM} =$

% ash × %flyash × $CBR = 0.10 \times 0.70 \times 0.727\,kg/kWh = 51\,g/kWh$. To estimate this emission per MJ of

heat input divide emission rate by heat rate: $\dfrac{a_{PM}}{HR} = \dfrac{51\,g/kWh}{10.91\,MJ/kWh} \simeq 4.67\,g/MJ$. Now to reduce to

0.013 g/MJ, you have to reduce by $\dfrac{4.67 - 0.013}{4.67} = 0.997$ or 99.7% removal efficiency.

Sulfur is removed using some method of FGD, which includes a wet or dry scrubber, where a spray of slurry of a sorbent (e.g., limestone) precipitates the sulfur as hydrated calcium sulfite and calcium sulfate (similar to gypsum). The mixture of these compounds and limestone (the fraction that did not react) falls to the bottom as a wet sludge. The wet scrubber is more efficient (~90%) than the dry scrubber (~80%), but more difficult and expensive operationally (~10% of the cost of producing electricity).

Example 7.16

Use the same information of the last example. The coal content includes 2% sulfur. What would be the sulfur emission rate if it were not controlled (g per kWh)? What removal rate at the scrubber is needed to reduce S to 0.130 g per MJ of heat input? Answer: Use 2% sulfur and multiply by CBR from the previous example to obtain sulfur emission to be controlled: a_S = %S × CBR = 0.02 KgS/kg × 0.727 kg/kWh = 15 g/kWh. To estimate this emission per MJ of heat input, divide emission rate by heat rate: $\frac{a_S}{HR} = \frac{15\,\text{g/kWh}}{10,910\,\text{kJ/kWh}} = \frac{15\,\text{g}}{10.91\,\text{MJ}} = 1.375\,\text{g/MJ}$. Now to reduce to 0.130 g/MJ you have to reduce by $\frac{1.375 - 0.130}{1.375} = 0.905$ or 90.5% removal efficiency.

Figure 7.5 summarizes the configuration of equipment to implement emission control after combustion. First in line is the SCR. Its outflow downstream goes to an ESP that collects fly ash, and then goes to a scrubber, which produces a sludge. The sludge can be dried for disposal or stored in ponds, representing a solid waste issue and a potential risk in terms of water pollution. Quantification of the amount of this waste, the associated risk, and opportunities of utilization of this waste will be discussed in the next section.

7.5.4 SLUDGE DISPOSAL AND REPURPOSING

The solid waste from a coal-fired power plant is significant. To put it in perspective, consider a plant with fly ash production rate of 50 g/kWh (see Example 7.15) and with fly ash control equipment of 100% removal efficiency. Assuming for the sake of illustration a 1 GW plant where fly ash accumulates annually at a rate of 50 g/kWh × 10^6 kW × 8760 h/yr ≃ 438 × 10^9 g/yr, that is to say, almost half a million tons per year. A similar calculation for sulfur using the values from Example 7.16 yields 15 g/kWh × 0.905 × 10^6 kW × 8760 h/yr ≃ 119 × 10^9 g/yr, which is nearly a fifth of a million tons per year.

From the results of these calculations, we can appreciate that solid waste from a coal-fired power plant is significant, and managing the sludge is of great concern. The sludge from the scrubber is sent to a landfill or stored in ponds. There is an opportunity of utilization instead of wasting, since fly ash can replace cement when making concrete (and thus saving CO_2 emissions), and the calcium sulfate can be converted to gypsum. However, currently only a small amount is utilized, primarily for concrete in the construction industry. A recent example is its use for the expansion of O'Hare International Airport in Chicago.

There are loose federal standards and regulations regarding the sludge. It is not classified as hazardous waste, but it can contaminate water (groundwater and rivers) by leaching. Furthermore, there are potential risks of spills into bodies of waters such as rivers. Recent examples are a spill at the Kingston plant in Tennessee in 2012, and the Elk River tank spill in January 2014 in South Carolina.

7.6 CARBON SEQUESTRATION

Long-term sustainable use of coal requires *sequestering* the CO_2 and preventing its emission to the atmosphere [4]. Carbon sequestration can be divided in two categories. *Indirect* consists of enhancing natural sinks of CO_2 increasing the uptake rate by nonatmospheric reservoirs (we will discuss indirect methods in Chapter 9 along with biomass energy conversion processes). *Direct* is the separation of CO_2 from the products of combustion and sending it to a nonatmospheric reservoir. We will cover direct methods in this chapter following the presentation in Vanek et al. [4].

7.6.1 ABSORPTION/ADSORPTION REMOVAL

Separation of the CO_2 from the flue gas requires an absorption or adsorption chemical agent that will eventually become saturated with CO_2 and requires regeneration to be able to continue removing

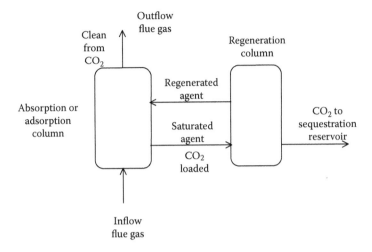

FIGURE 7.6 Direct carbon sequestration: separation from flue gas.

CO_2 (Figure 7.6). In absorption, the CO_2 in the flue gas is assimilated into the mass of the absorbing agent. In adsorption, the CO_2 adheres to the surface of the adsorption agent. Pilot facilities use monoethanolamine (MAE) as an absorbing agent. The regeneration process requires energy and therefore reduces the overall efficiency of power production. This technology is mature at small scales, but new approaches are needed for larger scales; research is looking at membranes to substitute absorption–regeneration, algae grown under controlled conditions, and nature-inspired absorbers such as carbonic anhydrase.

7.6.2 POTENTIAL RESERVOIRS, OIL, GAS, AND COAL SEAMS

Potential reservoirs are spent oil and gas deposits, coal seams, saline aquifers, and the bottom of the ocean (Figure 7.7). The first two of these processes include a by-product of the injection and represent an opportunity for financing the sequestration. When using partially depleted oil and gas deposits, injecting the CO_2 represents a form of recovery, which helps offset costs of sequestration. However, the recovered oil and gas when burned will emit CO_2 and this amount would have to be lower than CO_2 injected for net positive sequestration.

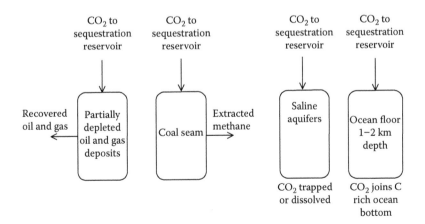

FIGURE 7.7 Direct carbon sequestration: storage in nonatmospheric reservoirs.

Methane gas is adsorbed to coal, in coal seams lining the surfaces of the pore space of the coal. This represents a potential fuel in coal as well as a potential risk when mining coal. Injection of CO_2 into coal seams is a viable sequestration option when CH_4 is extracted from the coal because of the CO_2 injection. Methane is then recovered as a fuel and can help fund the sequestration. As with oil and gas fields, the recovered methane when burned will emit CO_2 and this amount would have to be lower than CO_2 injected for net positive sequestration.

Example 7.17

CO_2 is injected in a coal seam and is able to displace and recover methane in the pores, amounting to 25% of the mass of CO_2 injected. The methane is burned as a fuel later. Assume all CO_2 injected remains in the pores and that this coal is not mined. How much is the net sequestration of CO_2 per kg injected? Answer: We would have sequestered 1 kg of CO_2 and recovered 0.25 kg of methane. When burning all of this methane we would release $\frac{44}{16} \times 0.25 \approx 0.69$ $kgCO_2$ (see Chapter 1). Net sequestration is $1.00 - 0.69 = 0.31$ $kgCO_2$ per kg of CO_2 injected or 31% of injected.

7.6.3 Potential Reservoirs: Saline Aquifers

Many coal-fired power plants are located on top or near saline aquifers. It is typically assumed that high salinity reservoirs are of little interest for irrigation or drinking. It is estimated that saline aquifers may be able to store 10,000 Gt or of CO_2, which is greater than the CO_2 amount that would result from burning all known fossil fuels reserves [4]. Projects around the world include plants in Europe and Canada ranging from a few Mt/yr to 1 Gt/yr sequestration rate. Current efforts are to sequester a fraction of the CO_2 emitted in the plant, and as the technology matures, move to 100% sequestration.

Sequestered CO_2 in saline aquifers may become trapped as a gas bubble (not desirable) or simply dissolve in the water. Long-term, the aquifer's caprock layer should prevent the CO_2 from returning to the environment. Dissolved CO_2 may precipitate out of the solution if reacting with minerals in the aquifer to form, for instance, calcite ($CaCO_3$). These processes will help to increase stability of the sequestration reservoir. Capacity for sequestration V_{CO2} in m^3 can be estimated from reservoir volume V_r and porosity f. As an approximation one can use $V_{CO2} = fV(1 - sw)$, where sw is the fraction of the pores saturated with water. The CO_2 will remain as a gas at subcritical pressure (<7.4 MPa). The CO_2 mass m_{CO2} sequestered can be calculated from V_g, pressure P, and temperature T of the aquifer using the ideal gas law $m_{CO2} = \frac{MV_{CO2}P}{RT}$, where M is the molecular mass of CO_2.

Example 7.18

Assume a saline reservoir is 10^{13} m^3 with 10% porosity and 0.2 water saturated pore space. Average conditions are 5 bar and 40°C. For how many years can it sequester 100% CO_2 from five 1 GW coal-fired power plants averaging 0.8 $kgCO_2$/kWh? Answer: First calculate storage volume $V_{CO2} = fV(1 - sw) = 0.1 \times 10^{13} \times (1 - 0.2) = 0.08 \times 10^{13}$ m^3 and evaluate mass storage capacity $m_{CO2} = \frac{V_g PM}{RT} = \frac{0.08 \times 10^{13} \times 5 \times 10^5 \times 44 \times 10^{-3}}{8.314 \times (273 + 40)} \approx 6.76 \times 10^{12}$ $kgCO_2$. Assuming all plants produce at full capacity 8760 h/yr, then CO_2 emissions would be 5×0.8 $kgCO_2$/kWh $\times 10^6$ kW $\times 8760$ h/yr $= 35 \times 10^9$ $kgCO_2$/yr. Now divide $t = \frac{6.76 \times 10^{12} \text{ kgCO}_2}{35 \times 10^9 \text{ kgCO}_2/\text{yr}} \approx 193$ yrs.

Environmental risks are associated with potential leakage, which can be divided into acute (sudden large release) and chronic (occurring gradually under normal conditions). Acute exposure could occur due to seismic events or unintended trapping of the gas to form a large bubble. Research is needed to characterize all risks.

7.6.4 Potential Reservoirs: Ocean Bottom

The ocean floor has carbonaceous material estimated at 30,000 to 40,000 Gt (or 30 to 40 Eg) of C. Storing additional C from CO_2 sequestration is thought to unlikely change the dynamics of this reservoir. Methods to transport sequestered CO_2 to this layer are being researched. One method is to pump it at high pressure in the water at depths of 1 to 2 km where it would be denser than the surrounding water and let it fall to the bottom. There are environmental risks associated with this approach (e.g., effect on marine life) and depending on depth and other conditions, currents may transport the gas back to shore or shallower areas and seep to the atmosphere. Other proposals include injecting CO_2 as a liquid into the seafloor using technology associated with deep-sea oil wells. At high-pressure and low temperature, the CO_2 would then form hydrate crystals (recall Chapter 2) sealing the rock pores and preventing diffusion of CO_2 to the ocean water.

7.6.5 Conversion to Stable Materials

Carbon is part of stable compounds such as calcite ($CaCO_3$) and magnesite ($MgCO_3$). Carbon dioxide precipitates to these compounds naturally in the ocean or in caves following

$$CaO + CO_2 \rightarrow CaCO_3$$
$$MgO + CO_2 \rightarrow MgCO_3$$

(7.4)

which are exothermic and therefore do not require energy to occur. The amount that precipitates naturally is much smaller than the increase of atmospheric CO_2. However, if these precipitation processes were to be enhanced to rates that keep up with emissions, then it offers an alternative to sequestration.

Increasing the rate requires higher temperature and pressure, thus requiring energy. This energy can be discounted from the energy produced by combusting coal. Both reactions in sequence (combustion and precipitation) have a total net favorable enthalpy. Calcium and magnesium oxides (CaO and MgO), reactants in the precipitation reaction, are not as common as silicate materials containing calcium and magnesium.

Example 7.19

How much calcium oxide is needed to sequester 100% CO_2 from a 1 GW coal-fired power plant averaging 0.8 kg CO_2/kWh? How much calcite is produced? Assume CaO is preheated to 1161 K to enhance precipitation rate and that at this temperature the reaction has $dH = -167$ kJ/mol. It takes 87 kJ/mol to heat CaO to this temperature from ambient conditions [4]. Answer: At full capacity in a year 0.8 kg CO_2/kWh $\times 10^6$ kW $\times$ 8760 h/yr $= 7 \times 10^9$ kg CO_2/yr. The molar mass of CaO is 40 + 16 = 56 g/mol, then the mass of CaO required to sequester this mass of CO_2 is $\frac{56 \text{ g/mol CaO}}{44 \text{ g/mol CO}_2} \times 7 \times 10^9$ kgCO$_2$/yr = 8.9 $\times 10^9$ kgCaO/yr). The calcite molar mass is 40 + 12 + 16 $\times$ 3 = 100 g/mol and therefore the mass of calcite produced is $\frac{100 \text{ g/mol CaO}}{44 \text{ g/mol CO}_2} \times 7 \times 10^9$ kgCO$_2$/yr = 15.9 $\times 10^9$ kgCaCO$_3$/yr. The net energy released after preheating is 167 − 87 = 80 kJ/mol, then the total energy is $80 \frac{\text{kJ}}{\text{mol}} \times \frac{7 \times 10^9 \text{ kg CO}_2/\text{yr}}{44 \text{g/mol CO}_2 \times 10^{-3} \text{kg/g}} = 1.272 \times 10^{15}$ kJ/yr converting to W $\frac{1.272 \times 10^{15} \text{ kJ/yr}}{8760 \times 3600 \text{ s/yr}} = 40.33$ MW.

The example illustrates that sequestering by precipitation is energetically favorable and that mass of reactant and products is considerable, 9 and 16 Mt of calcium oxide and calcite, respectively, in a year. Assessment of the environmental consequences of implementing these processes at large scale is needed.

7.7   OTHER STEAM-BASED SYSTEMS

The emission problems associated with burning coal have led to development of the Rankine cycle using other sources of heat to convert liquid water to steam. Of these, the most used is nuclear-fueled power plants that account for 10.6% of the world's electricity production (Chapter 1), followed by geothermal. In this section, we cover the fundamentals of these processes. An additional option is to use concentrated solar power to produce steam. We will discuss this option in Chapter 14 when we study solar resources.

7.7.1   NUCLEAR

The proportion of electricity produced by nuclear-fueled power plants varies by country, departing significantly from the world average. Organisation for Economic Cooperation and Development (OECD) countries reach 18% of electricity generated by nuclear contrasted with only 4.3% for non-OECD countries [12]. For example, in the United States, nuclear accounts for ~20%, in France for ~80%, and Japan for 30% of electricity production.

Nuclear-fueled power plants in operation are based on fission reactions. Since nuclear energy was harnessed in the 1950s, research has been underway to develop fusion energy, but there is not yet a fusion-based power plant. It has not been possible to control the reaction at reasonable temperature. Nuclear-fueled plants do not use fossil fuels, which is a great advantage in terms of environmental effects, but nuclear energy has risks of adverse effects on human health and the environment due to radioactivity of the fuel through mining, processing, and disposal. Reprocessing of spent fuel also has implications in terms of proliferation of nuclear weapons. Energy is derived from fission of uranium, thorium, and plutonium isotopes. Nuclear power plants actually have a short history, just half a century. Overall, nuclear plants have demonstrated safe operation, but several accidents have highlighted the risks, including Three Mile Island, Pennsylvania (1979); Chernobyl, Ukraine (1986); and the most recent in Fukushima, Japan, in 2011 due to a large tsunami.

A nuclear reactor of a nuclear-fueled power plant is a pressure vessel that contains the nuclear fuel in *fuel rods* that undergo a chain reaction (Chapter 1) and providing the heat to water to perform a Rankine cycle for power generation. The energy comes from fission of a heavy nucleus, for example, ^{235}U in controlled reaction. Most of the energy is in the kinetic energy of fission products: some is released as γ and β rays, and some is released gradually due to radioactive decay. Controlling the chain reaction requires additional reactor components: a moderator, control rods, and a coolant.

Natural uranium contains a small amount of ^{235}U and is insufficient to sustain a chain reaction; therefore, fuel rods are made by enrichment. Fuel rods are spent in about 2 to 3 years and need replacement. Moderators (e.g., water, heavy water, and graphite) slow down the neutrons from fission to become *thermal neutrons* and help sustain the chain reaction. Thermal neutrons have most probable energy of $0.5kT$, with k the Boltzmann constant and T the absolute temperature of the reaction site. Moderators are good at scattering neutrons and poor at absorbing them. Control rods on the contrary are made of material, for example, cadmium, which absorbs thermal neutrons and thereby controls or stops the chain reaction. The position of the control rods thereby determines the power output of the reactor and its stability. Nuclear reactors are run at full capacity and therefore are baseload power generating units.

The coolant removes heat from the reactor; in many power plants, the coolant is pressurized water (at about 15 MPa or 150 bar, preventing it to change phase to steam at that typical temperature of 300°C). See Figure 7.8. This hot water flows into a heat exchanger where the feed water is converted to steam for the Rankine cycle, whose other elements are located outside the reactor. The reason to keep coolant water in the liquid phase is that its moderator (neutron scattering and absorption) properties can be adjusted. In addition, it keeps coolant water separate from power generation steam; thus, radioactivity is confined to the coolant and away from the power generating units. The thermal efficiency of this heat engine is about 30%.

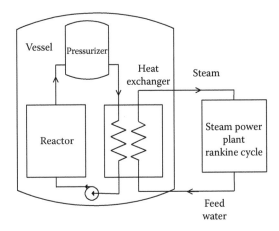

FIGURE 7.8 Nuclear-fueled power plant. Pressurized water reactor (PWR) type. The Rankine cycle on the right is implemented as discussed earlier (see Figure 7.4).

Spent fuel requires permanent disposal or it can also be reprocessed to recover some fresh fuel. Reprocessing waste requires permanent disposal. Many countries reprocess fuels; the United States does not reprocess spent fuel due to concern over weapons proliferation. A major environmental and human health issue of nuclear-fueled power plants is the disposal of spent fuel and waste from reprocessing fuel. Radioactive waste can take 1000 years to decay to levels of the originally mined fuel. In the United States, nuclear waste is stored in a stable geological formation under Yucca Mountain.

7.7.2 GEOTHERMAL

The energy flow from the interior of the Earth is greater than our current energy consumption, but except for a small fraction, it is very difficult to access and harness. This energy consists of two parts; one is "primordial" energy was captured during the gravitational collapse of interplanetary material when the Earth was formed, and the other is heat from radioactive decay [9,13].

Earth has a radius of ~6,371 km and its interior can be succinctly described by a few major layers; the *core* composed of the inner (solid, ~1220 km) and outer (liquid, ~3400 km) core; the *mantle*, which flows at very low velocities, and a thin *crust* (tens of km) (Figure 7.9). The core is at very high temperature (thousands of °C), and heat flows outward by conduction in its inner part and by convection in the outer part. Heat flow is by convection in the mantle (Figure 7.9). *Magma* sources occur in the upper mantle and lower crust. Large convection heat flow from the mantle powers geologic processes and plate tectonic movements at the crust (Chapter 2). The outward heat flow by conduction across the crust is small.

Example 7.20

Heat flow from the interior of the Earth is estimated to be 47 TW. Assuming this flow is spread over the surface of the Earth, what is the heat flow per unit area? Answer: Calculate surface of the Earth from its radius $4\pi r^2 = 5.76 \times 10^{14}\,m^2$, then divide total heat flow by the surface area $\dfrac{47 \times 10^{12}\,W}{5.76 \times 10^{14}\,m^2} = 81.5\,mW/m^2$.

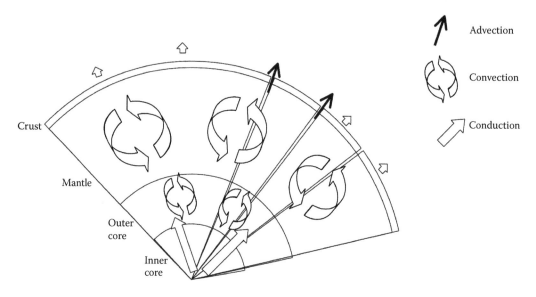

FIGURE 7.9   Heat flow processes in the Earth's interior.

Heat flow includes advection across the crust at the edge of tectonic plates or at locations of hotspots or plumes in the mantle (reaching 400 mW/m^2). These areas amount to a small fraction of the Earth's land area. Therefore, opportunities for geothermal power generation are not widespread around the world. Nevertheless, it is implemented in more than 20 countries and a total installed capacity of ~13 GW. Some countries have installed capacity to cover more than 15% of their demand. Geothermal is considered renewable because power generation actually uses a very small amount of the energy in the Earth's interior.

It is relatively easy to use a geothermal reservoir that is close to the surface of the Earth. It only requires tapping on the steam and running it to a Rankine cycle heat engine. This type of power production has been implemented since the early 1900s. However, near-surface reservoirs are not very abundant and geothermal generation requires drilling deep wells (1–5 km) to find the heat source to produce water or steam at high temperature for power generation.

As shown in Figure 7.10, the pressurized hot water and steam extracted from the geothermal reservoir is separated so that just the steam runs through a typical Rankine cycle power plant. The water is returned to the reservoir, together with the water condensed from steam once it runs through the plant. Depending on the temperature of the reservoir, the plant would have a higher or lower efficiency. Recall that the Carnot limit depends on the temperature difference between the hot and cold reservoirs. Geothermal plants tend to have lower temperature and pressure (for instance, only 180°C and 10 bar) and therefore much lower efficiency than coal-fired power plants.

To increase efficiency, geothermal power production employs *binary cycle*, in which the heat from the geothermal source is used to transfer heat to a different working fluid (binary fluid), which has a much lower boiling point than water (Figure 7.11). The heat transferred from the geothermal source vaporizes the working fluid. This vapor runs through a turbine and is afterward condensed to complete the thermodynamic cycle.

Heat removed from the reservoir in the form of hot water and steam is replenished by heat flow coming from deeper in the reservoir. Therefore, for sustainable geothermal power generation, the reservoir should sustain heat for the lifetime of the plant. Water ran through geothermal reservoirs contains substances that can cause adverse environmental and human health effects when released to surface waters or air. As we can see from this short discussion, making geothermal power generation renewable and clean requires favorable conditions and costly deep drilling.

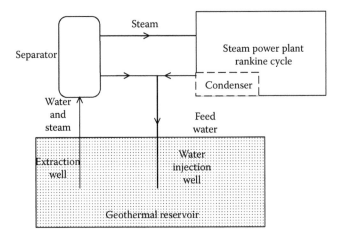

FIGURE 7.10 Geothermal power plant. Pressurized water and steam are extracted from the reservoir. The Rankine cycle on the right is implemented as discussed earlier (see Figure 7.4).

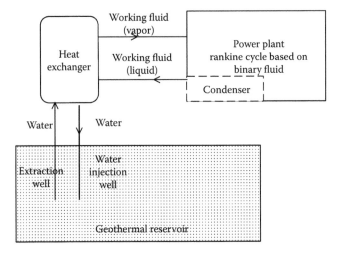

FIGURE 7.11 Binary-cycle geothermal power plant. Hot water is extracted from the reservoir to heat a working fluid with a low boiling point, which becomes the fluid for the Rankine cycle on the right (see Figure 7.4).

EXERCISES

7.1. Suppose water flows at 2 m/s carrying away heat from a source of 30 kJ/kg. Assume this process occurs at 30°C. What is the heat flux density?

7.2. Assume we burn 1 kg of coal with *HV* 24 MJ/kg and 40% C. What is the energy produced in kWh if only 30% of the released energy is converted to electricity? What is the mass of CO_2 produced? What is the mass of CO_2 per kWh of electricity?

7.3. Calculate time to deplete coal reserves if consumption rate continues to increase exponentially but only at 0.6% per year.

7.4. Calculate the system efficiency of ultra-supercritical plants at 8500 kJ/kWh. Compare its CO_2 emission to a typical 33% efficiency using the same coal with HV 20 MJ/kg and C content 40%.

7.5. Take a typical 33% efficient 1 GW coal-fired power plant and *HV* coal of 24 MJ/kg. What is the coal burn rate? What is the daily coal consumption?

7.6. Consider a coal-fired power plant with 33% system efficiency, thermal losses of 10%, thermal efficiency of 40%. Assume it uses a cooling tower. Find the mass of cooling water per kWh that is evaporated and removed from the river or lake.

7.7. Consider a 33% efficient coal-fired power plant that burns coal with a heat value of 24 MJ/kg and containing 15% mineral matter ("ash") by weight (assume that 70% of ash is PM as fly ash) and 2.5% sulfur. What would be the uncontrolled PM emission (g per kWh)? What would be the sulfur emission rate if it were not controlled (g per kWh)?

7.8. We inject CO_2 in a NG gas deposit and are able to displace and recover NG amounting to 25% of the mass of CO_2 injected. The NG is burned as a fuel later. Assume all CO_2 injected remains in the deposit. How much is the net sequestration of CO_2 per kg injected?

7.9. Assume a saline reservoir is 10^{14} m^3 with 15% porosity and 0.1 water saturated pore space. Average conditions are 8 bar and 40°C. For how many years can it sequester 100% CO_2 from three 1 GW coal-fired power plants averaging 0.8 kgCO$_2$/kWh?

7.10. Assume a geothermal gradient has a Carnot efficiency of 30%. What is the actual efficiency if heat and work losses were 8% and 2%, respectively, and the turbine is 90% efficient?

REFERENCES

1. Hemond, H.F., and E.J. Fechner, *Chemical Fate and Transport in the Environment*. 1994: Academic Press. 338 pp.
2. Christopherson, R.W., and G.E. Birkeland, *Geosystems: An Introduction to Physical Geography*. 10th edition. 2018: Pearson. 610 pp.
3. Graedel, T.E., and P.J. Crutzen, *Atmospheric Change: An Earth System Perspective*. 1993, New York: Freeman. 446 pp.
4. Vanek, F., L. Albright, and L. Angenent, *Energy Systems Engineering: Evaluation and Implementation*. Third edition. 2016: McGraw-Hill. 704 pp.
5. Crutzen, P.J., The influence of nitrogen oxides on the atmospheric ozone content. *Quarterly Journal of the Royal Meteorological Society*, 1970. 96: 320–325.
6. U.S. EPA. Clean Air Act. 2014. Accessed October 2014. Available from: http://www.epa.gov/air/caa/.
7. U.S. EPA. Summary of the Clean Water Act. 2014. US EPA. Accessed October 2014. Available from: http://www2.epa.gov/laws-regulations/summary-clean-water-act.
8. U.S. EPA. National Ambient Air Quality Standards (NAAQS). 2014. Accessed October 2014. Available from: http://www.epa.gov/air/criteria.html.
9. Fay, J.A., and D.S. Golomb, *Energy and the Environment. Scientific and Technological Principles*. Second edition. MIT-Pappalardo Series in Mechanical Engineering, M.C. Boyce and G.E. Kendall. 2012, New York: Oxford University Press. 366 pp.
10. International Energy Agency. Coal Information, Database Documentation. 2017. Accessed June 2017. Available from: http://wds.iea.org/wds/pdf/Coal_Documentation.pdf.
11. International Energy Agency. Key World Energy Statistics 2016. 2017. Accessed June 2017. Available from: https://www.iea.org/publications/freepublications/publication/key-world-energy-statistics.html.
12. International Energy Agency. Key Electricity Trends. 2017. Accessed June 2107. Available from: https://www.iea.org/publications/freepublications/publication/KeyElectricityTrends.pdf.
13. Turcotte, D.L. and G. Schubert, *Geodynamics*. Third edition. 2014: Cambridge University Press. 612 pp.

8 Alternating Current (AC) Circuits and Power

This chapter describes the fundamentals of power calculations in AC circuits. We start defining the concept of impedance as an AC analog to resistance, but that depends on frequency when including storage elements (capacitors and inductors). Using impedance and its reciprocal (admittance) we generalize nodal and mesh analysis methods that are practical for complicated AC circuits and an excellent option when the circuits are very large, as it occurs in an electric power system. Subsequently, we define AC power, both its instantaneous and average calculations, and focus on complex power and the important notion of power factor and its correction. We finalize the chapter discussing DC to AC converters or *inverters*, which play an important role in modern electric power systems since some renewable power harvesters are DC, and there is a need to connect AC systems operating at different frequencies. The chapter provides a review of material typically contained in introductory circuit analysis textbooks [1–3] as well as electric power systems textbooks [4].

8.1 ALTERNATING CURRENT (AC) CIRCUIT ANALYSIS USING IMPEDANCE

As we finished Chapter 5, the expressions $\mathbf{I} = \dfrac{\mathbf{V}}{R}$, $\mathbf{I} = j\omega C\mathbf{V}$, and $\mathbf{I} = \dfrac{\mathbf{V}}{j\omega L}$ were rearranged as $\dfrac{\mathbf{V}}{\mathbf{I}} = R$, $\dfrac{\mathbf{V}}{\mathbf{I}} = j\omega L$ and $\dfrac{\mathbf{V}}{\mathbf{I}} = \dfrac{1}{j\omega C}$ as a generalized Ohm's law to define the concept of impedance $\mathbf{Z} = \dfrac{\mathbf{V}}{\mathbf{I}}$ as an AC analog to resistance given in units of Ω. Impedance is a complex number, and the impedance of circuits with storage elements C and L depend on frequency. In this chapter, we make use of impedance to analyze AC circuits.

8.1.1 AC VOLTAGES AND CURRENTS FOR SIMPLE CIRCUITS

Applying voltage $v(t) = V_m \cos(\omega t + \theta_v)$ to a circuit with R and L in series (Figure 8.1, left-hand side) produces a current $i(t) = I_m \cos(\omega t + \theta_i)$ with I_m and θ_i unknown. The time-domain analysis is long, so instead use frequency-domain analysis. The voltage is $\mathbf{V} = V_m \angle \theta_v$ and we can write $\mathbf{V} = \mathbf{I}R + j\omega L\mathbf{I} = \mathbf{I}(R + j\omega L)$; therefore, solving for current we get $\mathbf{I} = \dfrac{\mathbf{V}}{R + j\omega L} = \dfrac{V_m \angle \theta_v}{R + j\omega L}$. Convert the denominator to polar and calculate

$$\mathbf{I} = \frac{V_m \angle \theta_v}{\sqrt{R^2 + \omega^2 L^2}\,\angle\left(\tan^{-1}\frac{\omega L}{R}\right)} = \frac{V_m}{\sqrt{R^2 + \omega^2 L^2}}\,\angle\left(\theta_v - \tan^{-1}\frac{\omega L}{R}\right)$$

The term $\mathbf{V} = \mathbf{I}(R + j\omega L)$ can be rearranged as $\dfrac{\mathbf{V}}{\mathbf{I}} = R + j\omega L$ as impedance $\mathbf{Z} = \dfrac{\mathbf{V}}{\mathbf{I}}$. Note from this simple circuit that the sum of impedances in series R and $j\omega L$ yields an equivalent impedance $\mathbf{Z} = R + j\omega L$. Note that current lags the voltage by the same angle as the impedance angle. A common practice is to draw circuit diagrams using a generic rectangular box to represent an impedance or an admittance and label it accordingly (Figure 8.1, right-hand side). Calculations of impedance and current as well as visualizing the response are expedited by programming.

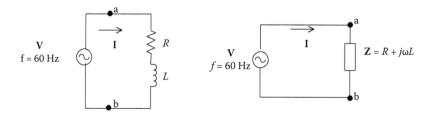

FIGURE 8.1   A simple RL circuit and its impedance representation.

Example 8.1

Consider a RL circuit at $f = 60$ Hz built with elements $R = 1\ \Omega$ and $L = 1$ mH in series powered with
$\mathbf{V} = 170\angle 0°$ V. Calculate impedance and current. Draw the time-domain and phasor plot. Answer:
The RL circuit has impedance $\mathbf{Z} = R + j\omega L$. We calculate it in rectangular and then convert to polar
so that we can divide voltage into impedance to obtain current. We can do all these using a script:

```
> w= 377; V.s=c(170,0)
> v.units <- c("V","A")
> y.lab <- "v(t) [V] or i(t) [A] "
> vt.lab <- c("v(t)","i(t)")
> vp.lab <- c("V","I")
>
> # RL
> R=1; L=1*10^-3
> Z.r <- c(R,w*L); Z.p <- polar(Z.r)
> I.p <- div.polar(V.s,Z.p)
> print(Z.r); print(Z.p); print(I.p)
[1] 1.000 0.377
[1] 1.069 20.656
[1] 159.027 -20.656
> ac.plot(waves(list(V.s,I.p)),vt.lab,v.units,y.lab)
> phasor.plot(list(V.s,I.p),vp.lab,v.units)
```

This yields $\mathbf{Z} = 1 + j0.377\ \Omega = 1.069\angle 20.656°\ \Omega$ and a current $\mathbf{I} = 159.027\angle -20.656°$ A. Note
current lags the voltage by the same angle as the impedance angle. The resulting graphs are shown
in Figure 8.2.

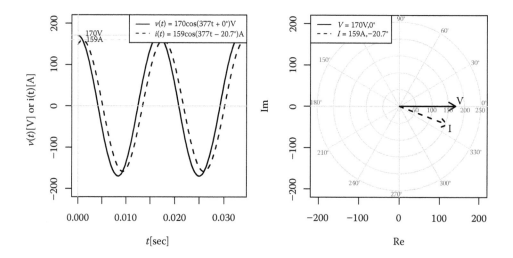

FIGURE 8.2   AC response of RL circuits.

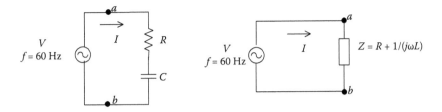

FIGURE 8.3   A simple RC circuit and its impedance representation.

Applying voltage $v(t) = V_m \cos(\omega t + \theta_v)$ to a circuit with R and C in series (Figure 8.3) produces a current $i(t) = I_m \cos(\omega t + \theta_i)$ with I_m and θi unknown. Use frequency-domain as we did for the RL circuit. The voltage is $\mathbf{V} = V_m \angle \theta_v$ and we can write $\mathbf{V} = \mathbf{I}R + \dfrac{\mathbf{I}}{j\omega C} = \mathbf{I}(R + \dfrac{1}{j\omega C})$. Therefore solving for current

$$\mathbf{I} = \frac{\mathbf{V}}{R + \dfrac{1}{j\omega C}} = \frac{V_m \angle \theta_v}{R - j\dfrac{1}{\omega C}}$$

Convert the complex number in the denominator to polar to get

$$\mathbf{I} = \frac{V_m \angle \theta_v}{\sqrt{R^2 + \omega^{-2}C^{-2}} \angle \left(-\tan^{-1}\frac{1/\omega C}{R}\right)} = \frac{V_m}{\sqrt{R^2 + \omega^{-2}C^{-2}}} \angle \left(\theta_v + \tan^{-1}\frac{1/\omega C}{R}\right)$$

The term $\mathbf{V} = \mathbf{I}(R + \dfrac{1}{j\omega C})$ can be rearranged as $\dfrac{\mathbf{V}}{\mathbf{I}} = R + \dfrac{1}{j\omega C}$ as impedance $\mathbf{Z} = \dfrac{\mathbf{V}}{\mathbf{I}}$. Note from this simple circuit that the sum of impedances in series yields an equivalent impedance $\mathbf{Z} = R + \dfrac{1}{j\omega C}$ (Figure 8.3). These calculations and visualizing the response are expedited by programming.

Example 8.2

Consider a RC circuit at $f = 60$ Hz build with elements $R = 1\ \Omega$ and $L = 1000\ \mu F$ in series powered with $\mathbf{V} = 170\angle 0°$ V. Calculate impedance and current. Draw the time-domain and phasor plot.

Answer: The RC circuit has impedance $\mathbf{Z} = R + \dfrac{1}{j\omega C}$. We calculate it in rectangular form and then convert to polar so that we can divide voltage into impedance to obtain current. We can do all these using a script:

```
> w= 377; V.s=c(170,0)
> v.units <- c("V","A")
> y.lab <- "v(t) [V] or i(t) [A]"
> vt.lab <- c("v(t)","i(t)")
> vp.lab <- c("V","I")
> # RC

> R=1;C=1000*10^-6
> Z.r <- c(R,-1/(w*C)); Z.p <- polar(Z.r)
> I.p <- div.polar(V.s,Z.p)
> print(Z.r); print(Z.p); print(I.p)
[1]  1.00000 -2.65252
[1]   2.835 -69.344
[1] 59.965 69.344
> ac.plot(waves(list(V.s,I.p)),vt.lab,v.units,y.lab)
> phasor.plot(list(V.s,I.p),vp.lab,v.units)
>
```

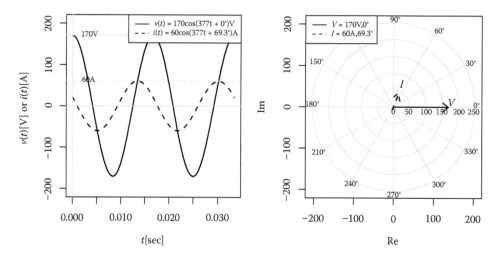

FIGURE 8.4 AC response of RC circuits.

This yields $\mathbf{Z} = 1 - j2.652 \ \Omega = 2.835\angle -69.344° \ \Omega$ and a current $\mathbf{I} = 159.027\angle -20.656°$ A. Note current lags the voltage by the same angle as the impedance angle. The resulting graphs are shown in Figure 8.4.

8.1.2 IMPEDANCE AND ADMITTANCE

The impedance concept is very useful for generalizing the analysis of circuits powered by AC. As we just saw in the last two sections, we can think of an AC "Ohm's law" using impedance $\mathbf{Z} = \dfrac{\mathbf{V}}{\mathbf{I}} = \dfrac{V_m \angle \theta_v}{I_m \angle \theta_i}$. An intuitive concept is that impedance opposes flow of current in the same manner as a pure resistor.

The real and imaginary parts of an impedance receive special names and symbols, *resistance* (R) and *reactance* (X), respectively. Therefore, we can write in general, using rectangular coordinates

$$\mathbf{Z} = R + jX \qquad (8.1)$$

The sign of the imaginary term is determined by the storage elements. The negative imaginary component corresponds to capacitive reactance, and the positive imaginary component corresponds to an inductive reactance. In the case of a simple RL circuit, the resistance is the resistor R and the reactance is $X = \omega L$. In the case of a simple RC circuit, the resistance is the resistor R and the reactance is $X = -\dfrac{1}{\omega C}$. This negative sign will make the impedance have a negative imaginary component.

Example 8.3

For example, consider a RL circuit and a RC circuit at $f = 60$ Hz built with elements $R = 1 \ \Omega$, $C = 1000 \ \mu F$, and $L = 1$ mH, as in Example 8.1 and Example 8.2. What are the resistance and reactance of these impedances? Answer: The RL circuit has impedance $\mathbf{Z} = 1 + j0.377 \ \Omega$. The resistance is $1 \ \Omega$ and the reactance is $0.377 \ \Omega$ (inductive). The RC circuit has impedance $\mathbf{Z} = 1 - j2.652 \ \Omega$. The resistance is $1 \ \Omega$ and the reactance is $-2.652 \ \Omega$ (capacitive).

In general, circuits include all three elements: resistors, inductors, and capacitors. The impedance is determined by the circuit topology and the values of R, L, and C. For example, a simple

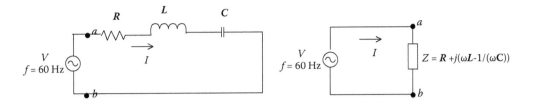

FIGURE 8.5 A simple RLC circuit and its impedance representation.

series combination of resistor, capacitor, and inductor (Figure 8.5) will have an impedance given by the sum of individual impedances $\mathbf{Z}_R$, $\mathbf{Z}_L$, and $\mathbf{Z}_C$:

$$\mathbf{Z} = \mathbf{Z}_R + \mathbf{Z}_L + \mathbf{Z}_C = R + j\omega L + \frac{1}{j\omega C} = R + j\left(\omega L - \frac{1}{\omega C}\right) \tag{8.2}$$

Example 8.4

Consider a RLC series circuit powered by $\mathbf{V} = 170\angle 0°$ V at $f = 60$ Hz and built with elements $R = 1\ \Omega$, $C = 1000\ \mu F$, and $L = 10$ mH. Calculate impedance and determine whether reactance is capacitive or inductive. Calculate current and determine whether it is lagging or leading the voltage, an draw the time-domain and phasor diagrams. Answer: $\mathbf{Z}_R = R = 1\ \Omega$, $\mathbf{Z}_L = j\omega L = j3.77\ \Omega$, and $\mathbf{Z}_C = -j2.652\ \Omega$. Thus $\mathbf{Z} = 1 + j(3.77 - 2.652) = 1 + j1.118$. The reactance is inductive. To calculate current, use

$$\mathbf{I} = \frac{\mathbf{V}}{\mathbf{Z}} = \frac{170\angle 0°}{1.5\angle 48.176°} = 113.33\ \angle -48.176°. \text{ Current lags the voltage.}$$

All of these calculations and the plots can be expedited by using the following script:

```
> w= 377; V.s=c(170,0)
> v.units <- c("V","A")
> y.lab <- "v(t) [V] or i(t) [A] "
> vt.lab <- c("v(t)","i(t)")
> vp.lab <- c("V","I")
>
> R=1;C=1000*10^-6;L=10*10^-3
> Z.r <- c(R,w*L-1/(w*C)); Z.p <- polar(Z.r)
> I.p <- div.polar(V.s,Z.p)
> print(Z.r); print(Z.p); print(I.p)
[1] 1.00000 1.11748
[1] 1.500 48.176
[1] 113.333 -48.176
> ac.plot(waves(list(V.s,I.p)),vt.lab,v.units,y.lab)
> phasor.plot(list(V.s,I.p),vp.lab,v.units)
>
```

This script confirms the values obtained earlier. Figure 8.6 shows the resulting plots, which illustrate that current lags the voltage.

Note that the reactance may be positive (inductive) or negative (capacitive) depending on the relative magnitude of ωL and $-\frac{1}{\omega C}$. A special case would be

$$\omega L - \frac{1}{\omega C} = 0 \text{ or } \omega L = \frac{1}{\omega C} \tag{8.3}$$

which makes the impedance purely resistive. Note that for any values L and C, this would occur at a frequency $\omega = \frac{1}{\sqrt{LC}}$ that satisfies Equation 8.3. However, for the purposes of analysis of power systems that have a fixed frequency (say, 60 Hz), satisfying Equation 8.3 is accomplished by

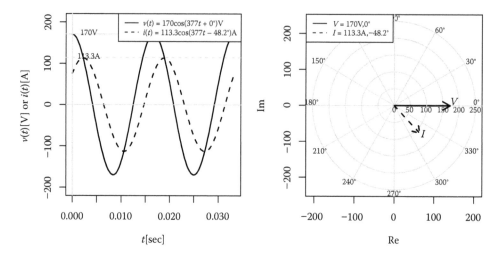

FIGURE 8.6 AC response of RLC circuits.

selecting elements. For example for a given L, we may select a capacitor C that satisfies Equation 8.3. That is to say, $C = \dfrac{1}{\omega^2 L}$.

Example 8.5

Consider a RLC series circuit powered at $f = 60$ Hz and built with elements $R = 1\ \Omega$ and $L = 10$ mH and an unknown capacitance. What capacitor value would make the impedance purely resistive? Answer: Find impedance of inductor at 60 Hz. It should be $\mathbf{Z}_L = j\omega L = j3.77\Omega$, then for purely resistive impedance $\mathbf{Z} = R + j0$ the capacitor impedance should be $\mathbf{Z}_C = -\mathbf{Z}_L = -j3.77\Omega$. Therefore $3.77 = \dfrac{1}{\omega C} = \dfrac{1}{377C}$ and solve for C to get $C = \dfrac{1}{377 \times 3.77} = 704\mu F$. This can be obtained directly by using $C = \dfrac{1}{\omega^2 L} = \dfrac{1}{(377)^2 \times 10 \times 10^{-3}} = 704\mu F$.

This calculation has importance in later chapters because as we will see a purely resistive impedance has benefits in terms of power consumption.

Another useful generalization is the concept of *admittance*, which generalizes conductance, using the inverse of impedance $\mathbf{Y} = 1/\mathbf{Z}$. For single elements (R, C, L), the admittance is easy to calculate. For a resistor $\mathbf{Y}_R = \dfrac{1}{\mathbf{Z}_R} = \dfrac{1}{R} = G$, for a capacitor $\mathbf{Y}_C = \dfrac{1}{\mathbf{Z}_C} = \dfrac{1}{{}^{1}\!/_{j\omega C}} = j\omega C$, and for an inductor $\mathbf{Y}_L = \dfrac{1}{\mathbf{Z}_L} = \dfrac{1}{j\omega L}$.

For an impedance with both real and imaginary parts, the admittance will also have real and imaginary parts. It is direct to calculate the admittance of an impedance given in polar form. Say we are given $\mathbf{Z} = Z_m \angle \theta$, the admittance will be $\mathbf{Y} = \dfrac{1}{\mathbf{Z}} = \dfrac{1}{Z_m \angle \theta} = Z_m^{-1} \angle -\theta$, that is, the inverse of the magnitude and the negative of the angle.

Now, many times we want to express admittance in rectangular coordinates. The real and imaginary terms are denoted by conductance (G) and *susceptance* (B), respectively:

$$\mathbf{Y} = G + jB \tag{8.4}$$

We can demonstrate relationships between these components and those of impedances $Z = R + jX$ by inverting the complex number for impedance and equating real and imaginary terms $G + jB = \dfrac{1}{R+jX}$. We get $G = \dfrac{R}{R^2 + X^2}$ and $B = \dfrac{-X}{R^2 + X^2}$. When the reactance is zero $X = 0$, the susceptance is also zero $B = \dfrac{-0}{R^2 + 0} = 0$, and the conductance is just the reciprocal of the resistance $G = \dfrac{R}{R^2 + 0} = 1/R$.

All these operations are facilitated by programming. We will use the function admit of renpow to calculate admittance in rectangular coordinates given impedance in rectangular coordinates. The following example illustrates how to use this function.

Example 8.6

Suppose $R = 1\ \Omega$, $C = 100\ \mu F$, $L = 10$ mH, and $f = 60$ Hz. What is the admittance of a simple series RC circuit? What is the admittance of a simple series RL circuit? What is the admittance of a simple series RLC circuit? Answer: Input the parameter values

```
> R=1; L=10*10^-3; C=100*10^-6; w= 377
```

Calculate impedance and use the function admit for an RC circuit:

```
> Z <- c(R,-1/(w*C))
> Y <- admit(Z)
> Z; Y
[1]  1.0000 -26.5252
[1] 0.001 0.038
>
```

The impedance is $Z = 1 - j26.525$ and the admittance is $Y = 0.001 + j0.038$. We see how a capacitive impedance has a negative reactance corresponding to a positive susceptance. Now calculate impedance and use the function admit or a RL circuit:

```
> Z <- c(R,w*L)
> Y <- admit(Z)
> Z; Y
[1] 1.00 3.77
[1]  0.066 -0.247
>
```

The impedance is $Z = 1 + j3.77$ and the admittance is $Y = 0.066 - j0.247$. We see how an inductive impedance has a positive reactance corresponding to a negative susceptance. Finally, for an RLC circuit

```
> Z <- c(R,w*L-1/(w*C))
> Y <- admit(Z)
> Z; Y
[1]  1.0000 -22.7552
[1] 0.002 0.044
>
```

The impedance is $Z = 1 - j22.756$ and the admittance is $Y = 0.002 + j0.044$. We see how a capacitive impedance has a negative reactance corresponding to a positive susceptance.

We can combine impedance in series by adding individual impedances $Z = Z_1 + Z_2$ and impedance in parallel by using admittance $Y = Y_1 + Y_2$ and then inverting Y to get Z. As an illustration, Figure 8.7 shows this process for parallel RLC circuit. We calculate each admittance, then add them up, and finally invert to calculate impedance. This process is further explained using the following example.

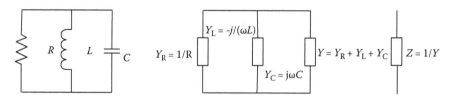

FIGURE 8.7 Impedance and admittance, example using parallel RLC circuit.

Example 8.7

Find the equivalent impedance of a parallel RLC circuit. Suppose $R = 1\ \Omega$, $C = 100\ \mu F$, $L = 10$ mH, and $f = 60$ Hz. Answer: Find three impedances $\mathbf{Z}_R = 1$, $\mathbf{Z}_C = -j26.525$, and $\mathbf{Z}_L = j3.77$ all in ohms. Find three admittances by inverting each impedance and add to obtain equivalent admittances $\mathbf{Y}_R = 1$, $\mathbf{Z}_C = j0.0377$, and $\mathbf{Z}_L = -j0.2652$ all in S. Add these admittances to obtain $\mathbf{Y} = 1 - j0.2275$ S. Now, invert this admittance to obtain impedance

$$\mathbf{Z} = \frac{1}{\mathbf{Y}} = \frac{1}{1.025\ \angle -12.789°} = 0.976\ \angle\ 12.789° = 0.952 + j0.216$$

This entailed converting to polar, inverting in polar, and converting back to rectangular form. All of these calculations could have been done by the following lines of code:

```
> R=1; L=10*10^-3; C=100*10^-6; w= 377
> Y <- admit(c(1,0))+admit(c(0,-1/(w*C)))+admit(c(0,w*L))
> Z <- admit(Y)
> print(Y);print(Z)
[1] 1.000 -0.227
[1] 0.952 0.216
>
```

Most circuit analysis laws and methods, such as Kirchhoff's circuit law (KCL), Kirchhoff's voltage law (KVL), Thévenin theorem, voltage and current dividers, and nodal and loop analysis, apply to AC circuits for impedance and admittance in an analogous manner as for resistance and conductance when using DC.

8.1.3 VOLTAGE DIVIDER

To illustrate, take the concept of a voltage divider (Figure 8.8, left-hand side) and use it in AC. The voltage $\mathbf{V}_2$ across impedance $\mathbf{Z}_2$ is $\mathbf{V}_2 = \mathbf{V}_s \dfrac{\mathbf{Z}_2}{\mathbf{Z}_1 + \mathbf{Z}_2}$. This calculation entails complex number algebra. In general, $\mathbf{V}_2 = V_m \angle \theta_s \dfrac{Z_2 \angle \theta_2}{(R_1 + R_2) + j(X_1 + X_2)}$, where the denominator is the sum of impedances in rectangular form. Then converting the denominator to polar form $Z_3\angle\theta_3$ we can write

$$\mathbf{V}_2 = V_m \angle \theta_s \frac{Z_2 \angle \theta_2}{Z_3 \angle \theta_3} = \frac{V_m Z_2}{Z_3} \angle (\theta_s + \theta_2 - \theta_3).$$

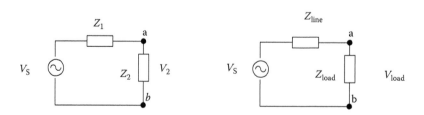

FIGURE 8.8 Voltage divider and its use to model line and load problems.

The voltage divider is useful as a model for a common problem in power systems consisting of source, transmission line impedance, and load impedance (Figure 8.8, right-hand side). We will discuss this situation later in Chapters 10 and 11.

Example 8.8

Suppose a voltage divider has source $\mathbf{V}_s = 170\angle10°$V, $\mathbf{Z}_1 = 2 - j1$ Ω, which is a capacitive reactance, and $\mathbf{Z}_2 = 1 + j2$ Ω, which has an inductive reactance. Calculate current $\mathbf{I}$ and $\mathbf{V}_2$. Answer:

$\mathbf{V}_2 = \mathbf{V}_s\dfrac{\mathbf{Z}_2}{\mathbf{Z}_1 + \mathbf{Z}_2} = 170\angle10°\dfrac{1 + j2}{2 - j1 + 1 + j2}$. Convert the numerator and denominator to polar $\mathbf{V}_2 \approx$

$170\angle10°\dfrac{2.24\angle63.44°}{3.16\angle18.44°} = 170\angle10° \times 0.71\angle45° = 120.19\angle55°$. The current is $\mathbf{I} = \dfrac{\mathbf{V}_s}{\mathbf{Z}_1 + \mathbf{Z}_2} =$

$\dfrac{170\angle10°}{3.16\angle18.44°} = 53.76\angle-8.44°$. Note that the current lags the voltage, and $\mathbf{V}_2$ leads the voltage source $\mathbf{V}_s$. Using R we can perform the calculation and draw a phasor plot of the result (Figure 8.9)

```
Vs <- c(170,10); Z1.r <- c(2,-1); Z2.r <- c(1,2)
Z2.p <- polar(Z2.r); Z3.r <- Z1.r+Z2.r; Z3.p <- polar(Z3.r)
Ip <- div.polar(Vs,Z3.p)
V2 <- mult.polar(Vs,div.polar(Z2.p,Z3.p))
print(V2)
phasor.plot(list(Vs,Ip,V2),c("Vs","Ip","V2"),c("V","A","V"))
```

which confirms the results obtained by hand.

```
> print(V2);print(Is)
[1] 120.19 55.00
[1] 53.763 -8.435
>
```

The graphic result (Figure 8.9) shows how the current lags the voltage source, and the resulting $\mathbf{V}_2$ leads the voltage source $\mathbf{V}_s$.

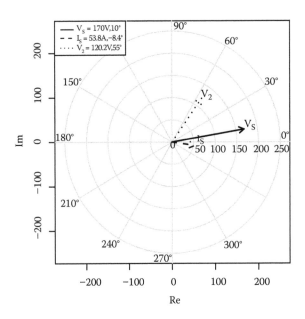

FIGURE 8.9 Response of a voltage divider.

8.1.4 AC NODAL AND MESH ANALYSIS

Recall from Chapter 3 that two systematic methods to analyze a circuit that lead to implementable algorithms are *nodal analysis* and *mesh analysis*. These methods are based on KCL and KVL, respectively, and both consist of setting up a system of linear equations solvable by matrix algebra. In nodal analysis, the unknowns in the system of equations are the independent node voltages, whereas in mesh analysis the unknowns are the independent loop (mesh) currents. These methods are practical for complicated AC circuits and an excellent option when the circuits are very large as it occurs in an electric power system or grid with thousands of nodes.

To explain nodal analysis, look at the circuit in Figure 8.10, where the elements are written as admittances. Assign unknowns $\mathbf{V}_1$ and $\mathbf{V}_2$ to the independent nodes (nodes 1 and 2), and assume you know the sources $\mathbf{I}_{s1}$ and $\mathbf{I}_{s2}$. Apply KCL and Ohm's law to each one of these nodes $\mathbf{V}_1\mathbf{Y}_1 + (\mathbf{V}_1 - \mathbf{V}_2)\mathbf{Y}_2 = \mathbf{I}_{s1}$ and $\mathbf{V}_2\mathbf{Y}_3 + (\mathbf{V}_2 - \mathbf{V}_1)\mathbf{Y}_2 = \mathbf{I}_{s2}$. Now, rearrange as a system of equations

$$\mathbf{V}_1(\mathbf{Y}_1 + \mathbf{Y}_2) + \mathbf{V}_2(-\mathbf{Y}_2) = \mathbf{I}_{s1}$$
$$\mathbf{V}_1(-\mathbf{Y}_2) + \mathbf{V}_2(\mathbf{Y}_3 + \mathbf{Y}_2) = \mathbf{I}_{s2}$$

(8.5)

or in matrix form

$$\begin{bmatrix} \mathbf{Y}_1 + \mathbf{Y}_2 & -\mathbf{Y}_2 \\ -\mathbf{Y}_2 & \mathbf{Y}_3 + \mathbf{Y}_2 \end{bmatrix} \begin{bmatrix} \mathbf{V}_1 \\ \mathbf{V}_2 \end{bmatrix} = \begin{bmatrix} \mathbf{I}_{s1} \\ \mathbf{I}_{s2} \end{bmatrix}$$

(8.6)

In the same manner as we did for DC, an easy way to think of the setup of the matrix equation is by main diagonal and off-diagonal entries. The main diagonal has sum of all admittances connected to a node, whereas an off-diagonal entry corresponds to the negative of shared admittance between a pair of nodes.

Now when using matrix notation for Equation 8.6, we run into a small difficulty because we have used boldface for matrices (Chapter 3) and also for complex numbers and phasors (Chapter 5 and this chapter). What we will do for simplicity is just use boldface for matrices even if their entries are complex numbers. Therefore, we rewrite Equation 8.6 as $\mathbf{YV} = \mathbf{I}_s$, where the matrix $\mathbf{Y}$ is an **admittance matrix** and vectors $\mathbf{V}$ and $\mathbf{I}_s$ are complex numbers and have obvious correspondence to Equation 8.6 terms. The solution for the unknown $\mathbf{V}$ is

$$\mathbf{V} = \mathbf{Y}^{-1}\mathbf{I}_s$$

(8.7)

In this simple case, the resulting matrix $\mathbf{Y}$ is symmetric. This is not necessarily true always, but when it is, the matrix setup is simpler.

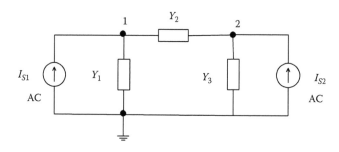

FIGURE 8.10 Simple circuit to explain AC nodal analysis.

Example 8.9

Take the circuit in Figure 8.10, with $\mathbf{Z}_1 = \mathbf{Z}_2 = 5 + j5\Omega$, $\mathbf{Z}_3 = 10 + j10\Omega$, and sources $\mathbf{I}_{s1} = 1\angle 0°$, $\mathbf{I}_{s2} = 0\angle 0°$. Set up and solve the matrix equation. Answer: The admittance of each element is

$$\mathbf{Y}_1 = \mathbf{Y}_2 = \frac{1}{\mathbf{Z}_1}S = \frac{1}{5+j5}S = \frac{1}{5 \times \sqrt{2}\angle 45°}S$$

and

$$\mathbf{Y}_3 = \frac{1}{\mathbf{Z}_3}S = \frac{1}{10+j10}S = \frac{1}{10 \times \sqrt{2}\angle 45°}S$$

The admittance matrix is

$$\mathbf{Y} = \begin{bmatrix} \dfrac{1}{5 \times \sqrt{2}\angle 45°} + \dfrac{1}{5 \times \sqrt{2}\angle 45°} & -\dfrac{1}{5 \times \sqrt{2}\angle 45°} \\ -\dfrac{1}{5 \times \sqrt{2}\angle 45°} & \dfrac{1}{5 \times \sqrt{2}\angle 45°} + \dfrac{1}{10 \times \sqrt{2}\angle 45°} \end{bmatrix}$$

Let us simplify

$$\mathbf{Y} = \frac{1}{5 \times \sqrt{2}\angle 45°} \times \begin{bmatrix} 2 & -1 \\ -1 & 1+1/2 \end{bmatrix} = (0.1414\angle -45°) \times \begin{bmatrix} 2 & -1 \\ -1 & 3/2 \end{bmatrix} = (0.1 - j0.1) \times \begin{bmatrix} 2 & -1 \\ -1 & 3/2 \end{bmatrix}$$

The matrix equation is

$$(0.1 - j0.1) \times \begin{bmatrix} 2 & -1 \\ -1 & 3/2 \end{bmatrix} \begin{bmatrix} \mathbf{V}_1 \\ \mathbf{V}_2 \end{bmatrix} = \begin{bmatrix} 1\angle 0° \\ 0\angle 0° \end{bmatrix} = \begin{bmatrix} 1+j0 \\ 0+j0 \end{bmatrix}$$

The easiest way in practice is to use a computer to solve the matrix equation. Using R we will code it as

```
Y1 <- 1/(5+5i) ; Y2 <- 1/(5+5i) ; Y3 <- 1/(10+10i)
Y <- matrix(c(Y1+Y2,-Y2,-Y2,Y3+Y2),ncol=2,byrow=TRUE)
Is <- c(1+0i,0+0i)
Vn <- solve(Y,Is)
```

Note the matrix, current, and voltage vectors are

```
> print(list(Y=Y,Is=Is,Vn=Vn))
$Y
        [,1]       [,2]
[1,]  0.2-0.2i -0.10+0.10i
[2,] -0.1+0.1i  0.15-0.15i

$Is
[1] 1+0i 0+0i

$Vn
[1] 3.75+3.75i 2.50+2.50i

>
```

The solution is $\begin{bmatrix} \mathbf{V}_1 \\ \mathbf{V}_2 \end{bmatrix} = \begin{bmatrix} 3.75 + j3.75 \\ 2.50 + j2.50 \end{bmatrix}$ V. We can represent the results of **V** given $\mathbf{I}_s$ as phasor diagrams by using function vector.phasor of renpow:

```
VpIp <- vector.phasor(Vn,Is)
phasor.plot(VpIp$VI,c("V1","V2","Is1","Is2"),c("V","V","A","A"),lty.p=c
(2,2,1,1))
```

The result is shown in Figure 8.11, where we can see the node voltages leading the current source. This is expected because the voltages have positive angle and the source is 0°, which results from all impedances being inductive.

Mesh analysis is the dual of nodal analysis when applied to independent loops (i.e., meshes). Think of the circuit in Figure 8.12 and setting up a system of equations for the unknown currents using KVL and Ohm's law to each one of these meshes $\mathbf{I}_1\mathbf{Z}_1 + (\mathbf{I}_1 - \mathbf{I}_2)\mathbf{Z}_2 = \mathbf{V}_{s1}$ and $\mathbf{I}_2\mathbf{Z}_3 + (\mathbf{I}_2 - \mathbf{I}_1)\mathbf{Z}_2 - \mathbf{V}_{s2}$. Now, rearrange as a system of equations

$$\mathbf{I}_1(\mathbf{Z}_1 + \mathbf{Z}_2) + \mathbf{I}_2(-\mathbf{Z}_2) = \mathbf{V}_{s1}$$
$$\mathbf{I}_1(-\mathbf{Z}_2) + \mathbf{V}_2(\mathbf{Z}_3 + \mathbf{Z}_2) = \mathbf{V}_{s2}$$

(8.8)

or in matrix form

$$\begin{bmatrix} \mathbf{Z}_1 + \mathbf{Z}_2 & -\mathbf{Z}_2 \\ -\mathbf{Z}_2 & \mathbf{Z}_3 + \mathbf{Z}_2 \end{bmatrix} \begin{bmatrix} \mathbf{I}_1 \\ \mathbf{I}_2 \end{bmatrix} = \begin{bmatrix} \mathbf{V}_{s1} \\ \mathbf{V}_{s2} \end{bmatrix}$$

(8.9)

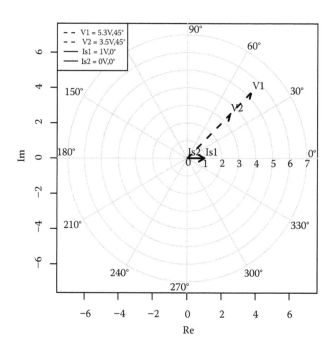

FIGURE 8.11 Phasor diagram of node voltages and current sources.

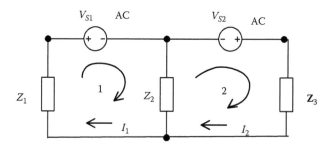

FIGURE 8.12 Simple circuit to explain AC mesh analysis.

The main diagonal has sum of all resistance encountered traversing a mesh, whereas an off-diagonal entry corresponds to the negative of shared resistance by a pair of meshes.

Now use matrix notation for Equation 8.9 keeping in mind that a boldface symbol may mean a matrix with complex entries. We write $\mathbf{ZI} = \mathbf{V}_s$, where the **impedance matrix Z** and vectors **I** and **V**$_\mathbf{s}$ have obvious correspondence to terms in Equation 8.9. The solution for the unknown **I** is

$$\mathbf{I} = \mathbf{Z}^{-1}\mathbf{V} \tag{8.10}$$

In this simple case, the resulting matrix **Z** is symmetric. This is not necessarily true always, but when it is, the matrix setup is simpler.

You should realize now that equations Equations 8.6 and 8.9 are matrix generalizations of Ohm's law. To see this, rewrite Ohm's law as $V = RI = \dfrac{1}{G}I = G^{-1}I$ and $I = \dfrac{V}{R} = \dfrac{1}{R}V = R^{-1}V$. Now match these scalar equations to Equations 8.6 and 8.9 recognizing the matrix generalization!

Example 8.10

Take the circuit in Figure 8.12 with $\mathbf{Z}_1 = \mathbf{Z}_2 = 1\ j1\ \Omega$, $\mathbf{Z}_3 = 2 + j2\ \Omega$, and sources $\mathbf{V}_{s1} = 12\angle 0°$, $\mathbf{V}_{s2} = 0\angle 0°$. Set up and solve the matrix equation. Answer: The impedance matrix is

$$\mathbf{Z} = \begin{bmatrix} (1+j1)+(1+j1) & -(1+j1) \\ -(1+j1) & (1+j1)+(2+j2) \end{bmatrix} = \begin{bmatrix} 2+j2 & -1-j1 \\ -1-j1 & 3+j3 \end{bmatrix} = (1+j1) \times \begin{bmatrix} 2 & -1 \\ -1 & 3 \end{bmatrix}, \text{ then the matrix}$$

equation is $(1+j1) \times \begin{bmatrix} 2 & -1 \\ -1 & 3 \end{bmatrix}\begin{bmatrix} \mathbf{I}_1 \\ \mathbf{I}_2 \end{bmatrix} = \begin{bmatrix} 12\angle 0° \\ 0 \end{bmatrix}$. Using R we will code it as

```
Z1 <- 1+1i; Z2 <- 1+1i; Z3 <- 2+2i
Z <- matrix(c(Z1+Z2,-Z2,-Z2,Z3+Z2),ncol=2,byrow=TRUE)
Vs <- c(12+0i,0+0i)
Im <- solve(Z,Vs)
```

Note the matrix, current, and voltage vectors are

```
> print(list(Z=Z,Vs=Vs,Im=Im))
$Z
     [,1]  [,2]
[1,] 2+2i -1-1i
[2,] -1-1i 3+3i

$Vs
[1] 12+0i 0+0i

$Im
[1] 3.6-3.6i 1.2-1.2i
```

The solution is $\begin{bmatrix} \mathbf{I}_1 \\ \mathbf{I}_2 \end{bmatrix} = \begin{bmatrix} 3.6 - j3.6 \\ 1.2 - j1.2 \end{bmatrix}$ A. We can represent the results of **I** given **V**$_s$ as phasor diagrams by using function vector.phasor of renpow:

```
VpIp <- vector.phasor(Vs,Im)
phasor.plot(VpIp$VI,c("Vs1","Vs2","I1","I2"),c("V","V","A","A"),lty.p=c
(1,1,2,2))
```

The result is shown in Figure 8.13 where we can see the mesh currents lag the source voltages. This is expected because the current have negative angle and the source is 0°, which results from all impedances being inductive.

These are extremely simple circuits, but the process is the same for complex circuits. Once computerized, the system can easily be calculated repeatedly when changing the parameters (matrices **Y** and **Z**). We will employ node analysis in Chapter 11 to analyze the grid voltages.

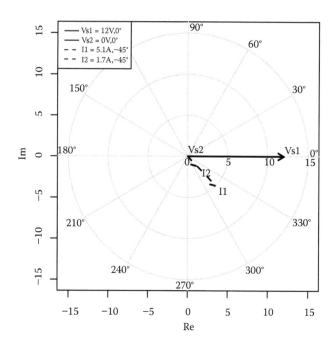

FIGURE 8.13 Phasor diagram of mesh currents and voltage sources.

8.2 INSTANTANEOUS AND AVERAGE POWER

As we know from Chapter 1, power is $p(t) = v(t)i(t)$ and therefore using a cosine wave for AC voltage and current, each with its own amplitude and phase

$$p(t) = v(t)i(t) = V_M \cos(\omega t + \theta_v) \times I_M \cos(\omega t + \theta_i) \tag{8.11}$$

To handle the multiplication of two cosine functions, we may use a trigonometric relation to convert it into a sum of cosine functions $\cos x \cos y = \dfrac{1}{2}[\cos(x+y) + \cos(x-y)]$ with $x = \omega t + \theta_v$ and $y = \omega t + \theta_i$ to obtain the *instantaneous* power for AC

$$p(t) = v(t)i(t) = \frac{V_M I_M}{2}[\cos(\theta_v - \theta_i) + \cos(2\omega t + \theta_v + \theta_i)] \tag{8.12}$$

This is an expression in the time domain. As we can see, the product of the magnitudes divided by 2, is multiplied by a constant term $\cos(\theta_v - \theta_i)$, which is the cosine of the phase difference between voltage and current, and a term $\cos(2\omega t + \theta_v + \theta_i)$ that oscillates at double the frequency as the voltage and the current.

It is of interest to calculate the *time average* of the instantaneous power. Recall that the time average of variable $x(t)$ is $\langle x \rangle = \dfrac{1}{T} \int\limits_0^T x(t)dt$. Because the time average of a cosine function is zero $\langle [\cos(2\omega t + \theta_v + \theta_i)] \rangle = 0$, the average power is simply

$$\langle p(t) \rangle = P = \frac{V_M I_M}{2} \cos \theta \tag{8.13}$$

where the angle θ is the angle between voltage and current phasors $\theta = \theta_v - \theta_i$. Here we should not confuse P used to denote electric power with P used to denote pressure.

When voltage and current are in phase $\theta_v = \theta_i$ the power waveform has a phase angle of $\theta_v + \theta_i = 2\theta_v$, the phase difference is $\theta = \theta_v - \theta_i = 0°$, then $\cos\theta = 1$ and therefore $P = \dfrac{V_M I_M}{2}\cos\theta = \dfrac{V_M I_M}{2}$.

Example 8.11

Suppose AC 60 Hz voltage of $\mathbf{V} = 170\angle 0°$ V applied to a resistor $R = 10\ \Omega$. Calculate current, instantaneous power, average power, and draw time-domain plots. Answer: The current is $\mathbf{I} = \dfrac{170}{10}\angle 0°$ A $= 17$ A. Power is $p(t) = \dfrac{170 \times 17}{2}[\cos(0°) + \cos(2\omega t)] = 1.45(1 + \cos(754t))$kW, which comprises an average power of $P = \dfrac{V_M I_M}{2}\cos\theta = 1.45$kW and an oscillatory component of double the frequency $1.45\cos(754t)$kW. We can use function inst.pow.plot of renpow to obtain time-domain plots:

```
vm=170; R=10
x <- list(c(vm,0),c(vm/R,0))
inst.pow.plot(x)
```

The result is illustrated in Figure 8.14. Note that current and voltage are in phase, and the oscillatory component of power oscillates at double the frequency.

We can see from Equation 8.13 that the average AC power for a capacitor and for an inductor is zero because $\cos(\pi/2) = \cos(-\pi/2) = 0$. This tells us that on the average, these elements do not dissipate power; in other words whatever is stored in part of the cycle is dissipated in the remainder of the cycle. To gain insight let us use a capacitor.

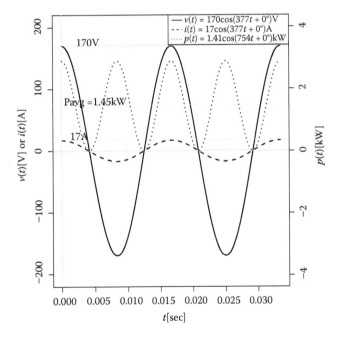

FIGURE 8.14 Instantaneous AC power when voltage and current are in phase.

Example 8.12

Suppose AC 60 Hz voltage of $\mathbf{V} = 170\angle 0°$ V applied to a capacitor C = 1000μF. Calculate current, instantaneous power, average power, and draw time-domain plots. Answer: The current is

$\mathbf{I} = j\omega CV = 377 \times 1000 \times 10^{-6} \times 170\angle(0 + 90°) = 64.09\angle 90°$. Power is $p(t) = \dfrac{170 \times 64.09}{2} \times$

$[\cos(90°) + \cos(2 \times 377t + 90°)] = 5.45(0 + \cos(754t + 90°))$kW, which comprises an average

power of $P = \dfrac{V_M I_M}{2} \cos\theta = 0$kW. The oscillatory component has double the frequency 5.45 cos (754t + 90°)kW and phase angle of 90°. We can use function powplot to obtain time-domain plots:

```
# power capacitor
w <- 377; v.s <- c(170,0)
C=1000*10^-6
# current response
i.res <- c(v.s[1]*(w*C),v.s[2]+90)
x <- list(v.s,i.res)
inst.pow.plot(x)
```

The result is illustrated in Figure 8.15. Note that current and voltage are out of phase by 90°, and the oscillatory component of power oscillates at double the frequency.

8.3   ROOT MEAN SQUARE (RMS) VOLTAGES AND CURRENTS

Using Equation 8.11 for a resistor power is $p(t) = V_M I_M \cos^2(\omega t + \theta v)$. For brevity we write $\cos^2\alpha = (\cos\alpha)^2$. We know that $I_M = V_M/R$ and therefore $p(t) = \dfrac{V_M{}^2}{R}\cos^2(\omega t + \theta_v)$ or $p(t) = I_M{}^2 R\cos^2(\omega t + \theta_v)$. In other words, power is proportional to the square of the current or the square of the voltage. To obtain average power implies calculating the time average of the square of the cosine function.

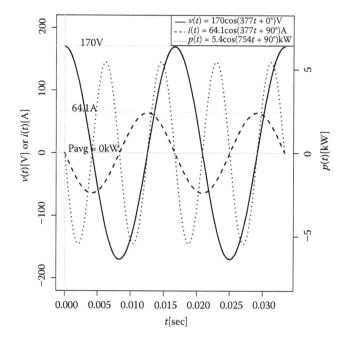

FIGURE 8.15   Instantaneous AC power when voltage and current are out of phase by 90°.

Let us use $x(t) = X_M \cos(\omega t)$ for a generic waveform instead of voltage or current. We calculate the time average of the square waveform and recall that $\cos^2\alpha = \dfrac{1}{2}(\cos 2\alpha + 1)$:

$$\left\langle x(t)^2 \right\rangle = \frac{1}{T}\int_0^T X_M{}^2\cos^2(\omega t)dt = \frac{X_M{}^2}{2T}\int_0^T (\cos 2\omega t + 1)dt = \frac{X_M{}^2}{2}$$

The last result occurs because the integral of cosine over a cycle is zero. It is convenient to work with the square root of this average, which is called the *root mean square* (*RMS* or *rms*):

$$V_{rms} = \sqrt{\left\langle v(t)^2 \right\rangle} \tag{8.14}$$

Therefore, for sinusoidal AC we can write the rms of voltage and current as

$$V_{rms} = \frac{V_M}{\sqrt{2}} \quad \text{or} \quad I_{rms} = \frac{I_M}{\sqrt{2}} \tag{8.15}$$

Example 8.13

Suppose AC 60 Hz voltage of $\mathbf{V} = 170\angle 0°$ V applied to a resistor $R = 10\ \Omega$. What is the RMS of voltage and current? Answer: $V_{rms} = \dfrac{170}{\sqrt{2}}$ V $\simeq 120$ V. We see that the rms value is the voltage for the U.S. grid. When we refer to the grid as providing 120 V, we really mean the rms value. The current is simply $I_{rms} = \dfrac{120}{10}$ A $\simeq 12$ A. We can plot and obtain Figure 8.16. In this case we use function ac.plot with optional argument rms=TRUE.

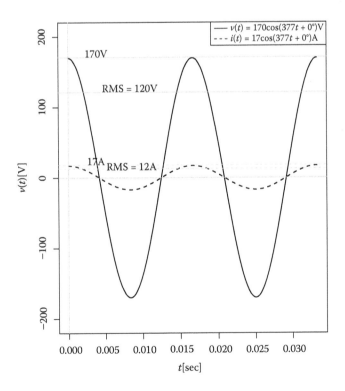

FIGURE 8.16 Root mean square (RMS).

```
vm=170; R=10
x <- list(c(vm,0),c(vm/R,0)); v.t <- waves(x)
v.lab <- c("v(t)","i(t)"); v.units <- c("V","A")
ac.plot(v.t,v.lab,v.units,rms=TRUE)
```

Using the rms value, we can rewrite Equation 8.12 for power as

$$p(t) = \frac{V_M I_M}{\sqrt{2}\sqrt{2}} [\cos\theta + \cos(2\omega t + \theta_v + \theta_i)] = V_{rms}I_{rms}[\cos\theta + \cos(2\omega t + \theta_v + \theta_i)]$$

In particular, Equation 8.13 for average power reduces to

$$\langle p(t) \rangle = P = V_{rms}I_{rms}\cos\theta \qquad (8.16)$$

For brevity, from now on we will simply write V and I to refer to the rms values, and thus write power as

$$p(t) = VI[\cos\theta + \cos(2\omega t + \theta_v + \theta_i)] \qquad (8.17)$$

and average AC power as

$$P = VI\cos\theta \qquad (8.18)$$

Note that if voltage and current are in phase (as in resistor) $\cos\theta = 1$ and average power is just $P = VI$, which is a similar expression to the one for DC power. This is why the rms value is referred to as the *effective* value, meaning that it is the equivalent to what the value would be if it were to be DC.

Example 8.14

Suppose AC 60 Hz voltage of $\mathbf{V} = 170\angle 0°$ V applied to a resistor $R = 10\ \Omega$. What is the voltage (rms)? What is the current (rms)? What is the instantaneous power? What is the average power? Answer: The voltage is $V = \frac{170}{\sqrt{2}}$ V $\simeq 120$ V rms. The current is $I \simeq \frac{120}{10}$ A $= 12$ A rms. Power is $p(t) \simeq (120 \times 12)\ [\cos(0°) + \cos(2\omega t) = 1.44\ (1 + \cos(754t))$kW, which has an average power of $P = VI\cos\theta = 1.44$ kW. We can use function powplot() of renpow to obtain time-domain plots together with the rms values (see Figure 8.17).

```
> vm=170; R=10
> x <- list(c(vm,0),c(vm/R,0))
> inst.pow.plot(x,rms=TRUE)
>
```

8.4 INSTANTANEOUS AND AVERAGE POWER OF A CIRCUIT

In general, the power consumed by a circuit given a voltage source $\mathbf{V} = V_m\angle 0°$ V is calculated from Equation 8.17 once we determine the current $\mathbf{I} = \dfrac{\mathbf{V}}{\mathbf{Z}}$ using the impedance of the circuit $\mathbf{Z} = Z\angle\theta$. Likewise, the average power is calculated from Equation 8.18. We will learn this using an example.

Example 8.15

Consider a RLC series circuit powered by $\mathbf{V} = 170\angle 10°$ V at $f = 60$ Hz and build with elements $R = 1\ \Omega$, $C = 1000\ \mu$F, and $L = 10$ mH. Calculate current and power. Answer: We first calculate the impedance.

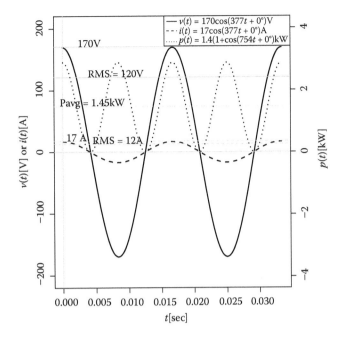

FIGURE 8.17 Voltage, current, and power in the time domain, showing rms values for voltage and current as well as average power.

$Z = 1 + j1.118 = 1.5\angle48.176°$, which has inductive reactance. To calculate current use $I = \dfrac{V}{Z} = \dfrac{170\angle10°}{1.5\angle48.176°} = 113.33 \angle -38.176°$. Current lags the voltage. The rms values of voltage and current are $V = \dfrac{170}{\sqrt{2}}$ V $\simeq 120$ V and $I = \dfrac{113.33}{\sqrt{2}}$ A $\simeq 80$ A. From Equation 8.17 $p(t) \simeq (120 \times 80)[\cos(-48.176°)] + \cos(2\omega t - 28.176°)] = 9.63 (0.667 + \cos(754t - 28.176°))$ kW. The average power is the constant term $P = 9.63 \times 0.667 = 6.42$kW. If we wanted to calculate directly from Equation 8.18 we have $P = 120 \times 80 \cos(-48.176°) = 9.63 \times 0.667 = 6.42$kW. All of these calculations can be expedited by using the following script, which produces the plots shown in Figure 8.18:

```
# RLC
w= 377; V.s=c(170,10)
v.lab <- c("v(t)","i(t)"); v.units <- c("V","A")
R=1;C=1000*10^-6;L=10*10^-3
Z.r <- c(R,w*L-1/(w*C)); Z.p <- polar(Z.r)
I.p <- div.polar(V.s,Z.p)
x <- list(V.s,I.p)
print(Z.r); print(Z.p); print(I.p)
inst.pow.plot(x,rms=TRUE)
```

Note that the average power is reduced by the factor cos θ with respect to the one for a purely resistive impedance $P = VI$.

8.5 COMPLEX POWER

Besides instantaneous power in the time domain and its average, it is of great importance to consider AC power in the frequency domain using phasors. *Complex power* is defined as a complex number

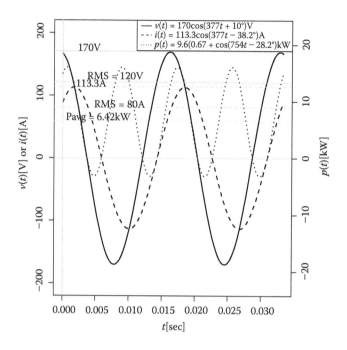

FIGURE 8.18 Instantaneous AC power for a RLC circuit.

obtained by multiplying phasors of voltage $\mathbf{V} = V\angle\theta_v$ and current $\mathbf{I} = I\angle\theta_i$. Here, by V and I, we mean the rms values. However, we can see that to express the result using the phase difference between voltage and current $\theta = \theta_v - \theta_i$, which is the angle of the impedance, we must make the angle θ_i negative. This is accomplished by using $\mathbf{I}^*$, which is the complex conjugate of the current phasor $\mathbf{I}$ or $\mathbf{I}^* = I\angle(-\theta_i)$. Then, we define complex power as

$$\mathbf{S} = \mathbf{V}\mathbf{I}^* \qquad (8.19)$$

In polar form,

$$\mathbf{S} = VI \angle(\theta_v - \theta_i) = VI \angle \theta \qquad (8.20)$$

where the magnitude VI, the product of rms values of voltage and current, and the angle is the phase difference $\theta = \theta_v - \theta_i$. The magnitude VI is named *apparent power*. Its units are not watts; rather they are defined as *volt-amp* or VA for short. It is common to denote apparent power VI by the symbol S, but we need to avoid confusion with the phasor symbol $\mathbf{S}$ or complex power. Thus, Equation 8.20 is rewritten as

$$\mathbf{S} = S\angle\theta \qquad (8.21)$$

In rectangular form, the complex power is written as the sum of real and imaginary parts

$$\mathbf{S} = VI\cos\theta + jVI\sin\theta = S\cos\theta + jS\sin\theta = P + jQ \qquad (8.22)$$

where P is *real power* (in W) and Q is *reactive power* (in volt-amp reactive or VAR). Figure 8.19 illustrates these relationships in the complex plane. This phasor relation of S, P, and Q can be represented as a simple geometric shape as shown in Figure 8.20, which is called the *power triangle*.

Figure 8.19 and Figure 8.20 were obtained for $\mathbf{V} = 170\angle10°$ V applied to an impedance of $\mathbf{Z} = 10\angle20°$ Ω, which determines a current of $\mathbf{I} = 17\angle(-10°)$ A (current lags the voltage), and that the rms values of voltage and current are $V \simeq 120$ V and $I \simeq 12$ A. From Equation 8.20 $\mathbf{S} = 120 \times 12\angle20° = 1.44\angle20°$kVA and converting to rectangular or using Equation 8.22 $\mathbf{S} = 1.44\cos(20°) + j1.44\sin(20°) = 1.36 + j0.49$. The real term is $P = 1.36$kW. The reactive power is $Q \simeq 0.49$kVAR.

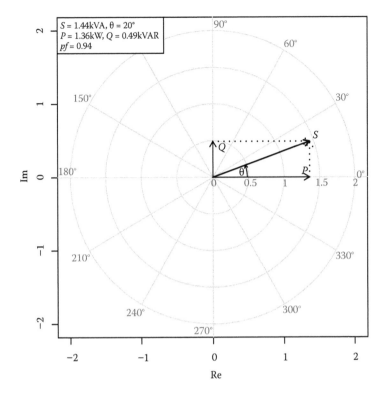

FIGURE 8.19 Complex power.

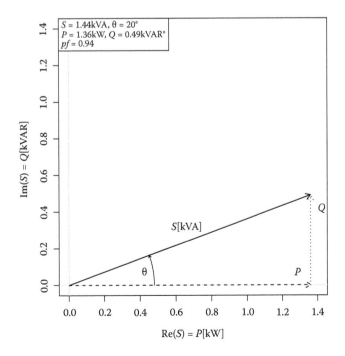

FIGURE 8.20 Power triangle.

Note that the average power is reduced by the factor cos θ = 0.94 with respect to the one for a purely resistive impedance $P = S = 1.44$. This term cos θ is the *power factor* (*pf*), which will be discussed in detail in the next section. All of these calculations can be expedited by using the following script that uses functions complex.pow.calc of package renpow. Its arguments are rms voltage V, rms current I, and angle θ:

```
# calc complex power
V.s=c(170,10); Z.p=c(10,20)
I.p <- div.polar(V.s,Z.p)
V <- V.s[1]/sqrt(2); I <- I.p[1]/sqrt(2)
theta <- V.s[2]-I.p[2]
cp <- complex.pow.calc(list(V,I,theta))
```

The results of cp confirm our calculations:

```
> cp
$units
[1] "kVA"  "kW"   "kVAR"
$S
[1] 1.44
$P
[1] 1.36
$Q
[1] 0.49
$theta
[1] 20
$pf
[1] 0.94
>
```

To visualize these results we can use two more functions complex.pow.plot and complex.pow.tri of package renpow, which produce Figure 8.19 and Figure 8.20:

```
complex.pow.plot(cp)
complex.pow.tri(cp)
```

Several observations can help us gain insight into the meaning of complex power. First, note that the real part P of Equation 8.22 is the same as the average power calculated in the time domain and given in Equation 8.18. This means that the average power is simply the real power. Second, note that the basic relations of the power triangle $\cos \theta = \frac{P}{S}$ and $\sin \theta = \frac{Q}{S}$ yield $\tan \theta = \frac{Q}{P}$. This means that the smaller the angle, the smaller the reactive power and the larger the real power.

Third, recall that for a purely capacitive impedance or a purely inductive impedance, the angle is −90° or +90°, respectively. Therefore, the complex power has zero real part because cos(90°) = cos(−90°) = 0, and it is purely imaginary or composed of purely reactive power because sin(90°) = 1 and sin(−90°) = 1. That is, $S = 0 + jQ = jS$ for inductors and $S = 0 - jQ = -jS$ for capacitors. By convention, negative power represents a source and positive power represents a load. Therefore, we can consider a capacitor as a source of reactive power, whereas an inductor absorbs reactive power. In other words, capacitors and inductors do not absorb real power, and the reactive power is just a model of oscillatory storage and release of energy.

Fourth, whenever the reactive power Q is nonzero, S is larger than the real power, because using Equation 8.22 we can see that $S = \sqrt{P^2 + Q^2}$. Fifth, the real power consumed by an impedance $\mathbf{Z} = R + jX$ can be calculated simply as the square of the rms current through the impedance times the resistance, and the reactive power is the square of the rms current through the impedance times the reactance:

$$P = I^2 R \quad \text{and} \quad Q = I^2 X \tag{8.23}$$

Note that the square of the current times the magnitude of the impedance Z is the magnitude of the apparent power. In other words, $I^2 Z = I^2 \sqrt{R^2 + X^2} = \sqrt{(I^2 R)^2 + (I^2 X)^2} = \sqrt{P^2 + Q^2} = S$, with the last part of the right-hand side of this equation being the power triangle.

In a similar fashion, using the voltage, real power and reactive power are

$$P = V^2/R \quad \text{and} \quad Q = V^2/X \tag{8.24}$$

Example 8.16

Consider a RLC series circuit given in Example 8.15. Calculate current and complex power. Answer: We know from Example 8.15 that impedance is $Z = 1 + j1.118 = 1.5\angle 48.176°$ (has inductive reactance) and current $I = 113.33\angle -38.176°$ (current lags the voltage), and that the rms values of voltage and current are $V = 120$ V and $I = 80$ A. From Equation 8.20 $S = 120 \times 80\angle 48.176° = 9.6\angle 48.176°$kVA and converting to rectangular or using Equation 8.22, $S = 9.6 \cos(48.176°) + j9.6 \sin(48.176°) = 9.6 \times 0.667 + j9.6 \times 0.745 = 6.4 + j7.152$. The real term is $P = 6.4$kW, which is the same as average power we calculated in Example 8.15. The reactive power is $Q \approx 7.2$kVAR. Note again that the average power is reduced by the power factor $pf = \cos\theta = 0.667$ with respect to the one for a purely resistive impedance $P = VI = S$. All of these calculations can be expedited by using the following script:

```
w= 377; V.s=c(170,10)
R=1;C=1000*10^-6;L=10*10^-3
Z.r <- c(R,w*L-1/(w*C)); Z.p <- polar(Z.r)
V.s=c(170,10); I.p <- div.polar(V.s,Z.p)
V <- V.s[1]/sqrt(2); I <- I.p[1]/sqrt(2)
theta <- V.s[2]-I.p[2]
cp <- complex.pow.calc(list(V,I,theta))
```

The results in cp confirm our previous calculations:

```
S=19.2kVA, theta=60°
P=9.6kW, Q=16.63kVAR
pf=0.5
```

To visualize these results we can use

```
complex.pow.plot(vp)
complex.pow.tri(vp)
```

which produce Figure 8.21. The quantity pf given in the output and in the legend of the figures is the power factor, which we discuss in the next section.

8.6 POWER FACTOR

As we now know, average power or real power is $P = VI \cos\theta$, where V and I are rms values of voltage and current and $\theta = \theta_v - \theta_i$ is the phase angle between voltage and current. As we mentioned in the previous section, power factor pf is the cosine of the phase angle, $pf = \cos\theta$. Therefore average or real power is

$$P = VI \times pf = S \times pf \tag{8.25}$$

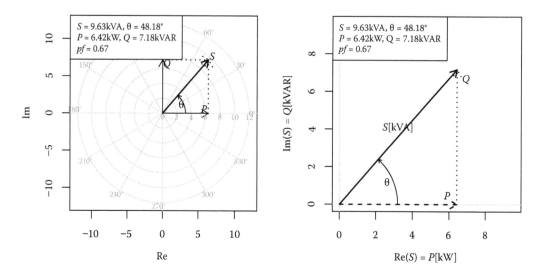

FIGURE 8.21 Complex power and power triangle.

Notice that *pf* will always be between 0 and 1, that is to say, $0 \le pf \le 1$, and therefore it always reduces the average power with respect to its maximum possible $S = VI$. A small phase angle yields a high *pf*, whereas a large phase angle yields a low *pf*. Moreover, $0 \le pf \le 1$, regardless of whether $\theta = \theta_v - \theta_i$ is positive (lagging current or inductive reactance) or negative (leading current or capacitive reactance). Therefore, when we mention *pf*, we would need to specify whether it corresponds to an inductive or capacitive impedance. However, because of the prevalence of motors in industry for very many applications, typical loads are inductive loads.

We have printed the power factor *pf* on the legend for complex power plots. For example, Figure 8.19 and Figure 8.20 illustrate a high *pf* of 0.94, which corresponds to $\theta = 20°$, whereas Figure 8.21 illustrates a low *pf* of 0.67, which corresponds to $\theta = 48.18°$.

A load impedance with a low *pf* can produce a large power loss in the transmission line that delivers power from a source to the load (Figure 8.22). To see this, note that to keep the same real power consumption at the load we would have a larger value of $S = VI$, and therefore the current would be higher for the same voltage. In other words, if *V* is the rms voltage at the load and *P* is the real power at the load, the rms current *I* (which is both the line current and the load current) will be $I = \dfrac{P}{V \times pf}$. We can see that the lower the *pf*, the higher the line current will be. Using Equation 8.23, the real power loss in the line is $P_{line} = I^2 R_{line} = \left(\dfrac{P}{V \times pf}\right)^2 R_{line}$, thus line loss increases as the square of the reciprocal of *pf*.

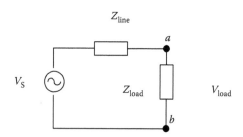

FIGURE 8.22 Source, line, and load to illustrate importance of power factor.

By the way, given the *pf*, we can calculate the reactive power Q using the power triangle, that is, $\tan \theta = \dfrac{Q}{P}$. Solving for the reactive power, $Q = P \tan \theta = P \tan (\cos^{-1}(pf))$. Similarly, we can calculate the apparent power S using $S = \dfrac{P}{\cos \theta} = P/pf$.

Example 8.17

Suppose we have 120V rms across a load of 1.2 kW with angle of $\theta = 60°$ (assume the load is inductive). The transmission line impedance has a real part $R_{line} = 1\ \Omega$. What is the power factor? What is the reactive power? What is the apparent power? What is the power loss in the line? Answer: The power factor is $\cos 60° = 0.5$. The reactive power is $Q = P \tan \theta = 1.2 \times \tan 60° = 2.08 kVAR$. This is a large reactive power! The apparent power is $S = P/pf = 1.2/0.5 = 2.4 kVA$, twice as large as the real power! The current is $I = \dfrac{P}{V \times pf} = \dfrac{1200}{120 \times 0.5} = 20A$. The loss is $P_{line} = I^2 R_{line} = 20^2 \times 1 = 0.4$ kW. This is a third of the power consumed by the load! We conclude that this is a very poor power factor. We can code this problem using R:

```
P=1200; V=120; theta=60; Rline=1
I <- P/(V*cos(theta*pi/180))
Pline <- I^2*Rline
```

Current and power loss in the line are 20 A and 400 W as calculated by hand:

```
> I;Pline
[1] 20
[1] 400
>
```

As well as the complex power results contained in the cp list:

```
>cp <- complex.pow.calc(list(V,I,theta))
S=2.4kVA, theta=60°
P=1.2kW, Q=2.08kVAR
pf=0.5
```

To visualize these results we can use

```
complex.pow.tri(vp)
```

which produces Figure 8.23.

8.6.1 POWER FACTOR CORRECTION

Large inductive (i.e., lagging) loads, such as those of motors, with poor power factor would cause high line loss. The source must provide the real power plus the reactive power, although the latter does not perform real work. To compensate for this, utilities include, in addition to the $ per kWh rate, an extra charge for reactive power as a charge per peak kVA. This in principle encourages the customers to improve their power factor.

Power factor correction is a process to make *pf* as close to 1 as possible or $pf = \cos \theta \approx 1$, that is, reduce the phase difference between voltage and current as much as possible or $\theta \approx 0°$. Let us assume, as is typically the case, that the load is inductive (lagging). In order to reduce *pf* to a corrected value pf_c, we need to reduce Q (recall the power triangle). The example shown in Figure 8.24 helps to visualize the process. The required corrected apparent power is $S_c = \dfrac{P}{pf_c}$ and the new reactive power is $Q_c = S_c \sin \theta_c = \dfrac{P}{pf_c} \sin (\cos^{-1} pf_c)$. The required reactive power reduction is then $\Delta Q = Q - Q_c$.

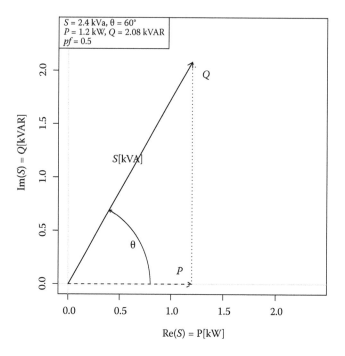

FIGURE 8.23   Power triangle for a poor *pf* example.

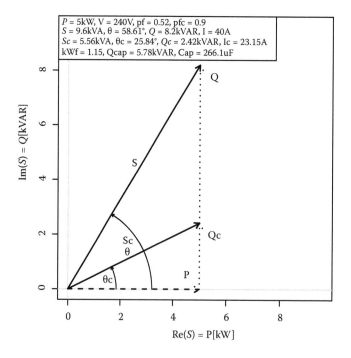

FIGURE 8.24   Results of power factor correction: from poor *pf* to a better *pf*.

Another way of seeing this is to use the tangent of both angles $\Delta Q = P(\tan(\cos^{-1}\theta) - \tan(\cos^{-1}\theta_c))$ where the quantity $\tan(\cos^{-1}\theta) - \tan(\cos^{-1}\theta c)$ is sometimes called the kW-factor, kWf, then $\Delta Q = P \times kWf$.

This required reduction ΔQ in kVAR is provided by a capacitor, which has a leading reactance. Thus we set $\Delta Q = Q_{cap}$. Recall that for a capacitor $X = 1/(\omega C)$ and that from Equation 8.24 $Q = V^2/X$. Therefore, $Q_{cap} = V^2/X = V^2(\omega C)$. In other words, solving for C we can calculate the capacitance C needed to correct the power factor:

$$C = \frac{Q_{cap}}{\omega V^2} \qquad\qquad (8.26)$$

These calculations and graphics are implemented as functions pf.corr() and pf.corr.tri() in the package renpow taking arguments of load in kW, rms voltage V in V, existing pf, and desired pfc. We will see how to apply these using an example.

Example 8.18

A plant draws 40 A from a 240 V line (RMS) at 60 Hz to supply a 5 kW lagging load. Calculate apparent power, power factor, phase angle, reactive power, and capacitance to improve power factor to $pf = 0.9$. Answer: Apparent power $S = 40 \times 240\text{VA} = 9.6\text{kVA}$. Power factor $pf = \dfrac{P}{S} = \dfrac{5\text{ kW}}{9.6\text{ kVA}} = 0.52$. The phase angle is $\theta = \cos^{-1} 0.52 = 58.61°$. Reactive power is $Q = S \sin\theta = 9.6\text{kVA} = 8.2$ kVAR. To improve pf to 0.9 we need a new phase angle of $\theta_c = \cos^{-1} 0.9 = 25.842°$. The new apparent power is $S_c = \dfrac{P}{pf_c} = \dfrac{5\text{ kW}}{0.9} = 5.56$ kVA and the new reactive power of $Q_c = S_c \sin\theta_C = 5.55\text{kVA} \times \sin 25.842° = 2.42$ kVAR.

Therefore, we need a reactive power reduction of $Q_{cap} = 8.2 - 2.42 = 5.78\text{kVAR}$. Use Equation 8.26 to get $C = \dfrac{5.78 \times 1000}{377 \times 240^2} \simeq 266.1\ \mu\text{F}$.

We can solve the example using functions of renpow in the following manner:

```
P=5; V=240; I=40; pfc=0.9
pf <- P*1000/(V*I)
# call pf correction function
pfcorr <- pf.corr(P,V,pf,pfc)
```

In this particular example, we know the current and use it to calculate existing pf so that we can use it as argument to pf.corr function. The output pfcorr has the relevant results of the correction:

```
> pfcorr <- pf.corr(P,V,pf,pfc)
P=5kW, V=240V, pf=0.52, pfc=0.9
S=9.6kVA, theta=58.61°, Q=8.2kVAR, I=40
Sc=5.56kVA, theta=25.84°, Qc=2.42kVAR, Ic=23.15
kWf=1.15, Qcap=5.78kVAR, Cap=266.1uF
>
```

In addition, we can visualize the changes in the power triangle yielding Figure 8.24, which we already described while discussing the aforementioned process.

```
pf.corr.tri(pfcorr)
```

8.6.2   POWER FACTOR CORRECTION AND PEAK DEMAND

Suppose monthly utility rates are given in $ per kWh consumed in a month plus $ per peak kVA. The latter is the highest demand of apparent power during the month. We will discuss this matter following an example based on problem 3.12 of Masters [5].

Example 8.19

Consider utility rates $0.04/kWh plus $7/month per peak KVA. Suppose that consumption is 750 kVA during 720 hours in the month 24h/d × 30d/mo = 720h/mo with peak of 1000 kVA. Calculate monthly bill for $pf = 0.7$ and then calculate savings by correcting pf to 0.9. Answer: Real power is $P = S \times pf = 750 \text{ kVA} \times 0.7 = 525 \text{ kW}$. In addition, peak kW power with pf 0.7 is $P_{peak} = S_{peak} \times pf = 1000 \text{ kVA} \times 0.7 = 700 \text{ kW}$. Calculate your real power monthly payment 0.04 $ / kWh × 525 kW × 720 h/mo = $15,120 and the peak kVA monthly 7 $ / kVA × 1000 kVA = $7,000. Add these two amounts to get the total monthly bill of $22,120. Now, recalculate with corrected $pf = 0.9$. The new kVA needed to meet the real power above $S_c = \dfrac{P}{pf_c} = \dfrac{525 \text{ kW}}{0.9} = 583.3 \text{ kVA}$ and the peak $S_{peak} = \dfrac{P_{peak}}{pf_c} = \dfrac{700 \text{ kW}}{0.9} = 777.78 \text{ kVA}$. With these new values, recalculate your bill. The base remains the same because the real power is the same $15,120. However, the peak changes to 7$ / kVA × 777.78KVA = $5,444. Money saved $7,000 − $5,444 = $1,555 per month.

8.7 COMPLEX POWER LOSS IN THE LINE

We discussed the real power loss in the line knowing the power factor, the voltage, and the real power delivered to the load. However, we would like to also know the voltage drop in the line, the complex power consumed by the line, and the effect on the pf as seen as the source. For this purpose, we would need to know the reactance X_{line} of the line in addition to the resistance R_{line}. Note that in reality a line has inductances and capacitances, and could be modeled as shown in Figure 8.25.

In general, we would have $\mathbf{V}_S = V_s\angle\theta_s$ $\mathbf{V}_{load} = V_{load}\angle\theta_v$ and current $\mathbf{I} = I\angle\theta_i$. Current is $\mathbf{I} = \dfrac{\mathbf{V}_s}{\mathbf{Z}_{line} + \mathbf{Z}_{load}}$ and from voltage divider $\mathbf{V}_{line} = \dfrac{\mathbf{V}_s\mathbf{Z}_{line}}{\mathbf{Z}_{line} + \mathbf{Z}_{load}}$ or $\mathbf{V}_{line} = V_s \angle\theta_s \dfrac{Z_{line}\angle\theta_{line}}{(R_{line} + R_{load}) + j(X_{line} + X_{load})}$, where the denominator is the sum of impedances in rectangular form. Then converting the denominator to polar form $Z_c\angle\theta_c$, we can write $\mathbf{V}_{line} = V_s\angle\theta_s \dfrac{Z_{line}\angle\theta_{line}}{Z_c\angle\theta_c} = \dfrac{V_sZ_{line}}{Z_c}\angle(\theta_s + \theta_{line} - \theta_c)$. The complex power consumed by the line $\mathbf{S}_{line} = \mathbf{V}_{line}\mathbf{I}^*$ and by the load is $\mathbf{S}_{load} = \mathbf{V}_{load}\mathbf{I}^*$. The complex power seen at the source is $\mathbf{S}_s = \mathbf{S} + \mathbf{S}_{line}$. From $\mathbf{S}_c$ we can determine the pf seen from the source.

Often, we know P, V, and pf for the load. Then we can determine its apparent power $S_{load} = \dfrac{P}{pf}$, current $I = \dfrac{S_{load}}{V}$, and the complex power $\mathbf{S}_{load} = P + jQ = P + jS_{load}\sin(\cos^{-1} pf)$, assuming that the phase angle of the voltage at the load is $\cos^{-1} pf$ (this is just an arbitrary convenient reference).

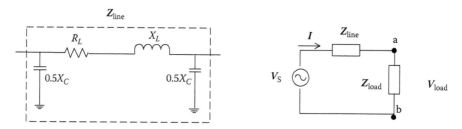

FIGURE 8.25 Simple model of a transmission line impedance.

Because the current is lagging by $-\cos^{-1} pf$, the current would have an angle of $\theta = \cos^{-1} pf - \cos^{-1} pf = 0°$. Since this current is the same one flowing through the line, the voltage drop at the line is $\mathbf{V}_{line} = \mathbf{I}_{line}\mathbf{Z}_{line} = I\angle 0°(R_{line} + jX_{line})$. The complex power of the line is $\mathbf{S}_{line} = I^2 (R_{line} + jX_{line})$. The complex power seen at the source is $\mathbf{S}_s = \mathbf{S}_{load} + \mathbf{S}_{line}$.

Example 8.20

Assume 60 Hz and a load of 2kW with $pf = 0.8$ lagging at 120 V. The line impedance is inductive $\mathbf{Z}_{line} = 0.1 + j0.2\Omega$. What is complex power supplied at the source? What is the voltage at the source? Answer: Assume voltage at the load has phase $\cos^{-1}(pf)$, the voltage at the load is $\mathbf{V}_{load} = 120\angle\cos^{-1}(0.8) = 120\angle 36.87°$ and the load apparent power is $S_{load} = \dfrac{P}{pf} = \dfrac{2}{0.8} = 2.5\text{kVA}$. The complex power consumed by the load $\mathbf{S}_{load} = 2 + j2.5 \sin(\cos^{-1} 0.8) = 2 + j1.5$ kVA. The current is $I = \dfrac{2.5}{120} \simeq 20.8\text{A}$ with phase $\cos^{-1}(0.8) - \cos^{-1}(0.8) = 0°$, that is to say, $\mathbf{I} = 20.8\angle 0°$A. The line impedance in polar is $\mathbf{Z}_{line} = 0.224\angle 63.435°$, and the voltage drop in the line is $\mathbf{V}_{line} = (20.8\angle 0°) \times (0.224\angle 63.435°) = 4.667\angle 63.435° = 2.087 + j4.174$ V. The voltage at the source is $\mathbf{V}_s = \mathbf{V}_{load} + \mathbf{V}_{line}$. To add these voltages, first convert $\mathbf{V}_{load}$ to rectangular $\mathbf{V}_{load} = 120\angle 36.87° = 96 + j72$V, then add $\mathbf{V}_s = \mathbf{V}_{load} + \mathbf{V}_{line} = (96 + j72) + (2.087 + j4.174) = 98.087 + j76.174$ in polar $\mathbf{V}_s = 124.192\angle 37.833°$. The complex power at the source is $\mathbf{S}_s = \mathbf{V}_s\mathbf{I}^* = (124.192\angle 37.833°) \times 20.8\angle 0° = 2.587\angle 37.833°$ kVA and taking the cosine of the angle we obtain $pf_s = 0.79$.

8.8   INVERTERS AND BACK-TO-BACK CONVERTERS

In Chapter 5, we studied AC to DC converters (rectifiers), and DC-DC converters. In this chapter, we will discuss DC to AC converters or *inverters*. These devices play an important role in modern electric power systems since some renewable power harvesters, such as photovoltaic (PV) cells (Chapter 14), are DC. In addition, there are needs to connect AC systems operating at different frequencies and these can be matched by converting to DC and then back to AC at another frequency; this is called a *back-to-back converter*.

8.8.1   INVERTER

Figure 8.26 shows a basic inverter consisting of a bridge of four switches (S1, S2, S3, and S4) each driven by a PWM signal and working in tandem. Each switch is composed of an IGBT transistor (see Chapter 5) with a diode in parallel to eliminate transients. Switches S1 and S2 are on (or off) simultaneously, when switches S3 and S4 are simultaneously in the opposite state off (or on). See the pulse trains shown in the lower part of Figure 8.26. When S1 = S2 = on and S3 = S4 = off, the positive side of the DC source is connected to the output and thus $V_{out} = V_{in}$. Conversely, when S1 = S2 = off and S3 = S4 = on, the negative side of the DC source is connected to the output and thus $V_{out} = V_{in}$.

To achieve a sinusoidal output, the PWM signal is produced by a modulating sine wave of the same frequency as the desired AC output frequency and a high-frequency carrier (Figure 8.27 top panel). Whenever the modulating value is larger than the carrier, the PWM output is positive (and will activate the gates of S1 and S2); otherwise the PWM is negative and changed sign (in order to activate the gates of S3 and S4). The resulting activation of the switches by these PWM signals is that $V_{out} = V_{in}$ or $V_{out} = V_{in}$ (Figure 8.27, bottom panel). You can see that the pulses have a larger duty cycle as the sine wave reaches a positive peak and a shorter duty cycle as the wave reaches a negative peak. The last step is filtering out the carrier to obtain a clean sine wave at the output of the inverter (Figure 8.27, bottom panel).

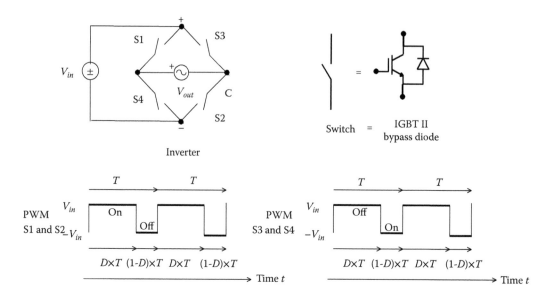

FIGURE 8.26   Inverter circuit, switches, and pulse width modulation (PWM).

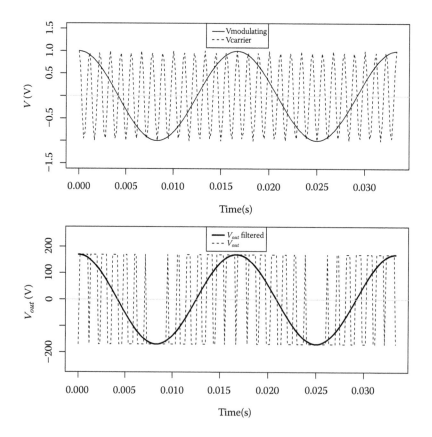

FIGURE 8.27   Top: Inputs to PWM, modulating and carrier waves. Bottom: V_{out} as response to the PWM signal and V_{out} once the carrier is filtered out.

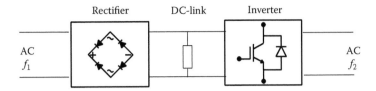

FIGURE 8.28   Back-to-back converter. AC power of frequency f_1 is rectified and put on a DC link or DC line, to deliver it as DC to an inverter, which converts it to AC at frequency f_2.

Example 8.21

Figure 8.27 is produced with function inverter, which takes the AC of desired frequency and amplitude. It calculates and plots the PWM signals needed and the output voltage. We can specify f = 60 Hz, V = 170 V, and nc = 2 two cycles to plot

```
x <- list(f=60,vin=170,nc=2)
inverter(x)
```

to obtain Figure 8.27.

8.8.2   BACK-TO-BACK CONVERTER

As shown in Figure 8.28, a *back-to-back converter* is composed of a rectifier (Chapter 5) that takes AC power at frequency f_1 and converts it to DC. This DC power is passed to an inverter that converts it back to AC at frequency f_2. These frequencies may be different, and in this case the back-to-back converter serves to couple two grids of different frequencies or an *asynchronous* generating unit to a grid. When the frequency is the same, the back-to-back converter may be serving as the equipment needed to transmit power via a DC link or to adapt the power factor of a load.

An important application of back-to-back converters that we will see in Chapters 10 and 11 is to adapt AC on each side of a *high-voltage direct current* (*HVDC*) transmission line.

EXERCISES

8.1. Consider a RL circuit at $f = 60$ Hz built with elements $R = 10\ \Omega$ and $L = 2$ mH in series powered with $\mathbf{V} = 170\angle 0°$ V. Calculate impedance and identify resistance and reactance. Calculate the admittance and identify conductance and susceptance. Calculate current. Draw the time-domain and phasor plots.

8.2. Consider a RC circuit at $f = 60$ Hz built with elements $R = 2\ \Omega$ and $L = 100$ μF powered with $\mathbf{V} = 170\angle 0°$ V. Calculate impedance and identify resistance and reactance. Calculate the admittance and identify conductance and susceptance. Calculate current. Draw the time-domain and phasor plots.

8.3. Consider a RLC series circuit powered by $\mathbf{V} = 170\angle 0°$ V at $f = 60$ Hz and built with elements $R = 10\ \Omega$, $C = 100$ μF, and $L = 1$ mH. Calculate impedance and determine whether reactance is capacitive or inductive. Calculate current and determine whether it is lagging or leading the voltage, and draw the time-domain and phasor diagrams.

8.4. Consider a voltage divider with source $\mathbf{V}_s = 170\angle 0°$ V, $\mathbf{Z}_1 = 1 + j2\ \Omega$ and $\mathbf{Z}_2 = 2 - j1\ \Omega$. Calculate current $\mathbf{I}$ and $\mathbf{V}_2$.

8.5. Consider the circuit in Figure 8.10 with $\mathbf{Z}_1 = \mathbf{Z}_2 = 5 - j5\Omega$, $\mathbf{Z}_3 = 10 - j10\Omega$, and sources $\mathbf{I}_{s1} = 1\angle 10°$, $\mathbf{I}_{s2} = 0\angle 0$. Set up and solve the matrix equation using nodal analysis. Plot a phasor diagram.

8.6. Suppose AC 60 Hz voltage of $\mathbf{V} = 170\angle 0°$ V applied to an inductor L = 10 mH. Calculate current, instantaneous power, average power, and draw the time-domain plots.

8.7. Suppose AC 60 Hz voltage of $\mathbf{V} = 170\angle 0°$ V applied to a resistor $R = 100$ Ω. What is the voltage (rms)? What is the current (rms)? What is the instantaneous power? What is the average power?

8.8. Consider a RLC series circuit powered by $\mathbf{V} = 170\angle 10°$ V at $f = 60$ Hz and built with elements $R = 10Ω$, $C = 100$ μF, and $L = 1$ mH. Calculate current, complex power, and real power.

8.9. A plant draws 40 A from a 240 V line (RMS) at 60 Hz to supply a 3 kW lagging load. Calculate apparent power, power factor, phase angle, reactive power, and capacitance to improve power factor to $pf = 0.9$.

8.10. Consider utility rates \$0.03/kWh plus \$10/month per peak KVA. Suppose that consumption is 800 kVA with peak of 1100 kVA. Calculate monthly bill for $pf = 0.7$ and then calculate savings by correcting pf to 0.9.

8.11. Assume 60 Hz and a load of 1 kW with $pf = 0.8$ lagging at 120V. The line impedance is inductive $\mathbf{Z}_{line} = 0.01 + j0.01Ω$. What is complex power supplied at the source? What is the voltage at the source?

8.12. Use function inverter to produce plots of required modulating, carrier, as well as output signals for a desired AC of 120 Hz and 340 V amplitude.

REFERENCES

1. Irwin, J.D., and R.M. Nelms, *Basic Engineering Circuit Analysis*. 11th edition. 2011: Wiley. 688 pp.
2. Hayt, W., J. Kemmerly, and S. Durbin, *Engineering Circuit Analysis*. Eight edition. 2012: McGraw Hill. 880 pp.
3. Alexander, C.K., and M.N.O. Sadiku, *Fundamentals of Electric Circuits*. Third edition. 2007: McGraw Hill. 1056 pp.
4. El-Hawari, M., *Introduction to Electrical Power Systems*. IEEE Press Series on Power Engineering, M. El-Hawari. 2008: Wiley. 394 pp.
5. Masters, G.M., *Renewable and Efficient Electric Power Systems*. Second edition. 2013: Wiley-IEEE Press. 690 pp.

9 Gas and Liquid Fuels
Gas Turbines and Combustion Engines

Natural gas has become a major fuel to generate electricity because it has a high heat value and cleaner emissions than coal. We examine the reserves and consumption of this resource, followed by a study of the Brayton cycle, which is used to model the process of gas-fired power plants. Moreover, gas-based conversion can be combined with coal into what are called combined cycle plants. Internal combustion engines, well known in transportation, are also employed for electric power generation. For this reason, we study engines based on the Otto cycle and Diesel cycle. Alternatives to natural gas and oil-based liquid fuels include biomass, landfill gas, manure, biodiesel, biofuels, and solar fuels. After examining these, we study alternative turbines and combustion engines, emphasizing microturbines and Stirling engines that have potential for distributed generation. At the end of the chapter, we briefly cover combined heat and power, which proposes that exhaust heat be recovered and used for a variety of purposes.

9.1 NATURAL GAS

9.1.1 THE RESOURCE

As explained in Chapter 2, oil and natural gas (NG) are contained in *reservoir rocks* where they were trapped between grains of the sediments that gave origin to the rock together with other substances, such as water and salt. In contrast to crude oil that has complex hydrocarbons, NG contains simple hydrocarbons. NG is increasingly becoming a major fuel to generate electricity for several reasons, which we will describe in detail. The combustible part of NG is mainly methane (CH_4), which is an atmospheric trace gas (Chapter 2), and some heavier alkanes (ethane, propane, and butane). The noncombustible gases in NG include N_2 and CO_2, in some cases in great proportion (e.g., 70% CO_2 by volume) [1].

NG is carbon rich and has a high heat value (*HV*); it can contain 74% C by weight and have a *HV* of 55 MJ/kg. Compare to subbituminous coal with 35% to 45% C and 20 to 24 MJ/kg *HV*. Several characteristics of NG make it desirable for electricity generation. Some reasons relate to the conversion process technology; NG mixes easily with air and thus the furnace/boiler can be smaller than the one for coal. The exhaust from burning NG can directly run a turbine, instead of requiring the intermediate step of producing steam [1]. Some other reasons for favoring NG are related to cleaner emissions from combustion: NG emissions do not contain PM and sulfur (reducing the negative impact on air quality), and CO_2 emission per kWh is ~1/2 that of coal and ~3/4 that of oil. This means a large reduction in impact on climate change. We will see how to calculate emission from NG later in the chapter.

However, one disadvantage of using NG for electricity is that NG reserves compared to consumption rate is smaller compared to other fossil fuels. This makes electricity generation by NG nonrenewable. Please recall our terminology of Chapter 1. Compared to coal-fired, NG-fired power plants would still be carbon based, cleaner, and still nonrenewable (probably more so than coal).

In order to examine the latter issue, let us look at reserves and consumption. Reserves of NG in conventional reservoirs are estimated to be 7350 EJ, and annual consumption in 2008 was 120 EJ. However, this rate was increasing rapidly at ~2.45% annually [1].

Example 9.1

Estimate time to depletion of NG reserves under scenarios of continued consumption at the current rate and increased consumption. Answer: Use formulas as in Chapter 7. Divide reserves by energy used in one year $t_f = \dfrac{X(0)}{c(0)} = \dfrac{7,350EJ}{120EJ/yr} \simeq 61$ yr, or NG reserves could last ~60 years. But under the increased consumption scenario, the time to depletion will shorten to $t_f = \dfrac{1}{0.024yr^{-1}} \ln \left(0.024yr^{-1} \dfrac{7,350EJ}{120EJyr^{-1}} + 1 \right) \simeq 38$yr, therefore NG reserves could last 38 years.

In this example we worked with conventional reserves; we did not account for increased avail- ability due to new and potential extraction processes. There is now technology to extract NG from unconventional reservoirs such as gas trapped in sandstone and *shale* rock, representing possibly an additional 10% to reserves in conventional reservoirs. Shale gas, particularly in the Unites States, has changed the outlook on NG availability. Unintentionally, new challenges to air quality have emerged making the "clean" dimension of NG-based conversion more complex to assess. This new tech- nology and air quality challenges are the subject of the next section.

9.1.2 SHALE GAS AND FRACKING

Clastic rocks are made up from fragments of broken older rocks that were eroded and weathered [2]. An easy-to-imagine situation would be a river carrying sediments of various grain size originating from erosion of rocks upstream. Upon reaching low-energy conditions, such as flats and lakes, these sediments deposit and eventually form new sedimentary rock; the resulting rock may be anywhere from coarse grain to fine grain depending on the sediment grain size. For example, sand sediments would form sandstone, a relatively coarse-grain clastic rock, and clay sediments would form *shale*, a *fine-grain clastic* rock. Thus, shale is made from clay mud and contains flakes of clay together with silt-size mineral particles and has breaks between layers [3].

Shales with organic matter can produce oil and gas. As pressure within the rock increases, oil and gas can escape into adjacent strata of a *conventional* reservoir. However, part of the gas remains entrapped or tightly bound inside the rock. This is the *unconventional* shale gas. Hydraulic fracturing or *fracking* is used to make this gas available. Fracking consists of drilling a well to the target formation, deeper than the aquifers, and pumping a fracturing fluid at a given flow and pressure, which will fracture the rock and free the gas. This fluid is mostly water but has chemical additives and sand; the latter is used to keep the fractures open once the pumping stops [4].

Importantly, the additives are a variation of 10 compounds representing ~30 chemicals, including hazardous air pollutants as well as compounds that imply risk to human health. Besides water consumption, hydraulic fracturing causes environmental concerns. These involve groundwater contamination, venting of contaminants, and disposal of fluids [4]. Emissions from various phases of shale gas development and production include volatile organic compounds (VOCs), nitrogen oxides (NO$_x$), and PM. Of these, VOCs and NO$_x$ are precursors of ground-level ozone (Chapter 7).

Shale gas production has significantly shifted the energy outlook of the United States. Following a decline in NG supply, the shale "boom" was boosted by the development of the Barnett, the Haynesville, and the Marcellus shale deposits. Shale gas production is emerging as an important source of natural gas in the United States, and expected to reach about half of the total supply by 2035 [5]. Canada and China have also added shale gas as an important component to their NG production and many countries have shale gas deposits.

9.2   GAS-BASED CONVERSION

9.2.1   BRAYTON CYCLE: GAS TURBINE

As we know from Chapter 7, the Rankine cycle employed in coal-fired power plants is based on steam, and it requires repeatedly vaporizing and condensing water. Recall that coal-based plants constitute a major source of power generation, represent a large investment, and run continuously providing *baseload* generation, that is, covering the bulk of the demand on a continuous basis.

In this section, we study the *Brayton cycle*, which is used to model the process of gas-fired power plants. These plants do not have to run continuously and can cover daily peaks in demand (*peaking* power plants). Moreover, gas-based conversion can be combined with coal into what are called *combined cycle* plants. Gas-fired power plants are based on gas turbines, which operate by combustion of the fuel gas. Shortly, we will focus on NG as fuel, but the principle of operation applies to liquefied petroleum gas (LPG), landfill gas, and other alternatives to NG, which we discuss at the end of the chapter.

Figure 9.1 summarizes the workings of a gas turbine. It consists of a turbine and compressor on the same shaft, along with a combustion chamber in between. The process can be modeled as a heat engine cycle as shown in Figure 9.2, which shows an idealized Brayton cycle on the *P-v* plane and the *T-s* plane. Work is extracted by the turbine from the shaft, but about half of the turbine power is used to drive the compressor. Air comes in at the inlet (state 1) and goes through the compressor blades changing to state 2. This process is isentropic and shows as an adiabat line on the *P-v* plane and as a vertical line on the *T-s* plane (Figure 9.2).

The combustion chamber burns gas and heats the compressed air, raising its temperature from state 2 to state 3 in an isobaric process, a horizontal line on the *P-v*, and an exponential determined by c_p on the *T-s* plane (Figure 9.2). This is heat input flow Q_{in}. At point 3 of the cycle, the fluid is air mixed with fuel-air and its temperature is highest (e.g., 1150°C). Because the amount of fuel-air compared to air is small, we will assume the fluid to be just air.

Now, the air goes through the turbine, which lowers its temperature and pressure (state 4) at constant entropy, that is, it follows a vertical line on the *T-s* plane and an adiabat on the *P-v* plane (Figure 9.2). Work is extracted from the turbine providing mechanical work *W* that is used to run the compressor and to drive an electrical generator. Air at state 4 is at a lower temperature (e.g., 550°C) and goes out as exhaust, representing heat flow Q_{out}.

At this point, the idealized cycle shows a return to state 1 following an isobaric process, horizontal line on the *P-v* plane, and exponential (with c_p) on the *T-s* plane. However, in reality a practical

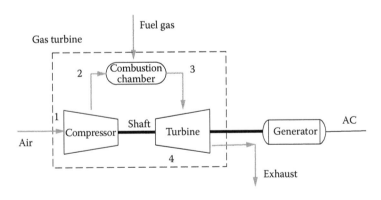

FIGURE 9.1   Simplified gas turbine.

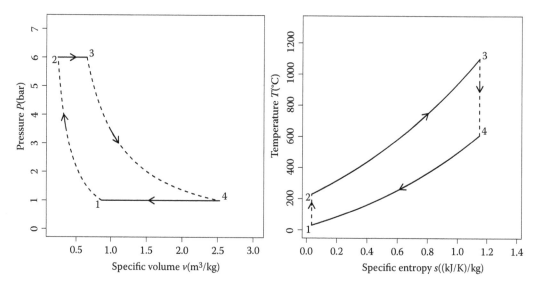

FIGURE 9.2   Idealized Brayton cycle showing curves as adiabats in the *P-v* plane and isobars in the *T-s* plane. Left: 1–2: adiabat, compression 30°C–> 229.3°C, *W* = 145.8 kJ/kg, *Q* = 0 kJ/kg; 2–3: isobar, expansion 229.3°C–> 1100°C, *W* = –249.9 kJ/kg, *Q* = 977.6 kJ/kg; 3–4: adiabat, expansion 1100°C–> 607.9°C, *W* = –430.5 kJ/kg, *Q* = 0 kJ/kg; 4–1: isobar, compression 607.9°C–> 30°C, *W* = 165.9 kJ/kg, *Q* = –608.8 kJ/kg. Right: 1–2: adiabat, compression 30°C–> 229.3°C, *W* = 145.8 kJ/kg, *Q* = 0 kJ/kg; 2–3: isobar, expansion 229.3°C–> 1100°C, *W* = –249.9 kJ/kg, *Q* = 977.6 kJ/kg; 3–4: adiabat, expansion 1100°C–> 607.9°C, *W* = –430.5 kJ/kg, *Q* = 0 kJ/kg; 4–1: isobar, compression 607.9°C–> 30°C, *W* = 165.9 kJ/kg, *Q* = –608.8 kJ/kg.

The Brayton cycle is an *open cycle* and does not include this cooling at constant pressure going from 4 to 1, but rather hot exhaust air is released to the environment. This is in contrast to a Rankine cycle, which is a *closed cycle*. Note that the lower temperature of this process is in the same order of magnitude as the highest temperature in a Rankine cycle, namely, that of the superheated steam.

Example 9.2

Plot the Brayton cycle as in Figure 9.2 for both the *P-v* and *T-s* planes for one mole of air with $T_H = 1100°C$, $T_L = 30°C$ and a compression ratio of 6, from 1 bar to 6 bar assuming that the initial entropy is 1 J/K. What is the work performed by the turbine and work done on the compressor? What is the heat in and heat out? Answer:

```
# brayton cycle
x <- list(TH=1100,TL=30,P1=1,P2=6,S1=1,cty='brayton')
y <- path.cycles(x,plane='Pv')
y <- path.cycles(x,plane='Ts')
```

   The results are in Figure 9.2, which reports work 145.8 kJ/kg done on the compressor (adiabatic compression path 1–2) and –430.5 kJ/kg work performed by the turbine (adiabatic expansion path 3–4). Heat input is 977.6 kJ/kg (isobaric expansion path 2–3) and heat output is –608.8 kJ/kg (isobaric compression path 4–1).

9.2.2   BRAYTON CYCLE EFFICIENCY

We will analyze the Brayton cycle using enthalpy. Each point of the cycle is a state, and the fluid at that state has a value of enthalpy. At state 3, the fluid has enthalpy H_3 that was gained from the inlet by adiabatic compression (to get from state 1 to state 2) plus heat added by the combustor at constant pressure (to get from state 2 to 3). At 4, the enthalpy of the fluid H_4 corresponds to a loss from state 3 due to work done by the turbine. Therefore, the difference in enthalpy $H_3 - H_4$ corresponds to the work done by the turbine W_t, or $W_t = H_3 - H_4$.

A similar argument applies to the compressor, therefore the work done by the compressor is $W_c = H_2 - H_1$. The net amount of work done by the turbine and compressor is then $W_n = W_t - W_c = (H_3 - H_4) - (H_2 - H_1)$. This is the same as the integral $\oint Tds$ for the entire cycle, which is the area in the polygon bounded by the cycle $W_n = \oint Tds$.

Now, the heat Q_{in} added at constant pressure by the combustor must equal the difference in enthalpy from state 3 to 2 $Q_{in} = (H_3 - H_2)$. This is the path integral $H_3 - H_2 = \int_2^3 Tds$ or area under the curve for the isobar from 2 to 3.

We can calculate efficiency using the ratio of the net amount of work to heat added

$$\eta = \frac{W_n}{Q_{in}} = \frac{(H_3 - H_4) - (H_2 - H_1)}{H_3 - H_2} = 1 - \frac{H_4 - H_1}{H_3 - H_2} = 1 - \frac{Q_{out}}{Q_{in}} = \frac{\oint TdS}{\int_2^3 TdS}$$

Geometrically, this is the area in the polygon bounded by the cycle divided by the area under the isobar from 2 to 3.

Example 9.3

Use the results of Example 9.2 given in Figure 9.2. These are for a Brayton cycle for one mole of air with $T_H = 1100°C$, $T_L = 30°C$ and a compression ratio of 6 from 1 bar to 6 bar assuming that the initial entropy is 1 J/K. What is the thermal efficiency? What percent of the net amount work is used to run the compressor? Answer: We can query the summary of the cycle

```
> y <- path.cycles(x,plane='Ts')
> path.cycles.summary(y)
$WQnet
 Qin(kJ/kg) Qout(kJ/kg)  Eff
1    977.6    -608.8 0.38

$end.state
   v(m3/kg) V(l) P(b)  T(C)  s(J/Kkg) S(J/K) W(kJ/kg) Q(kJ/kg) cv(kJ/Kkg)
1-2  0.240 7.0   6  229.3   0.035  1.0   145.8    0.0    0.747
2-3  0.657 19.0  6 1100.0   1.148  33.3  -249.9  977.6   0.915
3-4  2.528 73.2  1  607.9   1.148  33.3  -430.5    0.0    0.826
4-1  0.870 25.2  1   30.0   0.033  1.0   165.9  -608.8   0.718
   cp(kJ/Kkg) cp/cv
1-2    1.034 1.384
2-3    1.202 1.313
3-4    1.113 1.348
4-1    1.005 1.400

$pts;
[1] 1001

$nM
 n(mol) M(g/mol)  m(kg)
1    1   28.97 0.02897

>
>
```

The results indicate heat input q_{in} = 977.7 kJ/kg and heat output q_{out} = −608.8 kJ/kg. Efficiency is $\eta = 1 - \dfrac{q_{out}}{q_{in}} = 1 - \dfrac{608.8}{977.7} \simeq 0.38$ or 38% work as indicated in the output. We also see that w_c = 145.9 kJ/kg done on the compressor and w_t = −430.2 kJ/kg work performed by the turbine. The percent work done on the compressor is $\dfrac{w_c}{w_t} = \dfrac{145.9}{430.2} \simeq 0.34$ or 34%.

A practical approach to calculate efficiency of a Brayton cycle is to use the compression ratio and the specific heat ratio. Using the pressure at the inlet P_1 and the pressure just after the compressor as P_2, we can calculate the compression ratio as $r = P_2/P_1$. For example, a compression ratio of $r = 6$ means that compressed air has 6 times higher pressure than the inlet. The specific heat ratio is $\gamma = c_p/c_v$ (Chapter 4).

In an ideal cycle, pressures at 2 and 3 are the same $P_3 = P_2$, and pressures at 1 and 4 are the same $P_1 = P_4$. The compression ratio is $r = \dfrac{P_2}{P_1} = \dfrac{P_3}{P_4}$. Processes 1–2 and 3–4 are adiabatic and reversible (isentropic), and therefore $\dfrac{T_2}{T_1} = r^{\frac{\gamma-1}{\gamma}} = \dfrac{T_3}{T_4}$. The efficiency is $\eta = 1 - \dfrac{Q_{out}}{Q_{in}} = 1 - \dfrac{c_p(T_4 - T_1)}{c_p(T_3 - T_2)}$. Rearrange $\eta = 1 - \dfrac{T_1(T_4/T_1 - 1)}{T_2(T_3/T_2 - 1)} = 1 - \dfrac{T_1}{T_2}$ as in the Carnot efficiency. Using $\dfrac{T_2}{T_1} = r^{\frac{\gamma-1}{\gamma}}$ we can rewrite the cycle efficiency as a function of compression ratio

$$\eta = 1 - \frac{1}{r^{\frac{\gamma-1}{\gamma}}} \tag{9.1}$$

This is just an approximation because as we know γ is a function of temperature and it may vary in the range of temperatures involved in a Brayton cycle. For instance, at 30°C $\gamma \simeq 1.4$ and at 1000°C $\gamma \simeq 1.32$.

Example 9.4

Assume the compression ratio is $r = P_2/P_1 = 6$ and temperature varies between 30°C and 1000°C. What is the efficiency? What are the efficiencies for compression ratio between 3 and 10? Answer: First estimate γ at the middle of the temperature range 1030/2 = 515°C:

```
> cp.cv(TC=(1000+30)/2)$cp.cv
[1] 1.357
>
```

Thus use $\gamma = c_p/c_v = 1.357$ and then for $r = 6$ $\eta = 1 - \dfrac{1}{6^{(1-1/1.357)}} \simeq 0.376$. We can write a few lines of code to assist in this calculation and explore the effect of changing compression ratio on efficiency:

```
P.r <- seq(3,12,1)
gamma<- cp.cv(TC=(1000+30)/2)$cp.cv
eta <- 1-1/(P.r^(1-1/gamma))
P.r; round(eta,3)
plot(P.r,eta,type="b", xlab="Compression ratio r", ylab="Efficiency",lwd=2)
```

The plot is in Figure 9.3, where we can see that efficiency increases nonlinearly with the compression ratio and that achieves a value as high as ~0.46 for a 10:1 compression ratio.

This is an idealized value. Practical Brayton cycle implementations for electrical generation have ~33% efficiency. An increased efficiency is achieved by recovering heat from the exhaust air as

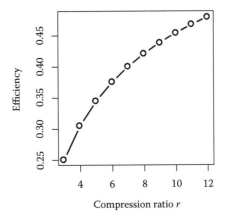

FIGURE 9.3 Efficiency versus compression ratio for an ideal Brayton cycle and assuming constant γ.

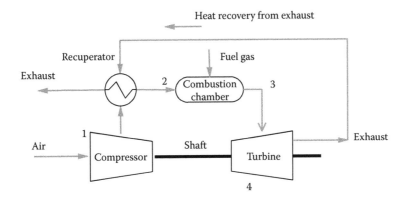

FIGURE 9.4 Heat recovery from exhaust to increase thermal efficiency.

illustrated in Figure 9.4. The hot air from the exhaust is sent to a heat exchanger or "recuperator," which raises the temperature of the compressed air before reaching the combustor. The increase in efficiency can be a few percent.

9.2.3 COMBINED CYCLES: INCREASED EFFICIENCY

Further use of heat recovery is to use hot (500°C) exhaust air from a gas turbine to boil water and drive a steam cycle without burning fuel for such a purpose. See Figure 9.5. This is called a *combined cycle* (CC) for power generation. The two generators shown in the figure can be mounted on the same shaft or reduced to only one generator. Because the temperature used to generate steam is lower than what would be in a normal steam-only plant, the Rankine cycle in a CC would operate at lower efficiency than a Rankine cycle in a steam-only plant. The gas turbine will also lower its efficiency compared to a gas-only system. However, overall, this combined system can generate more work from the same amount of fuel and achieve a high combined efficiency that can get to about 50%.

Let us follow Fay and Golomb [1] to see how to calculate CC efficiency η_{cc} from the components efficiencies η_g for gas and η_s for steam when the only source of heat is from burning gas, called Q_g. The work output is the sum of both systems $W_{cc} = W_g + W_s$, where subscript g is for gas and s for

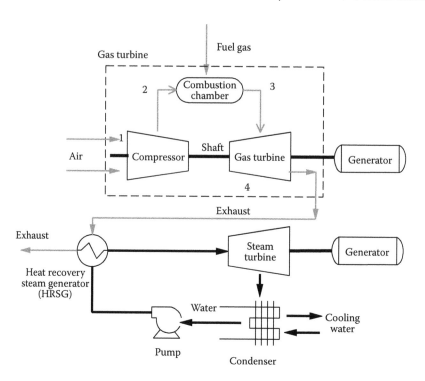

FIGURE 9.5 Combined cycle.

steam. The heat input for steam is what is left from Q_g after performing work $Q_s = Q_g(1 - \eta_g)$ and the work is given by the steam efficiency $W_s = Q_s\eta_s = Q_g(1 - \eta_g)(\eta_s) = Q_g(\eta_s - \eta_s\eta_g)$. Therefore combined efficiency is

$$\eta_{cc} = \frac{W_g + W_s}{Q_g} = \frac{W_g}{Q_g} + \frac{Q_g(\eta_s - \eta_s\eta_g)}{Q_g} = \eta_g + \eta_s - \eta_s\eta_g \qquad (9.2)$$

Example 9.5

Consider a CC process with gas system efficiency of 30% (assume lower than the typical 33%) and steam system efficiency of 28% (assumed lower than the usual 32%). What is the CC efficiency? Answer: $\eta_{cc} = \eta_g + \eta_s - \eta_s\eta_g = 0.30 + 0.28 - 0.30 \times 0.28 = 0.496$.

There are additional advantages of a combined cycle. Some relate to the environment, since overall the CO_2 and other emissions are reduced by decreasing the amount of coal burned. We will discuss emissions in the next section.

Some CC plants include burning supplementary coal to drive the Rankine cycle in addition to the heat provided by the exhaust gas. This increases the temperature of the steam-based process allowing the plant to cover variations in load demand; however, it lowers the CC efficiency and decreases the benefits of reduced emissions.

9.2.4 Gas-Fired and Combined Cycle Plants: Environmental Considerations

As we mention at the beginning of the chapter, NG has higher C content by weight (in %) and HV (in MJ/kg) compared to coal. For the sake of illustration, we can use 75% C by weight and a HV of

55 MJ/kg. NG has cleaner emissions from combustion (no PM and sulfur). In this section, we see how to calculate emissions from gas-fired plants using the carbon content in gas per unit energy C_e, as kg of C in the equivalent mass of gas that would produce a kJ of energy. For example, if 1 kg gas with 75% C by weight produces 55 MJ, then $C_e = 0.75\,\text{kgC}/55\,\text{MJ} = 1.36 \times 10^{-5}\,\text{kgC/kJ}$. Recall from Chapter 3 that heat rate (HR) is related to efficiency and using $\eta = 0.33$ for a Brayton cycle, we get a HR of $HR = \dfrac{3600\,\text{kJ/kWh}}{\eta} = \dfrac{3600\,\text{kJ/kWh}}{0.33} = 10909\,\text{kJ/kWh}.$

We can estimate carbon emission to the atmosphere a_C by multiplying the C content per unit energy C_e by heat rate HR, $a_C = C_e \times HR = 1.36 \times 10^{-5}\,\text{kgC/kJ} \times 10{,}909\,\text{kJ/kWh} \simeq 0.15\,\text{kgC/kWh}$.

This mass emission rate of carbon can be converted to CO_2 mass emission rate a_{CO2} by using the ratio 44:16 of molecular weight of CO_2 to atomic weight of C (Chapter 1). Therefore, $a_{CO2} = a_C \times \dfrac{44}{12} = 0.15\,\text{kgC/kWh} \times 3.666\,\text{kgCO}_2/\text{kgC} = 0.55\,\text{kgCO}_2/\text{kWh}$. We can see that for this gas and plant efficiency we emit about one half kg of CO_2 per kWh electrical energy generated. This is almost half of coal at typically 0.8 kg of CO_2 per kWh.

Consider a combined cycle power plant with 100% NG fuel and efficiency $\eta_{cc} = 50\%$. Assume the same C content and HV. The HR of the CC plant is $HR = \dfrac{3600\,\text{kJ/kWh}}{\eta_{cc}} = \dfrac{3600\,\text{kJ/kWh}}{0.5} = 7200\,\text{kJ/kWh}$. Carbon emission to the atmosphere a_C is $a_C = C_e \times HR = 1.36 \times 10^{-5}\,\text{kgC/kJ} \times 7{,}200\,\text{kJ/kWh} \simeq 0.098\,\text{kgC/kWh}$. Converted to CO_2 mass emission rate a_{CO2} it is $a_{CO2} = a_C \times \dfrac{44}{12} = 0.098\,\text{kgC/kWh} \times 3.666\,\text{kgCO}_2/\text{kgC} = 0.36\,\text{kgCO}_2/\text{kWh}$.

We can see that for this gas and CC plant efficiency, we emit about one third kg of CO_2 per kWh electrical energy generated. This is less than half (45%) of coal at typically 0.8 kg of CO_2 per kWh.

Example 9.6

Currently CC plants are trying to achieve 60% efficiency. What would be the reduced CO_2 emissions for NG compared to coal? Assume NG with C 75% by weight, an HV of 55 MJ/kg, and typical coal emissions of 0.8 kg of CO_2 per kWh. Answer:

$$\text{Heat rate: } HR = \frac{3600\,\text{kJ/kWh}}{\eta_{cc}} = \frac{3600\,\text{kJ/kWh}}{0.6} = 6000\,\text{kJ/kWh}$$

Carbon emission to the atmosphere:
$a_C\ a_C = C_e \times HR = 1.36 \times 10^{-5}\,\text{kgC/kJ} \times 6000\,\text{kJ/kWh} \simeq 0.082\,\text{kgC/kWh}$
Converted to CO_2 mass emission rate a_{CO2}:
$$a_{CO2} = a_C \times \frac{44}{12} = 0.082\,\text{kgC/kWh} \times 3.666\,\text{kgCO}_2/\text{kgC} = 0.3\,\text{kgCO}_2/\text{kWh}$$
Compared to coal at 0.8 this is 0.3 / 0.8 = 0.375 or 37.5% of coal.

9.3   INTERNAL COMBUSTION ENGINES

Steam-based processes rely on transferring heat from an external combustor to boil water and use the steam to perform work on the turbine, which in turn spins the generator. In contrast, *internal combustion* processes rely on burning the fuel internally and using the expanding gas from combustion to do work directly on the generator. Internal combustion engines are well known for their applications in transportation, including automobiles, trucks, locomotive, and vessels. They were employed in early aircraft engines. They are widely used for electric power generation with installed capacity ranging from fractions of a kW to tens of MW.

Internal combustion engines are based on compression and expansion *strokes* in a *cylinder* performed by a *piston* and controlled by inlet and outlet valves (Figures 9.6 and 9.7). The compression

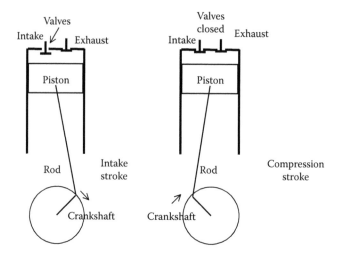

FIGURE 9.6 Internal combustion engine. Piston, cyclinder, crankshaft, and valves. Intake and compression strokes.

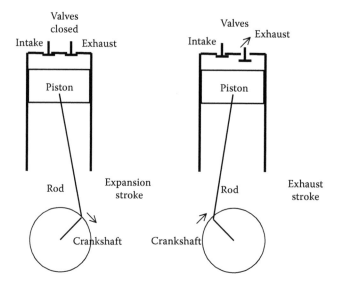

FIGURE 9.7 Internal combustion engine: piston, cylinder, crankshaft, and valves. Expansion and exhaust strokes.

or suction stroke occurs with only the inlet valve open, whereas the expansion or power stroke occurs with only the outlet valve open. To convert reciprocating motion to rotary motion, these engines use a slider crank mechanism based on the crankshaft (Figures 9.6 and 9.7). To extract useful work, the expansion work must exceed the work used for compression. Air is taken in, compressed, and heated by combustion of fuel; the resulting exhaust is released to the environment making the process an open cycle (Figure 9.8).

The following terminology relates to geometry and dimensions of the engine and position of the piston in the cylinder.

- TC, top center (position of the piston farthest from shaft)
- BC, bottom center (position of piston nearest the shaft)
- Bore (d), inner diameter of the cylinder
- Stroke (L), distance between dead centers

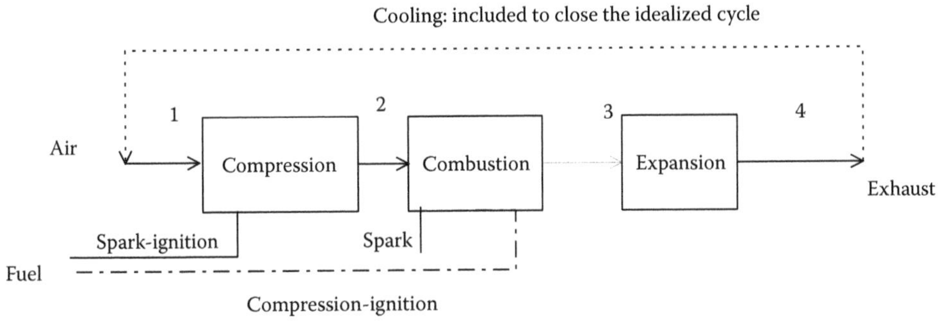

Cooling: included to close the idealized cycle

FIGURE 9.8 Internal combustion engine: spark-ignition versus compression ignition.

- Engine capacity is the swept volume or stroke volume (V_{sw}), volume swept by the piston as it moves from one dead center to the other dead center, calculated by $V_{sw} = (\pi d^2/4)L$
- Clearance volume (V_c), volume above the piston at TC
- Total volume (V_t), the volume above the piston at BC

Then, the following relation applies $V_t = V_{sw} + V_c$. The *volumetric compression ratio* is the ratio of total volume to clearance volume $r = Vt/Vc$.

Internal combustion engines can use a variety of fuels: gasoline, diesel, fuel oil, natural gas, and waste gas. Therefore, emissions of CO_2 and air pollutants vary according to the fuel used. Besides fuel characteristics, emissions vary for example by engine type; *four-stroke engines* are cleaner than *two-stroke engines*. In a two-stroke, there are only two movements of the piston in the cylinder during one crankshaft revolution. One stroke compresses air in the cylinder, and the other is caused by expansion. Exhaust and intake occur simultaneously, that is, the end of the expansion stroke coincides with air intake at the beginning of the compression stroke. In a four-stroke, it takes two revolutions of the crankshaft to complete a compression–expansion cycle.

Two-stroke engines are simpler and lighter but require mixing fuel with oil for lubrication, which burns together with the fuel, causing emissions and smoke in the exhaust. These emissions have air-quality effects. Nowadays, the oil used for two-stroke is synthetic oil. Two-stroke lubrication oil must have lower ash content than regular oil to avoid deposits in the engine and excessive PM emissions.

An internal combustion engine can be based on *spark-ignition* and modeled using the *Otto cycle*, or based on *compression-ignition* and modeled using the *Diesel cycle* (Figure 9.8). The volumetric compression ratio is typically 6:1 to 10:1 for Otto cycle–based engines (typically gasoline fueled) and higher (14:1 to 22:1) for Diesel cycle–based engines (diesel fueled). Efficiency varies from less than 30% (for spark-ignition engines) to about 40% (compression-ignition engines).

9.3.1 Otto Cycle

An idealized Otto cycle is a heat engine model for an internal combustion engine based on spark-ignition. Figure 9.9 shows a simplified Otto cycle for the purposes of describing how it works. Air is compressed from state 1 to 2, whereas expansion occurs from 3 to 4. Both processes are idealized as isentropic compression and expansion with given volumetric compression ratio idealized as $r = \dfrac{V_1}{V_2} = \dfrac{V_4}{V_3}$. They display as vertical lines in the *T-s* plane and curves in the *P-v* plane. The compressed air in state 2 is heated by ignition of the fuel at constant volume to state 3 (isochoric vertical line in *P-v* and curve in *T-s* plane). This heated air-combustion gas mix expands adiabatically (isentropic) from 3 to 4, performing output work (this is the power stroke). At 4 air is exhausted to the environment, and for the purposes of closing the cycle it is thought of as a return of fresh air at state 1 (air intake at ambient temperature and pressure). Thus, this path from 4 to 1 is modeled as isochoric cooling.

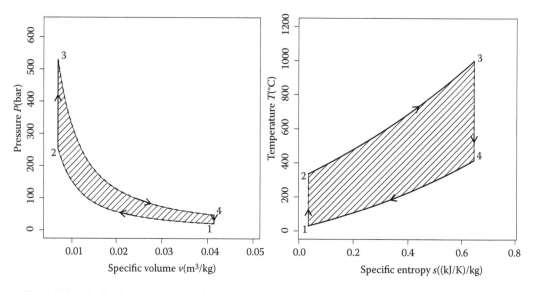

FIGURE 9.9 Idealized Otto cycle showing curved lines as adiabats in the *P-v* plane and isochors in the *T-s* plane. Example of compression ratio of 6. Left: 1–2: adiabat, compression 30°C–> 335.5°C, W = 226.3 kJ/kg, Q = 0 kJ/kg; 2–3: isochor, 335.5°C–> 1000°C, W = 0 kJ/kg, Q = 556.4 kJ/kg; 3–4: adiabat, expansion 1000°C–> 416.9°C, W = –492.9 kJ/kg, Q = 0 kJ/kg; 4–1: isochor, 416.9°C–> 30°C, W = 0 kJ/kg, Q = –289.2 kJ/kg. Right: 1–2: adiabat, compression 30°C–> 335.5°C, W = 226.3 kJ/kg, Q = 0 kJ/kg; 2–3: isochor, 335.5°C–> 1000°C, W = 0 kJ/kg, Q = 556.4 kJ/kg; 3–4: adiabat, expansion 1000°C–> 416.9°C, W = –492.9 kJ/kg, Q = 0 kJ/kg; 4–1: isochor, 416.9°C–> 30°C, W = 0 kJ/kg, Q = –289.2 kJ/kg.

In terms of piston position in the cylinder, these states correspond to the following: 1 to 2 compression, piston moving from TC to BC; 2 to 3 combustion, piston at BC; 3 to 4 expansion, piston moving from BC to TC; 4 to 1 cooling, piston at the TC position.

Example 9.7

Draw the idealized Otto cycle of Figure 9.9 in the *P-v* and *T-s* planes for one mole of air using volumetric compression ratio of 6, starting from 1.2 l at 30°C and reaching 1000°C. State heat input and output, and work output. Answer: We can use path.cycles as follows

```
x <- list(TH = 1000, TL = 30, V1 = 1.2, V2 = 0.2, S1 = 1, cty = 'otto')
y <- path.cycles(x, plane='Pv', shade.cycle = TRUE)
y <- path.cycles(x, plane='Ts', shade.cycle = TRUE)
path.cycles.summary(y)
```

resulting in Figure 9.9. As reported in this figure, heat input is 556.4 kJ/kg, heat output is 289.2 kJ/kg, and work output is 492.9 kJ/kg.

9.3.2 OTTO CYCLE EFFICIENCY

Efficiency of the idealized cycle is calculated as work output divided by heat input. Geometrically in the *T-s* plane, efficiency is the area inside the cycle (net amount of work, W_n) divided by area under the isochoric curve from 2 to 3 (heat input, Q_{in}). Using path integrals

$$\eta = 1 - \frac{Q_{out}}{Q_{in}} = \frac{\oint T ds}{\int_2^3 T ds}$$

We can query path.cycles.summary for component WQnet

```
x <- list(TH=1000,TL = 30,V1 = 1.2,V2 = 0.2,S1 = 1,cty = 'otto')
y <- path.cycles(x,plane = 'Pv',shade.cycle = TRUE)
> path.cycles.summary(y)$WQnet
$WQnet
 Qin(kJ/kg) Qout(kJ/kg)  Eff
1   556.4      -289.2 0.48
```

which shows an efficiency of 48%.

A practical way of evaluating efficiency is

$$\eta = 1 - \frac{1}{r^{\gamma-1}} \tag{9.3}$$

where $r = V_1/V_2$ is the volumetric compression ratio and $\gamma = c_p/c_v$ as before.

Example 9.8

Assume the compression ratio is $r = V_1/V_2 = 9$ and use a temperature between 30°C and 1000°C. What is the efficiency? What are the efficiencies for compression ratio between 3 and 10? Answer: First estimate γ at the middle of temperature range

```
> cp.cv(TC = (1000 + 30)/2)$cp.cv
[1] 1.357
>
```

Use $\gamma = c_p/c_v = 1.357$ and calculate $\eta = 1 - \frac{1}{9^{(1.357-1)}} = 0.544$.

We can use similar code as employed for Brayton cycle to assist in this calculation and explore the effect of changing compression ratio on efficiency

```
V.r <- seq(7,10,1)
gamma<- cp.cv(TC = (1000 + 30)/2)$cp.cv
eta <- 1-1/(V.r^(gamma-1))
V.r; round(eta,3)
plot(V.r,eta,type = "b", xlab = "Compression ratio r", ylab = "Efficiency",lwd = 2)
```

which produces Figure 9.10.

Several technological factors limit practical efficiency. Otto cycle implementations for electrical generation have ~30% efficiency.

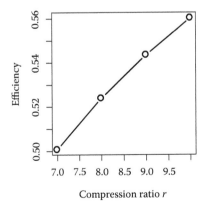

FIGURE 9.10 Efficiency versus compression ratio for idealized Otto cycle.

9.3.3 DIESEL CYCLE

An internal combustion engine based on compression-ignition is different with respect to an Otto cycle because the ignition occurs due to the high pressure and temperature of the fuel and without the addition of a spark. An idealized Diesel cycle is a heat engine model for an internal combustion engine based on compression-ignition (Figure 9.11). Air is compressed from state 1 to 2, whereas expansion occurs from 3 to 4. Both processes are idealized as isentropic compression and expansion. They display as vertical lines in the *T-s* plane and curves in the *P-v* plane. The compressed air in state 2 is heated by ignition of the fuel at constant pressure to state 3 (isobaric horizontal line in *P-v* and curve in *T-s* plane). This heated air-combustion gas mix expands adiabatically (isentropic) from 3 to 4 performing output work (this is the power stroke). At 4, air is exhausted to the environment, and for the purposes of closing the cycle it is thought of as a return of fresh air at state 1 (air intake at ambient temperature and pressure). Thus, this path from 4 to 1 is modeled as isochoric cooling.

Note from the *P-v* diagram that the volumetric compression ratio $r = V_1/V_2$ is different from the volumetric expansion ratio $r_e = V_4/V_3 = V_1/V_3$. The quotient of both ratios $\dfrac{r}{r_e} = \dfrac{V_1/V_2}{V_1/V_3} = V_3/V_2$ is called the cut-off ratio and denoted by $\alpha = V_3/V_2$. This is the ratio of the end to start volume before combustion.

Example 9.9

Draw the idealized Diesel cycle of Figure 9.11 in the *P-v* and *T-s* planes for one mole of air using a volumetric compression ratio of 15 and expansion ratio of 5, starting from 1.5 l at 30°C. State heat input and output. What is the efficiency? Answer: We can use path.cycles as follows

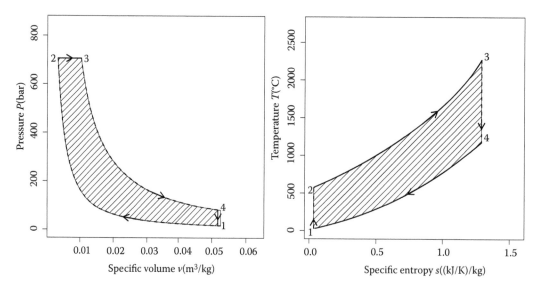

FIGURE 9.11   Idealized Diesel cycle showing adiabat curves in *P-v* plane and isobaric curves in the *T-s* plane. This is an example with compression ratio of 15 and expansion ratio of 5. Left: 1–2: adiabat, compression 30°C–> 575.9°C, $W = 416.8$ kJ/kg, $Q = 0$ kJ/kg; 2–3: isobar, expansion 575.9°C–> 2273.8°C, $W = -487.3$ kJ/kg, $Q = 1922.1$ kJ/kg; 3–4: adiabat, expansion 2273.8°C–> 1175.6°C, $W = -880.1$ kJ/kg, $Q = 0$ kJ/kg; 4–1: isochor, 1175.6°C–> 30°C, $W = 0$kJ/kg, $Q = -942.9$ kJ/kg. Right: 1–2: adiabat, compression 30°C–> 575.9°C, $W = 416.8$ kJ/kg, $Q = 0$ kJ/kg; 2–3: isobar, expansion 575.9°C–> 2273.8°C, $W = -487.3$ kJ/kg, $Q = 1922.1$ kJ/kg; 3–4: adiabat, expansion 2273.8°C–> 1175.6°C, $W = -880.1$ kJ/kg, $Q = 0$ kJ/kg; 4–1: isochor, 1175.6°C–> 30°C, $W = 0$ kJ/kg, $Q = -942.9$ kJ/kg.

```
x <- list (TL = 30, V1 = 1.5, V2 = 0.1, V3 = 0.3,  S1 = 1, cty = 'diesel')
y <- path.cycles (x, plane= 'Pv', shade.cycle = TRUE)
y <- path.cycles (x, plane= 'Ts', shade.cycle = TRUE)
path.cycles.summary (y) $WQnet
```

resulting in Figure 9.11. Input and output heat are 1922.1 and 942.9 kJ/kg, respectively. Efficiency is 51%. We see that from path.cycles.summary

```
> path.cycles.summary (y) $WQnet
 Qin (kJ/kg) Qout (kJ/kg)  Eff
1   1922.1    -942.9 0.51
>
```

9.3.4 DIESEL CYCLE EFFICIENCY

Efficiency of the idealized Diesel cycle is calculated as work output divided by heat input. Geometrically in the T-s plane, efficiency is the area inside the cycle divided by area under the isochoric curve from 2 to 3 (heat input, Q_{in}). Using path integrals $\eta = 1 - \frac{Q_{out}}{Q_{in}} = \dfrac{\oint Tds}{\displaystyle\int_2^3 Tds}$.

We can query path.cycles.summary WQnet as before

```
> path.cycles.summary (y) $WQnet
 Qin (kJ/kg) Qout (kJ/kg)  Eff
1   1922.2    -942.7 0.51
>
>
```

which shows an efficiency of 51%.

This efficiency is more complicated to calculate because the compression and expansion ratio are different. A closed formula is

$$\eta = 1 - \frac{1}{r^{\gamma-1}} \left(\frac{\alpha^{\gamma} - 1}{\gamma(\alpha - 1)} \right) \tag{9.4}$$

Or equivalently

$$\eta = 1 - \frac{1}{\gamma} \left(\frac{r_e^{-\gamma} - r^{-\gamma}}{r_e^{-\gamma} - r^{-1}} \right)$$

Example 9.10

Assume the compression ratio is $r = V_1/V_2 = 15$, the expansion ratio is $r_e = V_1/V_3 = 5$, and use a temperature between 30°C and 2000°C. What is the efficiency? What are the efficiencies for a compression ratio between 10 and 20? Assume $\gamma = c_p/c_v = 1.4$. Answer: Note that $\alpha = r/r_e = 3$, then

$$\eta = \eta = 1 - \frac{1}{r^{\gamma-1}} \left(\frac{\alpha^{\gamma} - 1}{\gamma(\alpha - 1)} \right) = 1 - \frac{1}{15^{1.4-1}} \left(\frac{3^{1.4} - 1}{1.4 \times (3 - 1)} \right) = 0.56$$

or using the alternative expression

$$\eta = 1 - \frac{1}{\gamma} \left(\frac{r_e^{-\gamma} - r^{-\gamma}}{r_e^{-1} - r^{-1}} \right) = 1 - \frac{1}{1.4} \left(\frac{5^{-1.4} - 15^{-1.4}}{5^{-1} - 15^{-1}} \right) = 0.56$$

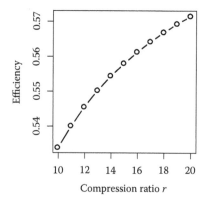

FIGURE 9.12 Efficiency versus compression ratio for idealized Diesel cycle.

As we did for the Brayton and Otto cycles, we can write simple code to assist in this calculation and explore the effect of changing compression ratio on efficiency:

```
V.r <- seq(10,20,1);V.r.e <- 5
gamma <- 1.4
cut.off <- V.r/V.r.e
fac <- (cut.off^gamma -1)/(gamma*(cut.off-1))
eta <- 1-(1/(V.r^(gamma-1)))*fac
# alternative equation
#eta <- 1-(1/gamma)*(V.r.e^-gamma-V.r^-gamma)/(V.r.e^-1-V.r^-1)
V.r; round(eta,2)
plot(V.r,eta,type = "b", xlab = "Compression ratio r", ylab="Efficiency",lwd=2)
```

The result is given in Figure 9.12. Technological factors limit practical efficiency. Diesel cycle implementations for electrical generation have ~40% efficiency.

9.4   OIL AS FUEL FOR POWER GENERATION

9.4.1   Electricity from Oil

As we discussed in Chapter 1, oil represents only a minor percentage (4.3%) of the world's electricity production [6]. Liquid fuels derived from refining crude oil include gasoline (the lightest), diesel fuel, and fuel oils (the heaviest). Gasoline fuels are mostly used for engines employed in transportation, such as automobiles. Diesel fuels are used in engines for transportation (automobiles, trucks, and locomotives), and agricultural machinery. Fuel oils are used primarily for space heating and electric power generation. Power plants have used heavy fuel oil or #6 Bunker C fuel oil. Diesel is used in some small generators, such as those for emergency backup systems.

Heavy fuel oil is the left over after the lighter fuels (gasoline, diesel) have been refined. Therefore, it has substantially more ash and impurities, for example, sulfur and heavy metals. Use of heavy fuel oil has decreased and is being substituted by natural gas, but it is still in use for power generation. Fuel oil is heavy because the molecules are larger. Heating value of residual oil is ~40 MJ/kg, which is lower than natural gas and approximately 85% C by weight. Fuel oil is used in a furnace to generate steam for a Rankine cycle or used directly in internal combustion engines using the diesel cycle.

9.4.2   Oil as a Resource

Petroleum reserves are estimated to about 10,000 EJ and is consumed at a rate of 180 EJ/y, which is increasing at a rate of ~1.1%/y [1]. We can estimate times to depletion as we did for natural gas.

Example 9.11

Estimate time to depletion of oil reserves under scenarios of continued consumption at current rate and increased consumption. Answer: Use formulas as in Example 9.1. Divide reserves by energy used in one year $t_f = \dfrac{X(0)}{c(0)} = \dfrac{10,000EJ}{180EJ/yr} \simeq 55.55$ yr, or oil reserves could last ~50 to 60 years (or about the same as NG) but under increased consumption scenario the time to depletion will shorten to $t_f = \dfrac{1}{0.011yr^{-1}} \ln\left(0.011yr^{-1} \times 55.55\,yr + 1\right) \simeq 43$ yr.

Therefore, oil reserves could last 43 years.

This estimate does not account for processes of *secondary and tertiary recovery* and the exploitation of unconventional oil resources, such as *oil shale* and *oil sands*. Such factors would increase the time to depletion of oil resources, but would represent further environmental challenges. Recovery consists of injecting gases and chemicals into the reservoirs to fracture the rock and promote the flow of oil, thus potential water contamination is an issue. Petroleum extraction from oil shale and oil sand is disruptive to ecosystems and water resources. The petroleum in sands is very viscous, and chemicals are added to facilitate flow.

An additional oil recovery process that will aid in increasing time to depletion is to inject liquid CO_2 into the reservoir. In the future, when CO_2 sequestration becomes more prevalent and coal-fired power plants include CO_2 capture, the CO_2 captured could be employed for recovery.

9.4.3 CO_2 Emissions from Oil

Oil-derived fuels have important pollutant emissions from combustion, including PM, ozone precursors, and sulfur. Diesel fuel is a particularly important contributor to PM contamination.

Heavy oil has higher C content by weight (85%) and lower HV (~40 MJ/kg) compared to NG. The carbon content per unit energy C_e, as kg of C, in the equivalent mass of oil that would produce a kJ of energy. For example, if 1 kg with 85% C by weight produces 40 MJ, then $C_e = 0.85$ kgC/40 MJ = 2.12×10^{-5} kgC/kJ. Using $\eta = 0.40$ for a Diesel cycle we get a HR of $HR = \dfrac{3600\,kJ/kWh}{\eta} = \dfrac{3600\,kJ/kWh}{0.40} = 9000\,kJ/kWh$.

Carbon emission to the atmosphere a_C is obtained by multiplying the C content per unit energy C_e by heat rate HR: $a_C = C_e \times HR = 2.12 \times 10^{-5}$ kgC/kJ $\times$ 9,000 kJ/kWh $\simeq 0.19$ kgC/kWh.

This mass emission rate of carbon can be converted to CO_2 mass emission rate a_{CO2} by using the ratio 44:16 of molecular weight of CO_2 to atomic weight of C. [3]Therefore,

$$a_{CO2} = a_C \times \frac{44}{12} = 0.19\,kgC/kWh \times 3.666\,kgCO_2/kgC \simeq 0.70\,kgCO_2/kWh.$$

We can see that for heavy oil used as a fuel and this plant efficiency we emit almost ¾ of kg of CO_2 per kWh electrical energy generated. This is almost as high as coal at typically 0.8 kg of CO_2 per kWh.

9.4.4 Fuel Consumption

Power output P in kW is a function of thermal efficiency η_t of an engine, generator efficiency η_g, the fuel heat value HV in MJ/kg, and the fuel consumption rate f_c in g/s:

$$P = \eta_t \times \eta_g \times HV \times f_c.$$

Example 9.12

Gasoline has a heat value of 36 MJ/kg. Assume a fuel consumption of 2.0 g/s with thermal and generator efficiencies of 0.25 and 0.9, respectively: $P = 36 \dfrac{MJ}{kg} \times 0.25 \times 0.9 \times 2 \dfrac{g}{s} = 16.2$ kW.

9.5 ALTERNATIVE OR SUBSTITUTE GAS AND LIQUID FUELS

9.5.1 Converting Coal to Liquid and Gas Fuels

Coal is converted to other types of fuels, liquid and gas, by *pyrolysis*, which is an endothermic chemical reaction occurring at high temperature in the absence of oxygen [7]. Pyrolysis of coal produces liquid, gas, and solid reaction products. Among the gases, we find H_2, CH_4, CO_2, and CO. The liquids include tar and light oils. The solid is *char*, which is carbon rich, also called *coke*, and employed in the metallurgic industry to make steel.

Coal gasification refers to converting coal to gas [7]. There is a variety of reactions to model coal gasification. When using steam and oxygen, gas products are H_2, CH_4, CO_2, and CO and synthesis gas, or *syngas*, which is a mixture of hydrogen and carbon monoxide. Syngas can then be chemically converted into other fuels. Coal liquefaction refers to converting coal to liquid fuels. In general, coal is deficient in hydrogen when compared to petroleum-derived liquid fuels; thus, coal liquefaction involves rejection of carbon or addition of hydrogen. Pyrolysis is a type of coal liquefaction, but the liquid products require further treatment.

9.5.2 Biomass

For a long time, humans have used biomass energy from plants and plant-derived materials for protection, cooking food, and keeping warm. Evidence indicates that *Homo erectus* began controlled burning of wood possibly ~0.6 Mya. For reference, this is a similar time span to the ice-core record 0.8 Mya from Antarctica (Chapter 2). Wood is still the largest biomass energy resource today in some parts of the world. In the context of power systems, the term *biomass* energy is very broad. It applies to all living plant matter as well as organic wastes from plants, animals, and humans, including sewage, garbage, wood construction residues, sludge from wastewater treatment plants, and municipal solid waste [1,8].

In Chapter 2, we studied the basics of ecosystems, and how energy flows from solar radiation through a food chain, and that fossil fuels formed from geologic forces operating for a long time on remains of energy captured by ecosystems on land (coal formed from plants) and water (petroleum formed by algae and zooplankton).

Biomass energy is carbon based and burning it releases CO_2 to the atmosphere in the same manner as burning fossil fuels. Combustion of biomass with typical carbohydrate composition CH_2O is modeled by

$$CH_2O + O_2 \rightarrow CO_2 + H_2O \tag{9.5}$$

However, fossil fuels release CO_2 captured by photosynthesis millions of years ago and essentially adding greenhouse gas from carbon matter that was previously sequestered. In contrast, biomass from living matter releases CO_2 that may be balanced by the CO_2 more recently captured by its own growth.

Biomass qualifies as a renewable resource because it can be regenerated in time scales that are comparable to or less than typical time scales for human use of the resource [7]. In quantitative terms, the ratio $\tau = \dfrac{t_r}{t_u}$ of time for resource regeneration t_r to utilization time t_u is used as a numerical measure of renewability of the resource. When $\tau \ll 1$, the resource is renewable; conversely, when $\tau \gg 1$, the resource is nonrenewable. When $\tau \approx 1$, the resource is marginally renewable. This ratio is also determined as the quotient of rates of regeneration and rate of utilization.

Example 9.13

Calculate renewability for two contrasting cases. (a) Suppose it took fossil fuels hundreds of million years to form and we deplete most reserves within 100 years. (b) Suppose methane from solid waste builds up in one year and we use it in 10 years. Answer: (a) We have $\tau \approx \dfrac{100\mathrm{My}}{100\mathrm{y}} \approx 10^6$, which means it is nonrenewable. (b) $\tau \approx \dfrac{1}{10} \approx 10^{-1}$, which means it is renewable.

Biomass could be burned directly as a fuel to generate electricity, using, for instance, wood pellets made from compacted sawdust, a by-product of sawmilling. Pellets are very dense and have high combustion efficiency.

However, more often biomass is converted into liquid or gas *biofuels* that burn more efficiently and have fewer contaminants.

9.5.3 LANDFILL GAS

Landfill gas (LFG) is a product of bacterial decomposition of organic matter in the solid waste of landfills [9]. The amount of gas and its composition depend on the type and age of the waste, the organic matter content, moisture, and temperature. LFG is mostly CH_4 (~50%), CO_2 and water vapor (~50%), small amounts of other compounds (oxygen, hydrogen, and nitrogen), hazardous air pollutants, and volatile organic compounds (VOCs). Composition changes by stages of decomposition starting from *aerobic* (i.e., under oxygen and little generation of methane) and followed by *anaerobic* (i.e., lack of oxygen and higher generation of methane). Depending on conditions, it takes about a year to reach a steady ~50% to 50% CH_4 and CO_2 composition. A model reaction for anaerobic digestion is

$$CH_2O \rightarrow \frac{1}{2}CH_4 + \frac{1}{2}CO_2 \tag{9.6}$$

which is an exothermic reaction.

As discussed in Chapter 2, methane itself is an important greenhouse gas; it can absorb and emit heat ~30 times more effectively than CO_2 [9]. In many parts of the world, emission of methane is mostly due to agricultural activities, for instance, crops grown under inundation as it occurs with rice paddies. Overall, LFG (being mostly CH_4 and CO_2) released to the atmosphere contributes to the planetary greenhouse effect (GH) and global warming. In some countries, LFG emissions can be a large source of methane emission to the atmosphere. For instance, in the United States, LFG is the third largest source of methane released to the atmosphere due to human activities. The Clean Air Act requires that large landfills implement systems for collection and control of LFG; the most basic control is flaring the LFG. It is common practice to monitor the surroundings of the landfill with gas analyzers to understand the impact on air quality for the sake of the environment and human safety.

Recall that CH_4 is the main gas of NG and that power production using NG represents a reduction in CO_2 emissions when compared to coal and oil. Capturing LFG and using it as fuel for electricity production by means of gas turbines is then a reasonable step in the process of reducing emissions from landfills, reducing waste accumulating in landfills, and helping to expand the gas resource for electricity production. Being about 50% methane, the *HV* of LFG is about half of the *HV* of NG. Once purified, it is often considered a renewable or green energy conversion process (Chapter 1). The valuable LFG resource then pays for the required controls and provides an additional revenue source.

Example 9.14

Calculate emission of CO_2 and potential contribution to the greenhouse effect from atmospheric release of 1 kg of LFG and compare to the contribution if you would burn the LFG for electricity production. Answer: Assume 50% CH_4 and 50% CO_2 for the LFG composition, and 30 times radiative effect of CH_4 to CO_2. Simple atmospheric release will contribute $0.5 \, kgCO_2/kgLFG$, but the impact on the GH effect $0.5 \, kgCH_4/kgLFG \times 30 + 0.5 \, kgCO_2/kgLFG \times 1 = 15.5$ times. Now if we burn the LFG, for half of it $\frac{44}{16} \, kgCO_2/kgCH_4$, then $0.5 \times \frac{44}{16} \, kgCO_2/kgLFG$. Assume the other half contributes $0.5 \, kgCO_2/kgLFG$. Combining, $0.5 \times \left(1 + \frac{44}{16}\right) \, kgCO_2/kgLFG = 1.875 \, kgCO_2/kgLFG$. Thus, we emit almost twice as much CO_2 than for simple release. However, the impact on the GH is $0.0 \, kgCH_4/kgLFG \times 30 + 1.875 \, kgCO_2/kgLFG \times 1 = 1.875$, which is much less than simple release.

This example illustrated the convenience of using landfill gas for electricity production instead of simply releasing it to the atmosphere. Now, let us compare how much power we can generate and emissions compared to NG. Since LFG has 50% CH_4, which can be assumed to be similar to NG, it takes twice as much LFG as NG to produce 1 kWh, but the CO_2 emission from combustion would be similar, $a_{CO_2} = 0.55 \, kgCO_2/kWh$ having to add the CO_2 contained in twice that mass.

Example 9.15

Calculate how many kg of LFG it takes to produce 1 kWh and CO_2 emissions per kWh. Assume that the half of LFG that is CO_2 becomes part of the emissions. Answer: Assuming 33% efficiency, and emissions from the methane part of the LFG would be $a_{CO_2} = 0.55 \, kgCO_2/kWh$ and the mass of methane would be $0.55 \frac{kgCO_2}{kWh} \times \frac{16}{44} \frac{kgCH_4}{kgCO_2} = 0.2$ $kgCH_4/kWh$. The mass of LFG is twice as much but half of this is CO_2, resulting in a total of $a_{CO_2} = (0.55 + 0.2) \, kgCO_2/kWh = 0.75 \, kgCO_2/kWh$. From this example, we see that emissions are higher than NG and still a little lower than coal.

9.5.4 BIOMASS CONVERSION PROCESSES

Although biomass itself is a fuel, with certain HV and potential emissions (nitrogen, sulfur, ash) when combusted, it is common to *convert* that biomass into a less amount of liquid or gas *biofuel*, which may have a higher HV and cleaner emissions. For instance, cow manure can be dried and burned as a fuel (with HV ~20 MJ/kg and will emit air contaminants), and it can be converted to methane with a higher HV (~55 MJ/kg) and cleaner emissions.

Energy recovery for a particular fuel product of a biomass conversion process is the ratio of the product yield W_p (in kg) times its HV_p divided by the biomass weight W_b (kg) times its HV_b [10]. In other words, $E_R = \dfrac{W_p \times HV_p}{W_b \times HV_b}$. This ratio could also be interpreted as an energy efficiency obtained by dividing energy output to energy input. When mixing biomass for pyrolysis (say, manure, agricultural trash, and algae) [10], the efficiency calculation can be extended by summation over products and reactants $E_R = \dfrac{\sum W_p \times HV_p}{\sum W_b \times HV_b}$.

Example 9.16

Suppose we convert 1 kg of biomass with HV_b = 20 MJ/kg to 0.2 kg of a fuel with HV_p = 50 MJ/kg. What is the energy recovery? Answer: $E_R = \dfrac{W_p \times HV_p}{W_b \times HV_b} = \dfrac{0.2 \times 50}{1 \times 20} = 0.5$ or 50%.

Some of the major processes to convert biomass to fuel involve anaerobic processes or low-oxygen chemical transformation. For instance, as we have already discussed, organic matter is decomposed to methane anaerobically by bacterial action. This occurs in wetlands and in landfills, but we can make it occur under controlled conditions in a reactor designed for this purpose: an *anaerobic digester*.

Pyrolysis can also be used for biomass conversion. Pyrolysis of organic matter produces liquid (e.g., tar), gas reaction products (e.g., methane and syngas), and a biochar. For instance, pyrolysis of wood at 200°C to 300°C yields a char (charcoal) that is mostly carbon and mineral ash. Gasification is also used to convert biomass at high temperature and a low-oxygen environment into a mixture of gases including syngas, which is a mixture of hydrogen and carbon monoxide.

Another form of biomass conversion is *fermentation* of carbohydrates to obtain ethanol C_2H_6O

$$CH_2O \rightarrow \frac{1}{3}C_2H_6O + \frac{1}{3}CO_2 \tag{9.7}$$

Ethanol is mostly used as a blending agent with gasoline to increase octane and to cut down carbon monoxide and other smog-causing emissions. Some vehicles are designed to run on E85, an alternative fuel with much higher ethanol content than regular gasoline.

9.5.5 MANURE

In small farms, manure from grazing herbivores (e.g., cows) rarely represent an environmental concern and is typically employed as fertilizer. However, when a large number of animals are concentrated in a small area, the manure on the land can be excessive and have the potential for methane emissions (and therefore contribution to the greenhouse effect), while at the same time excess of nitrogen and other contaminants affect water quality. Therefore, manure management becomes problematic. An option is to convert this waste into an energy resource, for instance, converting it to biogas by using anaerobic digesters (Figure 9.13), and purifying the gas to methane, which is used to generate electricity to power the facility, or to pipe it to market.

Biogas yield per kg of cow manure vary according to several factors such as moisture (~80%–90%) and organic content (80%–85% of dry matter), anaerobic digestion process, and the feeding and age of the cows. The organic fraction of totals solids (TS) is given as volatile solids (VS) and it is the contributor to biogas. Thus, a typical way of expressing biogas production from manure is in m³ of biogas per kg of VS or m3/kgVS and a percent of methane in the biogas.

Example 9.17

Suppose we produce 0.30 m3/kgVS of biogas, which is 60% methane, and that the VS/TS ratio is 80%. Calculate biogas production per kg of TS (or ~kg of dry manure DM) and methane production per kg of DM. Answer: $\dfrac{0.60 \times 0.33\,\text{m}^3}{1\,\text{kgVS} \times \frac{1}{0.8}\frac{TS}{VS}} \simeq 0.16\dfrac{\text{m}^3\text{CH}_4}{\text{kgDM}}$. Methane's density is 0.656 kg/m³, thus we produce 0.656 × 0.16 kgCH₄/kgDM = 0.105 kgCH₄/kgDM.

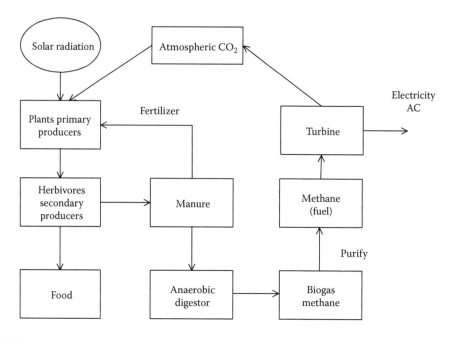

FIGURE 9.13 Methane from manure.

When applied to large concentrated animal feed operations (CAFOs), large capacity anaerobic digesters are required. One example in the United States is the facility at Huckabay Ridge, Stephenville, Texas, which has eight anaerobic digester tanks with capacity of 3500 m^3 each. Biogas is purified to methane and fed to a NG pipeline. The scheme in Figure 9.13 is not limited to large feeding operations and can provide a renewable power system for a small farm in combination with other options.

An alternative to anaerobic digestion of the manure to obtain methane is pyrolysis of the manure to yield syngas and biochar. The latter has potential applications in the farm and provides an opportunity for carbon sequestration.

9.5.6 Biodiesel

Biodiesel is a cleaner (lower emissions) diesel fuel replacement made from vegetable oil, animal fat, or recycled cooking oil. Feedstock for vegetable oil includes soybean, rapeseed or canola, and algae. Glycerin (a valuable product itself) is separated from the oil to yield *methyl esters* that constitute the biodiesel. Biodiesel can be used as an additive to reduce vehicle emissions or in its pure form as a renewable alternative fuel for a diesel engine. Algae are produced in open circulation ponds or in closed photobioreactors.

9.5.7 Growing Crops for Fuel

When the biomass is grown as a *crop* to be used for *biofuel*, the atmospheric CO_2 is constantly assimilated to the biomass and released to the atmosphere by respiration. Converting the standing biomass to biofuel will release CO_2 to the atmosphere once the biofuel undergoes combustion. In essence, we bypass the long geologic time needed to form fossil fuels. Biofuels are used as an important source of an energy carrier but mostly for transportation. For instance, ethanol from fermentation of sugar cane is an important fuel in Brazil. It is also a supplemental fuel in some countries, for instance, in the United States ethanol is derived from corn.

A concern with growing crops for fuels is is derived that it competes with food production and thus represents a risk to global food security. There are ~800 million undernourished people in the world (~11% of the population), and great efforts are needed to produce and provide food for all. Options to increase production by *extensification* (expanding cultivation area) are limited because there is no more land available for agricultural expansion without considerable environmental risks. Increasing production by *intensification* (increased crop yield for the same area) is still an option and demands technological advances to reduce environmental adverse effects. Biofuels compete with food in several forms. For instance, some of the crops used for fuel (maize, sugarcane, and soybeans) are also food crops, and their yields are partly diverted to fuels. Land and water that would be used to grow other food crops are diverted to growing plants for fuels.

9.5.8 CARBON SEQUESTRATION: FOREST GROWTH

As we have seen, carbon-based fuels, both fossil and biomass derived, inevitably lead to CO_2 emission with consequences for the greenhouse effect. Long-term sustainable use of carbon-based fuels requires sequestering the CO_2 and preventing its emission to the atmosphere [11]. We covered direct sequestration methods in Chapter 7, and those direct methods are applicable not only to coal but to any other carbon-based fuel. Indirect sequestration methods are based on enhancing natural carbon sinks and apply to sequestering C already in the atmosphere, not just preventing CO_2 release into the atmosphere. Currently, the most often used indirect sequestration method is *forest growth*.

Two major opposing processes affecting forest cover are *deforestation* and *afforestation*. In deforestation, forests are removed by fire or bulldozing to clear land for agriculture or urbanization. In afforestation, forest is allowed to regrow in previously occupied agricultural land. In some parts of the world, deforestation has been the prevalent process, whereas in other parts of the world, afforestation has become prevalent in the last half century. Overall, the balance is favorable to deforestation, which means that in total deforestation/afforestation currently constitutes a net source of atmospheric CO_2. Sequestering CO_2 by forest growth would require a significant increase in afforestation rates.

A forest stand is composed of N trees per unit area (per ha or 10000 m^2) of various diameters D_i and height H_i, $i = 1, \ldots, N$ and belonging to different species. The diameter D is measured at a standard height above the ground (breast height). Tree diameter increases through time according to a complicated nonlinear differential equation $\dfrac{dD(t)}{dt} = f(D(t))$, which is a function of site conditions, species-specific parameters, and other trees [12]. Biomass is estimated empirically from diameter $B(t) = aD^b(t)$, where a and b are species-specific coefficients.

Combining these equations, and taking into account all trees, biomass dynamics can be modeled in very simplified terms by one of several growth equations. For instance, the Richard's growth equation

$$\frac{dB(t)}{dt} = r\left(1 - \left(\frac{B(t)}{B_{\text{max}}}\right)^v\right) \tag{9.8}$$

where B_{max} is the maximum biomass of the stand and v is a coefficient adjusting for the shape of growth. This differential equation has a solution

$$B(t) = \frac{B_{\text{max}}}{(1 + A \exp(-rvt))^{1/v}} \tag{9.9}$$

where $A = \left(\dfrac{B_{max}}{B_0}\right)^v - 1$ is determined from the initial condition B_0. We can translate biomass to CO_2 sequestered assuming carbon content to be ~50% of dry biomass and the 44/12 ratio of molecular weight CO_2 to C, as $X(t) = 0.5\dfrac{44}{12}B(t)$, where $X(t)$ is CO_2 sequestered. There is point of inflexion on the growth curve at which the growth rate is maximum. Pass this point the growth rate decreases. The maximum growth rate is $\max(\dot{B}) = \dfrac{B_{max}rv}{(1+v)^{\frac{v+1}{v}}}$, and it occurs at a value of biomass equal to $B_i = \dfrac{B_{max}}{(1+v)^{1/v}}$. In other words, we would have to harvest the forest (extract timber for nonenergy uses) to a value B_i to sustain a max rate of sequestration $0.5\dfrac{44}{12}\max(\dot{B})$.

Example 9.18

Assume a forest with $B_{max} = 150$ t/ha; $r = 0.3\,\text{yr}^{-1}$; $v = 0.5$. How many ha of this forest kept at maximum growth rate conditions are needed to sequester all CO_2 from a 1 GW NG plant emitting 0.4 kgCO_2/kWh? Answer: First, calculate maximum sequestration rate $0.5 \times \dfrac{44}{12} \times \dfrac{B_{max}rv}{(1+v)^{\frac{v+1}{v}}} = 12.12\ tCO_2/(\text{ha yr})$. Operating at full capacity 8760 h/yr, CO_2 emissions from the plant would be 0.4 kgCO_2/kWh $\times 10^6$ kW $\times$ 8760 h/yr = $3.5 \times 10^6\ tCO_2$/yr. Now divide to obtain the area $\dfrac{3.5 \times 10^6\ tCO_2/\text{yr}}{12.12\ tCO_2/\text{yr/ha}} \approx 2.9 \times 10^5$ ha.

This example illustrates that indirect CO_2 sequestration by forest growth would require a considerable amount of land.

These calculations are facilitated by function forest.seq of renpow, which will produce rates, area, and a plot of sequestration dynamics (Figure 9.14):

```
forest <- list(t = seq(0,100),B0 = 1,Bmax = 150,nu = 0.5,r = 0.3)
plant <- list(kgCO2.kWh = 0.4,P = 10^9,C = 1.0)
forest.seq(x = forest,y = plant)
```

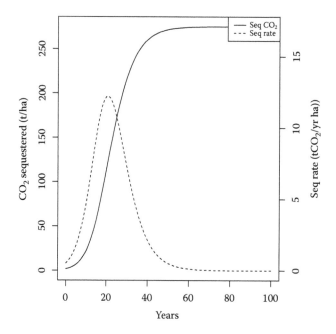

FIGURE 9.14 CO_2 sequestration dynamics by forest growth.

We can examine the output

```
> forest.seq(x=forest,y=plant)
$yr.max
[1] 21
$B.i
[1] 66.66667
$max.dCO2
[1] 12.21676
$tCO2.emiss
[1] 3504000
$area
[1] 286819.2
```

to realize that it would take 21 years (see Figure 9.14) to grow a forest from low biomass value to the required biomass of 66.66 t/ha to sequester CO_2 at the maximum rate.

9.5.9 SOLAR FUELS

The *solar fuel* approach is to emulate nature and use sunlight to derive fuels but in a shorter time scale. This idea is to produce a synthetic chemical *energy carrier* (fuel) from sunlight through photochemical reactions. In this process, solar radiation is converted to chemical energy by reduction of protons to *hydrogen* (H_2) or of CO_2 into *organic compounds*. This differs from converting sunlight to electricity using photovoltaics (PV) (Chapter 14) because the carrier can be stored and transported, very much like fossil fuels. For example, hydrogen can be used later in a fuel cell to produce electricity. Further connections with biology have been researched by using microorganisms to produce the fuel.

Solar fuels are produced by direct or indirect methods. In the latter, sunlight is converted first to an intermediate form of energy and then this intermediate carrier is converted to fuel. Direct methods do not use this intermediate step and are then more efficient.

An example of production of hydrogen (the solar fuel in this case) using indirect methods is a *photoelectrochemical cell* (PEC). An electrochemical cell generates electricity from chemical reactions, or conversely promotes chemical reactions by providing electrical power. A PEC is an electrochemical cell with a photosensitive anode. Solar radiation impinging on this anode generates electricity, which then splits water into hydrogen and oxygen. This is the same as electrolysis, but driven by sunlight and not by an external source of electric power. Because this process harvests sunlight, it is called *artificial photosynthesis*. A related PEC technology is a *dye-sensitized solar cell* or Grätzel cell, which has a photosensitive anode and an electrolyte. An example of direct methods is production of hydrogen from electrons released from a photosensitive electrode by using a *catalyst* that reduces protons to hydrogen. This more efficient process has not yet been scaled for commercial use.

Starting from CO_2, at atmospheric concentrations, researchers have attempted to produce organic compounds (such as hydrocarbons) by using photocatalysts. Success of this direct method has been relatively limited and is not yet ready for commercial use. Note that if successful, using these solar-produced hydrocarbons (solar fuel) would be the analog of using plant biomass as fuel. In the case of organic solar fuel, the process takes in atmospheric CO_2, which is then released when we burn this fuel. This then becomes a pathway in the carbon cycle with a much shorter time to return to fuel as compared to fossil fuels.

9.6 ALTERNATIVE TURBINES AND COMBUSTION ENGINES

9.6.1 MICROTURBINES

Microturbines are compact gas turbines that incorporate the major principles of a power generation plant including compressor, turbine, and generator, plus heat recuperation for increased efficiency.

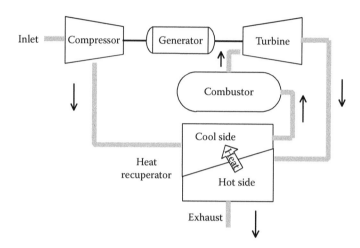

FIGURE 9.15   Microturbine.

Figure 9.15 schematically illustrates the components and workings of a microturbine. Air comes in at the inlet, is compressed, flows to the cool side of the recuperator (heat exchanger), where it increases its temperature due to heat flow from the hot side, which receives the hot expanded combustion gas from the turbine, then sending it to the exhaust. Raising the temperature of the gas before combustion increases efficiency, as we discussed in the section dealing with Brayton cycle efficiency. The compressor, turbine, and generator rotate together at fast speed.

Microturbines may operate continuously, on demand, stand-alone, or grid connected. Their capacity covers the range of a few kW to hundreds of kW. Units on the smaller side of the range are used for *distributed generation* (DG) and microgrids. We will discuss this topic in Chapters 13 and 14. For now, you can think of DG as an alternative to providing electricity by a centralized grid, with the advantage of moving generation closer to the consumer or providing electricity off-grid.

This type of compact power plant may run on a variety of fuels, such as NG, biogas (landfill and wastewater treatment plants), and propane. The heat from the exhaust can be used in applications of *combined heat and power* (CHP) where the hot air can be employed for space heating. We will discuss this topic later in this chapter.

9.6.2   EXTERNAL COMBUSTION ENGINES: STIRLING ENGINES

In an external combustion engine, the heat is supplied to a working fluid from an external source instead of internal combustion of the fuel. It is based on reciprocating, piston-driven action. This type of engine was developed with the intent of reducing the working pressure of steam engines. A Stirling engine is a *closed cycle* and *regenerative* external combustion engine, consisting of two separate spaces: one for compression (and cool) and one for expansion (and hot). External to the cylinder, these engines have a heater and a cooler. Importantly, there is a *regenerator* in between spaces, capable of thermal storage and maintaining the thermal difference between spaces (Figure 9.16, left-hand side and middle panels).

Stirling engines can be configured as *alpha*, *beta*, and *gamma* type engines [13]. An alpha type is the simplest; it has two pistons corresponding to compression and expansion spaces and two cylinders, one for each piston. The beta type is the classic Stirling engine, consisting of a power piston and a *displacer* (Figure 9.16, left-hand side). The latter displaces the working fluid at constant

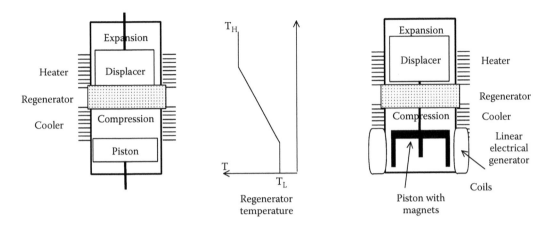

FIGURE 9.16 Stirling engine. Left: Beta type. Middle: Temperature gradient in the regenerator. Right: Free-piston and linear electrical generator.

volume between the compression and expansion spaces through the heater, regenerator, and cooler. The gamma type is based on a piston and displacer as the beta type, but housed on different cylinders. For all configurations, mechanical linkages are needed between the piston and displacer.

The beta type design has evolved into a *free piston* Stirling engine that gets rid of the mechanical linkages between piston and displacer. In this configuration, a *linear electrical generator* can be added to generate AC voltage (Figure 9.16, right-hand side). A linear generator works the same way as a rotating alternator, except that variation of the magnetic field that interacts with the coil is on a linear displacement instead of a rotational movement. The moving piston is equipped with permanent magnets, and therefore the magnetic field links with the stator coil; an emf is induced in the coils (Figure 9.16, right-hand side).

There are several advantages of Stirling engines in regard to power production versus internal combustion engines. Because of external combustion, this engine can use any fuel and take advantage of landfill gas or other resources. It is also possible to use concentrated sunlight (CSP) as the heat source (Chapter 14). Stirling engines are quiet and vibration free, being ideal for small portable power systems. They could play a role in distributed generation and micro combined heat and power.

An illustration of a Stirling engine cycle is shown in Figure 9.17. We will use this idealized cycle to explain the principle of operation. As in any of the cycles we have studied, the system returns repetitively to state 1 by four sequential paths. From 1 to 2, heat is added at constant volume V_1; this is a vertical line on the *T-s* plane and an isochoric curve on *T-s*. Then, there is isothermal expansion at the HTR temperature from 2 to 3, reaching V_3. This is followed by an isochoric path to descend in pressure and temperature from 3 to 4, staying at volume $V_3 = V_4$. Finally, the fluid returns to state 1 by an isotherm at the LTR temperature.

Example 9.19

Draw the Stirling cycle of Figure 9.17 using path.cycles assuming $V_1 = 1 l$ and $V_4 = 6\ l$, with HTR at 300°C and LTR at 30°C. Answer:

```
x <- list(TH = 300,TL = 30,V1 = 1,V4 = 6,S1 = 1,cty ="stirling')
y <- path.cycles(x,plane ="Pv',shade.cycle=TRUE)
path.cycles(x,plane ="Ts',shade.cycle = TRUE)
```

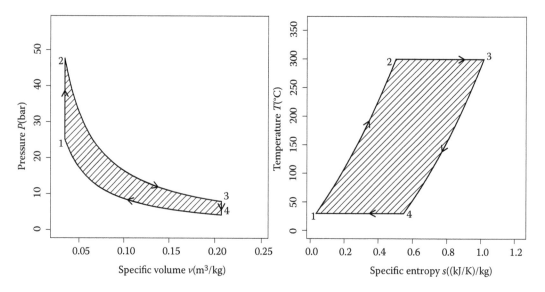

FIGURE 9.17 Stirling engine cycle in *P-v* and *T-s* planes. Left: 1–2: isochor, 30°C–> 300°C, W = 0 kJ/kg, Q = 199 kJ/kg; 2–3: isotherm, expansion 300°C–> 300°C, W = –294.6 kJ/kg, Q = 294.6 kJ/kg; 3–4: isochor, 300°C–> 30°C, W = 0 kJ/kg, Q = –199 kJ/kg; 4–1: isotherm, compression 30°C–> 30°C, W = 155.8 kJ/kg, Q = –155.8 kJ/kg. Right: 1–2: isochor, 30°C–> 300°C, W = 0 kJ/kg, Q = 199 kJ/kg; 2–3: isotherm, expansion 300°C > 300°C, W = –294.6 kJ/kg, Q = 294.6 kJ/kg; 3–4: isochor, 300°C–> 30°C, W = 0 kJ/kg, Q = –199 kJ/kg; 4–1: isotherm, compression 30°C–> 30°C, W = 155.8 kJ/kg, Q = –155.8 kJ/kg.

Let us look into more detail at this cycle and how to calculate efficiency. From state 1 to 2, we add heat at constant volume; the work is zero and the only contribution to internal energy is heat Q_{in} or

$$U_2 - U_1 = \int_1^2 \delta Q = Q_{in}.$$ This heat raises the temperature according to a ratio equal to ratio of pressure

$\dfrac{T_2}{T_1} = \dfrac{P_2 V_1/nR}{P_1 V_1/nR} = \dfrac{P_2}{P_1}$. Then, from state 2 to 3 the gas performs work while it expands at constant temperature. There is zero change in internal energy; therefore, $W = \int_2^3 PdV$ and it is the area under the curve of the upper trace of the cycle $W = nRT_2 \ln\left(\dfrac{V_3}{V_2}\right)$. From state 3 to 4, the work is zero.

Therefore, $U_4 - U_3 = -\int_3^4 \delta Q = -Q_{out}$. And as before for states 1 and 2, $\dfrac{T_4}{T_3} = \dfrac{P_4 V_4/nR}{P_3 V_4/nR} = \dfrac{P_4}{P_3}$. Finally, the working fluid will return to state 1 by isothermal compression; work is done on the fluid, and heat must be removed to avoid changing the temperature. The integral $Q_{out} = W = \int_4^1 PdV$ is the area under the curve of the lower trace $W = nRT_4 \ln\left(\dfrac{V_1}{V_4}\right)$. The area of the polygon between the upper and lower traces is equal to the net amount of work. Now, note that in this cycle $V_1 = V_2$ and $V_4 = V_3$, therefore $\dfrac{V_3}{V_2} = \dfrac{V_4}{V_1}$ and $\ln\left(\dfrac{V_3}{V_2}\right) = -\ln\left(\dfrac{V_1}{V_4}\right)$. Consequently, $\Delta U = Q_{in} - Q_{out} + nR \ln\left(\dfrac{V_1}{V_4}\right)(T_2 - T_4)$.

Example 9.20

Calculate efficiency for the Stirling cycle of Example 9.19 (Figure 9.17) assuming $V_1 = 1$ l and $V_4 = 6$ l, with HTR at 300°C and LTR at 30°C. Answer: We can query the path.cycles.summary component WQnet:

```
> path.cycles.summary(y)$WQnet
 Qin(kJ/kg) Qout(kJ/kg) Eff
1  294.6  -155.8 0.47
```

We see that ideal efficiency is 47%. This is a highly idealized cycle and such high efficiencies are not attained. We have to account for other losses, such as in the regenerator, which reduces efficiency by 20%. Increasing the temperature difference, increases efficiency; for instance, changing the hot temperature to 600°C in the aforementioned cycle yields 65% efficiency as we can see from the following:

```
> x <- list(TH = 600,TL = 30,V1 = 1,V4 = 6,S1 = 1,cty ="stirling')
> y <- path.cycles(x,plane ="Pv',shade.cycle = TRUE)
> path.cycles.summary(y)$WQnet
 Qin(kJ/kg) Qout(kJ/kg) Eff
1   448.9  -155.8 0.65
```

9.7 COMBINED HEAT AND POWER (CHP)

A recurring theme in our treatment of heat engines has been the generation of heat resulting from the thermodynamic cycle and accounting for reduction of efficiency. We have seen, except for combined-cycle plants, how heat is wasted or released to the environment. Combined heat and power (CHP) proposes that this heat be recovered and used for a variety of purposes, such as space heating, water heating, and process heating. CHP was called *co-generation* to mean that both heat and electricity were generated from the same fuel input; however, the term CHP is now more commonly used.

For CHP, the potential users of the heat should be preferably at or near the electrical power plant, since delivering heat for more than a few hundred meters is not efficient. Nevertheless, centralized or district heating is in use, for example, to heat greenhouses adjacent to the power plant. An important consideration in designing CHP applications is that the heat has utilization most of the year, for example, having use of heat besides the winter season. This is a reason for employing CHP in industrial processes and not just space heating.

One important CHP application is for "captive" power plants, that is to say, electrical generation plants associated with specific industrial complexes. In these cases, power generated is almost entirely consumed by that industrial plant, and the heat can be used for a variety of processes in the plant. An example could be a paper mill; a boiler produces steam that is used to produce electricity, and the exhaust heat is used to dry freshly made paper.

EXERCISES

9.1. Suppose NG reserves were to be increased by 50% due to new discoveries. Estimate time to depletion of NG reserves under scenarios of continued consumption at current rate and increased consumption.

9.2. Plot the Brayton cycle in both the P-v and T-s planes for one mole of air with $T_H = 1100$°C, $T_L = 30$°C and a compression ratio of 10 from 1 bar to 10 bar assuming that the initial entropy is 1 J/K. What is the work performed by the turbine and work done on the compressor? What is the heat in and heat out? What is the efficiency?

9.3. For an Otto cycle, assume the compression ratio is $r = P_2/P_1 = 10$ and temperature varies between 30°C and 900°C. What is the efficiency?

9.4. Draw an idealized Diesel cycle in the P-v and T-s planes for one mole of air using volumetric compression ratio of 12, and expansion ratio of 4, starting from 1.5 l at 30°C. State heat input and output.

9.5. Suppose we mix biomass from manure, agricultural trash, and algae in equal proportions. Their heat values are 20, 18, and 23 MJ/kg, respectively. By pyrolysis, we obtain 30% bio-oil, 20% syngas, and 40% char with HV 10, 20, and 0 MJ/kg, respectively. What is the energy recovery?

9.6. Calculate how many kg of LFG it takes to produce 1 kWh and CO_2 emissions per kWh. Assume that half of LFG, that is, CO_2 becomes part of the emissions. Assume efficiency improves to 40%.

9.7. Draw the Stirling cycle using path.cycles assuming $V_1 = 1$ l and $V_4 = 6$ l, with HTR at 250°C and LTR at 30°C. What is the efficiency?

9.8. Assume a forest with $B_{max} = 100$t/ha; $r = 0.35$yr^{-1}; $v = 0.5$. How many ha of this forest kept at maximum growth rate conditions are needed to sequester all CO_2 from a 100 MW NG plant emitting 0.5 kgCO$_2$/kWh but running only 0.8 of the year? Use renpow package.

REFERENCES

1. Fay, J.A., and D.S. Golomb, *Energy and the Environment. Scientific and Technological Principles.* Second edition. MIT-Pappalardo Series in Mechanical Engineering, M.C. Boyce and G.E. Kendall. 2012, New York: Oxford University Press. 366 pp.
2. Plummer, C., D. Carlson, and L. Hammersley, *Physical Geology.* 15th edition. 2015: McGraw-Hill Higher Education. 672 pp.
3. Wicander, R., and J.S. Monroe, *Essentials of Geology.* Fourth edition. 2006: Thomson, Brooks/Cole. 510 pp.
4. Massachusetts Geological Survey. Frequently Asked Questions About Shale Gas and Hydraulic Fracturing in Massachusetts. 2017. Accessed June 2017. Available from: https://mgs.geo.umass.edu /frequently-asked-questions-about-shale-gas-and-hydraulic-fracturing-massachusetts.
5. Stevens, P. The 'Shale Gas Revolution': Developments and Changes. 2012. Accessed June 2017. Available from: https://www.chathamhouse.org/publications/papers/view/185311.
6. International Energy Agency. Key Electricity Trends. 2017. Accessed June 2017. Available from: https://www.iea.org/publications/freepublications/publication/KeyElectricityTrends.pdf.
7. Tester, J.W., E.M. Drake, M.J. Driscoll, M.W. Golay, and W.A. Peters, *Sustainable Energy: Choosing Among Options.* Second edition. 2012: MIT Press. 1021 pp.
8. Capareda, S. *Introduction to Biomass Energy Conversions.* 2013: CRC Press. 645 pp.
9. US EPA. Landfill Methane Outreach Program (LMOP). 2017. Accessed August 2017. Available from: https://www.epa.gov/lmop.
10. Hanif, M.U., S.C. Capareda, H. Iqbal, R.O. Arazo, and M.A. Baig, Effects of Pyrolysis Temperature on Product Yields and Energy Recovery from Co-Feeding of Cotton Gin Trash, Cow Manure, and Microalgae: A Simulation Study. *PLoS ONE*, 2016. **11**(4): e0152230.
11. Vanek, F., L. Albright, and L. Angenent, *Energy Systems Engineering: Evaluation and Implementation.* Third edition. 2016: McGraw-Hill. 704 pp.
12. Acevedo, M.F. *Simulation of Ecological and Environmental Models.* 2012, Boca Raton, Florida: CRC Press, Taylor & Francis Group. 464 pp.
13. Ohio University. Stirling cycle machine analysis. 2017. Accessed August 2017. Available from: https:// www.ohio.edu/mechanical/stirling/.

10 Transformers and Three-Phase Circuits

Transformers are important components of electrical power systems, allowing to step up voltage for transmission and step it down for distribution. We start this chapter by studying magnetic circuits in order to gain a better understanding of transformers. After reviewing these devices, we continue our study of AC systems by looking into three-phase systems, that is, three AC voltages and currents at the same frequency but with different phase angles. Most electrical power systems are based on three-phase circuits. Among the major reasons for their use in power systems is that they provide constant total power. The chapter ends with a study of harmonics, or multiple values of the fundamental frequency, which can create load unbalance and detrimental current in the neutral. Harmonics can be created by a variety of reasons, including nonlinear electronic circuits to convert DC to AC, and vice versa such as rectifiers and inverters.

10.1 TRANSFORMERS

Together with electromechanical devices using magnetic fields such as generators, *transformers* are magnetic devices that are very important components of electrical power systems. In fact, the grid was made possible by use of transformers. To gain a better understanding of transformers, we first discuss magnetic circuits.

10.1.1 Magnetic Circuits

Recall from Chapter 5 that magnetic fields are the result of electric currents and moments of particles of magnetic materials. This vector field is visualized by lines of magnetic flux typically denoted by ϕ. A coil of N turns wrapped around a core of high magnetic permeability containing the flux ϕ creates an inductor. Figure 10.1 (left-hand side) shows an inductor made with an air-gapped core. Using reluctance $\mathfrak{R}$ and the magnetomotive force (mmf) given by current $\mathfrak{I} = Ni(t)$ and the number of turns, the flux was given by the analogy to Ohm's law

$$\phi(t) = \frac{\mathfrak{I}}{\mathfrak{R}} = \frac{Ni(t)}{\mathfrak{R}} \tag{10.1}$$

Inspired on this analogy, we can model interaction of current and magnetic flux as *magnetic circuits* with sources given by mmfs and circuit elements quantified by their reluctance. Figure 10.1 (right-hand side) shows how to model this example using one mmf and two reluctances in series, which corresponds to the core and the air gap. For simplicity, magnetic circuits use the same drawing symbol for a direct current (DC) source and resistor, but please realize these are magnetic elements. We can combine reluctances in series by adding them; therefore, for the magnetic circuit of Figure 10.1 (right-hand side) we have

$$\phi(t) = \frac{\mathfrak{I}}{\mathfrak{R}} = \frac{Ni(t)}{\mathfrak{R}_c + \mathfrak{R}_g} \tag{10.2}$$

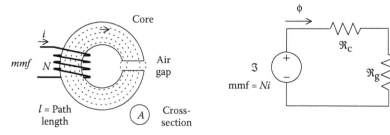

$\mathfrak{R}_c$ = Reluctance of core
$\mathfrak{R}_g$ = Reluctance of gap

FIGURE 10.1   Coil and magnetic circuit model.

We can easily see from Equation 10.2 that a much higher gap reluctance compared to the core reluctance would reduce the flux substantially with respect to what it would be with no gap. In fact, that is the purpose of the gap, namely, to reduce the flux in the core thereby avoiding saturation. Carrying this analogy further, we can calculate the mmf "drops" in each one of the reluctances using the analog of the voltage divider concept. For example, the mmf drop at the gap is $\mathfrak{F}_g = \mathfrak{F}\dfrac{\mathfrak{R}_g}{\mathfrak{R}_c + \mathfrak{R}_g}$. This can help us see that a much higher gap reluctance compared to the core would make most of the mmf be spent at the gap $\mathfrak{F}_g \simeq \mathfrak{F}$.

We can see that the permeability of the magnetic circuit is not just the permeability of the core or of the gap; instead the equivalent permeability is $\mu_e = \dfrac{l}{\mathfrak{R}A}$.

Example 10.1

Suppose we have a core of 1 cm^2 cross-sectional area, path length 6 cm, with $\mu_r = 1000$, and a 1 mm air gap. We wrap 100 turns of coil and inject 1 A of current. What is the mmf? What are the reluctances of the core and the gap? What is the magnetic flux? What is the equivalent permeability? Answer: The mmf is $\mathfrak{F} = 100 \times 1 = 100$A-turns. The reluctances of the core and gap are

$$\mathfrak{R}_c = \frac{l}{(\mu_r\mu_0)A} = \frac{6 \times 10^{-2}}{1000 \times 4\pi \times 10^{-7} \times (1 \times 10^{-4})} \simeq 0.48\,\text{MA-turns/Wb}$$

and

$$\mathfrak{R}_g = \frac{l}{(\mu_r\mu_0)A} = \frac{1 \times 10^{-3}}{1 \times 4\pi \times 10^{-7} \times (1 \times 10^{-4})} \simeq 7.96\,\text{MA-turns/Wb}$$

The equivalent reluctance is $\mathfrak{R} = \mathfrak{R}_c + \mathfrak{R}_g = 8.44$ MA-turns/Wb. Note that most of the total reluctance is due to the gap. The magnetic flux is $\phi(t) = \dfrac{\mathfrak{F}}{\mathfrak{R}} = \dfrac{100\,\text{A-turn}}{8.4\,\text{MA-turn/Wb}} = 11.9\,\mu\text{Wb}$. The equivalent permeability is $\mu_e = \dfrac{l}{\mathfrak{R}A} = \dfrac{6 \times 10^{-2}}{8.44 \times 10^6 \times 10^{-4}} = 7.11 \times 10^{-5}\,\text{Wb/(A-turns m)}$. Dividing by μ_0 we have the equivalent relative permeability $\mu_{re} = \dfrac{\mu_e}{\mu_0} = \dfrac{7.11 \times 10^{-5}}{4\pi \times 10^{-7}} \simeq 56.5$. You can see that it has been reduced substantially compared to the core by virtue of the gap.

We can expedite these calculations using functions of renpow:

```
mucore=1000; lcore=6*10^-2; Acore=1*10^-4
mugap=1; lgap=1*10^-3; Agap <- Acore
reluc <- reluctance(x=list(c(mucore,lcore,Acore),c(mugap,lgap,Agap)))
reluc$prnt
> reluc$prnt
[1] "Rel= 0.48 MA-turn/Wb" "Rel= 7.96 MA-turn/Wb"
>
```

We use this result to calculate flux:

```
> magckt <- flux(x=list(N=10,i=1,rel=rel.eq))
> magckt$prnt
[1] "mmf= 10 A-turn Rel= 8.44 MA-turn/Wb flux= 1.18 uWb L= 0.01 mH"
```

As we see from this code, the magnetic circuit calculation can include inductance. Thus, let us relate the concepts we just covered to inductance. Recall that using Faraday's law of induction, there is an induced electromotive force (emf), which, assuming no losses, is the same as the voltage across the coil:

$$v(t) = N \frac{d\phi(t)}{dt} \quad (10.3)$$

Using Equations 10.1 and 10.3 we have $v(t) = N \frac{d}{dt} \left(\frac{N}{\mathfrak{R}} i(t) \right) \simeq \frac{N^2}{\mathfrak{R}} \frac{di(t)}{dt}$ ignoring time variations

of reluctance. Recall that inductance is defined as $L = \frac{N^2}{\mathfrak{R}}$, the basis for the equation $v(t) = L \frac{di(t)}{dt}$, which we have used many times. Observe that this is an electric circuit relationship (i.e., voltage and current) and different from the reluctance, mmf, and flux of a magnetic circuit. It is very important to recall that an inductor behaves as a short circuit if there is no variation in current, $\frac{di(t)}{dt} = 0$. In essence, a steady magnetic flux $\frac{d\phi(t)}{dt} = 0$ will not induce a voltage.

10.1.2 FIELD INTENSITY AND FLUX DENSITY

For a better understanding of magnetic circuits and transformers, we need two additional field concepts: *magnetic field intensity* (H), which is the mmf per unit length or given in A-turns/m, and *magnetic flux density* (B), which can be thought of as a total field or an augmentation of the flux by permeability. Going into more detail, *Ampère's circuital law* (do not confuse this law with the same name related to force) states that the integration of magnetic field along a closed path is related to the current enclosed by that path or flowing through that path. Referring to Figure 10.1, the closed path follows the core and the current enclosed is the current $i(t)$ times the number of turns, that is to say, $Ni(t)$, which is the same as the mmf. Mathematically, using the field intensity

$$\oint H \cdot dl = Ni(t) = \mathfrak{I} \quad (10.4)$$

where dl is an infinitesimal of length along the closed path for flux. When H is constant along the path of total length l, Equation 10.4 reduces to $Hl = \mathfrak{I}$ or $H = \mathfrak{I}/l$. Intuitively, this says that the mmf establishes a field in the core. However, we have not yet invoked the magnetic properties of material that makes up the core. To include this, we bring in permeability. The field H produces alignment in the material according to its permeability. Take μH and calculate $\mu H = \frac{\mu \mathfrak{I}}{l} = \frac{\mu \mathfrak{I}}{l} \times \frac{A}{A} = \frac{\mathfrak{I}}{A} \times \frac{\mu A}{l}$.

Recognize the last term as the reciprocal of reluctance and rewrite $\mu H = \dfrac{\mathfrak{F}}{A} \dfrac{1}{\mathfrak{R}} = \dfrac{\phi}{A}$. This last term is defined as the *magnetic flux density* $B = \phi/A$ given in Wb/m² defined as tesla (T), and therefore $B = \mu H$ represents the effect of magnetic field given the permeability. Please realize that B and H are two views of the same magnetic field; H is created by the mmf, and B takes into account the material.

Example 10.2

What are the magnetic field intensity and flux density for the conditions of Example 10.1? Answer: The density is $B = \phi/A = \dfrac{1.19 \times 10^{-6}\,\text{Wb}}{1 \times 10^{-4}\,\text{m}^2} = 0.0119\,\text{T}$. The intensity is $H = \mathfrak{F}/l = \dfrac{10\,\text{A-turn}}{6 \times 10^{-2}\,\text{m}} = 166.66\,\text{A-turn m}^{-1}$. The permeability relating B and H is not just the permeability of the core or of the gap, but the equivalent permeability $\mu_e = \dfrac{B}{H} = \dfrac{0.0119\,\text{T}}{166.66\,\text{A-turn m}^{-1}} = 7.1 \times 10^{-5}\,\text{Wb/(A-turns m)}$, which corresponds to the result we obtained in Example 10.1.

10.1.3 TRANSFORMER BASICS

Armed with the concepts of the previous section, we tackle an examination of transformers and follow the presentation typically given in introductory circuit analysis textbooks (e.g., Irwin and Nelms [1]). A good exposition of transformers is also provided by Masters [2]. We keep the coil shown in Figure 10.1 with N_1 turns, voltage $v_1(t)$, and current $i_1(t)$, which we call the *primary*. As before, the mmf $N_1 i_1(t)$ creates a flux ϕ and field intensity H. Now we wind a second coil (Figure 10.2) to the core that has N_2 turns, current $i_2(t)$, and voltage $v_2(t)$. This coil is called the *secondary*. There is then a varying flux $\phi(t)$ in the core and it links the coils. This changing flux will induce a voltage $v_1(t)$ in the first coil given by Equation 10.3 $v_1(t) = N_1 \dfrac{d\phi(t)}{dt}$ and will induce a voltage $v_2(t) = N_2 \dfrac{d\phi(t)}{dt}$ on the second coil. Therefore, the two coils are magnetically coupled and share the same flux $\phi(t)$. Assume the voltages have the same sign as a response to $\dfrac{d\phi(t)}{dt}$.
    The ratio of voltages is obtained by division:

$$\frac{v_2(t)}{v_1(t)} = \frac{N_2 \dfrac{d\phi(t)}{dt}}{N_1 \dfrac{d\phi(t)}{dt}} \tag{10.5}$$

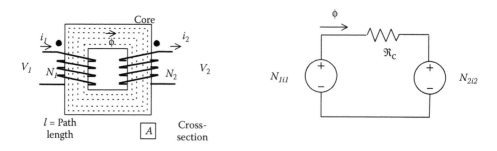

FIGURE 10.2 Transformer showing primary and secondary coils and magnetic circuit model.

As long as $\dfrac{d\phi(t)}{dt} \neq 0$, this ratio is not indeterminate and reduces to the ratio of the number of turns

of the coils, which is defined as the *turns ratio* $n = \dfrac{N_2}{N_1}$. Therefore, in the time domain and frequency

domain we have

$$\frac{v_2(t)}{v_1(t)} = \frac{N_2}{N_1} = n \quad \text{and} \quad \frac{V_2(t)}{V_1(t)} = \frac{N_2}{N_1} = n \tag{10.6}$$

Make sure you grasp the importance of the condition $\dfrac{d\phi(t)}{dt} \neq 0$. We must have varying $v_1(t)$ and $v_2(t)$ to be able to obtain the relations in 10.6. Therefore, this is only valid for alternating current (AC) or varying current, not for direct current (DC).

The current in the secondary flowing through N_2 turns of the coil generates another mmf $n_2 i_2(t)$. Applying Ampère's circuital law, the total mmf is the total current enclosed:

$$\oint H \cdot dl = N_1 i_1(t) \pm N_2 i_2(t) = \mathfrak{I} \tag{10.7}$$

Here we have used the plus-minus sign because the total current depends on the direction of current flow in the secondary, whether it contributes or opposes the primary current. Assume that the secondary current opposes the primary current, then $\oint H \cdot dl = N_1 i_1(t) - N_2 i_2(t) = \mathfrak{I}$. Assume H is constant along the path $Hl = N_1 i_1(t) - N_2 i_2(t) = \mathfrak{I}$. From a magnetic circuit point of view, we have two mmfs $N_1 i_1(t)$ and $-N_2 i_2(t)$ connected to the same reluctance $\mathfrak{R}$. To achieve a near-to-ideal transformer, we make the core with very low reluctance, such that $\mathfrak{I} = \mathfrak{R}\phi$ implies that the flux can be established by negligible mmf. In other words, $Hl = N_1 i_1(t) - N_2 i_2(t) \approx 0$. Therefore, we have $N_1 i_1(t) \approx N_2 i_2(t)$. Thus, for an ideal transformer

$$\frac{i_1(t)}{i_2(t)} \approx \frac{N_2}{N_1} = n \quad \text{and} \quad \frac{I_1(t)}{I_2(t)} \approx \frac{N_2}{N_1} = n \tag{10.8}$$

Actually, the relations in 10.6 and 10.8 can be negative $\dfrac{v_2(t)}{v_1(t)} = -n$ and $\dfrac{i_1(t)}{i_2(t)} = -n$ depending on the relative direction of current flow in the coils. A simple circuit model representation is to add a dot at either end of each coil. When the dots are aligned, the voltage and current relations are positive, whereas if the dots are on opposite ends of the coils, then the relations are negative. For the purposes of this book, we will assume that the relation is positive with aligned dots and that current in the primary flows into the coil and current in the secondary is defined as flowing out of the coil (Figure 10.3).

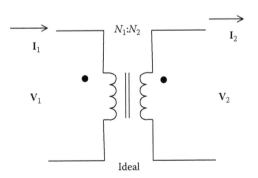

FIGURE 10.3 Transformer basic circuit representation and dot convention for signs.

Combining the relations in 10.6 and 10.8, we can derive an important conclusion about power. The power at the primary and secondary would be the product of voltage and current $p_1(t) = v_1(t)i_1(t)$ and $p_2(t) = v_2(t)i_2(t)$, respectively. Using the relations in 10.6 and 10.8, we can write the power at the secondary $p_2(t) = v_2(t)i_2(t) \approx (v_1(t)n)(i_1(t)/n) = v_1(t)i_1(t)$ and this is equal to the power at the primary. This is also valid for complex power, $\mathbf{S}_2 = \mathbf{V}_2\mathbf{I}_2^* \approx n\mathbf{V}_1\left(\dfrac{\mathbf{I}_1}{n}\right)^* = \mathbf{V}_1\mathbf{I}_1^* = \mathbf{S}_1$. In other words, assuming a near ideal transformer, the power loss from primary to secondary is very small

$$p_2(t) \approx p_1(t) \quad \text{and} \quad \mathbf{S}_2 \approx \mathbf{S}_1 \qquad (10.9)$$

Moreover, $\mathbf{S}_2 \approx \mathbf{S}_1$ is valid when separated into real and imaginary parts $P_1 \approx P_2$ and $Q_1 \approx Q_2$ since there is no change of phase. Therefore, the ideal transformer does not consume real nor reactive power. One more relation between primary and secondary corresponds to impedances. For this, we use phasors. As seen from the primary side, the impedance would be $\mathbf{Z}_1 = \dfrac{\mathbf{V}_1}{\mathbf{I}_1}$ and from the secondary $\mathbf{Z}_2 = \dfrac{\mathbf{V}_2}{\mathbf{I}_2}$. Using the relations in 10.6 and 10.8 we have $\mathbf{Z}_2 = \dfrac{\mathbf{V}_2}{\mathbf{I}_2} = \dfrac{n\mathbf{V}_1}{\mathbf{I}_1/n} = n^2\dfrac{\mathbf{V}_1}{\mathbf{I}_1} = n^2\mathbf{Z}_1$.

In summary, for an ideal transformer with turns ratio $n = \dfrac{N_2}{N_1}$ and for aligned dots, the voltage ratio goes directly with n, $\dfrac{\mathbf{V}_2}{\mathbf{V}_1} = \dfrac{N_2}{N_1} = n$; the current ratio with reciprocal of n, $\dfrac{\mathbf{I}_2}{\mathbf{I}_1} = \dfrac{N_1}{N_2} = \dfrac{1}{n}$; the impedance ratio with the square of n, $\dfrac{\mathbf{Z}_2}{\mathbf{Z}_1} = \left(\dfrac{N_2}{N_1}\right)^2 = n^2$; and power loss from primary to secondary is very low. Secondary voltage and current will change sign when dots are unaligned.

A *transformer is step-up* when $n > 1$ and therefore the voltage across the secondary is larger than at the primary $\mathbf{V}_2 > \mathbf{V}_1$. On the contrary, if $n < 1$, the voltage across the secondary is lower than at the primary $\mathbf{V}_2 < \mathbf{V}_1$, therefore we have a *step-down transformer*. The opposite is true for current; it is reduced at the secondary for a step-up transformer and augmented for a step-down transformer. Since the decrease (or increase for step-down) of current is the same as the increase (or decrease for step-down) in voltage, the power stays the same.

Example 10.3

Assume a step-up transformer from 120 V to 240 V rms. Connect a load of $R = 200\ \Omega$ to the secondary (Figure 10.4). What is the turns ratio? What is the impedance seen at the primary? What is the current entering the primary? What is the current leaving the secondary? What is the power at

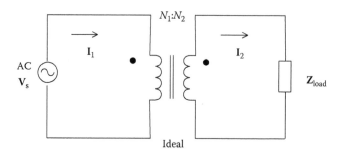

FIGURE 10.4 Transformer circuit with a simple load.

each side? Answer: Turns ratio $n = \dfrac{N_2}{N_1} = \dfrac{v_2}{v_1} = \dfrac{240}{120} = 2$ and at the primary $R_{in} = \left(\dfrac{N_1}{N_2}\right)^2 R =$

$\left(\dfrac{1}{2}\right)^2 200 = 50\ \Omega$. Current entering the primary is $i_1 = \dfrac{v_1}{R_{in}} = \dfrac{120}{50} = 2.4$ A, then reflect this current

to the secondary using the reciprocal of turns ratio $i_2 = \dfrac{N_1}{N_2} i_1 = \dfrac{1}{2} 2.4 = 1.2$ A or simply use Ohm's

law at the secondary $i_2 = \dfrac{v_2}{R} = \dfrac{240}{200} = 1.2$ A. Finally, power is the same for both sides $v_1 i_1 = 120$ V $\times$

2.4 A $= 288$ W $= 240$ V $\times 1.2$ A $= v_2 i_2$.

When considering more complicated circuits (i.e., more components connected to the transformer), we proceed similarly but have to take into account that the primary and secondary voltages (or currents) must obey Kirchhoff's voltage law (KVL) (or Kirchhoff's current law, KCL) for each of the circuits connected to these coils.

Example 10.4

Refer to Figure 10.5 with $\mathbf{V}_s = 12 \angle 30°$ V and $N_1 = 2$, $N_2 = 1$. Assume impedances connected to the transformer are $\mathbf{Z}_s = 1\ \Omega$, $\mathbf{Z}_o = 0\ \Omega$, $\mathbf{Z}_{load} = 1\ \Omega$. Is this transformer step-up or step-down? What are primary and secondary voltage, current, and power? What is the load voltage, current, and power? Answer: Turns ratio is $n = \dfrac{N_2}{N_1} = \dfrac{1}{2}$, therefore the transformer is step-down. Impedance seen from the secondary is $\mathbf{Z}_2 = \mathbf{Z}_o + \mathbf{Z}_{load} = 0 + 1 = 1\ \Omega$. Reflect secondary impedance (purely resistive) back to the primary side using $\mathbf{Z}_1 = \dfrac{\mathbf{Z}_2}{n^2} = \dfrac{1}{(1/2)^2} = 4\ \Omega$, which is also purely resistive. The impedance seen from the source is $\mathbf{Z}_s + \mathbf{Z}_1 = 1 + 4 = 5\ \Omega$, again purely resistive. Calculate primary current $\mathbf{I}_1 = \dfrac{\mathbf{V}_s}{5} = \dfrac{12}{5} \angle 30°$ A $= 2.4 \angle 30°$ A. This current times $\mathbf{Z}_1$ determines voltage drop across primary $\mathbf{V}_1 = 4 \times \mathbf{I}_1 = 4 \times 2.4 \angle 30°$ V $= 9.6 \angle 30°$ V. Now reflect primary current to the secondary using $\mathbf{I}_2 = \dfrac{1}{n}\mathbf{I}_1 = 2 \times 2.4 \angle 30°$ A $= 4.8 \angle 30°$ A and primary voltage using $\mathbf{V}_2 = n\mathbf{V}_1 = (1/2) \times 9.6 \angle 30°$ V $= 4.8 \angle 30°$ V. Complex power at primary $\mathbf{V}_1 \mathbf{I}_1^* = 9.6 \angle 30°$ V $\times 2.4 \angle (-30°)$ A $= 23.04$ W, which is all real power. Note that power supplied by the source is $\mathbf{V}_s \mathbf{I}_1^* = 10 \angle 30°$ V $\times 2.4 \angle (-30°)$ A $= 24$ W, also all real power. So there is a loss of $24 - 23.04 = 0.96$ W at $\mathbf{Z}_s$. The current through the load is the same as $\mathbf{I}_2$. Then voltage across the load is $\mathbf{V}_{load} = \mathbf{Z}_{load}\mathbf{I}_2 = 1\ \Omega \times 4.8 \angle 30°$ A $= 4.8 \angle 30°$ V, and real power consumed by the load is $P = I_{load}^2 R_{load} = 4.8^2 \times 1 = 23.04$ W.

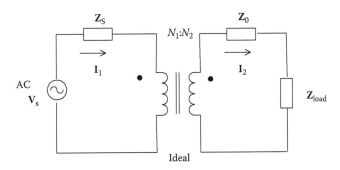

FIGURE 10.5   Transformer circuit showing primary and secondary circuits.

We can use function xformer.ckt of renpow to perform these calculations:

```
> x <- list(N=c(2,1),Vs.p=c(12,30),Zs.r=c(1,0),Zo.r=c(0,0),Zl.r=c(1,0))
> xf <- xformer.ckt(x)
Source: Vs.p=c(12, 30), Is.p=c(2.4, 30), Ss.p=c(28.8, 0), Ss.r=c(28.8, 0)
Transformer: n=0.5, Z2.r=c(1, 0), Z1.r=c(4, 0), I1.p=c(2.4, 30), V1.p=c(9.6,
30), I2.p=c(4.8, 30), V2.p=c(4.8, 30), S1.p=c(23.04, 0), S1.r=c(23.04, 0)
Load: V1.p=c(4.8, 30), I1.p=c(4.8, 30), S1.p=c(23.04, 0), S1.r=c(23.04, 0)
>
```

In a more general case, there will be non-zero reactance for $\mathbf{Z}_s$, $\mathbf{Z}_o$, and $\mathbf{Z}_{load}$, therefore there are phase differences between voltages and currents, as well as reactive power.

10.1.4 ROLE OF TRANSFORMERS IN TRANSMISSION AND DISTRIBUTION

In Chapter 11, we will discuss the typical grid or electric power system. As we will see, the major concept is to step up the voltage for transmission, and step it down for distribution. This is possible by using transformers. Take a simplified example to step up from a generator at 24 kV to 138 kV for transmission, step down for subtransmission (e.g., 4.8 kV), and step down to a consumer residential voltage (e.g., 240 V). Figure 10.6 illustrates this overly simplified electrical power system to focus on the roles played by the transformers. From generation to transmission, the step up has turns ratio greater than 1. From transmission to subtransmission, then to consumer, the voltage is stepped down twice. Note that to supply a consumer with power $\mathbf{S}_c$, it will require $\mathbf{I}_c$. The current draw seen at the primary of the step-down transformer $\mathbf{I}_s$ would be lower, this current is further reduced at the primary of the next transformer, and finally increased at the primary of the step-up for the generator. Neglecting losses, the power stays the same at all stages.

Example 10.5

Consider the system in Figure 10.6 and, assuming negligible line impedances, neglect all losses and assume perfect *pf*. Calculate currents and power at all stages to supply a consumer with 24 kVA of power. Answer: From generation to transmission $\dfrac{N_T}{N_g} = \dfrac{138}{24} \simeq 5.75$. From transmission to subtransmission $\dfrac{N_S}{N_T} = \dfrac{4.8}{138} \simeq \dfrac{1}{28.75}$. Then to the consumer $\dfrac{N_c}{N_s} = \dfrac{240}{4800} = \dfrac{1}{20}$. Note that to supply a consumer with about 24 kVA, it will require $I_c = 100$ A, assuming perfect *pf*. The current draw seen at the primary of the step-down transformer *Is* would be lower *Is* = 100/20 = 5 A. This current is further reduced at the primary of the next transformer $I_T = 5/28.75 = 0.173$ A, and finally increased at the primary of the step-up for the generator $I_T = 5.75 \times 0.173 \approx 1$ A. Neglecting losses, the power stays the same at all stages: $P_G = 24$ kV $\times 1$ A = 24 kVA; $P_T = 138$ kV $\times 0.173$ A $\simeq 24$ kVA; $P_S = 4.8$ kV $\times 5$ A $\simeq 24$ kVA; $P_C = 240$ V $\times 100$ A $\simeq 24$ kVA.

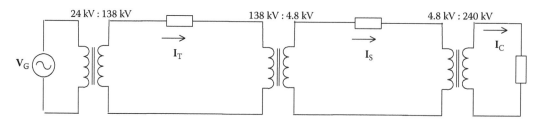

FIGURE 10.6 Role of transformers in transmission and distribution.

10.1.5 SINGLE-PHASE THREE-WIRE

Figure 10.7 depicts typical residential electrical power system drops (left-hand side) and circuits (right-hand side) [2]. A *single-phase three-wire* drop obtained by a center-tap on the secondary of the utility transformer on the pole, goes to the residence's meter, and from there to the breaker panel. The secondary's center-tap provides a "neutral" connection. The single-phase three-wire and its neutral should not be confused with three-phase and its neutral that is described in Chapter 4. High voltage feeds the primary and it is stepped down at the secondary to 240 V. Assuming $V_1 = 4.8$ kV, the turns-ratio of this step-down transformer is $n = \dfrac{N_2}{N_1} = \dfrac{240}{4800} = \dfrac{1}{20}$. Voltage $\mathbf{V}_{2B}$ is $180°$ out of phase with respect to $\mathbf{V}_{2A}$. Thus, the positive part of the wave $\mathbf{V}_{2A}$ (with respect to ground) corresponds to the negative part for $\mathbf{V}_{2B}$ (Figure 10.8). At the breaker panel, we have 120 V circuits (e.g., for loads $\mathbf{Z}_1$ and $\mathbf{Z}_2$) and 240 V circuits (e.g., for load $\mathbf{Z}_3$). The 120 V circuits are taken from $\mathbf{V}_{2A}$ and $\mathbf{V}_{2B}$. Each circuit feeds a load with a return by a "neutral" wire to ground. Because a load such as $\mathbf{Z}_3$ has double the value of voltage, its load current would be half for the same power, which means that it can be serviced with wires of lower ampacity. Therefore, larger loads (electric stove and oven, clothes dryer, and air-conditioning) are wired to 240 V circuits and lighter loads (light bulbs, small appliances) to the 120 V circuits.

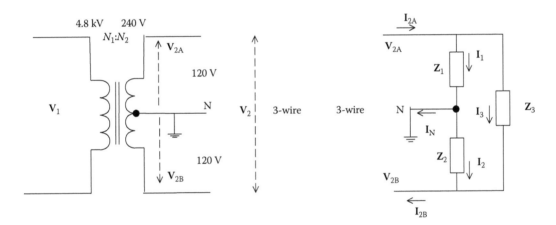

FIGURE 10.7 Single-phase three-wire power. Utility transformer and load circuits.

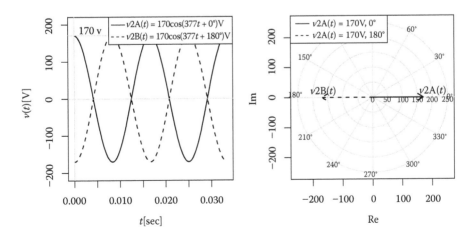

FIGURE 10.8 Single-phase three-wire. Phase inversion in time-domain and phasor plot. Amplitude is $V_{rms}\sqrt{2}$.

Example 10.6

Assume purely resistive impedances Z_1, Z_2, and Z_3. Assume real power demands for each load are 1.2, 2.4, and 4.8 kW, respectively. Calculate currents drawn by each load. Calculate total current drawn from each of the three-wire drop. What is this current if the loads are "balanced," i.e., both loads Z_1 and Z_2 are the same. Answer: Purely resistive loads mean $pf = 1$, thus currents are calculated using $I = \dfrac{P}{V}$. For each load we have $I_1 = \dfrac{1200\,W}{120\,V} = 10\,A$, $I_2 = \dfrac{2400\,W}{120\,V} = 20\,A$, and $I_3 = \dfrac{4800\,W}{240\,V} = 20\,A$. Note that the current for the heavy load of 4.8 kW remains 20 A, because the voltage is doubled. Otherwise, it would have been 40 A requiring heavier gauge wire. Now use KCL to calculate $I_{2A} = I_1 + I_3 = 10 + 20 = 30\,A$, $I_{2B} = I_2 + I_3 = 20 + 20 = 40\,A$, and $I_N = I_1 - I_2 = 10 - 20 = -10\,A$. This last current occurs because the loads are not balanced. If loads Z_1 and Z_2 were the same, then $I_1 = I_2$, $I_N = I_1 - I_2 = 0\,A$, and $I_{2A} = I_{2B}$.

10.1.6 Transformers and Converters

Recall from Chapter 5 that a full wave rectifier takes AC of a given amplitude V_m and converts it to a DC consisting of an average value close to V_m, plus a ripple that is minimized by increasing the capacitance of the filter (see Figures 5.28 and 5.31). Also, recall from that chapter that a switching power supply is made by stepping down the output of the rectifier using a DC-DC buck converter (see Figure 5.33).

An alternative design for a power supply is to use a transformer to step down the AC before it is input to the rectifier (see Figure 10.9). This is how most power supplies were made before the advent of modern transistorized switches that have allowed DC-DC converters. A four-diode bridge is used in the design shown on the left-hand side of Figure 10.9, whereas an alternative design is shown on the right-hand side of Figure 10.9 and is based on a center-tapped transformer similar to the one we used in the previous section to obtain a three-wire circuit.

A combination of transformer and rectifier in one end, plus a combination of inverter and transformer at the other end, is an alternative to make a DC link, while stepping up and down using transformers and not DC-DC converters (Figure 10.10).

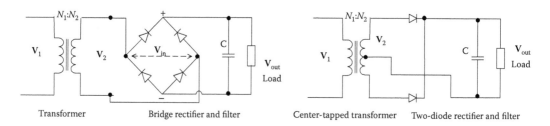

FIGURE 10.9 Power supply using a transformer. Center-tapped transformer and rectifier.

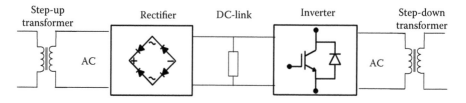

FIGURE 10.10 Transformers at each end of a DC link.

10.1.7 ENVIRONMENTAL IMPACTS OF POWER TRANSFORMERS

A power transformer contained in an enclosure is immersed in a dielectric fluid for cooling and electrical insulation. Petroleum-based mineral oil has been used for this purpose because of its good thermal and dielectric performance [3]. However, when a transformer fails, there is potential for explosion and fire, with potential risks to life, property, and mineral oil spills.

To reduce the risk of fire, PCBs (polychlorinated biphenyls) were used in transformers for several decades (since the late 1920s until the late 1970s) because of their low flammability properties. However, PCBs were found to be hazardous for human health and the environment, therefore manufacturing of PCBs was stopped in the United States as of 1977. Since PCBs accumulated in the environment, it is still found in aquatic ecosystems [4].

To replace PCBs, other fluids were researched and employed, such as silicone oils and synthetic esters. Nowadays, attention has turned to the potential of natural esters for transformer fluids, which are biodegradable and manufactured from crops such as rapeseed (canola) and corn (see Chapter 9). The use of these vegetable oil–based transformer fluids is being investigated, in particular their performance in reducing fire risks, impacts of spills, and electrical equipment performance [3,5].

10.2 THREE-PHASE POWER SYSTEMS

Most electrical power systems are based on three-phase circuits, that is, three AC voltages and currents at the same frequency but different phases. Among the major reasons for this are constant or uniform total power and lower costs. We discuss details in this section. Basic descriptions of three-phase systems are provided in textbooks of fundamental circuit analysis [1,6] and textbooks on electric power systems [7].

10.2.1 THREE-PHASE GENERATOR

Figure 10.11 illustrates a three-phase generator. Please refer back to Chapter 2 when we studied that the armature coil picks up an induced sinusoidal voltage as the rotor spins. The induced voltage changes polarity as the magnetic poles (N and S) face one side or the other side of the coil. In the

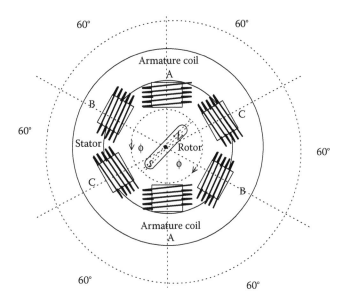

FIGURE 10.11 Three-phase generator.

three-phase generator, we have added two more coils separated by 60°. This way there is a phase angle of 120° between each pair of voltages induced. A circuit model is shown in Figure 10.12, which shows the coils (left-hand side) and the voltage induced in each as a single-phase generator (right-hand side). Each single-phase generator is connected between one of the phase nodes (a, b, or c) and the common node to all coils or neutral (n). This circuit configuration is called a *wye* (for its resemblance to the letter Y) or a *star*.

This particular one shown in Figure 10.12 is a letter Y flipped upside down (Figure 10.13) and it is still called a wye. A Y or star circuit can be converted to a *delta* circuit, named because of resemblance to the Greek letter Δ. A regular Y will convert to a Δ flipped upside down, whereas a flipped Y converts to a regular Δ. These configurations are also used for loads; substitute the circles of sources by boxes of impedances (Figure 10.13).

Recall that a convenient notation for voltage across a pair of nodes is to use two symbols, one for each node. For instance, to denote voltage of node a with respect to node b in the time domain, we can write $v_{ab}(t)$ or as a phasor $\mathbf{V}_{ab} \angle \theta$. Using this notation, v_{ab} is positive when a is at higher potential than b (half cycle of sine AC), and negative when a is at lower potential than b (the other half cycle of a sine wave AC). In other words, $v_{ba}(t) = -v_{ab}(t)$, which is the same as a phasor voltage at 180° out of phase $\mathbf{V}_{ba} = \mathbf{V}_{ab} \angle 180° = -\mathbf{V}_{ab}$.

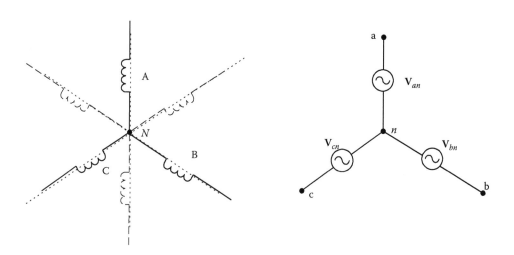

FIGURE 10.12   Circuit model of three-phase generator connected in a wye configuration.

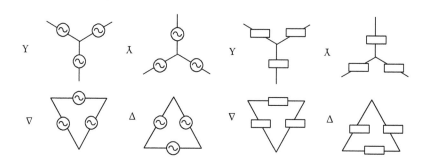

FIGURE 10.13   Wye and delta (Y and Δ) circuits. Used for three-phase sources and loads.

With the notation shown in Figure 10.12, the induced voltages are

$$\mathbf{V}_{an} = \sqrt{2}V_p \angle 0°, \quad \mathbf{V}_{bn} = \sqrt{2}V_p \angle -120°, \quad \mathbf{V}_{cn} = \sqrt{2}V_p \angle 120° \qquad (10.10)$$

where V_p is the rms of phase voltage. In the time domain $v_{an}(t) = \sqrt{2}V_p \cos(\omega t)$, $v_{bn}(t) = \sqrt{2}V_p \cos(\omega t - 120°)$, and $v_{cn}(t) = \sqrt{2}V_p \cos(\omega t + 120°)$. These voltages are in a, b, c sequence; or *positive phase sequence*. This means that a leads, b follows a by 120°, and c follows b by 120°.

Example 10.7

Use $V_p = 120$ V and functions ac.plot and phasor.plot of renpow to illustrate three-phase voltages.
Answer: $V_p = 120$ V corresponds to a magnitude of $Vm = \sqrt{2} \times 120 \simeq 170$ V

```
Vp <- 120; Vm <- 170; Van.p <-c(Vm,0); Vbn.p <- c(Vm,-120); Vcn.p <- c(Vm,120)
x <- list(Van.p,Vbn.p,Vcn.p); v3.t <- waves(x)
v3t.lab <- c("van(t)","vbn(t)","vcn(t)"); v.units <- rep("V",3)
ac.plot(v3.t,v3t.lab,v.units)
v3p.lab <- c("Van","Vbn","Vcn")
phasor.plot(x,v3p.lab,v.units,lty.p=rep(1,3))
```

The time-domain results are shown in Figure 10.14, whereas the phasors are illustrated in Figure 10.15.

The three-phase voltages add up to zero at all values of t. That is to say, $\mathbf{V}_{an} + \mathbf{V}_{bn} + \mathbf{V}_{cn} = 0$. We can see this in the time domain $v_{an}(t) + v_{bn}(t) + v_{cn}(t) = 0$ by using the trigonometric identity $\frac{1}{2} \times [\cos(x+y) + \cos(x-y)] = \cos x \cos y$ in $v_{bn}(t) + v_{cn}(t)$

$$v_{bn}(t) + v_{cn}(t) = \sqrt{2}V_p(\cos(\omega t - 120°) + \cos(\omega t + 120°))$$
$$= 2\sqrt{2}V_p \cos \omega t \cos(120°) = -\sqrt{2}V_p \cos \omega t$$

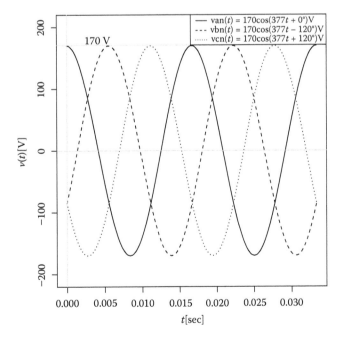

FIGURE 10.14 Time-domain plots for phase voltages. Three-phase sequence *abc*.

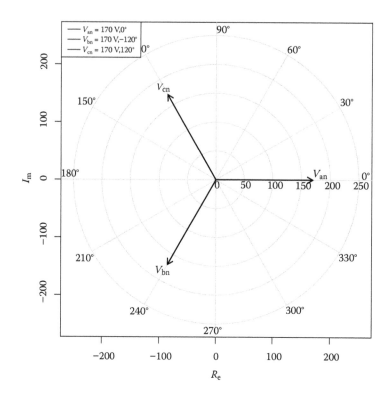

FIGURE 10.15   Phasor diagram for phase voltages. Three-phase sequence *abc*.

and then $v_{an}(t) + v_{bn}(t) + v_{cn}(t) = \sqrt{2}V_p[\cos \omega t - \cos \omega t] = 0$. In the frequency domain, convert the phasors to rectangular form

$$\mathbf{V}_{an} = \sqrt{2}V_p(\cos 0° + j \sin 0°)$$

$$\mathbf{V}_{bn} = \sqrt{2}V_p(\cos (-120°) + j \sin (-120°))$$

$$\mathbf{V}_{cn} = \sqrt{2}V_p(\cos (120°) + j \sin (120°))$$

and therefore,

$$\mathbf{V}_{an} = \sqrt{2}V_p + j0$$

$$\mathbf{V}_{bn} = \sqrt{2}V_p\left(-\frac{1}{2} - j\frac{\sqrt{3}}{2}\right)$$

$$\mathbf{V}_{cn} = \sqrt{2}V_p\left(-\frac{1}{2} + j\frac{\sqrt{3}}{2}\right)$$

(10.11)

We can see that $\mathbf{V}_{bn}$ and $\mathbf{V}_{cn}$ are complex conjugates; their imaginary part $\pm j\frac{\sqrt{3}}{2}$ cancels out, and their real part $-\frac{1}{2}$ is negative and half of the real part of $\mathbf{V}_{an}$, so they all cancel out, and the sum is zero:

$$\mathbf{V}_{an} + \mathbf{V}_{bn} + \mathbf{V}_{cn} = 0$$

(10.12)

Another insight about this comes from the identity

$$\cos(x + 0°) + \cos(x - 120°) + \cos(x + 120°) = 0 \qquad (10.13)$$

We can see that for $x = 0°$. This is the sum of the real parts of the complex voltages $\mathbf{V}_{an}$, $\mathbf{V}_{bn}$, $\mathbf{V}_{cn}$. This indicates that for any reference angle x as long as the angles are $0°$, $-120°$, $120°$, the voltages add to zero.

This type of three-phase source, where all three voltages add to zero is called *balanced*. A requirement for balanced source is that all three rms V_p voltages are equal. Note how in the time domain (Figure 10.14) the three voltages fill out the entire time interval, and at any particular time instant, the three voltages add up to a constant value, which is zero.

Example 10.8

Using renpow functions, calculate the sum of voltages in the time domain and frequency domain. Confirm that it is zero. Answer: To confirm this we can calculate out the three waves and add them up at all values of t:

```
y <- waves(v3p) ; nt <- length(y$y[,1])
y0 <- array()
for(i in 1:nt) y0[i] <- sum(y$y[i,])
round(y0,4)
```

The result is

```
> round(y0,4)
 [1] 0 0 0 0 0 0 0 0 0 0 0 0 0 0 0 0 0 0 0 0 0 0 0 0 0 0 0 0 0 0 0 0 0 0 0 0 0
[38] 0 0 0 0 0 0 0 0 0 0 0 0 0 0 0 0 0 0 0 0 0 0 0 0 0 0 0 0 0 0 0 0 0 0 0 0 0
... several lines of zeroes
>
```

Also, note that in the complex plane when we add up the vectors we get zero. To see this we can convert all three phasors to rectangular and add the vectors:

```
Van.r <- recta(Van.p)
Vbn.r <- recta(Vbn.p)
Vcn.r <- recta(Vcn.p)
> Van.r;Vbn.r;Vcn.r
[1] 170   0
[1] -85.000 -147.224
[1] -85.000 147.224
> Van.r+Vbn.r+Vcn.r
[1] 0 0
```

We can see that $\mathbf{V}_{bn}$ and $\mathbf{V}_{cn}$ are complex conjugates; their imaginary part $\left(\pm\dfrac{\sqrt{3}}{2}170 = \pm147.224\right)$ cancels out, and that their real part $\left(-\dfrac{1}{2}170 = -85\right)$ is negative and half of the real part of $\mathbf{V}_{an}$ $+170$, so they all cancel out, and the sum is zero.

10.2.2 WYE-WYE OR Y-Y CONNECTION

Let us connect the three-phase wye source in such a way that there is a load impedance connected to each phase. The load is then also configured as a wye. The overall circuit (Figure 10.16) thus obtained is a *wye-wye* or Y-Y three-phase circuit. A simplified schematic is shown in Figure 10.17, where we have moved the common node neutral (n) to the bottom of the circuit. This is still a wye circuit, for both source and load, but used often since it is just easier to draw and visualize.

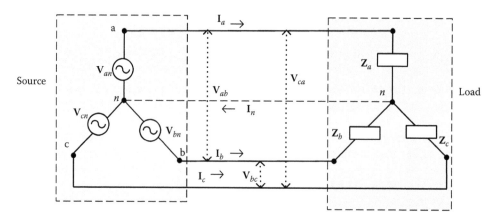

FIGURE 10.16 Wye-wye three-phase voltages sequence *abc*. Phase voltages and line-to-line voltages.

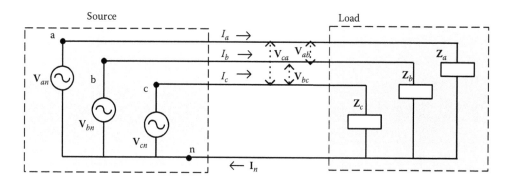

FIGURE 10.17 Wye-wye simplified schematic. Three-phase voltages and currents.

When all three impedances are the same $\mathbf{Z}_a = \mathbf{Z}_b = \mathbf{Z}_c = \mathbf{Z}$ and the sources comply with Equation 10.10, then we have a balanced circuit. We can also see that the *line currents* from source to load are $\mathbf{I}_a = \dfrac{\mathbf{V}_{an}}{\mathbf{Z}}$, $\mathbf{I}_b = \dfrac{\mathbf{V}_{bn}}{\mathbf{Z}}$, $\mathbf{I}_c = \dfrac{\mathbf{V}_{cn}}{\mathbf{Z}}$ and they are the same as the phase currents, that is, the current through the source of each phase. The sum of all currents in the neutral wire $\mathbf{I}_n$ will be $\mathbf{I}_n = \dfrac{1}{\mathbf{Z}}(\mathbf{V}_{an} + \mathbf{V}_{bn} + \mathbf{V}_{cn}) = 0$ because of Equation 10.12.

We can calculate each line current. For instance, current $\mathbf{I}_a = \dfrac{\sqrt{2}V_p \angle 0°}{\mathbf{Z}}$ assuming $\mathbf{Z} = Z \angle \theta = Z \angle (\theta_v - \theta_i)$ we have $\mathbf{I}_a = \dfrac{\sqrt{2}V_p \angle 0°}{Z \angle \theta} = \sqrt{2}I_p \angle -\theta$, where $I_p = \dfrac{V_p}{Z}$ is the rms of current in each phase. In summary for all phases $\mathbf{I}_a = \sqrt{2}I_p \angle -\theta$, $\mathbf{I}_b = \sqrt{2}I_p \angle (-120° - \theta)$, $\mathbf{I}_c = \sqrt{2}I_p \angle (120° - \theta)$.

Example 10.9

Assume a three-phase wye-wye connected balanced circuit (abc sequence) with $V_p = 120$ V and the same load in each phase $\mathbf{Z} = 1.7 \angle 20°$. What are the currents? Draw a phasor diagram of voltages and currents. Answer: The magnitude of the currents is $I_m = \dfrac{\sqrt{2}V_p}{Z} = \dfrac{170}{1.7} = 100$A.

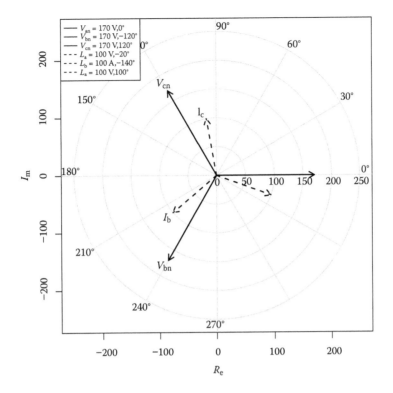

FIGURE 10.18 Phasor diagram of phase voltages and currents.

The angles are 0°, −120°,120° for voltages and −20°,−140°,100° for currents. To draw a phasor diagram use

```
Vp <- 120; Vm<-170; Van.p <-c(Vm,0); Vbn.p <- c(Vm,-120); Vcn.p <- c(Vm,120)
Z.p <- c(1.7,20)
Ia.p <- div.polar(Van.p,Z.p)
Ib.p <- div.polar(Vbn.p,Z.p)
Ic.p <- div.polar(Vcn.p,Z.p)
v3pl <- list(Van.p,Vbn.p,Vcn.p,Ia.p,Ib.p,Ic.p)
v.units <- c(rep("V",3),rep("A",3))
v3pl.lab <- c("Van","Vbn","Vcn","Ia","Ib","Ic")
phasor.plot(v3pl,v3pl.lab,v.units,lty.p=c(rep(1,3),rep(2,3)))
```

The result is shown in Figure 10.18. We can see how each current lags the corresponding voltage by 20° and it is reduced in magnitude according to impedance.

10.2.3 POWER

To calculate power, we can treat each phase independently of the others and then sum the results. Each phase is calculated as for single-phase circuits (Chapter 8). Taking into account that all impedances are the same, and the angles are 0°,−120°,120°, the instantaneous power in each phase is

$$p_a(t) = v_{an}(t)i_a(t) = V_pI_p[\cos\theta + \cos(2\omega t + 0° - \theta)]$$

$$p_b(t) = v_{bn}(t)i_b(t) = V_pI_p[\cos\theta + \cos(2\omega t - 120° - \theta)] \qquad (10.14)$$

$$p_c(t) = v_{cn}(t)i_c(t) = V_pI_p[\cos\theta + \cos(2\omega t + 120° - \theta)]$$

We add all three equations $p(t) = p_a(t) + p_b(t) + p_c(t)$ to obtain total instantaneous power $p(t) =$

$$V_p I_p \left[3 \cos \theta + \underbrace{\cos(2\omega t - \theta + 0°) + \cos(2\omega t - \theta - 120°) + \cos(2\omega t - \theta + 120°)}_{\text{use identity}} \right].$$ Now, use the

identity given by Equation 10.13 with $x = 2\omega t - \theta$ to obtain that the second term within the brackets is zero. Therefore,

$$p(t) = 3V_p I_p \cos \theta \qquad (10.15)$$

This is a very important result. It says that the *three-phase instantaneous power is constant*! It does not vary with time as in a single-phase AC system, which oscillates at twice the frequency. In other words, power delivery is smooth instead of varying. This is one important reason to use three-phase in electric power systems.

Furthermore, the three-phase balanced instantaneous power is three times the *real power of each phase* $P_p = V_p I_p \cos \theta$, and the term $pf = \cos \theta$ is the power factor of each phase. The apparent power of each phase is $S_p = V_p I_p$, and the reactive power $Q_p = V_p I_p \sin \theta$. Therefore, for each phase we can correct the power factor the same way as we did for single phase (Chapter 8).

Because the three-phase balanced instantaneous power is constant, its average is also constant and has the same value. This is the real power of the three-phase system

$$P_{3\phi} = 3V_p I_p \cos \theta \qquad (10.16)$$

and therefore, $3V_p I_p$ plays the role of the apparent power of the three-phase system

$$S_{3\phi} = 3V_p I_p \qquad (10.17)$$

The power triangle is completed by writing

$$Q_{3\phi} = S_{3\phi} \sin \theta \qquad (10.18)$$

Example 10.10

Assume a three-phase balanced circuit abc sequence with $V_p = 120$ V and the same load in each phase $Z = 1.7 \angle 20°$. What is the power factor of each phase? What is the average power in each phase? What is the three-phase instantaneous power? What is the three-phase average power?

Draw time-domain plots of phase power and total power. Answer: The rms of the currents is $I_p = \dfrac{V_p}{Z} = \dfrac{120}{1.7} = 70.59$A. The power factor is $pf = \cos 20° = 0.94$. The average power of each phase is $P_p = V_p I_p \cos \theta = 120 \times 70.59 \times 0.94 \approx 7.96$ kW, and the three-phase instantaneous power is $p(t) = 3 \times 7.96 = 23.89$ kW. The average three-phase power is $P_{3\phi} = 3V_p I_p \cos \theta = 23.89$ kW. To draw plots, use Van.p, Vbn.p, Vcn.p, and I.p as in the previous example, then

```
pa <- inst.pow.calc(list(Van.p,Ia.p))$p
pb <- inst.pow.calc(list(Vbn.p,Ib.p))$p
pc <- inst.pow.calc(list(Vcn.p,Ic.p))$p
ptot <- pa+pb+pc
y <- cbind(pa,pb,pc,ptot)
t <- inst.pow.calc(list(Van.p,Ia.p))$t
matplot(t,y,type="l",col=1,lty=1:4,ylim=c(0,1.3*max(ptot)),lwd=1.8,
    xlab="Time[sec]",ylab="Power[kW]")
```

```
legend("topright",leg=c("pa(t)","pb(t)","pc(t)","ptot"),col=1,lty=1:4)
text(0.1*max(t),1.05*max(ptot),"Ptot")
abline(h=mean(pa),col=1,lty=5)
text(0.1*max(t),1.05*mean(pa),"Pavg-phase")
```

The result is shown in Figure 10.19. We can see how each phase has a varying power and the total power is constant (~24 kW), and three times the average power in each phase (~8 kW).

10.2.4   LINE-TO-LINE VOLTAGES

Let us revisit the Y-Y circuit (Figures 10.16 and 10.17). The *phase* voltages are also called the *line-to-neutral* voltages. Now consider the voltages across each pair of nodes; these are called *line-to-line* voltages or *line* voltages for short. For instance, $\mathbf{V}_{ab}$ is a line-to-line voltage, whereas $\mathbf{V}_{an}$ and $\mathbf{V}_{an}$ are line-to-neutral voltages. We have three line voltages $\mathbf{V}_{ab}$, $\mathbf{V}_{bc}$, $\mathbf{V}_{ca}$. Quick inspection of the circuit reminds us that the currents $\mathbf{I}_a$, $\mathbf{I}_b$, and $\mathbf{I}_c$ correspond to the phases as well as the lines. The double subscript notation allows distinguishing between line and phase voltages in three-phase systems.

Let us calculate the line-to-line or line voltages assuming balanced conditions. This is easily done using KVL and phasors in rectangular form given in Equation 10.11. For instance, the difference between voltages of phases a and b, $\mathbf{V}_{an}$ and $\mathbf{V}_{bn}$, would be $\mathbf{V}_{ab} = \mathbf{V}_{an} - \mathbf{V}_{bn} = \sqrt{2}V_p(1+j0) - \sqrt{2}V_p(-\frac{1}{2}-j\frac{\sqrt{3}}{2})$, which simplifies to $\mathbf{V}_{ab} = \sqrt{2}V_p(\frac{3}{2}+j\frac{\sqrt{3}}{2}) = \sqrt{2}V_p\sqrt{3}(\frac{\sqrt{3}}{2}+j\frac{1}{2})$. In polar

$$\mathbf{V}_{ab} = \sqrt{2}V_p\sqrt{3}\angle 30° \qquad (10.19)$$

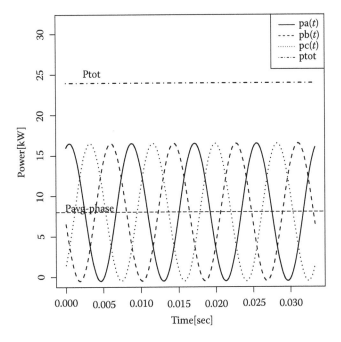

FIGURE 10.19   Instantaneous power per phase and total three-phase power (constant).

Note that this line-to-line voltage has a larger amplitude (factor of $\sqrt{3}$) and a phase of 30° with respect to voltage at phase a $\mathbf{V}_{an} = \sqrt{2}V_p \angle 0°$.

We can repeat this process for phases b and c, that is to say, the difference between $\mathbf{V}_{bn}$ and $\mathbf{V}_{cn}$, and for phases c and a, that is to say, difference between $\mathbf{V}_{cn}$ and $\mathbf{V}_{an}$. To make a long story short we would obtain

$$\mathbf{V}_{bc} = \sqrt{2}V_p\sqrt{3} \angle (-90°) \tag{10.20}$$

which is 30° out of phase with respect to $\mathbf{V}_{bn} = \sqrt{2}V_p \angle -120°$, and

$$\mathbf{V}_{ca} = \sqrt{2}V_p\sqrt{3} \angle (-210°) \tag{10.21}$$

which is 30° out of phase with respect to $\mathbf{V}_{cn} = \sqrt{2}V_p \angle 120°$.

In summary, defining $V_l = \sqrt{3}V_p$ as rms of line voltage, we have

$$\mathbf{V}_{ab} = \sqrt{2}V_l \angle 30°, \ \mathbf{V}_{bc} = \sqrt{2}V_l \angle -90°, \ \mathbf{V}_{ca} = \sqrt{2}V_l \angle -210° \tag{10.22}$$

Line voltages are out of phase by 30° with respect to phase voltages and magnified by $\sqrt{3}$. The phasor diagrams for phase and line voltages are in Figure 10.20, where it is easy to visualize the relationships between phase and line voltages. This plot was obtained using renpow as follows:

```
Vp <- 120; Vm <- 170; Van.p <-c(Vm,0); Vbn.p <- c(Vm,-120); Vcn.p <- c(Vm,120)
Vab.p <- Van.p*c(sqrt(3),1)+ c(0,30)
```

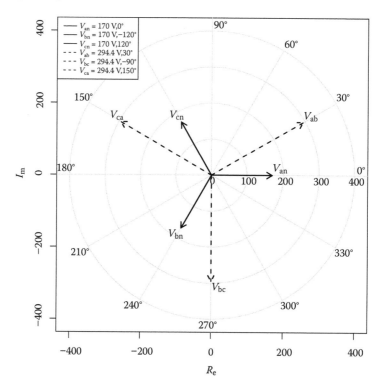

FIGURE 10.20 Wye-wye three-phase voltages sequence *abc* phasors: phase voltages and line-to-line voltages.

```
Vbc.p <- Vbn.p*c(sqrt(3),1)+c(0,30)
Vca.p <- Vcn.p*c(sqrt(3),1)+c(0,30)
v3pl <- list(Van.p,Vbn.p,Vcn.p,Vab.p,Vbc.p,Vca.p)
v.units <- rep("V",6)
v3pl.lab <- c("Van","Vbn","Vcn","Vab","Vbc","Vca")
phasor.plot(v3pl,v3pl.lab,v.units,lty.p=c(rep(1,3),rep(2,3)))
```

We can express power in terms of line voltages and currents. Take the line apparent power per phase, $S_l = V_l I_l$. To convert in terms of phase apparent power $S_p = V_p I_p$ use $V_l = \sqrt{3} V_p$ and $I_l = I_p$ for Y-connected. Then, $S_l = V_l I_l = \sqrt{3} V_p I_p = \sqrt{3} S_p$. To express the three-phase instantaneous three-phase power as a function of line apparent power $p(t) = 3S_p \cos\theta = \dfrac{3}{\sqrt{3}} S_l \cos\theta$ or

$$p(t) = \sqrt{3} S_l \cos\theta = \sqrt{3} V_l I_l \cos\theta \tag{10.23}$$

This is the real power of the three-phase system.

$$P_{3\phi} = \sqrt{3} V_l I_l \cos\theta \tag{10.24}$$

and therefore, $\sqrt{3} V_l I_l$ plays the role of the apparent power of the three-phase system

$$S_{3\phi} = \sqrt{3} V_l I_l \tag{10.25}$$

This is a practical equation because *rated* voltage and current in three-phase systems are given with respect to line voltages and current instead of phase.

Example 10.11

Consider a Y-Y connected three-phase balanced abc sequence. Assume phase voltage rms is $V = 120$ V. Total three-phase power is $P_{3\phi} = 360$ W and $pf = 1$ in all phases. What is the three-phase apparent power? What is the average power and the current in each phase? What are the phase currents and line voltages? Answer: The three-phase apparent power is $S_{3\phi} = \dfrac{P_{3\phi}}{pf} = \dfrac{360}{1} = 360$ VA.

The real power of each phase is $P_p = \dfrac{P_{3\phi}}{3} = \dfrac{360}{3} = 120$ W. Since $pf = 1$, the apparent power of each phase is the same as real power $P_p = V_p I_p \underbrace{\cos\theta}_{pf=1} = V_p I_p = S_p$. Thus, $S_p = 120$ VA. The rms of phase currents is $I_p = \dfrac{S_p}{V_p} = \dfrac{120}{120} = 1$ A. In addition, since $pf = 1$, there is no phase difference between current and voltage, thus the currents are $\mathbf{I}_a = \sqrt{2}\angle 0°$, $\mathbf{I}_b = \sqrt{2}\angle -120°$, $\mathbf{I}_c = \sqrt{2}\angle 120°$ in A. These are the same as line currents. The rms of line voltage is $V_l = \sqrt{3} V_p = 1.73 \times 120 = 208$ V or magnitude $208\sqrt{2} = 294.15$ V, then $\mathbf{V}_{ab} = 294.15 \angle 30°$, $\mathbf{V}_{bc} = 294.15 \angle -90°$, $\mathbf{V}_{ca} = 294.15 \angle -210°$. Note that after applying Equation 10.25, the rms of line currents is $I_l = \dfrac{S_{3\phi}}{\sqrt{3} V_l} = \dfrac{360}{\sqrt{3} \times 208} = 1$ A, which is the same as rms of phase current.

10.2.5 SYNCHRONOUS GENERATORS

Generators at the generating stations of the grid, except for solar PV plants many wind generators, and fuel cell plants, are *synchronous* AC generators, which produce a fixed frequency (e.g., 60 Hz in the United States) and after step-up are connected to the transmission line. A simple three-phase equivalent Y-Y balanced circuit is shown in Figure 10.21 (left-hand side), which includes an internal stator impedance $\mathbf{Z}_s$ for each phase. The right-hand side of Figure 10.21 shows a simple equivalent circuit for just one phase. $\mathbf{E}$ is the electromotive force, $\mathbf{Z}_s = R_s + jX_s$ is armature impedance, $\mathbf{V}_p$ is the

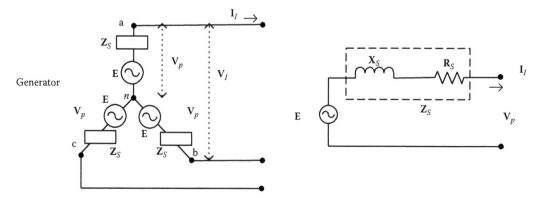

FIGURE 10.21   Synchronous generator model.

output voltage or phase voltage, and I_l is the line current delivering power at a given *pf*. Therefore, $\mathbf{E} = \mathbf{I}_p\mathbf{Z}_s + \mathbf{V}_p$.

The magnitude E of **E** is controlled by the amount of field current provided to the generator, whereas the angle δ of **E** is controlled by the torque provided by the prime mover to the generator. Recall from Chapters 1 and 5 that the power provided by the prime mover is torque multiplied by mechanical angular speed $P_{mech} = \tau\omega_{mech} \times \dfrac{\pi}{30}$ W, where torque τ is in Nm, and angular speed ω_{mech} is in revolutions per minute (rpm). For a fixed speed in rpm, to obtain 60 Hz increasing the torque will require more power from the prime mover. For a Y-Y connected three-phase generator with rated $S_{3\phi}$ and V_l, the real power is $P_{3\phi} = S_{3\phi}\cos\theta$. The mechanical power required for a given electrical generator efficiency η_e is $P_{mech} = \dfrac{P_{3\phi}}{\eta_e}$. We can calculate torque for a desired rpm.

As we mentioned in the previous section, the *rated* rms voltage and current are given as the line voltage V_l and current I_l. For Y-connected, $V_l = \sqrt{3}V_p$, $I_l = I_p$, and also $P_{3\phi} = S_{3\phi}\cos\theta = \sqrt{3}V_lI_l\cos\theta$. As a reference we can assume $\mathbf{V}_p = V_p \angle 0°$ and then $\mathbf{I}_p = I_p \angle \cos^{-1}(pf)$.

Example 10.12

Consider a 15 MVA, 13.8 kV, 1800 rpm, 60 Hz, 98% efficient, Y-connected three-phase generator with $\mathbf{Z}_s = 0.1 + j2\ \Omega$ per phase, delivering rated current at *pf* = 0.85 lagging.

What is the torque in Nm? What is the three-phase real power? What is the armature current? What is the phase voltage? What is the **E** required to provide this phase voltage? Note: The values given are the rated three-phase apparent power and line voltage. Answer: We are given rated $S_{3\phi}$ = 15 MVA and V_l = 13.8 kV. The real power is $P_{3\phi} = S_{3\phi}\cos\theta = 15 \times 0.85 = 12.75$ MW. The mechanical power required is $P_{mech} = \dfrac{P_{3\phi}}{\eta_e} = \dfrac{12.75\ \text{MW}}{0.98} = 13.01$ MW. Torque $\tau = \dfrac{P_{mech}}{\omega_{mech} \times \dfrac{\pi}{30}}$ and substituting values $\tau = \dfrac{13.01 \times 10^6}{1800 \times \dfrac{\pi}{30}}$ Nm = 69.02 kNm. We can use $S_{3\phi} = \sqrt{3}V_lI_l$ to solve for the rms of line current (which will be the same as the phase and the armature current) $I_l = \dfrac{S_{3\phi}}{\sqrt{3}V_l} = \dfrac{15000\ \text{kVA}}{\sqrt{3} \times 13.8\ \text{kV}} = 627.55$ A. Thus, $I_p = 627.55$ A. The phase voltage is solved from $V_l = \sqrt{3}V_p$ to get $V_p = \dfrac{13.8}{\sqrt{3}} = 7.967$ kV. Then, $\mathbf{E} = \mathbf{I}_p\mathbf{Z}_s + \mathbf{V}_p = 627.55$ A $\angle \cos^{-1}(0.85)(0.1 + j2\ \Omega) + 7967 \angle 0°$V. Now, convert impedance to polar, and multiply $\mathbf{E} = (627.55$ A $\angle 31.79°)(2.002 \angle 87.13°) + 7967 \angle$

$0° = 8.743 \angle 6.789°$ kV. We have expressed the phasors with magnitude given by the rms. We could multiply by $\sqrt{2}$ to express as amplitude of the wave.

The **E** result from this example is interpreted to be the induced voltage **E** required to provide the required real power at **V$_p$** for this line and load. The field current of the generator controls the magnitude of **E**, whereas the angle δ of **E** is controlled by the torque provided by the prime mover.

These calculations are expedited with function generator of renpow

```
x <- list(S3p = 15*10^6, Vl.rms = 13.8*10^3, pf=0.85, lead.lag=-1, Zs.r = c(0.1,2))
generator(x)
```

which gives us the **E**, **V$_p$**, and **I$_l$** in polar form and produces a phasor diagram (Figure 10.22)

```
> generator(x)
$E.p
[1] 8743.058   6.789
$Vp.p
[1] 7967.434   0.000
$Il.p
[1] 627.55464 -31.78833
```

This phasor diagram represents conditions when E is larger than V_p and the current lags **V$_p$**.

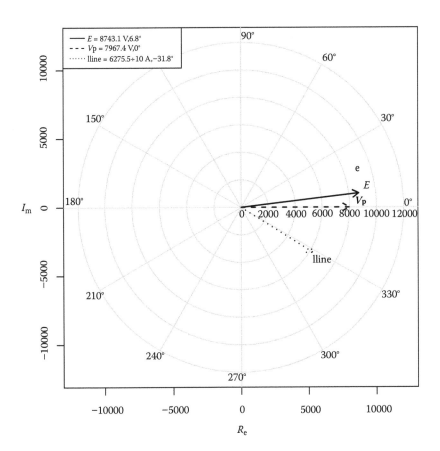

FIGURE 10.22   Phasor diagram of synchronous generator.

10.2.6 OTHER CONFIGURATIONS

We have studied the most common and basic wye-wye circuit. Several alternative circuit configurations are wye source and delta load, delta source and wye load, and delta source with delta load. It is often convenient when faced with those circuits to convert it to a wye-wye, perform the analysis, and then convert the results back to the original form.

Consider a delta-delta configuration as illustrated in Figure 10.23. A quick inspection shows that the line-to-line voltages are the same as the generator voltages. Using $0°, -120°, 120°$ as a reference, the line voltages are $\mathbf{V}_{ab} = \sqrt{2}V_l \angle 0°$, $\mathbf{V}_{bc} = \sqrt{2}V_l \angle -120°$, $\mathbf{V}_{ca} = \sqrt{2}V_l \angle 120°$, where V_l is the rms of the line voltage. However, also by quick inspection, we see that the line currents $\mathbf{I}_a$, $\mathbf{I}_b$, and $\mathbf{I}_c$ are no longer the same as the currents through the generators $\mathbf{I}_{ab}$, $\mathbf{I}_{bc}$, and $\mathbf{I}_{ca}$.

Current flow from a to b is represented as time domain $i_{ab}(t)$ or phasor $\mathbf{I}_{ab} \angle \theta$. The current i_{ab} is positive when flowing from a to b and negative in the opposite direction (from b to a) $i_{ba} = -i_{ab}$. In phasor form, negative is $180°$ out of phase $\mathbf{I}_{ba} = \mathbf{I}_{ab} \angle 180° = -\mathbf{I}_{ab}$.

We can use KCL to calculate the line currents. For instance, $\mathbf{I}_a = \mathbf{I}_{ca} - \mathbf{I}_{ab} = \sqrt{2}I_p(1 + j0) - \sqrt{2}I_p\left(-\frac{1}{2} + j\frac{\sqrt{3}}{2}\right)$, which can be rewritten $\mathbf{I}_a = \sqrt{2}Ip\left(\frac{3}{2} - j\frac{\sqrt{3}}{2}\right) = \sqrt{3}\sqrt{2}I_p\left(\frac{\sqrt{3}}{2} - j\frac{1}{2}\right) = \sqrt{3}\sqrt{2}I_p \angle 150°$.

Similarly, we would find the other two currents $\mathbf{I}_b = \mathbf{I}_{ab} - \mathbf{I}_{bc} = \sqrt{3}\sqrt{2}I_p \angle 30°$ and $\mathbf{I}_c = \mathbf{I}_{bc} - \mathbf{I}_{ca} = \sqrt{3}\sqrt{2}I_p \angle -90°$. In summary, the rms of each line current is $I_l = \sqrt{3}I_p$ and phased by $\mathbf{I}_a = \sqrt{2}I_l \angle 150°$, $\mathbf{I}_b = \sqrt{2}I_l \angle 30°$, $\mathbf{I}_c = \sqrt{2}I_l \angle -90°$.

We can express power in terms of line voltages and currents for a delta-connected circuit. Take line apparent power, $S_l = V_l I_l$. To convert in terms of phase apparent power $S_p = V_p I_p$ use $I_l = \sqrt{3}I_p$ and $V_l = V_p$ for delta-connected. Then, $S_l = V_l I_l = \sqrt{3}V_p I_p = \sqrt{3}S_p$, the same as it was for a wye-connected. In other words, regardless of whether we have delta or wye, $S_l = \sqrt{3}S_p$ and

$$P_{3\phi} = \sqrt{3}S_l \cos \theta = \sqrt{3}V_l I_l \cos \theta \qquad (10.26)$$

An often-used approach is to convert delta sources to wye sources and delta loads to wye loads. First, to convert the source use Equation 10.22 and then the wye generator voltages would be

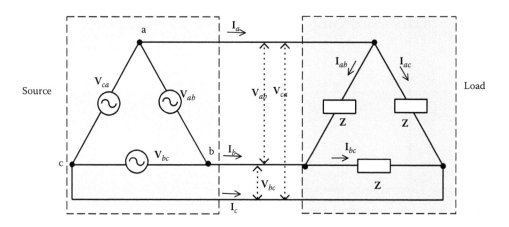

FIGURE 10.23 Delta-delta three-phase voltages sequence *abc*. Phase voltages and line-to-line voltages.

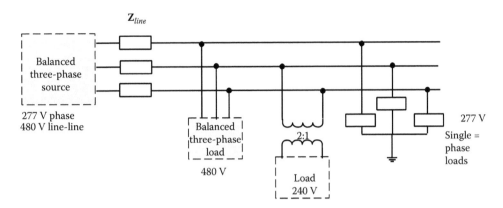

FIGURE 10.24 Three-phase distribution line and a variety of loads.

$\mathbf{V}_{an} = \sqrt{2}\dfrac{V_l}{\sqrt{3}} \angle -30°$, $\mathbf{V}_{bn} = \sqrt{2}\dfrac{V_l}{\sqrt{3}} \angle -150°$, $\mathbf{V}_{cn} = \sqrt{2}\dfrac{V_l}{\sqrt{3}} \angle 90°$. To convert a delta-balanced load where all impedances are the same $\mathbf{Z}_\Delta$ to a wye-balanced load simply use $\mathbf{Z}_Y = \dfrac{1}{3}\mathbf{Z}_\Delta$ to obtain the wye equivalent impedance $\mathbf{Z}_Y$.

Example 10.13

Consider a delta-delta circuit with $V_l = 208$ V generators $\mathbf{V}_{ab} = \sqrt{2}V_l \angle 0°$, $\mathbf{V}_{bc} = \sqrt{2}V_l \angle -120°$, $\mathbf{V}_{ca} = \sqrt{2}V_l \angle 120°$, and a delta load with impedance $\mathbf{Z}_\Delta = 3\Omega$ real ($pf = 1$). Calculate line currents. Answer: First, convert to a wye-wye circuit. The wye generators' rms voltage is $V_p = \dfrac{V_l}{\sqrt{3}} = \dfrac{208}{\sqrt{3}} =$ 120 V and the phase voltages are $\mathbf{V}_{an} = 170 \angle -30°$, $\mathbf{V}_{bn} = 170 \angle -150°$, $\mathbf{V}_{cn} = 170 \angle 90°$. The wye loads are $\mathbf{Z}_Y = \dfrac{1}{3}3 = 1\Omega$. Now, the line currents are $\mathbf{I}_a = \dfrac{\mathbf{V}_{an}}{\mathbf{Z}_Y}$, $\mathbf{I}_b = \dfrac{\mathbf{V}_{bn}}{\mathbf{Z}_Y}$, $\mathbf{I}_c = \dfrac{\mathbf{V}_{cn}}{\mathbf{Z}_Y}$, therefore, $\mathbf{I}_a = $ 170 $\angle -30°$, $\mathbf{I}_b = 170 \angle -150°$, $\mathbf{I}_c = 170 \angle 90°$.

   We can see from the example that the line currents have a phase difference of $-30°$ with respect to the generators in the delta source.

In general, a three-phase distribution line feeds a variety of loads. This is exemplified in Figure 10.24 for a phase voltage of 277 V. The figure illustrates three-phase (delta or wye) loads using the full line voltage (480 V), single-phase loads using the phase voltages (277 V), and single-phase loads using half of the line voltage (240 V) by means of a step-down transformer.

10.3 POWER QUALITY: HARMONIC DISTORTION

Ideally, AC is a sine wave $x(t) = X_m \cos(\omega t)$ of just a given *fundamental frequency* $\omega = 2\pi f$. However, due to a variety of reasons we will describe shortly, AC can include multiple values of this fundamental frequency or *harmonics*. Simply put, the electrical signal is the sum of sinusoids at this set of frequencies and can create load unbalance and detrimental current in the neutral.

   Among the reasons that AC is distorted by harmonics, are nonlinear effects of many loads and electrical equipment, for instance, hysteresis in the magnetic material of transformers, nonlinear electronic circuits to convert DC to AC, and vice versa such as rectifiers and inverters. When including renewable DC power systems (PV solar panels, fuel cells) in interaction with the grid (AC), we include more electronic equipment that produce harmonics.

Let us consider a current consisting of the sum of its harmonic components $i(t) = \sum_k i_k(t)$ where each component $i_k(t) = Im_k \cos(k\omega t)$ has a frequency $k\omega$ or a multiple k of the fundamental frequency. In other words,

$$i(t) = \underbrace{I\,m_0}_{DC} + \underbrace{I\,m_1 \cos \omega t}_{Fundamental} + \overbrace{Im_2 \cos 2\omega t}^{Even} + \overbrace{Im_3 \cos 3\omega t}^{Odd} + \ldots \qquad (10.27)$$

Example 10.14

Write a current with 10 A amplitude at the fundamental at 60 Hz, 2 A amplitude at 180Hz (third harmonic), and 1 A amplitude at 300 Hz (fifth harmonic). Assume no DC component. Answer: Using Equation 10.27, $i(t) = 10 \cos \omega t + 2 \cos 3 \omega t + 1 \cos 5 \omega t$.

Whenever the fundamental has a nonzero phase angle θ, the harmonics are affected by the angle and have a phase angle, which is a multiple of the angle of the fundamental, that is to say, $i_k(t) = Im_k \cos(k(\omega t + \theta))$.

Example 10.15

Write a current with 10 A amplitude at the fundamental at 60 Hz with phase angle 30°, 2 A amplitude at 180 Hz (third harmonic), and 1 A amplitude at 300 Hz (fifth harmonic). Assume no DC component. Answer: $i(t) = 10 \cos(\omega t + 30°) + 2 \cos(3(\omega t + 30°)) + 1 \cos(5(\omega t + 30°))$, which yields $i(t) = 10 \cos(\omega t + 30°) + 2 \cos(3 \omega t + 90°) + 1 \cos(5 \omega t + 150°)$.

10.3.1 HARMONIC DISTORTION MEASURES

The RMS of $i(t)$ is the square root of the average of the sum of all squares

$$I_\Sigma h = \sqrt{\langle i(t)^2 \rangle} = \sqrt{\left\langle \left(\sum_k i_k(t)^2 \right) \right\rangle} \qquad (10.28)$$

Each harmonic component has a RMS, for instance, $I_3 = \dfrac{Im_3}{\sqrt{2}}$ for the third harmonic. For a sum of sinusoids, the total RMS is the square root of the sum of the RMS of the components. The total RMS is

$$I_\Sigma h = \sqrt{\sum_k I_k^2} = \sqrt{\sum_k \left(\frac{Im_k}{\sqrt{2}} \right)^2} \qquad (10.29)$$

In order to quantify the effect of harmonics, it is common to use *total harmonic distortion (THD)*, which is a ratio of sum of the rms of all harmonics (except the fundamental) with respect to the rms of the fundamental.

$$THD = \frac{\sqrt{\displaystyle\sum_{k \neq 1} I_k^2}}{I_1} \qquad (10.30)$$

This is the same as

$$THD = \sqrt{\sum_{k \neq 1}\left(\frac{I_k}{I_1}\right)^2} \qquad (10.31)$$

In other words, THD is the square root of the sum of the squares of the ratios of the harmonic rms to the fundamental rms. The ratio $\frac{I_1}{I_1} = 1$ is excluded in the sum. Actually, because the terms are ratios, it can be done as ratio of amplitude or ratio of rms values. THD can have a value larger than 1 or greater than 100%. We want THD to be low for lower effect of harmonics. If we do not have any harmonics, then THD is zero.

When we limit our discussion to cosine waves only, it can be shown that we only have odd harmonics. For instance, for odd harmonics $k = 3, 5, 7,\ldots$

$$THD = \sqrt{(I_3/I_1)^2 + (I_5/I_1)^2 + (I_7/I_1)^2 + \ldots} \qquad (10.32)$$

Another measure to quantify the effect of harmonics is *distortion factor*, which is a ratio of the RMS of the fundamental to the rms of the signal (it does not exclude the fundamental in calculating the ratio)

$$DF = \frac{I_1}{I_\Sigma h} = \frac{I_1}{\sqrt{\sum_k I_k^2}} \qquad (10.33)$$

This measure is less than 1 because the numerator is part of the sum in the denominator. The closer the numerator (rms of the fundamental) is to the true rms, the closer DF is to 1.

Note that if we take the square of the THD, we get $THD^2 = \dfrac{\sum\limits_{k \neq 1} I_k^2}{I_1^2}$. Now add 1 to get $1 + THD^2 =$

$1 + \dfrac{\sum\limits_{k \neq 1} I_k^2}{I_1^2} = \dfrac{I_1^2 + \sum\limits_{k \neq 1} I_k^2}{I_1^2} = \dfrac{\sum\limits_k I_k^2}{I_1^2}$. If we invert this, and take the square root, we get $\sqrt{\dfrac{1}{1 + THD^2}} =$

$\sqrt{\dfrac{I_1^2}{\sum\limits_k I_k^2}} = \dfrac{I_1}{\sqrt{\sum\limits_k I_k^2}}$, which is the distortion factor. Therefore,

$$DF = \sqrt{\frac{1}{1 + THD^2}} \qquad (10.34)$$

The measures THD and DF do not account for phase angle. They are related only to how large the harmonics are in comparison to the fundamental.

Example 10.16

Suppose we have a current with odd harmonics 3, 5, and 7 with rms value equal to 20%, 10%, and 5% of the rms of the fundamental, respectively. What is the THD? What is the DF? Answer: $THD =$

$\sqrt{(I_3/I_1)^2 + (I_5/I_1)^2 + (I_7/I_1)^2} = \sqrt{0.2^2 + 0.1^2 + 0.05^2} = 0.229$. The DF is $DF = \sqrt{\dfrac{1}{1 + THD^2}} =$

$\sqrt{\dfrac{1}{1 + 0.23^2}} = 0.975$.

Renpow has a function harmonic to plot harmonics and calculate THD and DF. For instance, to visualize the effect of the harmonics of the previous example, we write the following code to produce Figure 10.25. Here x is the fundamental amplitude (10 A) and phase (0°); harm.odd is the fraction of odd harmonics with respect to the fundamental $I_3/I_1 = 0.2$, $I_5/I_1 = 0.1$, and $I_7/I_1 = 0.05$

```
x <- list(c(10,0)); harm.odd <- list(c(0.2,0.1,0.05)); lab.units <- "I [A]"
y <- harmonic(x,harm.odd,lab.units)
```

We can see the distortion of the wave due to these harmonic components. One of the exercises at the end of the chapter asks you to check, using this function, that adding an angle, say 30°, to the fundamental will produce a wave that is shifted differently, but the THD and DF are the same.

10.3.2 HARMONICS AND POWER

Recall that instantaneous power for a pure sinusoidal wave, say, the fundamental, is $p(t) = v(t)i(t) = V_1I_1[\cos\theta + \cos(2\omega t - \theta)]$, where θ is the phase angle between the voltage and current, V_1 and I_1 are rms values of fundamental voltage and current, and V_1I_1 is apparent power. Because the cosine term averages to zero, the average power is $P = V_1I_1\cos\theta$ and we call $pf = \cos\theta$ the power factor. What happens if $v(t)$ remains as a pure sinusoidal, but the load responds with a current that contains harmonics measured by DF? The answer is that instantaneous power, average power, and power factor are now more complicated. To see this, write the current as a sum of harmonic components $p(t) = v(t)i(t) = Vm\cos(\omega t) \times \sum_k Im_k\cos(k(\omega t + \theta))$ and expand $p(t) = v(t)i(t) = \underbrace{Vm\cos(\omega t) \times Im\cos(\omega t + \theta)}_{fundamental} + Vm\cos(\omega t)\sum_{k\neq1}Im_k\cos(k(\omega t + \theta))$.

The first term is the same as instantaneous power of the fundamental, whereas the second term will be all the contributions due to harmonics. The first term is $V_1I_1[\cos\theta + \cos(2\omega t - \theta)]$, and our familiar concepts of average power and power factor apply. The second term depends on the specifics of harmonic content.

One simple approach is to assume that a more general power factor pf_h relates average power to new apparent power $VI_{\Sigma h}$. Then, we write $pf_h = \dfrac{P}{V_1I_{\Sigma h}} = \dfrac{V_1I_1\cos\theta}{V_1I_{\Sigma h}} = \dfrac{I_1\cos\theta}{I_{\Sigma h}}$.

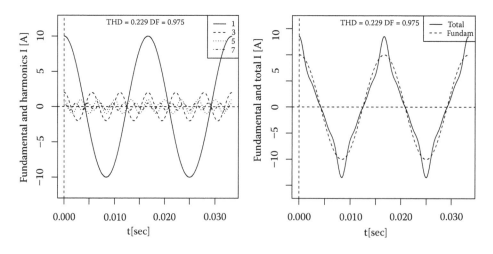

FIGURE 10.25 Effect of third, fifth, and seventh harmonics on a fundamental at 60 Hz. Left: Components. Right: Distorted wave compared with the fundamental.

Now, recognize the term $\dfrac{I_1}{I_{\Sigma h}}$ as the DF per Equation 10.33 and obtain

$$pf_h = \cos\theta \times DF \tag{10.35}$$

In other words, we can compound the effect of phase shift $\cos\theta$ with effect of harmonic distortion to account for reduction of apparent power by both effects.

Example 10.17

Suppose you have a current with odd harmonics 3, 5, and 7 with RMS value equal to 20%, 10%, and 5% of the rms of the fundamental (respectively) and a phase angle of 30° with respect to voltage. What is the compounded power factor pf_h? Answer: From the previous example, we know this current has DF = 0.975. Then, $pf_h = \cos 30° \times 0.975 = 0.844$.

10.3.3 HARMONICS AND THREE-PHASE SYSTEMS

Recall from earlier in this chapter that phase currents cancel out in balanced three-phase circuits and therefore the current in the neutral is zero. This statement remains true for the fundamental, but not all harmonics are canceled in balanced three-phase systems. Therefore, balanced three-phase systems can suffer from a total current in the neutral due to harmonics. Of particular relevance is the third harmonic, which can manifest in the neutral.

To see why the third harmonic does not cancel, consider the three currents and the third harmonic:

$$i_a(t) = \sqrt{2}(I_1 \cos(\omega t) + I_3 \cos(3\omega t))$$

$$i_b(t) = \sqrt{2}(I_1 \cos(\omega t - 120°) + I_3 \cos(3(\omega t - 120°)))$$

$$i_c(t) = \sqrt{2}(I_1 \cos(\omega t + 120°) + I_3 \cos(3(\omega t + 120°)))$$

Now, perform addition to obtain the current in the neutral $i_n(t) = i_a(t) + i_b(t) + i_c(t)$. We know the fundamental cancels out, therefore, $i_n(t) = \sqrt{2}I_3(\cos(3\omega t) + \cos(3\omega t - 360°) + \cos(3\omega t + 360°))$. Now, realize that $\cos(\omega t \pm 360°) = \cos(\omega t)$, therefore, $i_n(t) = 3\sqrt{2}I_3 \cos(3\omega t)$. Which in terms of rms is simply $I_n = 3I_3$. In other words, the rms of the neutral current is 3 times the rms of the third harmonic present in each phase!

Actually, this occurs whenever we have $3 \times n \times 120°$, $n = 1, 2, 3,...$, and therefore all harmonics multiple of 3 will manifest at the neutral. One advantage of delta-connected circuits over wye-connected circuits is that the third harmonic cancels out.

Example 10.18

Using renpow, calculate the sum of currents in a balanced wye-wye circuit $I_a = 10 \angle 0°$, $I_b = 10 \angle -120°$, $I_c = 10 \angle 120°$. In each line (phase because it is wye connected), we have odd harmonics 3, 5, and 7 with RMS value equal to 20%, 10%, and 5% of the rms of the fundamental, respectively. What happens to the current returning from the neutral? Answer:

```
x <- list(c(10,0),c(10,-120),c(10,120))
harm.odd <- list(c(0.2,0.1,0.05),c(0.2,0.1,0.05),c(0.2,0.1,0.05))
lab.units <- "I [A]"
y <- harmonic(x,harm.odd,lab.units)
```

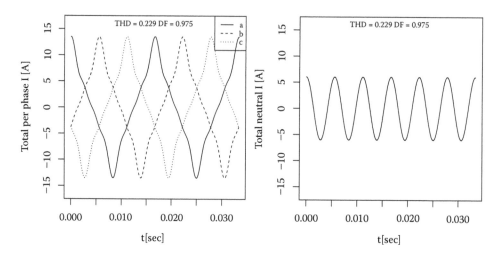

FIGURE 10.26 Effect of third, fifth, and seventh harmonics on a wye-wye balanced three-phase system. Left: Distorted phase (or line) currents. Right: Total current or current in the neutral.

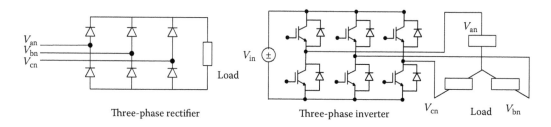

FIGURE 10.27 Left: Three-phase rectifier. Right: Three-phase inverter.

The result is shown in Figure 10.26. We can see that each line current is distorted by these harmonics (left-hand side) and the neutral current is nonzero, has a frequency 3 times the fundamental, and equal to 6 A in amplitude, which is precisely 3 times the magnitude of the third harmonic component (2 A). Note that the fifth and seventh harmonic are not seen in the neutral.

10.4   AC-DC AND DC-AC CONVERTERS IN THREE-PHASE SYSTEMS

Figure 10.27 (left-hand side) shows a *three-phase full wave rectifier*. Each one of the phases provides a full-wave rectified signal. Thus, when all three phases are present, the peaks come closer together and the ripple is small even without a filter. The right-hand side of Figure 10.27 shows a *three-phase inverter* based on two switches per phase that act in tandem. The simplest scheme is to apply a square wave to the gate of all switches. However, the AC signal would have fifth, seventh, and other harmonics. A better result is obtained by activating the gates using PWM signals as the ones described in Chapter 8.

EXERCISES

10.1. Suppose we have a core of 2 cm^2 cross-sectional area, path length 10 cm, with $\mu_r = 1000$ and a 1 mm air gap. We wrap 150 turns of coil and inject 2A of current. What is the mmf? What are the reluctances of the core and the gap? What is the magnetic flux? What is the equivalent permeability?

10.2. Refer to Figure 10.5 with $V_s = 170 \angle 30°$ V and $N_1 = 1$, $N_2 = 2$. Assume impedances connected to the transformer are $Z_s = 1\ \Omega$, $Z_o = 0\ \Omega$, $Z_{load} = 1\ 0\Omega$. Is this transformer step-up or step-down? What are primary and secondary voltage, current, and power? What is the load voltage, current, and power?

10.3. Consider Figure 10.7. Assume impedances $Z_1 = Z_2 = 10 + j1\ \Omega$ and $Z_3 = 20 + j2$ respectively. Calculate currents drawn by each load. Calculate total current drawn from each of the three-wire drop.

10.4. Assume three-phase wye-wye connected balanced circuit (abc sequence) with $V_p = 240$ V and the same load in each phase $Z = 1 \angle 10°$. What are the currents? Draw a phasor diagram of voltages and currents.

10.5. Consider a Y-Y connected three-phase balanced abc sequence. Assume the phase voltage rms is $V_p = 240$V. Total power is $P_{tot} = 720$ W and $pf = 1$ in all phases. What is the average power and the current in each phase? What are the phase currents and line voltages?

10.6. Consider a 1 MVA, 13.8 kV, 3600 rpm, 60 Hz, 97% efficient, Y-connected three-phase generator with $Z_s = 0.01 + j1\ \Omega$ per phase, delivering rated current at $pf = 0.80$ lagging. What is the torque in Nm? What is the three-phase real power? What is the armature current? What is the phase voltage? What is the **E** required to provide this phase voltage?

10.7. Suppose you have a current with odd harmonics 3, 5, and 7 with RMS value equal to 20%, 10%, and 5% of the rms of the fundamental, respectively, and a phase angle of 30°. Check using the harmonic function of renpow that the resulting distorted wave is shifted according to the phase angle, but the THD and DF are the same.

10.8. Using renpow harmonic function, display the current in the neutral when the third harmonic of each phase is zero, although the fifth and seventh are nonzero. Confirm that the current in the neutral is zero.

10.9. Using renpow, calculate the sum of currents in a balanced wye-wye circuit. $I_a = 10 \angle 0°$, $I_b = 10 \angle -120°$, $I_c = 10 \angle 120°$. In each line (phase because it is Wye), we have odd harmonics 3 and 9 with RMS values equal to 20% and 10% of the rms of the fundamental, respectively. What happens to the current returning from the neutral?

REFERENCES

1. Irwin, J.D., and R.M. Nelms, *Basic Engineering Circuit Analysis*. 11th edition. 2011: Wiley. 688 pp.
2. Masters, G.M., *Renewable and Efficient Electric Power Systems*. Second edition. 2013: Wiley-IEEE Press. 690 pp.
3. Asano, R., and S.A. Page, Reducing Environmental Impact and Improving Safety and Performance of Power Transformers with Natural Ester Dielectric Insulating Fluids. *IEEE Transactions on Industry Applications*, 2014. 50(1): 134–141.
4. EPA, PCBs Questions and Answers. 2017. Accessed August 2017. Available from: https://www3.epa.gov/region9/pcbs/faq.html.
5. Oommen, T.V., Vegetable Oils for Liquid-Filled Transformers. *IEEE Electrical Insulation Magazine*, 2002. 18(1): 6–11.
6. Alexander, C.K., and M.N.O. Sadiku, *Fundamentals of Electric Circuits*. Third edition. 2007: McGraw Hill. 901 pp.
7. El-Hawari, M., *Introduction to Electrical Power Systems*. IEEE Press Series on Power Engineering, M. El-Hawari. 2008: Wiley. 394 pp.

11 Power Systems and the Electric Power Grid

Now that we have studied the essentials of direct current (DC) and alternating current (AC) circuits, transformers, generators, three-phase circuits, and AC power, in this chapter we will describe the major components and workings of a typical *electrical power system* or *grid*. A typical power system consists of three major components: generation, transmission, and distribution. After reviewing these components, this chapter describes models of transmission lines. Then we introduce the bus admittance matrix, its use to calculate voltage and current, and to setup and solve power flow equations. We follow with a study of varying demand from the loads and the load-duration curve. We conclude emphasizing that a large grid is a complex dynamic system that requires tools from real-time control systems. We simplify the presentation substantially from material contained in textbooks devoted to electrical power systems [1,2] and the integration of renewable generation with power systems [3].

11.1 ELECTRIC POWER SYSTEMS: MAJOR COMPONENTS

A typical power system consists mainly of three major components: generation, transmission, and distribution (Figure 11.1). Traditionally these components have been mainly based on three-phase AC systems. However, a modern power system also includes DC generation and transmission components because more renewable power systems are available and the technology of high voltage DC-AC and AC-DC conversion has been greatly improved. Recall the concepts of AC-DC, DC-AC, and DC-DC converters that we covered in Chapters 5, 8, and 10.

The *generation* component includes many power plants or generator stations operating in *synchrony* at the system's frequency (e.g., 60 Hz in the United States). Generating stations include coal-fired and nuclear power plants (Chapter 7), gas-fired power plants (Chapter 9), traditional renewables such as hydroelectric power plants (Chapter 12), and possibly utility-scale renewables from solar and wind power plants (Chapters 13 and 14). Generated voltages are in the order of tens of kV (e.g., 14–24 kV) (Figure 11.1). Using *high voltage DC (HVDC)* equipment has made it possible to integrate generating units of different frequency (e.g., 60 Hz with 50 Hz units) and wind power systems that may have varying frequencies (Chapter 13). HVDC allows for AC at one frequency to be converted to DC and then DC converted back to AC at a different frequency (Chapter 8).

Most large generating facilities are located far away from the electricity consumers, and, therefore, *transmission* lines carry the power generated at these stations to areas closer to the consumers. Because line power loss decreases with increasing voltage, power is transmitted at high voltages in the order of hundreds of kV (e.g., from 138 to 765 kV; typical values are 230, 345, 500, and 765 kV). Therefore, near the generating station, at the switchyard or substation, the voltage is raised by a step-up transformer before transmission. Transformers are connected to AC-DC converters before transmission over a HVDC transmission line. Transmitting DC has several advantages over AC, particularly for underground or submarine lines. These advantages in many cases compensate the additional high cost of the converters.

At a *transmission substation*, voltage steps down from the hundred kV range to tens of kV (e.g., 34.5 kV–138 kV), and then power is transmitted by a *subtransmission* component to the *distribution substations*. Once the power is closer to the location of the consumers, it is brought to a distribution substation where it is decreased to the kV or tens of kV range (e.g., 4.16 kV or 34.5 kV) by a step-down transformer. If transmission occurred over a HVDC line, then DC-AC converters are included to integrate to the rest of AC power. From a distribution substation, power flows on a

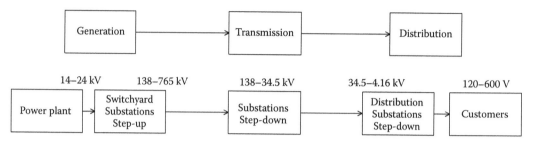

FIGURE 11.1   Major components of a power system.

distribution network using the kV or tens of kV range, and, finally, stepped down again for consumers to the hundreds of volts (120–600 V) using step-down transformers.

11.2 MAJOR COMPONENTS OF A SUBSTATION

We can identify several major components of a substation: incoming lines, switches to connect (or disconnect) incoming lines, transformers (step-up or step-down), and switches to connect (or disconnect) outgoing lines (Figure 11.2). The term *bus* is used often to refer to the lines bringing the power or sending the power. The bus lines are connected by a *bus bar*, which is a metal strip that can carry high current at short distances. Additional components include lighting arrestors, control systems, voltage regulators, and converters (AC-DC or DC-AC).

The incoming bus could be a *generator bus* (from the generator at a generating substation) or a transmission bus (from transmission lines at a distribution substation). Similarly, the outgoing bus could be the *feeder* to a transmission line (at a generating substation) or a feeder to a distribution network (at a distribution substation). For example, a generator bus and feeder bus at a generating substation could be a 14 kV bus and a 138 kV bus, respectively. The bus concept is very useful when visualizing the power system as a set of single lines or a *single-line diagram* (*SLD*) as shown in

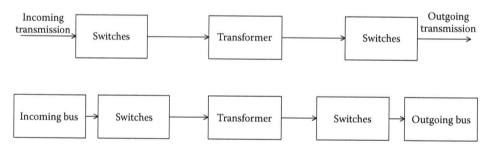

FIGURE 11.2   Major components of a substation.

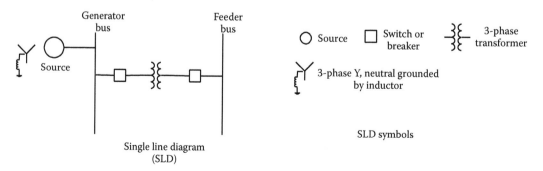

FIGURE 11.3   SLD of major components of a substation.

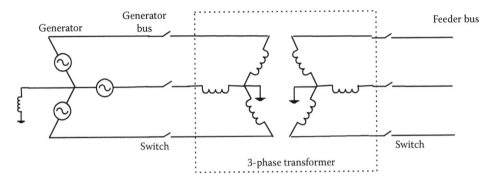

FIGURE 11.4 Circuit diagram of major components of a substation.

Figure 11.3. This is a conceptual or high-level diagram not showing detail of the electrical lines. Its equivalent is a circuit diagram, as shown in Figure 11.4.

Electrically, we will show the details of the generator, switches, and transformer. Generator neutrals are typically grounded via a high-resistance value or an inductor, and transformer neutrals are also grounded (Figure 11.4). The resistance to ground is added in order to avoid flow of current during a fault.

11.3 DISTRIBUTION BUS

At a distribution substation, the set of circuit points that would feed power to several circuits is the distribution feeder bus. This can be seen as a more general concept of interconnecting sources with loads. For DC and single-phase AC, we can think of this bus as a pair of wires that are used to power several circuits. This voltage is provided by a source. For example, a 24 V DC bus will provide 24 V to the loads. This general concept is visualized in a SLD as shown in Figure 11.5 together with its electrical description.

In the specific context of a distribution substation, we can say that a load is connected to a bus in a substation. The SLD is the same, but we also specify the source as Y or delta. See Figure 11.6, where the source is specified as Y and the loads are unspecified. A small Y symbol can be used in the SLD to denote the configuration of the source.

A SLD representation of a HVDC link is shown in Figure 11.7, which includes a back-to-back converter with special symbols for rectifier, DC line, and inverter. This is a simplified view of the more detailed diagram shown in Figure 10.10 (Chapter 10), connecting the back-to-back converter to transformers.

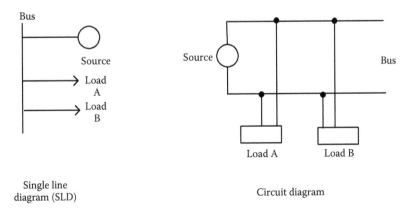

FIGURE 11.5 Source–load bus concept.

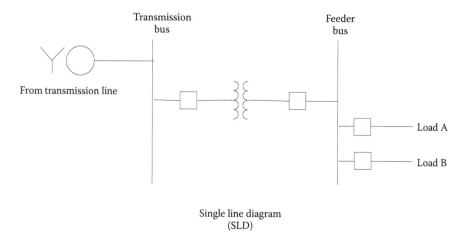

Single line diagram
(SLD)

FIGURE 11.6 Three-phase distribution feeder bus. Source is Y-connected and connection of loads is not specified.

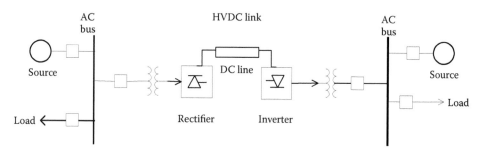

FIGURE 11.7 SLD showing HVDC link based on back-to-back converter including rectifier and inverter.

Electrically, we can see a diagram as shown in Figure 11.8, which does not go into details about the type of connection of the loads (whether it is Y or delta). In addition, we can specify the connection at the load as Y or delta. For example, see Y-connected loads shown in Figure 11.9. We can use what we have learned so far about SLD and buses to put together the SLD for a simple electrical power system, which shows the bus-to-bus interconnections (Figure 11.10).

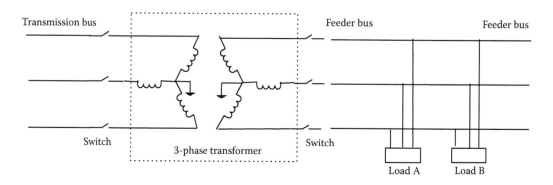

FIGURE 11.8 Concept of three-phase bus. Source is Y-connected and connection of loads is not specified.

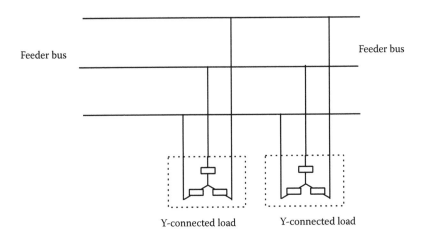

FIGURE 11.9   Concept of three-phase bus. Source and loads are Y-connected.

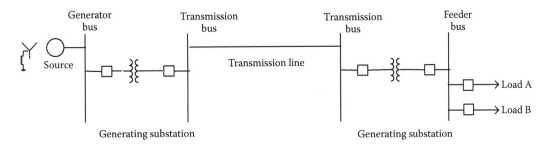

FIGURE 11.10   Electric power system, showing bus-to-bus interconnections.

11.4   TRANSMISSION LINE MODELS

There are four electrical parameters to consider when modeling a transmission line. These are resistance, inductance, capacitance, and leakage conductance [1]. The first two are modeled as occurring in-series and the last two are modeled as in shunt or diverted. *Resistance*, as in any wire, is related to the resistivity, diameter, and length of the line. *Inductance* captures the magnetic flux linkage due to the current flowing in the line. *Capacitance* represents the line charging current that is diverted from the line current, and last, *leakage conductance* diverts current from the line due to conduction.

These terms are modeled as lumped or distributed according to the length of the line. Recall from Chapter 1 that an electromagnetic (EM) wave of frequency v (in Hz or s^{-1}) has a wavelength λ that can be calculated from $c = v\lambda$, where c is the speed of light (3×10^8 m/s). At the grid frequency (60 Hz in the United States), the wavelength is $\lambda = c/v = \dfrac{3 \times 10^8}{60}$ m $= 5 \times 10^6$ m $= 5000$ km. Transmission lines that are much shorter than the wavelength can be modeled as if the electrical parameters are *lumped*. In other words, one can apply Kirchhoff's current law (KCL) or Kirchhoff's voltage law (KVL) to circuit elements representing the entire line. Typically, the length will be less than 5% of the wavelength, or ~250 km. Lumped models can be used for short (<80 km) and medium (<250 km) lines. For long lines, the electrical parameters are considered distributed along the line, in other words, taking values per differential of length dx. For some purposes not demanding high accuracy, the distributed model can be approximated by a lumped circuit for lines less than 320 km. We do not study long transmission line models in this introductory textbook.

A short transmission line is one of length less than ~80 km (and is typically of relatively low voltage <20 kV). It can be modeled by including only the in-series components, that is, resistance and inductance, and ignoring the shunt elements (Figure 11.11, left-hand side). A medium transmission has length less than ~250 km (and voltages up to 100 kV). Its model includes shunt capacitance, which may be allocated in equal parts to each end of the line (Figure 11.11, right-hand side). The line impedance is then written as resistive part R as well as reactive inductive X_L and capacitive X_C parts, as shown in Figure 11.11. For instance, the in-series impedance is $\mathbf{Z}_L = R + jX_L = R + jX_L$ and each one of the shunt impedances is $\mathbf{Z}_C = -j0.5X_C$. Typically, the reactive components of the line impedance are larger than the resistive component, and therefore, often $R \ll X_L$.

Example 11.1

Consider a 120 km line with $R = 0.01$ Ω/km, $L = 2$ mH/km, $C = 5$ nF/km. Select a model and write the impedances and admittances. Answer: Select medium-length model. Calculate lumped parameters $R = 0.01$ Ω/km $\times$ 120 km = 1.2 Ω, $X_L = j\omega L = j377 \times 2 \times 10^{-3}$ Ω/km = $j0.75$ Ω/km, or $X_L = j0.75$ Ω/km $\times$ 120 km = $j90$ Ω. Since $R \ll X_L$, the in-series impedance is approximated by just $\mathbf{Z}_L \simeq j90$ Ω, and the admittance is $\mathbf{Y}_L \simeq -j0.011$ S. The shunt impedance is $X_C = \dfrac{1}{j\omega C} = -j\dfrac{1}{377 \times 5 \times 10^{-9}} = -j530.5$ kΩ/km or $X_C = -j530.5$ kΩ/km $\times$ 120 km = $-j63.6$ MΩ. Each shunt impedance is assigned half $\mathbf{Z}_{sh} = 0.5X_C = -j31.8$ MΩ, and each admittance is $\mathbf{Y}_{sh} = j0.031$ μS.

Lumped models apply to lines used for transmission, subtransmission, and distribution according to their length. The actual resistive and reactive values depend on several factors such as length, conductor spacing, and conductor cross-sectional area. Inductive reactance decreases with cross-sectional area and increases with increasing distance between conductors. On the contrary, capacitive reactance increases with cross-sectional area and decreases with increasing distance between conductors. Capacitive reactance increases for underground lines because spacing decreases. The reactive components of the line are responsible for having a *charging current* flowing in the line regardless of whether the circuit is closed or open. This current is charging and discharging the line as AC voltage varies from positive to negative. Also the reactive components increase as the line voltage increases; so high-voltage transmission lines have higher X than lower-voltage transmission lines.

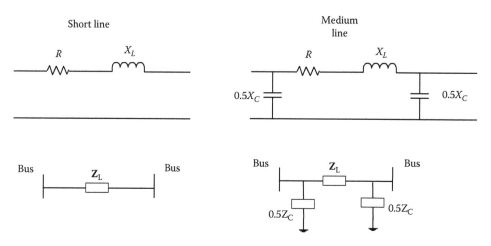

FIGURE 11.11 Transmission line lumped models: short and medium lines.

11.5 BUS ADMITTANCE MATRIX

A grid can be conceptualized as a collection of buses interconnected by transmission lines. Figure 11.12 illustrates using a simple example of three buses and two lines with the analogy between the grid consisting of buses and lines (and visualized by a SLD) to a circuit consisting of nodes and admittances interconnecting the nodes (and visualized as a circuit diagram). We studied the latter in Chapter 8 using nodal analysis, which hinges on the circuit admittance matrix $\mathbf{Y}$ to calculate the relation between currents and voltages by $\mathbf{I} = \mathbf{YV}$. Recall that the diagonal terms of $\mathbf{Y}$ was determined by the sum of admittances tied to a node, and the off-diagonal terms were determined by the negative of the "mutual" admittance between two nodes. We will employ the same concept to the grid and define a *bus admittance matrix*, which is determined by the interconnecting lines.

It is easier to explain this concept using a small network. For a three-bus system, the bus admittance matrix would be

$$\mathbf{Y} = \begin{bmatrix} \mathbf{Y}_{11} & -\mathbf{Y}_{12} & -\mathbf{Y}_{13} \\ -\mathbf{Y}_{12} & \mathbf{Y}_{22} & -\mathbf{Y}_{23} \\ -\mathbf{Y}_{13} & -\mathbf{Y}_{23} & \mathbf{Y}_{33} \end{bmatrix}$$

where $\mathbf{Y}_{ii}$ is the sum of admittances connected to bus i, and $\mathbf{Y}_{ij}$ is the mutual admittance or admittance connected between bus i and j. In determining the bus admittance matrix, we specify whether a line is modeled as a short, medium, or long line.

Example 11.2

Determine the bus admittance matrix $\mathbf{Y}$ for the simple network of Figure 11.12. Assume the line between buses 1 and 2 is short with $\mathbf{Z}_L = j50\ \Omega$ and the line between buses 1 and 3 is medium with $\mathbf{Z}_L = j90\ \Omega$ and $\mathbf{Z}_{sh} = -j20\ \text{M}\Omega$. Answer: First, set up a diagram as shown in Figure 11.13 identifying the known admittances to help visualize the problem. The in-series admittance of the short line between buses 1 and 2 is $\mathbf{Y}_1 = -j0.02\ \text{S}$. The in-series admittance for the medium line is $\mathbf{Y}_2 = -j0.011\ \text{S}$, and the shunt admittance of this line is $\mathbf{Y}_{sh} = j0.05\ \mu\text{S}$. Next, calculate the diagonal entries. For bus 1, $\mathbf{Y}_{11} = \mathbf{Y}_1 + \mathbf{Y}_2 + \mathbf{Y}_{sh} = -j0.02 - j0.011 + j0.05 \times 10^{-6} \simeq -j0.031\ \text{S}$. For bus 2, $\mathbf{Y}_{22} = \mathbf{Y}_1 = -j0.02\ \text{S}$. For bus 3, $\mathbf{Y}_{33} = \mathbf{Y}_2 + \mathbf{Y}_{sh} = -j0.011 + j0.05 \times 10^{-6} \simeq -j0.011\ \text{S}$. Now calculate the off-diagonal terms. Between bus 1 and 2 it is $-\mathbf{Y}_{12} = -\mathbf{Y}_1 = j0.02\ \text{S}$, between bus 1 and 3 it is $-\mathbf{Y}_{13} = -\mathbf{Y}_2 = j0.011\ \text{S}$, and between bus 2 and 3 it is $-\mathbf{Y}_{13} = 0$. We put it all together:

$$\mathbf{Y} = \begin{bmatrix} \mathbf{Y}_1 + \mathbf{Y}_2 + \mathbf{Y}_{sh} & -\mathbf{Y}_1 & -\mathbf{Y}_2 \\ -\mathbf{Y}_1 & \mathbf{Y}_1 & 0 \\ -\mathbf{Y}_2 & 0 & \mathbf{Y}_2 + \mathbf{Y}_{sh} \end{bmatrix} = \begin{bmatrix} -j0.031 & j0.02 & j0.011 \\ j0.02 & -j0.02 & 0 \\ j0.011 & 0 & -j0.011 \end{bmatrix} \text{S}$$

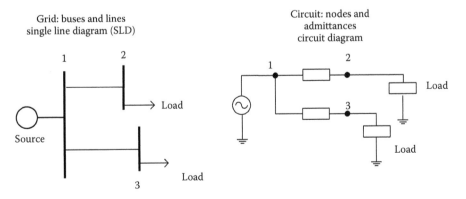

Grid: buses and lines
single line diagram (SLD)

Circuit: nodes and
admittances
circuit diagram

FIGURE 11.12 Grid model as buses and lines. Same concept as circuit modeled by nodes and admittances.

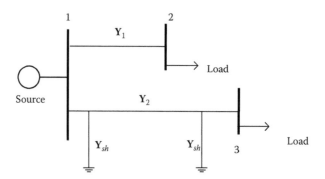

FIGURE 11.13 Grid model of three-bus example with two lines (one short and one medium) identifying known in-series and shunt admittances for each line.

This example illustrates a couple of points. First, it is straightforward to set up the bus admittance matrix using nodal analysis concepts. Second, when the shunt capacitance of a medium line leads to a small shunt admittance, its value may be very small compared to the inductive admittance of the lines and could be ignored when deriving the bus admittance. In these cases, it is easier to set up the bus admittance matrix by considering only the in-series inductive admittances.

11.6 VOLTAGE AND CURRENT

Once we have the bus admittance matrix, we can relate voltages and currents by $\mathbf{I} = \mathbf{YV}$. For instance, using a 3-bus example as in the previous section:

$$\begin{bmatrix} \mathbf{I}_1 \\ \mathbf{I}_2 \\ \mathbf{I}_3 \end{bmatrix} = \begin{bmatrix} \mathbf{Y}_{11} & -\mathbf{Y}_{12} & -\mathbf{Y}_{13} \\ -\mathbf{Y}_{12} & \mathbf{Y}_{22} & -\mathbf{Y}_{23} \\ -\mathbf{Y}_{13} & -\mathbf{Y}_{23} & \mathbf{Y}_{33} \end{bmatrix} \begin{bmatrix} \mathbf{V}_1 \\ \mathbf{V}_2 \\ \mathbf{V}_3 \end{bmatrix} \tag{11.1}$$

First, let us observe that for each bus the current can be partitioned as $\mathbf{I}_i = \mathbf{Y}_{ii}\mathbf{V}_i - \sum_{i \neq j}\mathbf{Y}_{ij}\mathbf{V}_j$. Since $\mathbf{Y}_{ii}$ itself corresponds to a sum of admittances $\mathbf{Y}_{ii} = \sum_k \mathbf{Y}_k$ of all lines connected to the bus (and shunt admittance if significant), the first term $\mathbf{Y}_{ii}\mathbf{V}$ corresponds to a sum of currents injected to the bus. The second term is the sum of currents leaving the node and flowing to other buses.

Second, observe that for each bus we have four quantities that could be known or unknown. These are magnitude and angle for voltage and current. At each bus, there could be a load or a source constraining the magnitude and phase of voltage or current. This observation means, for instance, that solving an equation like Equation 11.1 requires knowing 2×3 quantities for 3 buses. In general, $2 \times N$ for N buses.

Example 11.3

Round up the bus admittance from the previous example to two decimals and assume we know voltages of each bus to be 4.2, 3.8, 3.6 kV with angles 10°, 20°, 0°. Calculate the currents of all three buses. Answer: The matrix equation is

$$\begin{bmatrix} I_1 \\ I_2 \\ I_3 \end{bmatrix} = \begin{bmatrix} -j0.03 & j0.02 & j0.01 \\ j0.02 & -j0.02 & 0 \\ j0.01 & 0 & -j0.01 \end{bmatrix} \begin{bmatrix} 4200\angle 10° \\ 3800\angle 20° \\ 3600\angle 0° \end{bmatrix}$$

and we can calculate using R:

```
Y <- matrix(ncol=3, nrow=3)
Y[1,1] <- 0-3i; Y[1,2] <- 0+2i; Y[1,3] <- 0+i
Y[2,1] <- Y[1,2]; Y[2,2] <- 0-2i; Y[2,3] <- 0+0i
Y[3,1] <- Y[1,3]; Y[3,2] <- Y[2,3]; Y[3,3] <- 0-i
Y <- Y*0.01
Vm <- c(4.2,3.8,3.6)*1000 #kV
Va <- c(10,20,0)
V <- complex(mod=Vm, arg=Va*pi/180)
I <- Y%*%V # matrix multiplication
Im <- c(Mod(I))
Ia <- c(Arg(I)*180/pi) # convert to polar
```

We can then print the admittance, voltage, and current:

```
> print(Y);print(Vm);print(Va);print(round(Im,1)); print(round(Ia,1))
        [,1]    [,2]     [,3]
[1,] 0.00-0.03i 0+0.02i 0.03+0i
[2,] 0.00+0.02i 0-0.02i 0.00+0i
[3,] 0.03+0.00i 0+0.00i -0.03+0i
[1] 4200 3800 3600
[1] 10 20 0
[1] 116.5 16.1 27.2
[1] -26.9 44.7 53.7
> >
```

We see that the currents injected to the buses have magnitudes 116.5, 16.1, 27.2A with angles $-26.9°$, $44.7°$, $53.7°$.

11.7 BASICS OF PER UNIT (P.U.) SYSTEM

The *per unit* (p.u.) expression relates to a ratio of a voltage, current, power, or impedance to a chosen *base* quantity. For example, if the selected base is 120 kV, then a voltage of 108 kV is expressed as $\frac{108\,kV}{120\,kV} = 0.9$ p.u. This concept is useful in power systems dealing with large quantities, for instance kV and kW. A base in kW or kVA is the result of base voltage in kV and base current in A.

In three-phase, the use of p.u. results in the same line and phase voltages. For example, suppose we select a base apparent total power of 30,000 kVA and a base line-to-line voltage V_l of 120 kV. Note that dividing by 3, the base apparent power per phase is 10,000 kVA. Suppose we actually have $V_l = 108$ kV and total $S = 18,000$ kVA. Then, $Vl = \frac{108}{120} = 0.9$ p.u. But the line-to-neutral in p.u. would be $V_{an} = \frac{108/\sqrt{3}}{120/\sqrt{3}} = 0.9$ p.u. because the $\sqrt{3}$ cancels. Therefore, we see that the line and phase are the same in p.u. when the system is balanced. The apparent power in p.u. is

$S = \dfrac{18,000}{30,000} = 0.6$ p.u. and the apparent power per phase is $S = \dfrac{18,000/3}{30,000/3} = 0.6$ p.u. We can see

that total apparent power and apparent power per phase is the same when the system is balanced.

11.8   POWER FLOW

11.8.1   MATRIX EQUATION

Complex *power flow* into a bus is determined by the bus voltage and injected current. This current is determined by the bus admittance $\mathbf{I} = \mathbf{YV}$. Thus, complex power is $\mathbf{S} = \mathbf{VI}^* = \mathbf{V}(\mathbf{YV})^* = \mathbf{VY^*V^*}$. For each bus i the power is

$$\mathbf{S}_i = \mathbf{V}_i \sum_j \mathbf{Y}^*_{ji}\mathbf{V}^*_j \tag{11.2}$$

which will also be possible to write as

$$\mathbf{S}_i = P_i + jQ_i \tag{11.3}$$

Separating real and imaginary parts in Equation 11.2 and equating to their counterparts in Equation 11.3 constitutes the power flow equations. Note that in Equation 11.2, variables are magnitude and angle of voltage, and real and reactive power of the bus. Also, note that power in Equation 11.2 can be partitioned as $\mathbf{S}_i = \mathbf{V}_i\mathbf{Y}^*_{ii}\mathbf{V}^*_i + \mathbf{V}_i \sum_{i \ne j} \mathbf{Y}^*_{ij}\mathbf{V}^*_j$, as we did the current separating the

terms related to the mutual admittances. The power flow problem consists of solving Equation 11.2 combined with Equation 11.3 subject to restrictions on the voltages, currents, and power for each bus. The solution is found numerically and is computationally challenging for networks composed of thousands of buses. In this textbook, we will only cover simple examples of a few buses.

Example 11.4

Use base 1 kV, 0.01 Ω, 1 A, 1 kVA to express variables in p.u. The bus admittance from the previous example on a p.u. to $\mathbf{Y} = \begin{bmatrix} -j3 & j2 & j1 \\ j2 & -j2 & 0 \\ j1 & 0 & -j1 \end{bmatrix}$ and voltages $\mathbf{V} = \begin{bmatrix} 4.2\angle 10° \\ 3.8\angle 20° \\ 3.6\angle 0° \end{bmatrix}$ on p.u.

Calculate the complex power of all three buses. Answer: The matrix equation is

```
Y <- matrix(ncol=3, nrow=3)
Y[1,1] <- 0-3i; Y[1,2] <- 0+2i; Y[1,3] <- 0+1i
Y[2,1] <- Y[1,2]; Y[2,2] <- 0-2i; Y[2,3] <- 0+0i
Y[3,1] <- Y[1,3]; Y[3,2] <- Y[2,3]; Y[3,3] <- 0-1i
# base 1 kV 1MW 1 MVA
Vm <- c(4.2,3.8,3.6)
Va <- c(10,20,0)
V <- complex(mod=Vm, arg=Va*pi/180)
S <- V*Conj(Y)%*%Conj(V) # matrix multiplication
round(S,1)
> round(S,1)
      [,1]
[1,] -2.9+6.6i
[2,]  5.5-2.6i
[3,] -2.6-1.9i
>
```

We have obtained real power $\begin{bmatrix} P_1 \\ P_2 \\ P_3 \end{bmatrix} = \begin{bmatrix} -2.9 \\ 5.5 \\ -2.6 \end{bmatrix}$ and reactive power $\begin{bmatrix} Q_1 \\ Q_2 \\ Q_3 \end{bmatrix} = \begin{bmatrix} 6.6 \\ -2.6 \\ 1.9 \end{bmatrix}$. There

is import of real power at bus 2 and export at bus 1 and 3; vice versa, bus 2 exports reactive power and the other two buses import it.

11.8.2 POWER FLOW FOR A TWO-BUS SYSTEM

Consider an extremely simple case of two buses connected by a short line with impedance $\mathbf{Z}_1 = jX$ or admittance $\mathbf{Y}_1 = -jX^{-1}$. The bus admittance matrix is $\mathbf{Y} = \begin{bmatrix} \mathbf{Y}_{11} & \mathbf{Y}_{12} \\ \mathbf{Y}_{21} & \mathbf{Y}_{22} \end{bmatrix} = \begin{bmatrix} -jX^{-1} & jX^{-1} \\ jX^{-1} & -jX^{-1} \end{bmatrix}$

and then we write the power flow equations

$$\mathbf{S}_1 = \mathbf{V}_1 \mathbf{Y}_{11}^* \mathbf{V}_1^* + \mathbf{V}_1 \mathbf{Y}_{12}^* \mathbf{V}_2^*$$
$$\mathbf{S}_2 = \mathbf{V}_2 \mathbf{Y}^* \mathbf{V}_1^* + \mathbf{V}_2 \mathbf{Y}_{22}^* \mathbf{V}_2^*$$

Substitute admittance $\mathbf{Y}_{ii}$ according to the matrix and assume $\mathbf{V}_i = V_i \angle \theta_i$ for $i = 1, 2$

$$\mathbf{S}_1 = jX^{-1}V_1^2 - jX^{-1}V_1(\cos\theta_1 + j\sin\theta_1)V_2(\cos\theta_2 - j\sin\theta_2)$$
$$\mathbf{S}_2 = jX^{-1}V_2^2 - jX^{-1}V_1(\cos\theta_1 - j\sin\theta_1)V_2(\cos\theta_2 + j\sin\theta_2).$$

Expand

$$\mathbf{S}_1 = jX^{-1}V_1^2 - jX^{-1}V_1V_2[(\cos\theta_1\cos\theta_2 + \sin\theta_1\sin\theta_2) + j(\sin\theta_1\cos\theta_2 - \cos\theta_1\sin\theta_2)]$$
$$\mathbf{S}_2 = jX^{-1}V_2^2 - jX^{-1}V_1V_2[(\cos\theta_1\cos\theta_2 + \sin\theta_1\sin\theta_2) + j(-\sin\theta_1\cos\theta_2 + \cos\theta_1\sin\theta_2)].$$

Collect real and imaginary terms in each equation

$$\mathbf{S}_1 = X^{-1}V_1V_2(\sin\theta_1\cos\theta_2 - \cos\theta_1\sin\theta_2) + j[X^{-1}V_1^2 - jX^{-1}V_1V_2(\cos\theta_1\cos\theta_2 + \sin\theta_1\sin\theta_2)]$$
$$\mathbf{S}_2 = X^{-1}V_1V_2(\sin\theta_1\cos\theta_2 - \cos\theta_1\sin\theta_2) + j[X^{-1}V_2^2 - jX^{-1}V_1V_2(\cos\theta_1\cos\theta_2 + \sin\theta_1\sin\theta_2)]$$

For each equation, use Equation 11.3; make the real part equal to P_i and the imaginary part equal to Q_i:

$$X^{-1}V_1V_2(\sin\theta_1\cos\theta_2 - \cos\theta_1\sin\theta_2) = P_1$$
$$X^{-1}V_1{}^2 - X^{-1}V_1V_2(\cos\theta_1\cos\theta_2 + \sin\theta_1\sin\theta_2) = Q_1$$
$$X^{-1}V_1V_2(\sin\theta_1\cos\theta_2 - \cos\theta_1\sin\theta_2) = P_2 \qquad (11.4)$$
$$X^{-1}V_2{}^2 - X^{-1}V_1V_2(\cos\theta_1\cos\theta_2 + \sin\theta_1\sin\theta_2) = Q_2$$

Observe that $P_1 = P_2 = X^{-1}V_1V_2(\sin\theta_1\cos\theta_2 - \cos\theta_1\sin\theta_2)$ and that each equation can be partitioned into a term $X^{-1}V_i^2$ determined by the main diagonal terms and a term determined by the sum of mutual conductance terms $-X^{-1}V_1V_2(\cos\theta_1\cos\theta_2 + \sin\theta_1\sin\theta_2)$. Rewrite Equation 11.4 as

$$X^{-1}V_1V_2(\sin\theta_1\cos\theta_2 - \cos\theta_1\sin\theta_2) = P_1 = P_2$$
$$X^{-1}V_1[V_1 - V_2(\cos\theta_1\cos\theta_2 + \sin\theta_1\sin\theta_2)] = Q_1 \qquad (11.5)$$
$$X^{-1}V_2[V_2 - V_1(\cos\theta_1\cos\theta_2 + \sin\theta_1\sin\theta_2)] = Q_2$$

Let us consider the following observations regarding Equation 11.4 or Equation 11.5. First, these equations are nonlinear because they contain cross multiplication of voltage variables. Second, we have a total of four voltage variables (magnitude and angle for each bus) and three power variables, and only three equations; thus, $7 - 3 = 4$ variables have to be defined or constrained. You can now picture that the power flow equation for one bus connecting to several or many buses would have similar structure as Equation 11.5 but will include a summation of terms.

Example 11.5

Use base 1 kV, 0.01 Ω, 1 MW, 1 MVA, 1 MVAR to express variables in p.u. Suppose $X = 1$ and we constrain $P_1 = 1, V_1 = 13.8, \theta_1 = 6°, \theta_2 = 0°$ in p.u. How would you solve the system of equations? Answer: Using known values in the first equation of Equation 11.5 $13.8 \times V_2 \sin 6° = 1$, we can solve for V_2 to get $V_2 = \dfrac{1}{13.8 \sin 6°} = 0.69$ p.u. Substitute this value in the second and third equations to find Q1 and Q2 as

$$13.8 \times (13.8 - 0.693 \cos 6°) = Q_1 = 180.93 \text{ p.u}$$

$$0.693 \times (0.693 - 13.8 \cos 6°) = Q_2 = -9.03 \text{ p.u}$$

The results indicate that we inject reactive power into bus 1 and that reactive power for bus 2 is negative, indicating power imported from that bus.

This example illustrates the power flow problem in a simple case, how to specify the constraints to solve it, and gives us a glimpse that it would have to be computerized in order to find the solution for thousands of buses. In order to do that, our next step is to consider restrictions to some of the variables in Equation 11.2 based on the type of bus and fixing two variables per bus. For instance, for a bus connected to a generator, we could assume that real power and terminal voltage amplitude are known, or that angle and voltage are fixed. For a bus feeding a load, we can specify both the real and reactive power. An *infinite bus* assumes that voltage is fixed in magnitude and angle. For instance, if bus 2 were to be an infinite bus, then V_2 and θ_2 are given. We will expand on the concept of infinite bus in the next section.

Example 11.6

Write the power flow equations for two buses such that bus 1 is a generator bus with $P_1 = P$ and $\mathbf{V}_1 = V_1 \angle \theta_1$. Bus 2 is an infinite bus with $\mathbf{V}_2 = V_2 \angle 0°$. Answer: Substitute these conditions in Equation (11.5):

$$\begin{aligned}
X^{-1} V_1 V_2 \sin \theta_1 &= P_1 \\
X^{-1} V_1 (V_1 - V_2 \cos \theta_1) &= Q_1 \\
X^{-1} V_2 (V_2 - V_1 \cos \theta_1) &= Q_2
\end{aligned} \qquad (11.6)$$

We now see that if P and V_1 are set by the generator, and bus 1 and V_2 are set by infinite bus 2, we can solve for θ_1 from the first equation, and substitute in the second and third equations to find reactive power Q_1 and Q_2. Thus, the power flow problem is solved. In the next section, we further analyze this important case of a generator bus connected to an infinite bus.

11.8.3 GENERATOR CONNECTED TO AN INFINITE BUS

In Chapter 10, we solved a generator problem assuming that the generator is stand-alone or independent of other generators that may be connected to the load. However, when a generator is connected to the grid, its behavior is constrained by all other generators. Consider the circuit shown in Figure 11.14. This model is called an infinite bus or infinite bus–bar model and is applied to a generator connected to a large grid via a short line with reactance X. It is assumed that the grid has large generating capacity such that the voltage V and frequency are not affected if we change the mechanical power driving the generator. To simplify notation with respect to the previous section, we will use $\mathbf{V}_2 = \mathbf{V}$ and $\mathbf{V}_1 = \mathbf{E}$.

Taking the voltage $\mathbf{V}$ at the infinite bus as a reference in terms of phasor angle, we can write $V \angle 0°$. Then, the voltage from the generator is denoted as phasor $E \angle \delta$, where angle δ is called the *power angle*. As we explained in Chapter 10, the magnitude E of $\mathbf{E}$ is controlled by the field current or excitation provided to the generator, and the angle δ of $\mathbf{E}$ is controlled by the torque exerted by the prime mover.

The line impedance is $\mathbf{Z} = jX = X \angle 90°$. For this simple bus admittance, we can calculate current using the difference between $\mathbf{E}$ and $\mathbf{V}$ and dividing by the impedance $I \angle \phi = \dfrac{E \angle \delta - V \angle 0°}{jX}$. The calculation is conveniently done using rectangular form, since division by j is just $-j$. Then, $I \angle \phi = \dfrac{E(\cos \delta + j \sin \delta) - V}{jX} = \dfrac{-jE \cos \delta + E \sin \delta + jV}{X}$. Collecting real and imaginary parts, we have the current in rectangular form

$$I \angle \phi = \frac{E \sin \delta + j(V - E \cos \delta)}{X} \tag{11.7}$$

We can see that the relative magnitude of $E \cos \delta$ compared to V determines the angle ϕ of the current. The angle ϕ will be positive when $V - E \cos \delta > 0$ and negative when $V - E \cos \delta < 0$. The phasor diagram for these conditions are illustrated in Figure 11.15. Under the first condition, $E \cos \delta < V$, which gives $\phi > 0$, the current $\mathbf{I}$ leads the bus voltage $\mathbf{V}$, reactive power $Q = VI \sin \phi$ is positive, and thus the generator *imports reactive power*. This condition is called *underexcited*, implying that excitation by the field current is not large enough for $E \cos \delta$ to acquire the bus voltage V (Figure 11.15). On the contrary, when $E \cos \delta < V$, which gives $\phi < 0$, the current lags the voltage $\mathbf{V}$, reactive power $Q = VI \sin \phi$ is negative, and thus the generator *exports reactive power*. This condition is called *overexcited*, implying that excitation by the field current is large enough for $E \cos \delta$ to acquire the bus voltage V (Figure 11.15).

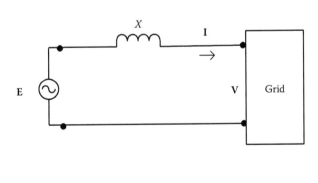

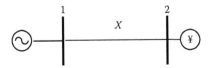

FIGURE 11.14   Generator bus and infinite bus.

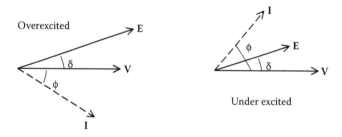

FIGURE 11.15 Phasor diagram of infinite bus conditions.

For a given E, the critical value δ_{cr} of the angle δ at which we change mode is found from $E\cos\delta_{cr} = V$ to get $\delta_{cr} = \cos^{-1}(V/E)$. For values $\delta < \delta_{cr}$, the generator is in overexcited mode, whereas for values $\delta < \delta_{cr}$, the generator is in underexcited mode. For a given δ, the critical value E_{cr} of E, in which the change mode is found from $E_{cr}\cos\delta = V$ to get $E_{cr} = V/\cos\delta$. For values $E > E_{cr}$, the generator is in overexcited mode, whereas for values $E > E_{cr}$, the generator is in underexcited mode.

Example 11.7

Assume infinite bus voltage V = 13.8 kV and consider two cases: (a) We increase the field current to achieve E = 15 kV. What is the critical δ? (b) We increase the torque of the prime mover to achieve δ = 10°. What is the critical E? Answer: (a) $\delta_{cr} = \cos^{-1}(V/E) = \cos^{-1}(13.8/15) = 23.073°$, we cannot get past this angle to stay in overexcited mode. (b) $E_{cr} = V/\cos\delta = 13.8/\cos(10°) = 14.01$ kV, we have to increase the field current to at least this value to stay in overexcited mode.

Of interest is to calculate the power flow or power at the sending (generator) and receiving (bus) ends. We will work with the conjugate of complex power because in this case it is easier to see the conjugate of voltage than the conjugate of current. In general for a voltage **V** and a current **I**, complex power is $\mathbf{S} = \mathbf{VI}^* = P + jQ$ and the conjugate of complex power is $\mathbf{S}^* = \mathbf{V}^*\mathbf{I} = P - jQ$. Note, that calculating the conjugate of complex power yields an opposite sign of reactive power.

The conjugate of complex power at the generator is $\mathbf{S}_1^* = \mathbf{E}^*\mathbf{I} = P_1 - jQ_1$.

$$\mathbf{S}_1^* = \mathbf{E}^*\mathbf{I} = (E(\cos\delta - j\sin\delta))\left(\frac{E(\cos\delta + j\sin\delta) - V}{jX}\right) = \left(\frac{E^2}{jX}\right) - \left(\frac{E(\cos\delta - j\sin\delta)V}{jX}\right)$$

therefore,

$$\mathbf{S}_1^* = \left(-j\frac{E^2}{X}\right) - \left(\frac{EVj\cos\delta - EV\sin\delta}{X}\right) = P_1 - jQ_1$$

The real part is $P_1 = \dfrac{EV}{X}\sin\delta$ and the imaginary part is $Q_1 = \dfrac{E^2}{X} - \dfrac{EV}{X}\cos\delta$. Likewise, the conjugate of complex power at the infinite bus is $\mathbf{S}_2^* = \mathbf{V}^*\mathbf{I} = P_2 - jQ_2$

$$\mathbf{S}_2^* = \mathbf{V}^*\mathbf{I} = (V + j0)\left(\frac{E(\cos\delta + j\sin\delta) - V}{jX}\right) = \frac{EV}{X}(\sin\delta - j\cos\delta) + (jV^2) = P_2 - jQ_2,$$

the real part is $P_2 = \dfrac{EV}{X}\sin\delta$ and the imaginary part is $Q_2 = \dfrac{EV}{X}\cos\delta - \dfrac{V^2}{X}$.

In summary, the real power

$$P_1 = P_2 = \frac{EV}{X} \sin \delta \qquad (11.8)$$

is the same for both ends, and the reactive power is

$$Q_1 = \frac{E(E - V \cos \delta)}{X} \qquad Q_2 = \frac{V(E \cos \delta - V)}{X} \qquad (11.9)$$

For overexcited conditions $Q_2 > 0$, which means the generator exports reactive power to the grid, whereas for underexcited conditions $Q_2 < 0$, which means there is reactive import from the grid. From Equation 11.8, there are two ways of increasing real power transfer: One is to increase $\frac{EV}{X}$ by increasing field current (and thus E) and the other is to increase $\sin \delta$ by increasing torque from the prime mover. The maximum real power is $\frac{E_{max} V}{X}$, where E_{max} is the maximum emf (determined by the maximum field current that can be applied). From Equation 11.9, the sign of the term $E \cos \delta - V$ is determined when we change mode from overexcited to underexcited. We already know the critical value δ_{cr} of the angle δ at which we change mode is $\delta_{cr} = \cos^{-1}(V/E)$. For values $\delta < \delta_{cr}$, the generator is in overexcited mode and Q_2 is positive, whereas for values $\delta < \delta_{cr}$, the generator is in underexcited mode and Q_2 is negative.

Example 11.8

Assume infinite bus voltage $V = 13.8$ kV. We increase the field current to achieve $E = 15$ kV. What is the critical δ? What is the real power and reactive power for two cases: (a) $\delta = 15°$ and (b) $\delta = 30°$. Answer: $\delta_{cr} = \cos^{-1}(V/E) = \cos^{-1}(13.8/15) = 23.073°$; we cannot get past this angle to stay in overexcited mode. (a) For $\delta = 15°$, the system would be in overexcited mode. The real power is $P_2 = \frac{EV}{X} \sin \delta = \frac{15 \times 13.8}{1} \sin 15° = 53.57$ MW, and the reactive power is $Q_2 = \frac{V(E \cos \delta - V)}{X} = \frac{13.8 \times (15 \times \cos 15° - 13.8)}{1} = 9.51$ MVAR. This is positive and therefore exported. (b) For $\delta = 30°$, the system would be in underexcited mode. The real power is $P_2 = \frac{EV}{X} \sin \delta = \frac{15 \times 13.8}{1} \sin 30° = 103.5$ MW, and the reactive power is $Q_2 = \frac{V(E \cos \delta - V)}{X} = \frac{13.8 \times (15 \times \cos 30° - 13.8)}{1} = -11.17$ MVAR. This is negative and therefore imported.

These calculations can be performed using function infinite.bus of renpow. For instance, for $V = 13.8$ kV, $X = 1$, we can explore the effect of a sequence of E (14, 15, 16 kV) and $\delta = 6, 7, ..., 30°$

```
x <- list(V=13.8,X=1,E=seq(14,16,1),delta=seq(6,30,1))
infinite.bus(x)
```

The resulting plots shown in Figure 11.16 contain real power, reactive power, and current (magnitude and angle) for this set of values of E and δ. We can see that for $E = 14$, 15, and 16 kV, the critical values of angle δ are 9°, 23°, and 30°, respectively. At these critical values, each one of the reactive power curves cross the y-axis and become negative. Similarly at these values, the angle of the current goes from negative (lagging V) to positive (leading V).

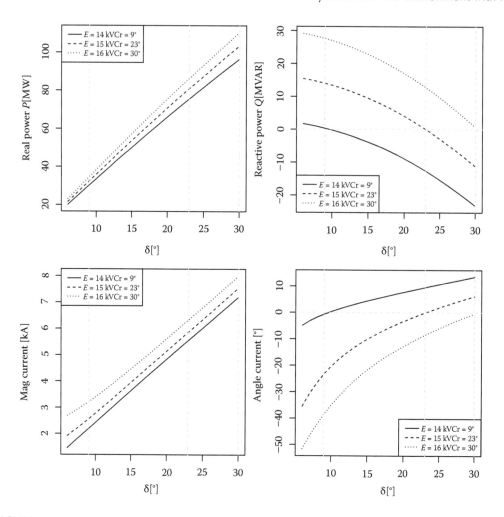

FIGURE 11.16 Infinite bus: power and current as a function of E and δ.

11.8.4 NEWTON-RAPHSON SOLUTION

As mentioned before, it is necessary to find a computer solution to the power-flow problem when we have thousands of buses. Here we will explain the Newton-Raphson method, and for simplicity, we will explain how it works for a simple two-bus system [1]. Let us revisit Equation 11.6 and consider only the two equations for bus 1:

$$X^{-1}V_1 V_2 \sin \theta_1 = P_1$$
$$X^{-1}V_2(V_2 - V_1 \cos \theta_1) = Q_2$$
(11.10)

Suppose V_1 and P_1 are fixed by the generator and Q_2 is a known power to bus 2. We have two remaining unknowns, V_2 and θ_1, which can be determined from two equations. However, these equations are nonlinear. One method to solve a system of nonlinear equations is to use the Newton-Raphson algorithm, which is based on the *Jacobian* matrix.

For simplicity, define vector of variables $\mathbf{x} = \begin{bmatrix} x_1 \\ x_2 \end{bmatrix} = \begin{bmatrix} \theta_1 \\ V_2 \end{bmatrix}$ and rewrite Equation 11.10 as

$$
\begin{aligned}
f_1(x_1, x_2) &= X^{-1} V_1 x_2 \sin x_1 - P_1 = 0 \\
f_2(x_1, x_2) &= X^{-1} x_2^{\,2} - X^{-1} V_1 x_2 \cos x_1 - Q_2 = 0
\end{aligned}
\tag{11.11}
$$

and in vector form

$$
\mathbf{f}(\mathbf{x}) = \begin{bmatrix} f_1(x_1, x_2) \\ f_2(x_1, x_2) \end{bmatrix} = \begin{bmatrix} X^{-1} V_1 x_2 \sin x_1 - P_1 \\ X^{-1} x_2^{\,2} - X^{-1} V_1 x_2 \cos x_1 - Q_2 \end{bmatrix} = \begin{bmatrix} 0 \\ 0 \end{bmatrix}
\tag{11.12}
$$

A Taylor expansion of $\mathbf{f}(\mathbf{x})$ is

$$
\mathbf{f}(\mathbf{x} + \delta\mathbf{x}) = \begin{bmatrix} f_1(\mathbf{x} + \delta\mathbf{x}) \\ f_2(\mathbf{x} + \delta\mathbf{x}) \end{bmatrix} = \begin{bmatrix} f_1(\mathbf{x}) \\ f_2(\mathbf{x}) \end{bmatrix} + \begin{bmatrix} \dfrac{\partial f_1}{\partial x_1} & \dfrac{\partial f_1}{\partial x_2} \\[2mm] \dfrac{\partial f_2}{\partial x_1} & \dfrac{\partial f_2}{\partial x_2} \end{bmatrix} \begin{bmatrix} \delta x_1 \\ \delta x_2 \end{bmatrix}
\tag{11.13}
$$

which can further be rewritten as

$$
\mathbf{f}(\mathbf{x} + \delta\mathbf{x}) \approx \mathbf{f}(\mathbf{x}) + \mathbf{J_f}(\mathbf{x})\delta\mathbf{x}
\tag{11.14}
$$

where $\mathbf{J_f}(\mathbf{x})$ is the Jacobian matrix, and given by partial derivatives of each function with respect to each one of the unknown variables

$$
\mathbf{J_f}(\mathbf{x}) = \begin{bmatrix} \dfrac{\partial f_1}{\partial x_1} & \dfrac{\partial f_1}{\partial x_2} \\[2mm] \dfrac{\partial f_2}{\partial x_1} & \dfrac{\partial f_2}{\partial x_2} \end{bmatrix}
\tag{11.15}
$$

In the case we are studying, the Jacobian based on Equation 11.12 is

$$
\mathbf{J_f}(\mathbf{x}) = \begin{bmatrix} X^{-1} V_1 x_2 \cos x_1 & X^{-1} V_1 \sin x_1 \\ X^{-1} V_1 x_2 \sin x_1 & X^{-1} 2x_2 - X^{-1} V_1 \cos x_1 \end{bmatrix}
\tag{11.16}
$$

We can solve for $\delta\mathbf{x}$ from Equation 11.14 to get $\delta\mathbf{x} = \mathbf{J_f}(\mathbf{x})^{-1}(\mathbf{f}(\mathbf{x} + \delta\mathbf{x}) - \mathbf{f}(\mathbf{x})) = -\mathbf{J_f}(\mathbf{x})^{-1}\mathbf{f}(\mathbf{x})$ and then $\mathbf{x} + \delta\mathbf{x} = \mathbf{x} - \mathbf{J_f}(\mathbf{x})^{-1}\mathbf{f}(\mathbf{x})$. Therefore we can find the roots iteratively by

$$
\mathbf{x}_{n+1} = \mathbf{x}_n - \mathbf{J_f}(\mathbf{x}_n)^{-1}\mathbf{f}(\mathbf{x}_n)
\tag{11.17}
$$

starting from an initial guess $\mathbf{x}_0$ until the difference $\mathbf{x}_{n+1} - \mathbf{x}_n$ is negligible or reduced below a predefined tolerance.

We will look at how to apply the Newton-Raphson method in R using package rootSolve for the case of two buses under study. First, we define the function $\mathbf{f(x)}$ and its Jacobian $\mathbf{J_f(x)}$ matrix. For this purpose we use the system 11.12 and matrix 11.16 reproduced here for easy reference

$$\mathbf{f(x)} = \begin{bmatrix} X^{-1}V_1x_2\sin x_1 - P_1 \\ X^{-1}x_2^2 - X^{-1}V_1x_2\cos x_1 - Q_2 \end{bmatrix}$$

$$\mathbf{J_f(x)} = \begin{bmatrix} X^{-1}V_1x_2\cos x_1 & X^{-1}V_1\sin x_1 \\ X^{-1}V_1x_2\sin x_1 & X^{-1}2x_2 - X^{-1}V_1\cos x_1 \end{bmatrix}$$

Define the f function:

```
f <- function(x,parms){
X <- parms$X; V1 <- parms$V1
P1 <- parms$P1; Q2 <- parms$Q2
y <- array()
y[1] <- X^-1*V1*x[2]*sin(x[1]*pi/180)-P1
y[2] <- X^-1*x[2]^2 -X^-1*V1*x[2]*cos(x[1]*pi/180)-Q2
return(y)
}
```

Define the Jacobian matrix:

```
Jf <- function(x,parms){
X <- parms$X; V1 <- parms$V1
P1 <- parms$P1; Q2 <- parms$Q2
nx <- length(x)
Y <- matrix(nrow=nx,ncol=nx)
Y[1,1] <- X^-1*V1*x[2]*cos(x[1]*pi/180)
Y[1,2] <- X^-1*V1*sin(x[1]*pi/180)
Y[2,1] <- X^-1*V1*x[2]*sin(x[1]*pi/180)
Y[2,2] <- X^-1*2*x[2] - X^-1*V1*cos(x[1]*pi/180)
return(Y)
}
```

Implement the solution by providing parameter values, initial guess, and running the solver:

```
parms <- list(X=1,V1=1,P1=1,Q2=1) # p.u.
x0 <- c(45,1.41)
f(x0,parms)
Jf(x0,parms)
require(rootSolve)
multiroot(f=f,start=x0,jacfunc=Jf,parms=parms,jactype="fullusr",
maxiter=1000)
```

Results:

```
> multiroot(f=f,start=x0,jacfunc=Jf,parms=parms,jactype="fullusr",
maxiter=1000)
$root
[1] 44.999991 1.414214
$f.root
[1] -1.535216e-07 -1.533620e-07
```

```
$iter
[1] 3
$estim.precis
[1] 1.534418e-07
>
```

From this simple example of two buses, we can appreciate that solving the power flow for a large network of buses is computationally demanding. Other methods are routinely used to solve the power flow, such as the *Gauss-Seidel method*. We will not go into more detail in this introductory book, but you will find excellent explanations in power system textbooks [1,3].

11.9 DEMAND

The loads to the electric power system, that is to say, the P and Q terms in the power-flow equations, are changing instantaneously all the time. Therefore, the network of generating plants have to produce electricity according to varying demand from the loads and dispatch the power according to demand. It is not difficult to imagine that operating a large grid requires tools to perform real-time control of a complex dynamic system.

In the United States, individual utilities are involved in electricity delivery including grid operation, transmission, and distribution. However, in some areas, regional organizations coordinate operation, transmission, and distribution, for instance, independent system operators (ISOs) and regional transmission organizations (RTOs). These are composed of utilities and federal and state regulators.

Grids are interconnected, making operation even more challenging. For instance, in the state of Texas, the grids are interconnected and managed by the Electric Reliability Council of Texas (ERCOT). It represents 85% of the Texas load covering 22 million customers, supplying electricity using 550 generation-units and 40,000 miles of transmission lines.

11.9.1 DAILY REGIME

Hourly demand within a day varies according to industrial, commercial, and residential activities. Demand shows a *peak hour* in the day that varies with location, season, and month. Depending on location, there is a peak day in the month and the season.

To provide concrete examples, we will use data from ERCOT [4]. From this website, one can download files that contain the demand on an hourly basis for the entire year [5]. Therefore, a file for one year has $24 \times 365 = 8760$ hourly records. The demand Hour_End is broken down by zone COAST, EAST, FAR_WEST, NORTH, NORTH_C, SOUTHERN, SOUTH_C, WEST, and ERCOT. The latter is the total. The _C designates as CENTRAL, for instance, NORTH_C is North Central Texas. Because each record has a time stamp consisting of date and hour, this type of data set is better analyzed as a *multivariate time series*, that is, a set of variables that are a function of real time, or equivalently a data object, where each multivalued record has a time stamp. As an example, renpow includes a file ERCOT2010.csv with hourly values for year 2010. You can read this from your local extdata folder or the system's by employing system.file

```
X <- read.table("extdata/ERCOT2010.csv",header=TRUE,sep=",")
```

or directly using the dataset using X <- ERCOT2010.

In R, there are several methods and packages to analyze time series. The simplest is ts, which is used by functions seg.ts and plot.series of renpow. For instance

```
x.t <- seg.ts(X,dh0="1/1/2010 1:00",dhf="1/1/2011 0:00",c(1:9))$x.t
```

converts the entire series contained in X.

A convenient package is xts, which in turn is based on package zoo. You can install and load xts for this part of the chapter. For instance we can convert X to a xts time series using

```
tt <- strptime(X[,1], format="%m/%d/%Y %H:%M")
x.ts <- xts(round(X[,2:10],2),tt)


filename <- "ERCOT2010.csv"
x.t <- read.ERCOT.make.xts(filename)
```

To examine the result, look at the first five records using:

```
> x.ts[1:5,]
                    COAST    EAST FAR_WEST    NORTH   NORTH_C SOUTHERN   SOUTH_C
2010-01-01 01:00:00 7775.68 1238.22 1237.78 950.70 12406.21 2467.65 5032.08
2010-01-01 02:00:00 7705.23 1236.12 1249.07 957.67 12486.48 2436.16 5111.74
2010-01-01 03:00:00 7651.33 1244.20 1251.42 954.24 12560.71 2414.68 5169.40
2010-01-01 04:00:00 7665.58 1253.03 1260.32 970.20 12669.99 2383.60 5241.12
2010-01-01 05:00:00 7745.08 1284.08 1264.76 982.34 12912.47 2498.22 5401.54
                    WEST    ERCOT
2010-01-01 01:00:00 1059.77 32168.09
2010-01-01 02:00:00 1064.67 32247.14
2010-01-01 03:00:00 1072.06 32318.04
2010-01-01 04:00:00 1086.68 32530.52
2010-01-01 05:00:00 1111.87 33200.37
>
```

The data set is in MW and the time stamp includes date and hour. In renpow, we also include a function series.plot to plot any segment of this data set. We can select a segment by specifying start date–hour and end date–hour, as well as column numbers to select variables.

Example 11.9

Let us examine the daily regime for two days, one in January (winter season) and one in July (summer season), for three zones of ERCOT. This function uses ts and does not require an additional package (such as xts). Answer: In the following, we select 00:00 to 23:00 hours for January 4 and July 5, and variables 1, 5, and 7 that correspond to COAST, NORTH_C, and SOUTH_C.

```
# week day Monday Jan 04 2010
x.t.wd.wt <- seg.ts(X,dh0="1/4/2010 0:00",dhf="1/4/2010 23:00",c(1,5,7))$x.t.seg
series.plot(x.t.wd.wt)
# week day Monday Jul 05 2010
x.t.wd.su <- seg.ts(X,dh0="7/5/2010 0:00",dhf="7/5/2010 23:00",c(1,5,7))$x.t.seg
series.plot(x.t.wd.su)
```

The results are illustrated in Figure 11.17. We can see how this winter day shows two peaks of demand in a day (morning and evening), whereas the summer day shows only one peak in the late afternoon.

11.9.2 Weekly Regime

Daily patterns are then part of weekly, monthly, and annual regimes. A weekly pattern is discernible, plotting hourly data. However, monthly and annual regimes can be examined by first summarizing the data at daily level, for example, taking the maximum and minimum demand of each day.

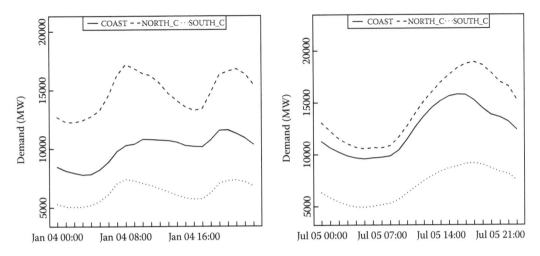

FIGURE 11.17   Examples of daily regime of load demand for winter and summer at three zones of ERCOT.

Example 11.10

Let us examine the weekly regime for two weeks—one in January (winter season) and one in July (summer season)—for three zones of ERCOT. Answer: In the following, we select Sunday, January 24 at 00:00 to Saturday, January 30 at 23:00 for winter. Then select Sunday, August 1 00:00 to Saturday, August 7 23:00, and variables 1, 5, and 7 that corresponds to COAST, NORTH_C, and SOUTH_C.

```
# week Sunday Jan 24 to Saturday Jan 30 2010
x.t.wd.wt <- seg.ts(X,dh0="1/24/2010 0:00",dhf="1/30/2010 23:00",c(1,5,7))$x.t.seg
series.plot(x.t.wd.wt)
# week Sunday Aug 1 to Saturday Aug 7 2010
x.t.wd.su <- seg.ts(X,dh0="8/1/2010 0:00",dhf="8/7/2010 23:00",c(1,5,7))$x.t.seg
series.plot(x.t.wd.su)
```

The results are illustrated in Figure 11.18. We can see how this winter day shows two peaks of demand in a day, increasing toward Saturday. The summer weeks show only one peak per day, which diminishes in the weekend. These patterns are not constant because they depend on actual weather for those days.

Next, we look at an entire year by extracting daily maximum and minimum for each day. For this task, we use the function daily.min.max of renpow. All we have to do is select a zone of ERCOT file as shown later. In addition, we can draw lines corresponding to installed capacity in MW and labels for these installed generation units.

Recall from Chapters 7 and 9 that coal-fired power plants constitute a major source of power generation and run continuously providing *baseload* generation, that is, covering the bulk of the demand running on a continuous basis and typically at lower cost. In contrast, natural gas (NG)–fired power plants do not have to run continuously and can cover daily peaks in demand (peaking power plants). Recall that natural gas combined cycle (NGCC) power plants are gas-based conversion combined with coal. Intermediate capacity is a bracket of generation between baseload and peaking. Other examples of baseload generation facilities are nuclear power plants (Chapter 7) and large hydroelectric power plants (which we study in the Chapter 12).

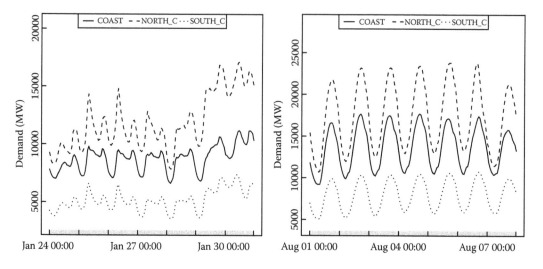

FIGURE 11.18 Examples of weekly regime of load demand for winter and summer at three zones of ERCOT.

Example 11.11

Draw a figure of NORTH_C zone for 2010 showing daily maximum and minimum, and add horizontal lines for hypothetical installed capacities of 10 GW (coal-steam), 8 GW (NGCC), and 8 GW (NG) corresponding to baseload, intermediate, and peaking loads. Answer: We select column 5 of x.t for NORTH_C and input the installed capacity information. Finally, call function daily.min.max. It should be noted that these capacities are for illustration only; they are not real capacities of ERCOT.

```
x.t.y <- x.t[,5];ylabel <- colnames(x.t)[5];inst.cap <- c(10000,8000,8000) # MW
inst.lab <- c("Baseload","Intermediate", "Peaking")
daily.min.max(x.t.y,ylabel,inst.cap,inst.lab)
```

The results are displayed in Figure 11.19. These curves correspond to a total of 24 h/d × 365 d/yr = 8760 h/yr.

An alternative manner of displaying this type of information is the load–duration curve, which is the subject of the next section.

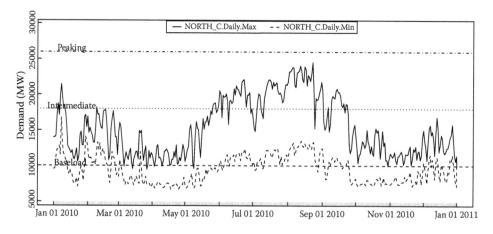

FIGURE 11.19 Example of daily minimum and maximum load for the entire year illustrating installed capacity.

11.9.3 LOAD–DURATION CURVE

A *load–duration curve* consists of a plot of hourly load values versus hours per year showing this value of load. In a year, we have 24 h/d × 365 d/yr = 8760 h/yr hourly demand or load values in MW. These values are ordered in descending order of magnitude and plotted versus the hours from 1 to 8760. An example is shown in Figure 11.20, which we have built using function load.duration of renpow. At the start of this function, a call is made to a function sort applied to the numeric values of the time series. These are the values of the time series excluding the time stamp.

```
load <- sort(as.numeric(x.t.y),decreasing=T)
```

Other aspects of function load.duration are to calculate the area under the curve. Note that the area under the curve is energy in MWh since we are integrating power with respect to time in hours. The total area under the curve is the MWh/year.

We can add horizontal lines to this graph to represent various installed generation capacity. In the example in Figure 11.20, the baseload line represents generation capacity of 10 GW, the second line intermediate represents an additional 8 GW for a total of 18 GW, and the last line for peaking represents an additional 8 GW for a total of 26 GW. Then, observe the following: The load is always between ~8000 MW and ~25000 MW. It is useful at this point to look again at Figure 11.19 for a better understanding of these limits, as the load is displayed chronologically. The load is greater than 10 GW (baseload line) for 6726 hours in the year. Going back to Figure 11.19, these will be all the hourly values above the baseload line that would be met by intermediate and peaking generation. Climbing lines in the graph, the load is greater than 18 MW (baseload plus intermediate) for 863 hours in the year. Going back to Figure 11.19, these will be all the hourly values above the intermediate line that would be met by peaking generation.

For each slice shown hatched in Figure 11.20, the area in the polygon A_i bounded by the line for capacity i and the curve represents the utilization of that capacity in MWh. The total possible utilization B_i is the area of the rectangle with height equal to capacity of segment i, and width equal to the total hours/year (8760 h). The ratio $CF_i = \dfrac{A_i}{B_i}$ is the *capacity factor* of that segment i of generation capacity. For instance, take the hatched for baseload; we can see that it is not utilized only for 8760 − 6726 = 2034 h in the year. In other words, only a small part of the rectangle for total possible utilization is above the curve beyond 6726 h. The ratio displayed CF = 0.97 indicates that baseload units are utilized 97% of the time in the year.

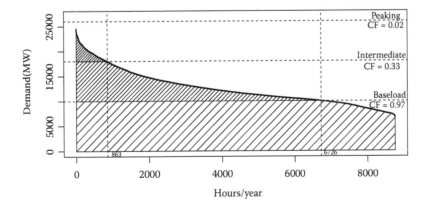

FIGURE 11.20 Example of load–duration curve illustrating installed capacity and CF values.

Example 11.12

Let us build a load–duration curve and calculate CF for the load data of zone NORTH_C and a set of hypothetical generation capacities as in the previous example. Answer: We use function load.duration of renpow:

```
x.t.y <- x.t[,5]
inst.cap <- c(10000,8000,8000) # MW
inst.lab <- c("Baseload","Intermediate", "Peaking")
load.duration(x.t.y, inst.cap,inst.lab)
```

The result is precisely Figure 11.20, which we have been discussing to explain the concepts of load duration and CF. Note from this figure that CFs are 0.97, 0.33, and 0.02 for baseload, intermediate, and peaking, respectively. Function load.duration returns relevant values for each generation bracket, hours possible, utilization energy in TWh, and CF as follows:

```
> load.duration(x.t.y, inst.cap,inst.lab)
$Hours
[1] 6726 863   1
$energy
[1] "TWh"
$possible
[1] 87.60 70.08 70.08
$utilization
[1] 85.01 23.15 1.75
$CF
[1] 0.97 0.33 0.02
>
```

Please recall that these portions of generation capacities are for illustration only; they are not real capacities of that zone of ERCOT. Capacity factor here was defined only on load duration. Other aspects influence the capacity factor. More generally, for generation unit, we can think of capacity factor as the ratio of energy delivered over its capacity. Energy delivered is lower than capacity for several reasons. For instance, a plant may be out of service for maintenance or failure, or a plant does not produce because the demand is low or the revenue is low compared to costs. As we study more renewables such as hydro, wind, and solar generation, capacity factor is also related to lack of the resource, for instance, no wind, no sun, or no water flow.

11.10   POWER DELIVERY: ENVIRONMENTAL RELATIONSHIPS

This chapter should have illustrated that the electric power system is a large and complex system. For instance, the U.S. power industry includes 275,000 miles of high-voltage transmission lines, millions of miles of distribution lines, 950 GW of generating capacity, and covers 300 million people. Such an enormous system has environmental consequences. For example, in terms of air pollution and carbon emissions alone, the U.S. power system is responsible for three-fourths of U.S. sulfur oxides (SO_x) emissions, one-third of U.S. CO_2 emissions, one-third of U.S. NO_x emissions, and one-fourth of U.S. PM and toxic heavy metals. Air pollution and carbon emissions are mostly associated with generation, and mainly because coal-fired power plants represent the bulk of generation capacity.

Besides emissions, other aspects of the electric power industry relate to the environment. These are associated with *electricity delivery* [6]. For instance, the hundreds of thousands of kilometers of transmission lines have an impact on land use and land cover, due to land for *right-of-way* (ROW) and construction. Depending on the ecosystems traversed, lines ROW impact wildlife habitat and may aid propagation of invasive species. Pollution is also possible, for example, ozone, heat, EM waves, and unintended leaks of sulfur hexafluoride (SF_6; an insulator used for high-voltage breakers, switches, and related equipment), which is a greenhouse gas. Also depending on landscapes traversed, a very important impact is on the scenic value of nature [7,8].

Increasingly, these impacts in turn constrain the increase of transmission infrastructure by increasing the time it takes to plan, site, and obtain permits for new lines [8]. In some cases, these times can be longer than the time to site and permit a plant based on renewables. Public perception of impacts of transmission lines assign higher concern in terms of aesthetics, property values, and health.

These factors represent a challenge to develop new technology and implement changes that would increase power transmission capacity in existing corridors, but maintain reliability without excessive cost increase [8]. Alternatives include reevaluation of line static rating, which is the maximum sustained current carried by the line, that does not exceed limiting temperature and clearance, subject to environmental conditions (wind speed, solar radiation, ambient temperature). More demanding in cost and technology is to implement dynamic rating, which uses real-time monitoring of ambient (meteorological) and line conditions (e.g., sagging sensors and mitigation devices) to adjust the rating while minimizing overload. In some cases, it is possible to replace existing conductors for greater size, ampacity, and lower sag, as well as reduce space between conductors, and additional phases; this requires new towers. Other alternatives are to increase voltage further in order to reduce current and to implement high-voltage DC lines (HVDC), requiring AC to DC conversion and vice versa. Visual impact on scenic landscapes could be reduced by taking into account topography and land cover, by integrating the transmission infrastructure with the landscape, and by creative tower design [8].

11.11 OTHER TOPICS RELEVANT TO THE GRID

Several other aspects of operating a grid are very important but are beyond the scope of this textbook. *Costs and revenue* are analyzed taking fixed costs (capital, tax, fixed maintenance and operation) and variable costs (fuel and variable maintenance and operation). A projection of total costs in a year allows calculating revenue and designing a rate structure in $/kWh. Screening curves are plots of levelized cost of energy versus capacity factor.

Faults and protection is a theme of great importance in power systems and treated at length in power system textbooks [1,2]. A short-circuit fault occurs when two conductors come into direct contact or through an arc. Faults can be balanced or symmetrical three-phase faults, line-to-ground and line-to-line faults, and double line-to-ground faults.

Monitoring is an important component of real-time control, particularly to implement demand side management. SCADA (supervisory control and data acquisition) systems are used for large complex networks. SCADA includes a human–machine interface (HMI), which an operator uses to monitor and control the processes, and encompasses many transducers and control points that are distributed over a large area. The remote units are data loggers and called RTU or remote terminal units, which communicate with the SCADA's HMI.

EXERCISES

11.1. Consider a 150 km line with R = 0.02 Ω/km, L = 3 mH/km, C = 4 nF/km. Select a model and write the impedances and admittances.

11.2. Determine the bus admittance matrix **Y** for the simple network of Figure 11.12. Assume the line between buses 1 and 3 is medium with $\mathbf{Z}_L = j80\,\Omega$ and $\mathbf{Z}_{sh} = -j10\,\mathrm{M}\,\Omega$, and that the line between buses 1 and 2 is short with $\mathbf{Z}_L = j40\,\Omega$.

11.3. Round up the bus admittance from the previous exercise to two decimals and assume we know voltages of each bus to be 4.8, 3.6, 3.2 kV with angles 15°,10°,0°. Calculate the currents of all three buses.

11.4. Use base 1 kV, 0.01 Ω, 1 A, 1 kVA to express variables in p.u. Express the bus admittance and voltages from the previous exercise on a p.u. basis. Calculate the complex power of all three buses.

11.5. Use base 1 kV, 0.01 Ω, 1 MW, 1 MVA, 1 MVAR to express variables in p.u. Suppose $X = 1$ and we constrain $P_1 = 1.5, V_1 = 13.8, \theta_1 = 10°, \theta_2 = 0°$ in p.u. How would you solve the system of two-bus power-flow equations?

11.6. Assume infinite bus voltage $V = 13.8$ kV. We increase the field current to achieve $E = 16$ kV. What is the critical δ? What is the real power and reactive power for $\delta = 20°$?

11.7. Examine the daily regime for two days, one in February (winter season) and one in August (summer season) year 2010 for EAST and SOUTHERN zones of ERCOT.

11.8. Draw a figure of COAST zone for 2010 showing daily maximum and minimum. Add horizontal lines for hypothetical installed capacities of 9 GW (coal-steam), 5 GW (NGCC), and 5 GW (NG) corresponding to baseload, intermediate, and peaking loads.

11.9. Draw a load–duration curve and calculate CF for the load data of zone COAST and a set of hypothetical generation capacities as in the previous exercise.

REFERENCES

1. Grainger, J.J., and W.D. Stevenson, *Power System Analysis*. McGraw-Hill Series in Electrical and Computer Engineering, S.W. Director. 1994: McGraw Hill Education. 787 pp.
2. El-Hawari, M., *Introduction to Electrical Power Systems*. IEEE Press Series on Power Engineering, M. El-Hawari. 2008: Wiley. 394 pp.
3. Keyhani, A., *Design of Smart Power Grid Renewable Energy Systems*. 2017: Wiley IEEE press. 566 pp.
4. ERCOT. Grid Information. 2017. Accessed October 2017. Available from: http://www.ercot.com /gridinfo.
5. ERCOT. Hourly Load Data Archives. 2017. Accessed October 2017. Available from: http://www.ercot .com/gridinfo/load/load_hist.
6. US EPA. Electricity Delivery and Its Environmental Impacts. 2017. Accessed October 2017. Available from: https://www.epa.gov/energy/electricity-delivery-and-its-environmental-impacts.
7. Public Service Commission of Wisconsin, Environmental Impacts of Transmission Lines. 2013.
8. Cybulka, L., Mitigating the Impacts of Electric Transmission Lines. 2012, California Institute for Energy and the Environment.

12 Hydroelectric Power Generation

Hydropower is a form of renewable power that has been employed for a long time. Before electricity became prevalent, water wheels or *water mills* were employed to provide mechanical power to a variety of industrial processes. With the advent of electricity, hydropower has been an important part of the generation capacity of many countries. This chapter begins with a brief review of water resources and clarifying important concepts such as watersheds, hydrology and hydrodynamics. Then, we use basic principles from fluid dynamics to develop calculations of power based on head and flow. We discuss a variety of turbine types and their applications for various conditions of head and flow. Hydrological data on streamflow is very important in planning and operation of hydroelectric facilities, be it reservoir based, run-of-river, and pumped storage. Other forms of hydroelectric generation involve the use of tides and waves in the coastal environment. We conclude this chapter by examining these forms of generation. Several textbooks treat hydroelectric power generation and can serve as supplementary reading to this chapter [1–4].

12.1 HYDROELECTRIC POWER: INTRODUCTION

12.1.1 TYPES AND PRODUCTION

This is a good time to recall our discussion of potential and kinetic energy in Chapter 1. A common form of hydroelectric generation is to build a dam on a river to create a *reservoir*, and use the potential energy of water stored in the reservoir by releasing it as needed to convert to kinetic energy and drive a hydraulic or water turbine (Figure 12.1). Another commonly employed form is *run-of-river*, which consists of diverting some flow of a running river for electric generation. In this case, only a small barrier is used to dam a small part of the river, and power is extracted from the kinetic energy of moving water (Figure 12.1). A third type includes *pumping* some of the water back up for storage when electricity demand is low and later utilizing it as needed when demand increases (Figure 12.1). Ultimately hydroelectric is solar, because sun evaporates water, then it rains in the watershed and creates flow.

As discussed in Chapter 1 (look again at Figure 1.15), hydroelectric is the third largest contributor of electricity production worldwide with 16.4%, after coal and natural gas. This means hydroelectric is the largest source of renewable power in use today. In fact, hydroelectricity with installed capacity of about one TW in 2016 accounted for 71% of all renewable generation. Hydroelectric plants range from tens of kW to ~20 GW [5].

Hydropower contributes 7% of the United States' electricity, or 100 GW, of which 10% is pumped. The 10 largest electric power plants in the world are hydro, including, Guri, Venezuela ~10 GW; Three Gorges, China ~17 GW; and Itaipu, Brazil-Paraguay ~10 GW.

As of 2014, the largest producer in the world was China (~1064 TWh, representing 18.7% of its production), while the United States with 282 TWh is the fourth largest producer, but with only 6.5% of its production. Other major producers such as Canada (second largest, 383 TWh), Brazil (third largest, 373 TWh), and Venezuela (ninth largest, 87 TWh) have a significant proportion of their production represented by hydro, 58.3%, 63%, and 68.3%, respectively. Going further in that direction, Norway is only the seventh producer with 129 TWh, but hydroelectric represents 96% of its production.

For hydropower, our explanation of capacity factor (CF) in Chapter 11 becomes very relevant. For hydroelectric, capacity factor is influenced by availability of the resource, whether stored in a

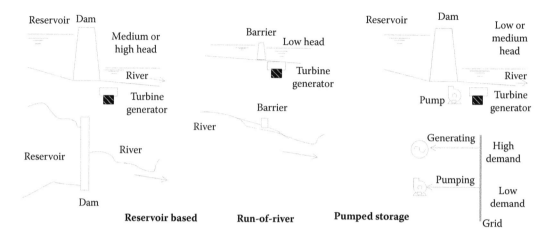

FIGURE 12.1   Types of hydroelectric plants.

reservoir or flowing. This resource ultimately depends on variability of precipitation (rainfall and snowfall), its runoff over land, and discharge as the river flows. For this reason, installed capacity is not necessarily correlated with demand. The United States has the second largest installed capacity of 102 GW, but its capacity factor is relatively low at 0.42. Brazil and Canada with lower installed capacity of 89 and 76 GW, respectively, produce more than the United States and are the third and second largest producers (their CF is 0.56 and 0.59, respectively). China has both the largest production and installed capacity; however, its CF is the lowest among top producers at ~0.37. Venezuela with the least installed capacity (15 GW) among the top 10 producers has the highest CF 0.67, achieving large production with low installed capacity.

Hydroelectric plants can be categorized in terms of installed capacity as large, small, and micro hydropower. As a generalization, *micro hydroelectric plants* are 1 to 100 kW, and designed for low flow and can be used off-grid or in microgrids (Chapters 13 and 14). A *small hydroelectric plant* is in between 100 kW and 20–30 MW operating with larger flow, and functioning as peaking and intermediate plants, and in few cases as baseload plants. Then, *large hydroelectric plants* have more than 20 to 30 MW of installed capacity using large flows and functioning as baseload plants.

12.1.2   Hydrologic Cycle and Watersheds

Water covers 70% of Earth's surface and almost all (97%) of the water is in the oceans and it is saline. The remainder 3% is freshwater of which 2% is iced in glaciers. The remaining 1% is surface (lakes, rivers), in the soil and groundwater, and water vapor [6].

As discussed in Chapter 6, water can exist as gas, liquid, and solid phases. Changes from one phase to another imply changes in energy, either gain or loss. Recall Figure 6.11 describing these three phases and the energy gain or lost when changing phase. Processes involved in changing between gas and liquid are condensation and evaporation, and between solid and liquid are freezing and melting. When liquid water becomes water vapor, energy is required as latent heat of vaporization. Increasing temperature promotes evaporation and increases vapor pressure of air.

These processes are involved in the water cycles on Earth. The hydrologic cycle consists of evaporation from surface water and soil, condensation of atmospheric water vapor into liquid, precipitation from atmosphere, *runoff* to surface water and infiltration into soil moisture, and back to the atmosphere by evaporation. Runoff to surface water is the most important for hydroelectricity; quantity and quality of freshwater runoff to streams and lakes represent important characteristics of the base resource for electricity generation. A basin or *watershed* is the land area contributing runoff to a river (Figure 12.2). The quantity and quality of water in a river are in large part dependent on the

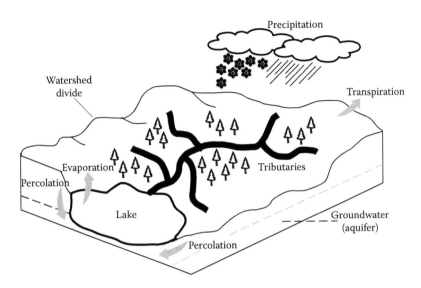

FIGURE 12.2   Watersheds, reservoirs, rivers.

conditions of its watershed. Sustainability of water resources requires monitoring of water quality and quantity over the entire watershed, as well as along the tributaries and the main river [6].

Precipitation is unevenly distributed on the continents. Those areas with high precipitation are likely to have the required flow conditions for hydroelectric generation. However, watershed area, relief, geology, soils, and land cover play major roles in making the resource adequate for hydroelectric generation. For instance, abundant rainfall over a large watershed area covered with vegetation and low ongoing soil erosional processes would yield higher sustained and regular water flows with lower *sediment load*. In contrast, a small watershed denuded of vegetation and with ongoing soil erosion would yield irregular water flow with higher sediment load.

12.1.3   HYDROLOGY

Planning and operating a hydroelectric facility requires modeling and quantitative analysis of the resource, be it reservoir based, run-of-river, or pumped storage. Starting with the watershed draining to the site, *precipitation-runoff* modeling allows analyzing historical behavior of streamflow in tributaries and the river, and predicting runoff for several planning and operating time horizons. These include at least daily precipitation, but in many cases hourly precipitation during storms, distributed over the watershed. Meteorological radar data together with ground-based meteorological stations are useful to provide rainfall data for analysis and prediction.

Next, models input the surface runoff into the various streams of the drainage network to track *streamflow* in tributaries and the river at hourly (for storm events) and daily scales that can be aggregated to monthly, weekly, and yearly scales. Streamflow data are required to test the rainfall-runoff models and the river network model, and to analyze dynamics of the drainage network, the river, or the reservoir. Several public domain models are available for hydrological modeling. One of these is the HEC-HMS maintained by the Hydrologic Engineering Center of the U.S. Army Corps of Engineers.

12.1.4   HYDRODYNAMICS AND HYDRAULICS

For large rivers and reservoirs, it is necessary to model the water movement within the water body in order to know the distribution of water velocity and water levels. These *hydrodynamic* models are built based on fluid dynamics equations. These serve as input to water quality, eutrophication, and sediment deposition models in reservoirs and inflow expected at run-of-river facilities in large rivers.

Depending on required level of detail, hydrodynamic models can be just 2-D (two-dimensional space and depth averaged) and 3-D (two surface axis and depth). Several hydrodynamic models in the public domain are available.

For design of dams, penstocks, barriers, and spillways, a more detailed view of fluid dynamics is necessary and is the subject of *hydraulic* models, which are less concerned with the overall reservoir or river and more focused on water velocity fields near and in the structures used for hydroelectric generation.

12.2 HYDROELECTRIC: CALCULATING POWER

Recall our discussion of potential and kinetic energy of Chapter 1. Potential energy of water stored at a substantial height is the basis of reservoir-based hydroelectric power generation. This potential energy is converted to kinetic energy by letting the water flow down a pipe or *penstock* as shown in Figure 12.3. The outflow of water acquires kinetic energy, which performs work by moving the blades of a turbine, which in turn spins the rotor of an electrical generator. Water flow rate and reservoir water level determine power. There are losses due to friction of flowing water against the walls of the pipe or penstock, plus the typical inefficiencies or losses of the turbine and the generator. Power depends on head and flow. We can achieve large power by combining high head with low flow or low head with large flow.

Kinetic energy can also perform work as long as there is force sustaining the momentum. Examples of renewable power systems that use this principle are hydroelectric power generation using the flow of water as in run-of-river (Figure 12.4). We will see this also in wind power generation using the kinetic energy in the wind (air is the fluid) (Chapter 13).

Hydroelectric facilities can use a variety of conditions thus requiring methods to calculate available power for electrical generation. Additional examples are increasing head using natural topography to obtain high head conditions (Figure 12.5) and pumping (Figure 12.6). Next, we will see how to model power available for both types of conditions using simplified fluid dynamics equations.

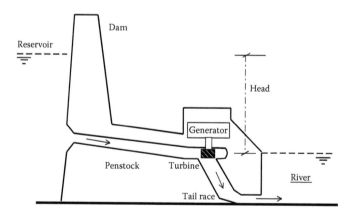

FIGURE 12.3 Reservoir-based, medium, and high-head hydroelectric power plant.

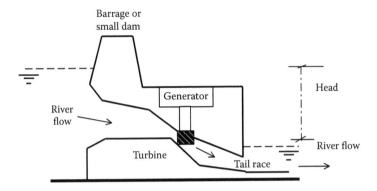

FIGURE 12.4   Run-of-river, low-head hydroelectric power plant.

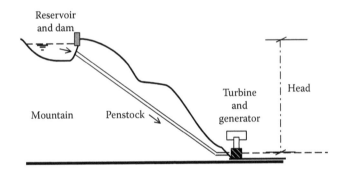

FIGURE 12.5   Reservoir-based, high-head hydroelectric power plant.

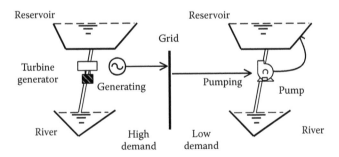

FIGURE 12.6   Pumped storage hydroelectric facility.

12.2.1   Bernoulli's Equation in Fluid Dynamics

Bernoulli's equation states conservation of energy for fluids in motion. A common form is

$$p + \frac{1}{2}\rho v^2 + \rho g z = \text{constant} \tag{12.1}$$

where p is pressure, z is elevation above a datum or reference, v is water velocity, ρ is density, and g is acceleration of gravity. The equation states that the sum of three forms of energy remains constant, one related to pressure "energy," another to kinetic energy, and the other to potential energy.

To see these terms as energy forms, let us examine the units of Equation 12.1, say, p in Pa. Recall from Chapter 4 that $1\,\text{Pa} = 1\,\dfrac{\text{N}}{\text{m}^2}$. Multiply both the numerator and denominator by m $\dfrac{\text{N} \times \text{m}}{\text{m}^2 \times \text{m}} = \dfrac{\text{W}}{\text{V}}$ to see that pressure can be seen as work or energy per unit volume, or *energy density*. In the second term, density is in kg/m^3 and v in m/s, so this is kinetic energy per unit volume, also energy density. Finally, in the third term, z is in m and g in m/s^2, therefore we have potential energy per unit volume and energy density.

In other words, Bernoulli's equation states that energy density of a moving fluid remains constant. This form of Bernoulli's equation applies to incompressible flows (density does not change with pressure), steady flow (fluid velocity at a point does not vary with time), and negligible loss by friction. Thus, it applies, for instance, to liquid water flowing in a pipe at constant velocity with very small friction against the pipe walls. The constant depends on the *streamline* (i.e., field line in a fluid flow), whereas p, v, and z may vary along the streamline.

Example 12.1

Consider an extremely simplified model of reservoir-based hydroelectric generation consisting of a pipe as shown in Figure 12.7 with both sides open to atmospheric pressure, say, $p_1 = 100$ kPa and $p_2 = 100$ kPa at points 1 and 2 of the streamline (centerline shown as dashed line). Recall from Chapter 4 that 1 atm = 101.325 kPa and 1 bar = 100 kPa. The water at point 1 has zero velocity $v_1 = 0$, and the streamline at 1 is $z_1 = 3$ m with respect to point 2, which is zero $z_2 = 0$. What is the velocity at point 2? Answer: Write Bernoulli's Equation 12.1 for both points $p_1 + \frac{1}{2}\rho v_1^2 + \rho g z_1 = p_2 + \frac{1}{2}\rho v_2^2 + \rho g z_2$. Substitute the values given; the pressure cancels on both sides, $\rho g z_1 = \frac{1}{2}\rho v_2^2$, and solve for v_2 to get $v_2 = \sqrt{2 g z_1} = \sqrt{2 \times 9.8 \times 3} = 7.67$ m/s. In other words, all potential energy density at 1 has been converted to kinetic energy density at 2. The pressure energy density does not change.

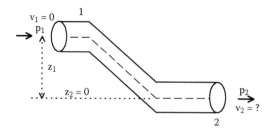

FIGURE 12.7   Illustration of Bernoulli's equation applied to water moving down a pipe.

This result should remind you of the simple example of complete conversion of potential energy to kinetic energy of a mass m at height z, $mgz = \frac{1}{2}mv^2$, which solves $v = \sqrt{2gz}$.

Bernoulli's equation can be normalized using the concept of *energy head* or energy per unit weight. Dividing Equation 12.1 by ρg, which is *weight density*, we obtain

$$\underbrace{\frac{p}{\rho g} + z +}_{static} \underbrace{\frac{v^2}{2g}}_{dynamic} = \frac{constant}{\rho g} = H \qquad (12.2)$$

where the constant, which was energy density in $\frac{Nm}{m^3}$, is now total energy head H in m. We can see this by looking at the units $\dfrac{\frac{Nm}{m^3}}{\frac{kg}{m^3}\frac{m}{s^2}} = $ m. In Equation 12.2, we have separated what we considered static and dynamic terms. The term $\underbrace{\frac{p}{\rho g} + z}_{static} = h$ is the hydraulic head or static head, or the sum of elevation and pressure head. Using these terms, we can rewrite Bernoulli's equation succinctly as

$$\underbrace{\frac{p}{\rho g} + z}_{h} + \frac{v^2}{2g} = h + \frac{v^2}{2g} = H \qquad (12.3)$$

Stating that the sum of static head and dynamic head is the total head H, which remains constant. In this parameterization of Bernoulli's equation, the units of head are m.

Example 12.2

Consider another simple analogy of hydroelectric generation consisting of a large tank full of water as in Figure 12.8 and water flowing out of the bottom. The tank is 3 m tall and both points 1 and 2 are open to atmospheric pressure. What is the water velocity flowing out of the bottom of the tank? Answer: We can solve this problem using Equation 12.1, but let us apply Equation 12.3 to practice calculations using hydraulic head. At points 1 and 2, we have $\frac{p_1}{\rho g} + z_1 + \frac{v_2^2}{2g} = \frac{p_2}{\rho g} + z_2 + \frac{v_2^2}{2g}$. Both pressure values are the same and cancel. We take the tank bottom as datum, and therefore $z_2 = 0$, $z_1 = 3$ m, thus head at 1 is $h_1 = z_1$ and $h_2 = 0$. At point 1, we can assume that velocity $v_1 = 0$ is zero

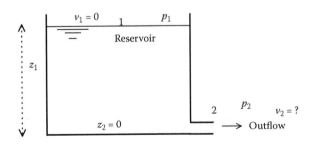

FIGURE 12.8 Illustration of Bernoulli's equation applied to water flowing out of a tank.

if the tank has a much larger volume than the outflow. Therefore, $z_1 = \frac{v_2^2}{2g}$. Solving for velocity $v_2 = \sqrt{2gz_1} = \sqrt{2 \times 9.8 \times 3} = 7.67$ m/s as in the previous exercise.

12.2.2 Power, Flow, and Head

We now use the aforementioned concepts to link head to flow and power. For simplicity, assume a situation as in the two previous examples, where all energy density at point 1 is potential and all energy density at point 2 is kinetic. In other words, head h at point 1 is only static and all is converted to dynamic head given by velocity v at point 2.

$$h = \frac{v^2}{2g} \tag{12.4}$$

In general, we can assume that h is the head difference between 1 and 2 by proper selection of the datum. In hydroelectric generation, the source is point 1 and the turbine is at point 2.

Assume the water flows out at velocity v through a cross-sectional area A. Flow Q is Velocity × Area or $Q = vA$. Now, importantly, assume that this *flow is small enough so that the head is unaffected*, or in other words that the area A can be increased as long as it is small enough to keep the *static head constant*. Now use the expression

$$P = \frac{1}{2}\rho A v^3 \tag{12.5}$$

for power in this moving fluid, which we derived in Chapter 1, and stating that power of the moving fluid is proportional to the cube of velocity of the fluid. Separate one v term to multiply A and get flow, that is, $P = \frac{1}{2}\rho A v \times v^2 = \frac{1}{2}\rho Q v^2$. Now, using Equation 12.4 ($v^2 = 2gh$) we get $P = \frac{1}{2}\rho A v \times v^2 = \frac{1}{2}\rho Q \times 2gh$ or finally

$$P = \rho g h Q \tag{12.6}$$

This is the *power equation* stating that power is the product of weight density ρg, water outflow, and head difference between points 1 and 2, which are the source and the turbine locations in hydroelectric generation. Let us look at units of Equation 12.6: ρg is in N/m^3, h is in m, and Q in m^3/s, therefore P is in Nm/s = W.

Equation 12.6 presumes that head and flow are independent. This is valid as long as flow Q is small enough to keep the head h constant. Intuitively, to meet this condition you need a lot of water in the reservoir so that the flow does not affect its level.

Equation 12.5 ($P = \frac{1}{2}\rho A v^3$) is also used for calculations of wind power generation after adjusting air density according to elevation above sea level, as we will see in Chapter 13.

Example 12.3

Consider the conditions of the previous example with head difference of 3 m, which will produce a velocity of 7.67 m/s and that it flows out of a pipe with cross-sectional area of $A = 2$ m^2. What is the power P? Answer: The flow is Q = 7.67 × 2 = 15.34 m3/s, the weight density is 1000 × 9.8 = 9800 N/m3, then the power is $P = \rho g h Q$ = 9800 × 3 × 15.34 ≈ 451 kW.

Function P.hA of renpow allows summarizing these calculations. For the conditions of this example, we would use it this way and obtain the results given in the example:

```
> x <- list(h=3,A=2)
> P.hA(x)
 Head(m) Vel(m/s) Area(m2) Flow(m3/s) Power(MW)
1    3    7.67      2      15.34      0.45
>
```

12.2.3 HEAD LOSS AND EQUIPMENT EFFICIENCY

So far, we have assumed no loss by friction of the flowing fluid. A way to include these losses in the power equation model is to assume that loss can be lumped as a given amount of head h_L subtracted from the head difference

$$P = \rho g(h - h_L)Q \qquad (12.7)$$

Head loss is a function of flow, pipe diameter, and material. Assume that loss goes as the square of flow Q, that is $h_L = kQ^2$. Then power

$$P = \rho g Q(h - h_L) = \rho g Q(h - kQ^2) = \rho g\left(hQ - kQ^3\right) \qquad (12.8)$$

Example 12.4

Suppose we have a reservoir with $h = 20$ m and can sustain flow of $Q = 1000$ m^3/s at constant head. Loss in the penstock can be modeled by $h_L = kQ^2$ with coefficient $k = 5 \times 10^{-6} \dfrac{m}{(m^3/s)^2}$. Answer: $\rho g = 1000$ kg/m$^3 \times 9.8$ m/s$^2 = 9800$ N/m^3. Loss is $h_L = kQ^2 = 5 \times 10^{-6} \times (1000)^2 = 5$ m. Then power is $P = \rho g Q(h - h_L) = 9800 \times 1000 \times (20 - 5) = 147 \times 10^6$ W $= 147$ MW.

12.2.4 MAXIMUM POWER

Equation 12.8 is a nonlinear function of flow. Power increases with flow, but then decreases as losses become larger. Thus, there is value of flow that yields maximum power. Take the derivative of Equation 12.8 with respect to flow and make it equal to zero $\dfrac{dP}{dQ} = h - 3kQ^2 = h - 3h_L = 0$. Maximum power occurs when $h = 3h_L$ or in other words when loss is a third of the head:

$$h_L = \frac{1}{3}h \qquad (12.9)$$

This gives us a handle on how much to increase the pipe diameter to reduce loss while keeping cost down.

Example 12.5

Suppose we have a reservoir with $h = 15$ m that can sustain flow of $Q = 1000$ m^3/s at constant head. What is the power in the fluid assuming that loss has been designed to yield maximum power? Answer: $\rho g = 1000$ kg/m$^3 \times 9.8$ m/s$^2 = 9800$ N/m^3. Loss is $h_L = \dfrac{1}{3}h = 5$ m. Then, power available in the water flow is $P = \rho g Q(h - h_L) = 9800 \times 1000 \times (15 - 5) = 98 \times 10^6$ W $= 98$ MW.

Function Pmax.Qh(x) of renpow allows for quick calculation of maximum power given flow and gross head. With the values of this example, we would use it as follows:

```
> x <- list(Q=1000,h=15)
> Pmax.Qh(x)
 Gross head (m) Net head (m) Flow (m3/s) Power(MW)
1         15         10      1000      98
>
```

12.2.5   HEAD-FLOW DIAGRAMS

A practical visualization tool is to plot Equation 12.12 in a graph with flow in the horizontal axis and net head in the vertical axis (Figure 12.9) using the logarithmic scale for both axes. This graph allows a quick estimate of the power expected for given flow and head regime. This graph was drawn taking into account head loss. In other words, it already discounts loss. This plot is obtained with function Pmax.Qh.plot of renpow:

```
x <- list(Q=1000,h=15,plab="A")
Pmax.Qh.plot(x)
```

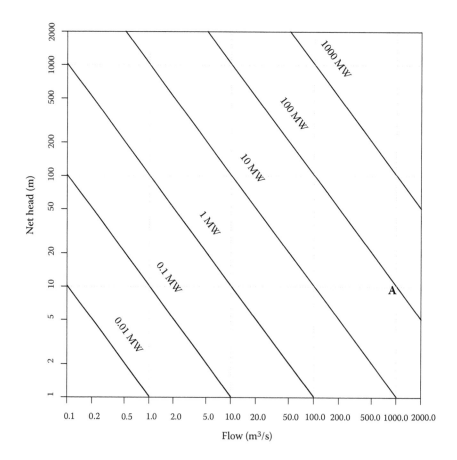

FIGURE 12.9   Power in flow–head plane including friction loss.

The point labeled A on the plot corresponds to the regime under examination of flow $Q = 1000$ m^3/s and $h = 15$ m (gross head), which yields almost 100 MW. We know from the previous example that it is 98 MW. Note that the net head is 10 m corresponding to $h - h_L = 15 - 5 = 10$m.

12.2.6   POWER MODEL FOR RUN-OF-RIVER

For run-of-river generation, we divert some of the river flow and use this for power generation by means of a low head h. The simplest assumption is that the diverted flow Q together with the head h enter into Equation 12.6 for power available in the fluid. We assume that head loss is negligible since for these low head situations typically you do not have a long penstock. Production is now more variable since it will depend on river flow.

Example 12.6

Assume we divert $Q = 100$ m^3/s from a river for electricity generation and the head is just $h = 1$ m. What is the power? Answer: Power is $P = \rho g h Q = 9800 \times 1 \times 100 \approx 980$ kW. We could also use function P.Qh of renpow:

```
> x <- list(h=1,Q=100)
> P.Qh(x)
 Head(m) Flow(m3/s) Power(kW)
1    1      100      980
>
```

12.2.7   PIPE LOSS FOR MICRO HYDROELECTRIC PLANTS

An alternative to calculation of head loss in pipes is the empirical Hazen-Williams equation

$$h_L = 10.67 \times \frac{L}{d^{4.87}} \times \left(\frac{Q}{C}\right)^{1.852} \tag{12.10}$$

where h_L is friction head loss in m, L is length of pipe (also in m), Q is flow (m^3/s), C is pipe roughness coefficient (given in tables, e.g., Engineering ToolBox [7]), and d is the inside pipe diameter (m). Some examples of typical values for C are 150 for polyvinyl chloride (PVC) pipes, 100 to 140 for concrete pipes, 90 to 110 for steel pipes, and 140 for polyethylene pipes. Other values can be consulted in the Engineering ToolBox [7].

Function pipe.loss of renpow implements the Hazen-Williams equation and includes values for some piping materials.

Example 12.7

From a creek, we take a flow of 0.01 m3/s to a 90% efficient turbine/generator located 30 m elevation below using 200 m of PVC pipe of $D = 0.075$ m. Calculate head loss in the pipe and power. Answer:

```
> x <- list(Q=0.01,d=0.075,L=200,mat='pvc')
> pipe.loss(x)
```

```
  Head loss(m) Roughness
1    11.84    150
>
```

Power is $P = \eta g \rho Q(h - h_L) = 0.9 \times 9800 \times 0.01 \times (30 - 11.84) = 1.6$ kW, or simply use P.Qh:

```
> hL <- pipe.loss(x)[1,1]
> x <- list(h=30-hL,Q=0.01,nu=0.9)
> P.Qh(x)
  Head(m) Flow(m3/s) Eff PowWater(kW) PowGen(kW)
1  18.16    0.01 0.9     1.78        1.6
>
```

12.2.8 Turbine: Converting Pressure to Kinetic Energy and Force

Let us look a little closer at what happens at the turbine. The water at the inlet of the turbine is at pressure p_i and velocity v_i. We can apply Bernoulli's equation assuming water at the surface of the reservoir is at atmospheric pressure p_a and zero velocity. Then,

$$p_i \simeq p_a + \rho g h - \frac{1}{2}\rho v_i^2 \tag{12.11}$$

Any pressure drop along the turbine $dp = p_i - p$ will produce an increase in kinetic energy given by Bernoulli's equation. This increased kinetic energy is an increased velocity that can exert a force on the turbine blades and cause rotational movement. To simplify the discussion, we will derive the equations for a specific type of turbine (impulse turbines) where the relationship between water velocity, and force and power is easy to see. In these turbines, pressure drop is converted to a water jet with constant velocity v.

First, recall from Chapter 1 that momentum is $\mathbf{p} = mv$ and $F = \dfrac{d\mathbf{p}}{dt} = m\dfrac{dv}{dt}$, which is to say that force is rate of change of momentum, or product of mass and rate of change of velocity for constant mass. The letter p is the usual symbol for momentum; we are using $\mathbf{p}$ boldface to avoid confusion with p for pressure. A water jet of constant velocity v (product of the pressure drop) applies a force to a turbine blade given by $F = \dfrac{d\mathbf{p}}{dt} = \dfrac{dm}{dt}v = \dot{m}v$, or the product of mass flow rate and velocity. Recall that $\dot{m} = \rho A v$ and therefore $F = \rho A v^2$.

But the blade is moving at velocity u, therefore the force applied to the blade is $F = 2\rho A (v - u)^2$. The factor 2 occurs in these turbines because the water jet returns at the same velocity but in opposite direction. Power given to the blade is the product of force and its velocity $P = Fu = 2\rho A(v - u)^2 u$. At $u = v$ (freewheeling) and at $u = 0$ (blade at rest), the power is zero. For a blade velocity between 0 and v, we should have maximum power. Taking the derivative of power $\dfrac{dP}{du} = 2\rho A \times (v - u)^2 - 2(v - u)u$, and making it equal to zero, we obtain $u = \dfrac{1}{2}v$ for maximum power. The rotational speed ω in rad/s will be $\omega = u/r$, where r is the radius of the turbine, or $N = \omega/2\pi$ in Hz. Also, power is $P = \tau\omega$, where τ is torque.

Example 12.8

Consider a turbine as described in this section. Assume a net high head such that velocity is 30 m/s from a nozzle of area $A = 10^{-3}$ m^2. The turbine has 2 m of radius. What is the mechanical power of the turbine, its angular velocity, and the torque if we achieve maximum power in the turbine? Answer: At maximum power, we have $u = \frac{1}{2}v$, therefore power is $P = Fu = 2\rho A(v/2)^2(v/2) = \frac{1}{4}\rho A v^3$.

Substitute values $P = \frac{1}{4}1000 \times 10^{-3} \times 30^3 = 6.75$ kW. The angular velocity is $\omega = 15/2 = 7.5$ rad/s, and torque is $\tau = P/\omega = 6750/7.5 = 900$ Nm.

Mechanical power output P_t, after discounting the energy losses in the turbine, is $P_t = \eta_t P_w$, where η_t is the turbine efficiency. Mechanical power output of the turbine is then used to spin the generator, which itself has an efficiency η_e. It is typical to include combined efficiency of the turbine and generator as a combined parameter $\eta = \eta_t \eta_e$ to obtain generated power as

$$P_e = \eta P_w = \eta \rho g(h - h_L)Q \qquad (12.12)$$

Example 12.9

Suppose we have a reservoir with $h = 15$ m that can sustain flow of $Q = 1000$ m^3/s at constant head. The combined efficiency of the turbine and generator is 90%. What is the power in the fluid assuming that loss has been designed to yield maximum power? What is the power generated? Answer:

```
> Pe.Pw(x)
  GrossHead(m) NetHead(m) Flow(m3/s) Press(kPa) Eff PowWater(MW) PowGen(MW)
1      15         10         1000      98 0.9      98            88.2
```

12.2.9 PUMPING

A centrifugal pump works by rotating an *impeller* and forcing water through a *volute*, which is a widening spiraling channel around an impeller. The water acquires exit velocity and thus kinetic energy. The impeller has curved vanes pointing out from the center or suction eye, and moves water radially outward from the suction eye due to centrifugal force. The displaced water creates lower pressure in the suction side of the pump; the liquid slows down in the volute, and then increases pressure at the discharge side.

The exit velocity needed to overcome a given net head h can be calculated using Bernoulli's equation as $v = \sqrt{2gh} = \sqrt{2 \times 9.8 \times 10} = 14$ m/s. The displacement x per revolution at a given rotational speed (rpm) is $x = v/\omega$. For instance, at 1800 rpm $\omega = \dfrac{1800\,\text{r/m}}{60\,\text{s/m}} = 30\,\text{rad/s}$ and then $x = v/\omega$ for 14 m/s, we have $x = \dfrac{14\,\text{m/s}}{30\,\text{rad/s}} = 0.46\,\text{m/rad}$, which means an impeller with diameter $D = \dfrac{x}{\pi} = 0.14$ m. In other words, an impeller of 14 cm at 1800 rpm would lift water to a head of 10 m.

12.3   TYPES OF TURBINES

Hydraulic turbines used for hydroelectric generation vary considerably in design, their operating principle, and scale. In terms of operating principles, turbines can be categorized using the direction of flow with respect to the axis of rotation and types of force.

Turbines can be of *axial-flow* type, that is, the flow lines are parallel to the axis of rotation. The water exerts a force on the blades, and the *reaction* causes rotational movement. Increasing the area to produce more power means a concomitant increase in radius of the blades and consequently of the velocity of the tip. Thus, the turbine rotates faster than the velocity of water. These are reaction turbines operating submerged, with water reaching the blades under pressure. As previously discussed, the pressure drop at the outlet with respect to the inlet is proportional to power. Many axial flow turbines resemble a propeller.

In contrast, the water can flow toward the center of the turbine along its radius as in *radial-flow* turbines. The water is conducted toward the center using guide vanes and a volute, and exerting a force on the blades, thus causing the rotational movement. Changing the angle of the guide vanes changes the force on the blades. As in axial-flow turbines, the water reaches the turbine under pressure, and the pressure drop relates to power.

Another major type is the *impulse turbine*, which does not operate submerged but exposed to air at atmospheric pressure. The impulse turbine is based on the principle of converting pressure to a force exerted by a water jet. Pressure drop is converted to kinetic energy in a water jet by Bernoulli's equation; water goes through to a nozzle, which shoots a water jet toward the blades, transferring velocity to the turbine. Almost the entire pressure drops at the nozzle.

These operating principles are used in various ways determining major turbine types. The *Kaplan turbine* is a representative of an axial-flow. It operates as a reaction force turbine and is designed as a propeller. It is widely used for large flow and low to medium head (Figure 12.10). *Francis turbines* are radial flow, covering a wide range of head from low to medium and high, and operate at medium flow (Figure 12.11). The *Pelton turbine*, based on a wheel with cups mounted on its circumference, is the most common impulse turbine and can operate at high head and low flows (Figure 12.12). A nozzle directs the flow to the cups, thus providing the impulse. Other impulse-based turbines are

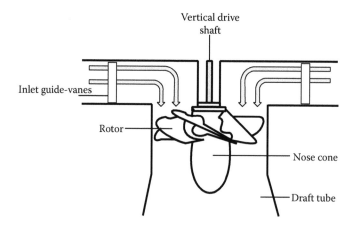

FIGURE 12.10   Kaplan turbines.

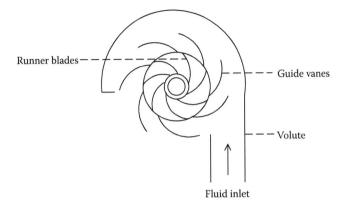

Runner blades

Guide vanes

Volute

Fluid inlet

FIGURE 12.11   Francis turbines.

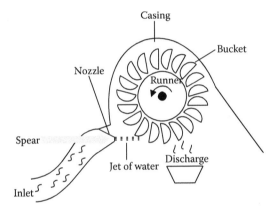

Casing

Bucket

Nozzle

Runner

Spear

Jet of water   Discharge

Inlet

FIGURE 12.12   Pelton turbines.

the *cross-flow* and the Natel Energy *hydroEngine*, which are suitable for low head and small hydropower applications.

Often, the turbine areas of operation are plotted on the same chart as an aid to select turbine types (see Figure 12.13).

Example 12.10

For four different sites (A, B, C, and D), we have the following conditions of flow Q (m³/s) and h gross head (m): A ($Q = 100$, $h = 5$), B ($Q = 30$, $h = 150$), C ($Q = 3$, $h = 200$), and D ($Q = 2$, $h = 5$). Determine expected power at each site and select a turbine type for each site. Answer: Use the power lines and turbine ranges graph in Figure 12.13. All units in m and m³/s. From the chart, we find A is a 2 MW Kaplan, B is a 20 MW Francis, C is a 8 MW Pelton, and D is a 80 kW hydroEngine. Plots shown in Figure 12.14 can be produced with function Pmax.Qh.plot and turbine.regions:

```
panels(wd,ht,2,2,pty="m")
x <- list(Q=100,h=5,plab="A")
```

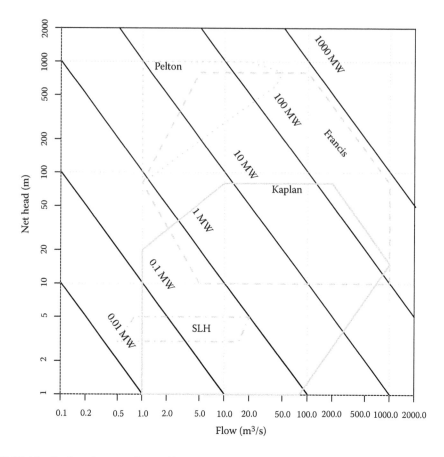

FIGURE 12.13   Design chart to select turbine type based on flow and head.

```
Pmax.Qh.plot(x)
turbine.regions(type='kaplan')
x <- list(Q=30,h=150,plab="B")
Pmax.Qh.plot(x)
turbine.regions(type='francis')
x <- list(Q=3,h=250,plab="C")
Pmax.Qh.plot(x)
turbine.regions(type='pelton')
x <- list(Q=2,h=5,plab="D")
Pmax.Qh.plot(x)
turbine.regions(type='slh')
```

Generators for large hydro plants require special consideration of design and construction because of large size.

12.4   HYDRO POWER DESIGN AND MANAGEMENT

As we explained at the beginning of this chapter, hydrological data on streamflow is very important in planning and operation of hydroelectric facilities, be it reservoir based, run-of-river, and pumped storage. Streamflow of all rivers coming to a reservoir, together with releases for generation and other

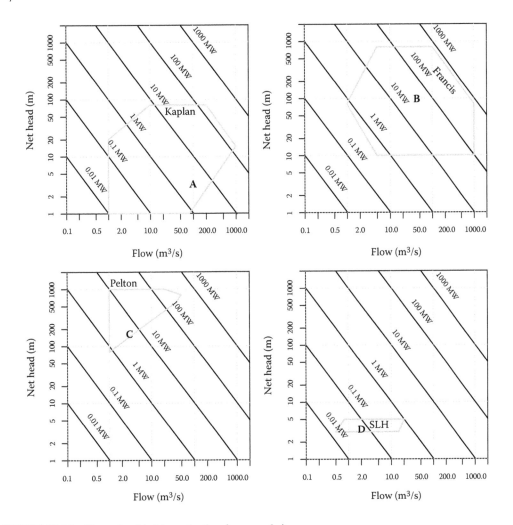

FIGURE 12.14 Example of turbine selection for several sites.

purposes (irrigation and water supply), and lake evaporation, will determine lake level. As we know, lake level will determine gross head and therefore power generation potential.

Streamflow varies daily, monthly, seasonally, and yearly; some small streams will respond hourly to storm events. Generally, a practical time scale for hydroelectric generation is daily, with data collected over many years, allowing long-term daily streamflow analysis. For this purpose, flows are time series. In this section, we examine various methods of analysis.

12.4.1 STREAMFLOW TIME SERIES AND FLOW–DURATION CURVE

We will focus on daily streamflow Q (in m^3/s) in a river using simulated values as a learning tool. For this purpose, we will use functions model.flow, exceed, and flow.exc.plot of renpow. The function model.flow simulates flow as an autoregressive model adding variability from a Gaussian variate. This means that for each day, a random variable x is calculated by $x(t) = a_1x(t-1) + a_2x(t-2) + a_3x(t-3) + a_4r(t)$, where a_i are coefficients of the autoregressive model, and $r(t)$ is a random variable with normal distribution. To this basic random process, we add seasonality, a peak flow, and a base flow. Object x in this short script contains all arguments passed to function model.flow:

```
x <- list(base.flow=10,peak.flow=100,day.peak=200,length.season=90,
     variab=c(0.005,1),coef=c(0.7,0.2,0.1))
flow <- model.flow(x)
flow.plot(flow,label="Simulated flow (m3/s)")
```

The flow time series is shown in the top panel of Figure 12.15. We will describe shortly the meaning of Q_{mean}, Q_{50}, and Q_{95} lines shown there.

A *flow–duration curve* (bottom panel of Figure 12.15) is built using daily streamflow rate Q. The vertical axis is streamflow, sorted into descending order, with the highest levels toward the left of the curve. The horizontal axis is the exceedance in probability units or percent of the time that the

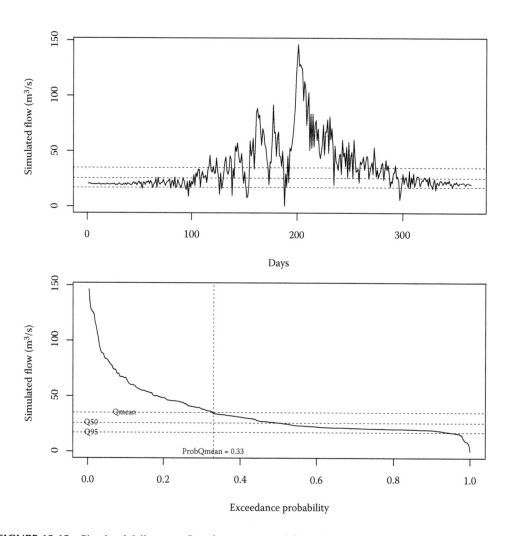

FIGURE 12.15 Simulated daily streamflow for one-year and flow–duration curve.

value of flow is larger than that value. This is similar to what we did in Chapter 11 with the load–duration curve using electricity demand data. In this particular case, we have plotted flow–duration for only one year for illustration, but in practice we employ the long-term multiple year flow record as discussed later.

The bottom panel of Figure 12.15 is drawn together with the flow time series using

```
exc <- exceed(flow)
flow.exc.plot(flow,exc,label="Simulated flow (m3/s)")
```

Function exceed, identifies several levels of exceedance probability that are typically considered of interest. For instance, 0.5 and 0.95. The flow values corresponding to these exceedance values are called the Q_{50} and Q_{95} values. They can be interpreted as a *low potential power* and *median potential power*. These levels can be queried from the exceedance function:

```
> exc$prob.Q
     Q50   Q95 Qmean
Prob 0.50  0.95  0.33
Q    25.65 17.09 35.00
> >
```

For example, the flow Q_{95} is 17.09 m^3/s, and Q_{50} is 25.65 m^3/s. As with load–duration curves, this means, for instance, that the flow rate is Q_{95} m^3/s or more for 95% of the year, that is, the curve is always greater than 17.09 m^3/s left of the 0.95 exceedance point. In addition, we can consider the average flow, called Q_{mean}, and its exceedance value.

The Q_{95} flow is used as a low-flow condition that must be maintained in the river when diverting flow through the hydroelectric plant. Comparing flow at various exceedance values, we can determine how variable the river is. The flow in Figure 12.16 has an average flow Q_{mean} of 35 m^3/s and a Q_{95} of 17.09 m^3/s, thus the $\dfrac{Q_{95}}{Q_{mean}} \approx 0.49$ or Q_{95} is 49% of the Q_{mean}. This high percent is typical of a high-baseflow river, meaning that the watershed has storage of runoff that is released gradually to the stream. A low percent of $\dfrac{Q_{95}}{Q_{mean}}$ would indicate a flashy river that would change flow fast in response to rain, meaning little storage of runoff in the watershed.

Example 12.11

Simulate a flashy river using model.flow. Run exceedance calculations using exceed and determine $\dfrac{Q_{95}}{Q_{mean}}$. Answer: Use a lower value for argument base.flow, say, 5 instead of 20. Run the model.

```
x <- list(base.flow=5,peak.flow=100,day.peak=200,length.season=90,
       variab=c(0.01,2),coef=c(0.4,0.3,0.2,0.1))
flow <- model.flow(x)
exc <- exceed(flow)
flow.exc.plot(flow,exc,label="Simulated flow (m3/s)")
```

Examine results in exc:

```
> exc$prob.Q
     Q50  Q95 Qmean
Prob 0.50 0.95  0.29
Q    10.06 2.59 27.17
>
```

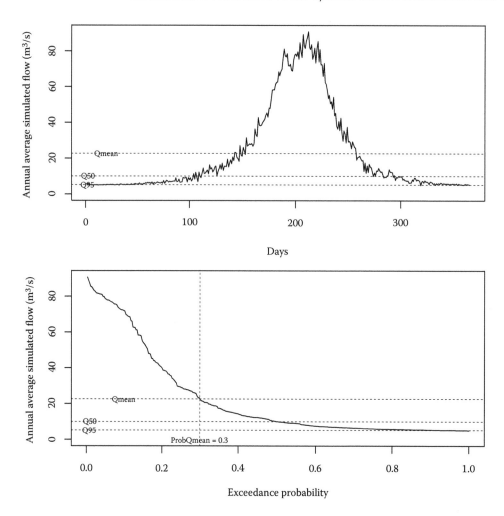

FIGURE 12.16 Example of multiyear annual averaged daily streamflow and flow–duration curve.

The ratio $\dfrac{Q_{95}}{Q_{mean}} = \dfrac{2.59}{27.17} \simeq 0.09$ or 9%. You may obtain slightly different values because the simulation generates random values.

A multiyear flow–duration curve can be obtained by taking the average over many years of the daily flow each day of the year. The series of 365 days of annual average flow values can be used to develop an annual average flow–duration curve. Calculations are facilitated by function annual.avg of renpow that simulates multiple years of flow and produces the annual average. With those results we employ exceed and flow.exc.plot as before:

```
flow.nyrs <- annual.avg(x,nyrs=20)
exc.nyrs <- exceed(flow.nyrs)
flow.exc.plot(flow.nyrs,exc.nyrs,label="Annual Average Simulated flow (m3/s)")
```

See results in Figure 12.16. Although we have performed these calculations with the flashy river of Example 12.11, you would notice that the spikes in flow and variability in exceedance has been reduced, because we have averaged them out. Examination of the exceedance values and their flows yields

```
> exc.nyrs$prob.Q
    Q50 Q95 Qmean
Prob 0.50 0.95 0.30
Q   10.04 5.15 22.62
```

The Q_{95} has been increased, and so has the ratio $\dfrac{Q_{95}}{Q_{mean}} = \dfrac{5.15}{22.62} \simeq 0.23$. You may obtain slightly different values because the simulation generates random values.

12.4.2 LAKE LEVEL AND LEVEL–DURATION CURVE

We look at a real data example based on a small reservoir in the headwaters of the Trinity River, Texas. Figure 12.17 (top panel) shows daily lake elevation as a time series in the interval 1989 to 2013 or 25 years of record. You can see periods of low level corresponding to years of reduced rainfall and short periods of high lake level corresponding to years of elevated rainfall.

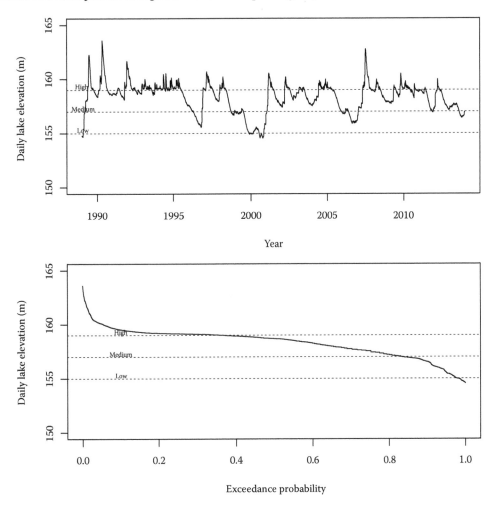

FIGURE 12.17 Example of multiyear daily reservoir level to derive level–duration curve.

Time series of reservoir level are used to determine a level–duration curve, built using the same process described for the flow–duration curve, but the vertical axis is the level in meters. The bottom panel of Figure 12.17 shows the level–duration curve for the data shown in the top panel. For illustration of an alternative approach to the annual average, we have plotted level–duration for the entire 25 years of record for the series shown in the top panel of Figure 12.17. When we plot it like this, the exceedance curve preserves more of the variability as compared to the annual average.

Flow–duration and level–duration curves are used to estimate capacity factor (CF) based on availability of the water resource. A plant with installed capacity scaled to use flow near the Q_{95} flow would have a high CF. In contrast, a plant with capacity that requires flow near the Q_{50} flow would have lower CF.

12.4.3 AREA–CAPACITY CURVE

A reservoir water level determines how much water surface area it covers and how much water volume it holds. Both surface area and volume depend on bottom topography underlain by the reservoir. This topography is the lake *bathymetry*, and it may be known before inundation from topographic maps or developed by depth surveys for multiple transects across the lake.

Volume is also called *capacity*. Area and capacity are given for a conceptual reservoir in Figure 12.18. Note that these curves are given in ha for area and ha × m for volume. Recall that 1 ha = 10,000 m^2. Therefore, 1 ha × m = 100,000 m^3. We have obtained these plots using function area.vol of renpow:

```
# H and B are pool Elevation (m) and bottom elevation at the dam (m)
# W is width (km), L is tail length (km)
x<- list(H=130, B=100,W=10,L=100)
area.vol(x)
```

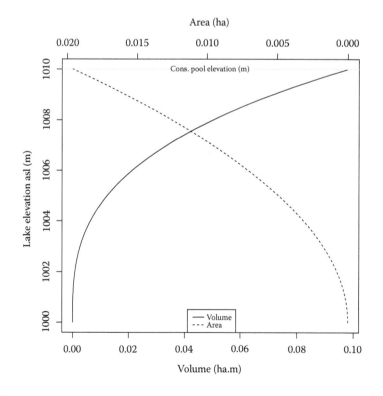

FIGURE 12.18 Area–capacity curves.

This function is not developed for a real reservoir but as a learning tool. It assumes a simplistic geometry of one linear water body, with bathymetry consisting of a V cross-sectional profile, and axial ramp from the dam to the reservoir tail water.

Example 12.12

Consider the conceptual reservoir with area-capacity of Figure 12.18. Assume we developed a hydroelectric plant drawing a flow of 1000 m^3/s for 24 h every day. Assume the turbine is at the base of the dam. Starting with a full lake (water at conservation pool elevation), what is maximum drought duration (in consecutive days of base river flow of only 1 m^3/s) that we can sustain before the gross head falls to two-thirds of its maximum? Assume evaporation rate is 5 mm/day. Answer: Each day we withdraw 1000 m^3/s × 24 h × 3600 s/h = 86400000 m^3 = 864 ha × m for electricity. We need to calculate gross evaporation loss, that is, over the entire lake. Thus, we need the surface area; we read maximum area from the graph to be ~5200 ha. Thus, evaporation loss is 0.005 m × A = 0.005 m × 5200 ha = 26 ha × m over the entire lake. Assume it to remain the same even though the area will decrease as the head decreases. The river inflow is 1 m^3/s × 24 h × 3600 s/h = 86400 m^3 = 0.864 ha × m. The balance per day is −864 − 26 + 0.864 ≈ −889.1 ha × m/d. For two-thirds gross head (120 m elevation) we read the chart of Figure 12.18 to get a capacity of ~25,000 ha × m, which means a decrease of 80000 − 25000 = 55000. Now, divide 55,000 ha × m/868.34 ha × m/d ≈ 62 d, or about 2 months.

Sediments carried by the river will deposit on the bottom as water velocity decreases, because the river goes from a riverine regime to a lake regime. Consequently, sediments including sand, silt, and clay accumulate as layers on the bottom, and therefore the capacity of the reservoir gradually decreases over years or decades depending on the river's sediment load. Periodically, the capacity is then updated by means of a bathymetric survey.

12.5   ENVIRONMENTAL INTERACTION

Environmental impacts of hydroelectric power generation depend strongly on which type of facility is examined. Large reservoirs have different impacts than small reservoirs, and run-of-river plants have different impacts than reservoirs based plants, since there is only flow diversion.

12.5.1   RESERVOIRS

Starting with reservoir-based generation, dams change the hydrology of a river. We will separate the discussion into effects *downstream* and *upstream* from the dam. A river becomes regulated in flow; downstream from the dam, the flows no longer show the pulsed nature of flows that aquatic eco-systems may have experienced for a long time. In particular, a flood pulse regime brings nutrients downstream. It is possible to design mitigation strategies; for example to release water periodically to simulate natural pulses of flow. This requires that in managing the reservoir, a portion of the flow is reserved for ecological releases.

Anadromous fish are those whose life cycle involves migration from the sea to freshwater; they are born in freshwater, migrate to the sea as juveniles, and then as adults migrate back to freshwater to spawn. An example is salmon. Dams on rivers that discharge to the sea interrupt anadromous fish migration. This is the case of many dams built in the Pacific Northwest region of the United States. This issue has been mitigated when possible by building runways parallel to the dam that fish can use to swim up the river.

Slowing down the water velocity has impacts on water quality that manifest itself downstream in the form of changed nutrients in solution, pH, and dissolved oxygen. We will discuss water quality in a separate section.

The impacts upstream are generally flooding of large areas with concomitant inundation of land, communities, and ecosystems. The larger the area, the greater the impact. For instance, large reservoirs in the Caroni River, Venezuela, have inundated many square kilometers of tropical forests, affecting this ecosystem and its habitat value for many wildlife species. The forest biomass is removed before inundation to reduce future CO_2 and CH_4 emissions from submerged biomass. These impacts have been mitigated by campaigns of salvaging wildlife during inundation.

Hydroelectric generation has several advantages with respect to fossil-based electricity generation. Besides electricity, a reservoir can have multiple uses, such as flood control, water provision for irrigation and drinking, and recreation. Comparing greenhouse emissions to fossil-based plants, greenhouse emissions are minimal. It should be noted that a reservoir for hydroelectric production, as like any other natural lake, emits CO_2 and CH_4 by decomposing of submerged biomass and organic matter. Aquatic organisms, algae and zooplankton, participate in the carbon cycle, as other lakes are part of the short-term exchange of carbon with the atmosphere.

It is typically said that hydroelectric plants are low maintenance and have a long life. This statement should be qualified because these plants do require judicious management of the watershed producing runoff that goes into flow of the river. Management includes curtailing extensive land-use change. An increase in impervious surface in the watershed would produce changes in the hydrological regime, increasing variability of the flows. Increases of areas producing sediment load to the river, such as deforestation and land denudation, lead to an increase of sediment load, which then deposits in the reservoir, shortening its useful life.

Many of the adverse environmental impacts of hydroelectric generation are associated with reservoirs for high-head generation. It is estimated that the United States has a potential 95 GW for high-head hydroelectric, of which 35 GW has been developed. It is unlikely that new high-head plants will be developed due to their high cost and environmental risks. However, there is potential for 70 GW of low-head hydroelectric, of which less than 2 GW has been developed. Small hydroelectric plants in low-head conditions represent a lower risk to the environment.

12.5.2 WATER QUALITY

Besides water quantity, which plays a major role in hydroelectric generation, an important aspect of streams and reservoirs is water quality. Although, only some parameters of water quality, for example, total suspended solids (TSS), may be of direct relevance to power generation, we should keep in mind that a stream and reservoirs have multiple purposes, and water quality parameters are relevant for irrigation, water supply, and recreational use.

In general, we can consider physical, chemical, and biological water quality parameters [6,8]. Major physical water quality parameters are temperature, TSS, and turbidity. Major chemical parameters are pH, dissolved oxygen (DO), total dissolved solids (TDS), major cations (Ca, Mg, Na, K, and NH_4), major anions (Cl, SO_4, HCO_3, CO_3, PO_4, H_2S, and NO_3), total organic carbon (TOC), and biological oxygen demand (BOD). Other variables of interest include chlorophyll and volatile organic hydrocarbons. Major biological parameters include bacteria (e.g., fecal coliforms), viruses, protozoans (e.g., giardia), helminthes (e.g., parasitic worms), and algae (e.g., blue-green) [6].

As part of the Clean Water Act [9], the U.S. Environmental Protection Agency (EPA) has criteria for 157 priority pollutants and 45 nonpriority pollutants. These guidelines are used by states and tribes to develop regulatory programs such as the National Pollutant Discharge Elimination System (NPDES) [10].

The STORET (short for STOrage and RETrieval) Data Warehouse is a repository for water quality, biological, and physical data and is used by state environmental agencies, the EPA and other federal agencies, universities, private citizens, and many others [11].

12.6 COASTAL HYDROELECTRIC: TIDAL AND WAVE POWER

Besides using water from overland runoff, which depends on the hydrologic cycle, it is also possible to generate electricity in coastal areas by using the changes in potential and kinetic energy from *waves* and *tides*. The reader is referred to several good references [1–4] for further reading on these topics. In this section, we will emphasize tidal power, whereas we will cover wave power generation towards the end of the chapter.

Worldwide, electricity production from these resources is much less than hydroelectric from overland surface runoff, mostly because the conditions to develop these facilities are constrained to fewer locations and the technology is still in development.

Using tidal power has a long history. *Tidal mills* installed on tidal streams and rivers were used in the Middle Ages in Britain and France to mill grains. These facilities captured the *flood* (incoming) tide in the river and then released at low or ebb tide (outgoing), making it flow to push a water wheel. It was not until the 1960s that a tidal power to generate electricity was built; this was a plant in the La Rance river estuary in France, with a 240 MW turbine.

Conventionally, tidal power plants are developed by building tidal *barrages* (or low dams, similar to run-of-river hydroelectric plants) to capture *vertical movement* of the tides as potential energy and force water through turbines installed on the barrage, thereby using kinetic energy (Figure 12.19). More recently, there is emphasis on direct use of the kinetic energy in *tidal currents* that occur with the ebb and flow of tides. The vertical movement of water level manifests as current, that is, *horizontal movement* toward and away from shore. Harvesting these currents for power production is by means of *free-standing turbines*. Infrastructure and technology varies; for example, there are small turbines on the seafloor and large turbines on submerged vertical towers (Figure 12.19).

Wave energy is transferred from *wind energy*, as winds flow over open water. In the next chapter, we will be discussing wind power generation, which consists of directly harvesting the kinetic energy in the wind. Wave energy has advantages in some areas, for instance, its increase in winter, and thus can cover peak electricity demand. In general, wave energy harvesting for electricity has lesser environmental costs than other coastal forms of hydroelectric production [12].

It is estimated that globally, tidal potential is ~3 TW [1], but realistically, constraints reduce it to 100GW. This is less than 12% of total surface-water hydroelectric potential, but still represents a significant resource [1]. Potential large-scale tidal sites are found in Russia, Canada, United States, Argentina, Korea, Australia, France, China, and India [1]. In some areas of the world, the tidal variation is very large and the resource is very abundant. For example, South Korea has potential for 2.4 GW [1].

In contrast to surface-water hydroelectric, which relies on the hydrologic cycle and hence on solar energy, tides are the result of gravitational pull by the moon, mostly, and in lesser extent by the sun. Thus, we can refer to tidal energy as combined *lunar energy* and solar energy [1]. Wave energy is driven by wind, and wind is the result of differential solar heating of the surface of the Earth, and thus a form of solar energy (we will discuss further in the next chapter).

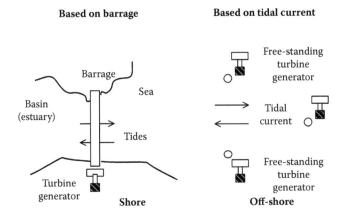

FIGURE 12.19 Approaches to tidal power: barrage and current.

12.6.1   TIDES: THE BASICS

The full process of tide formation is a complicated mechanism. Let us start with the simplest interaction of only the moon and Earth. As Figure 12.20 (top) illustrates, a combination of the relative rotation of moon and Earth and the gravitational pull of the moon will form a bulge on the Earth's ocean in the direction of alignment of the moon and the Earth. Relative rotation would increase the bulge on the side farthest from the moon, whereas gravitational pull would have the greatest effect on the side of the Earth nearest to the moon. This simple scenario, by itself, explains a *lunar tide* occurring twice daily (*semidiurnal*), since as the Earth rotates, any point will experience a bulge every 12 h and 24 min, which is half of a lunar day.

Now consider the gravitational effect of the sun, which will produce a solar bulge and another semidiurnal tide. The sun has a mass much larger than the moon, but is much farther away (390 times), and therefore the tidal bulge created by the sun is of lower height (about 46% that of the moon), and will occur every 12 h, which is half of a solar day [13].

Figure 12.20 (bottom) displays what happens when the moon and sun are aligned (full or new moon phases). The solar bulge is smaller than the lunar one, but they combine or superimpose, creating a larger bulge and producing the *spring tides* or the highest tides. In contrast, when the moon and the sun alignment with the Earth is perpendicular, the bulge created by the moon does not add up to the one from the sun, and in consequence the height decreases, producing *neap tides*, or the lowest tides. This occurs in first and third quarter phases of the moon. See Figure 12.21. The time in between neap and spring tides is one-quarter of the 29.5 days of the lunar cycle; this is 29.5/4 = 7.37 days or about 7 days. *Tidal range* for a particular tide cycle is the total excursion from high to low tide. For instance, if the high tide is 6 m and the low tide is 0 m, this cycle has a range of 6 m. The tidal range increases during spring tides and decreases during neap tides.

To these basic astronomical mechanisms, creating an oscillatory rising and falling of the sea surface, we have to add the amplification or attenuation of these oscillations due to the geometric arrangements of continental land surrounding the oceans. More locally, particular bathymetry conditions may amplify the oscillations and produce large tidal variation, such as it happens in the Bay of Fundy, Canada, where the tidal range can reach 16 m.

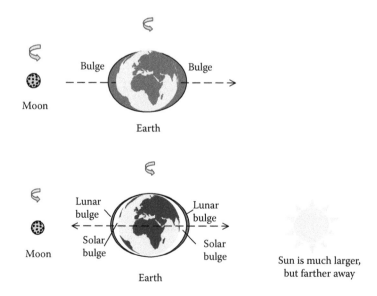

FIGURE 12.20   Tidal mechanisms. Top: Moon only. Bottom: Interaction of moon and sun. Spring tide.

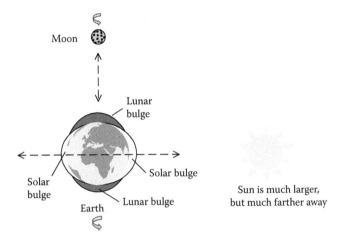

FIGURE 12.21 Tidal mechanisms. Interaction of moon and sun. Neap tide.

12.6.2 HARMONIC CONSTITUENTS

One way of modeling the oscillatory fall and rise of the sea at a particular location is by using *Fourier series*, that is to say, a summation of sine and cosine functions of several periods related to the astronomical process we just described. In the United States, this approach consists of adding *harmonic constituents*, which represents the periodic variation of the relative positions of the Earth, moon, and sun [14]. The water elevation $y_i(t)$ due to constituent i is written as

$$y_i(t) = A_i \cos(a_i t + \mu_i) \tag{12.13}$$

where t is hours given with respect to a specific initial time, A_i is amplitude (in m) or half the range for that constituent, $a_i t + \mu_i$ is the *phase* of the constituent (in degrees), a_i is the *speed* (which is rate of change of phase in degrees per hour), and μ_i is the phase of the constituent at the initial time. The period of the constituent is the time that it takes the phase to change through 360 degrees and is the cycle of the astronomical condition represented by the constituent. Values for these parameters are given by National Oceanic and Atmospheric Administration (NOAA) for stations throughout the United States [14]. The value for elevation is given as the sum of all the constituents

$$y(t) = \sum_i A_i \cos(a_i t + \mu_i) \tag{12.14}$$

As an example, consider the tide at Anchorage, Alaska. This site has parameter values for 124 constituents. The following list is only the first seven constituents. As you can see, the principal lunar semidiurnal (M2) and solar semidiurnal (S2) contribute the most in amplitude.

Cons#	Name	Amplitude	Phase	Speed	Description
1	M2	3.531	107.8	28.984104	Principal lunar semidiurnal constituent
2	S2	1	150.4	30	Principal solar semidiurnal constituent
3	N2	0.604	82	28.43973	Larger lunar elliptic semidiurnal constituent
4	K1	0.689	341.8	15.041069	Lunar diurnal constituent
5	M4	0.289	84.9	57.96821	Shallow water overtides of principal lunar constituent
6	O1	0.385	322.2	13.943035	Lunar diurnal constituent
7	M6	0.152	26.8	86.95232	Shallow water overtides of principal lunar constituent

Example 12.13

Use harmonic constituents to determine tides during a 29-day period for Anchorage, Alaska, and Port Aransas, Texas. You can visit NOAA's website [14] to study the parameter values from these locations. Package renpow includes one csv sample data file from each location that we can use them for this example. Use function harmonics.tide of renpow. Answer: First, we read the file containing the parameter values, and then call the function to calculate and plot values for a number of days, starting at a given t0. In this example, we plot 29 days.

```
x <- read.tide("extdata/AnchorageTide.csv")
harmonics.tide(x,days=29)
x <- read.tide("extdata/PortAransasTide.csv")
harmonics.tide(x,days=29)
```

Alternatively, we could have used

```
x <- read.tide(system.file("extdata","AnchorageTide.csv",package = "renpow"))
harmonics.tide(x,days=29)

x <- read.tide(system.file("extdata", "PortAransasTide.csv", package = "renpow"))
```

The resulting plots are shown in Figure 12.22. The top plot shows that the tidal range is ~12 m at spring tide and ~4 m at neap tide for this period at Anchorage, with both M2 and S2 showing strongly. The bottom plot for Port Aransas shows a much smaller tidal range of only 0.5 m at spring tide, and almost null range at neap tide. The S2 is attenuated with respect to other constituents and shows as a small dip instead of full semidiurnal fluctuation. These plots let you appreciate the variation of neap and spring tides within the period.

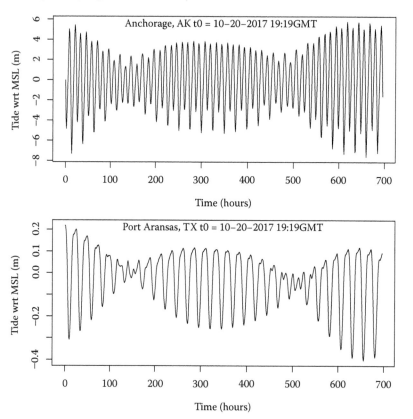

FIGURE 12.22 Example of modeling tides using harmonic constituents. Top: High tidal range. Bottom: Low tidal range and small semidiurnal S2.

12.6.3 SALINITY AND DENSITY OF SEAWATER

You know from our discussion of hydroelectric that water density is an important parameter in the power equation. For freshwater, density remains almost constant, and we did not worry about calculating it. However, in marine environments, density varies with *salinity*, temperature, and pressure. In coastal areas, and particularly at the interface between rivers and sea, such as estuaries, salinity experiences spatial and temporal gradients.

It is common to express salinity in parts per thousand ‰ or psu (practical salinity units) because it is approximately the same as grams of salt per kilogram of solution. Freshwater's salinity limit is 0.5‰, then water is considered for the 0.5‰ to 30‰ range; above that we have *saline* water in the 30‰ to 50‰ range, and *brine* with more than 50‰. Density is calculated from salinity, temperature, and pressure using an equation of state and polynomials of temperature and salinity. We will not go into details in this introductory book. For our examples and exercises, we will assume that *density of seawater* is

$$\rho_{sw} = 1025 \, \text{kg/m}^3 \tag{12.15}$$

12.6.4 BARRAGE-BASED TIDAL POWER

A barrage separates the sea from a tidal basin (Figure 12.23). A common tidal power plant operation is one way or *single effect*; during flood tide, the sluice in the dam is opened and water enters the basin freely. The basin fills and water acquires an elevation equal to the tidal range H at the end of the flood cycle. Water is then retained by closing the sluice until the tide recedes on the side of the sea and reaches the low tide level. At this point, there is a head equal to the tidal range, and water is released from the basin. Power is calculated by means of potential energy available per tidal cycle. For simplicity, assume that the volume of water behind the barrage is $V = \overline{A}H$, where $\overline{A}$ is the averaged surface area of the tidal basin. More precisely, this area could be calculated dynamically from the area–capacity curve.

A simple assumption is to lump the mass of water at the center of mass and assume that potential energy is given by all mass at that height. Thus, the amount of water to be released is not determined by H but by h, which is located at the center of mass of the pool when full at H. For simplicity, assume $h = aH$ with $1/2 < a < 1$. The potential energy Ep available to be converted in a cycle is $Ep = \underbrace{\rho \overline{A}H}_{mass} gh = \rho \overline{A}Hg \times aH = \rho g \overline{A}aH^2$. Then, if the tidal period is T, we have the following power available in a cycle:

$$P = \frac{a}{T} \rho g \overline{A} H^2 \tag{12.16}$$

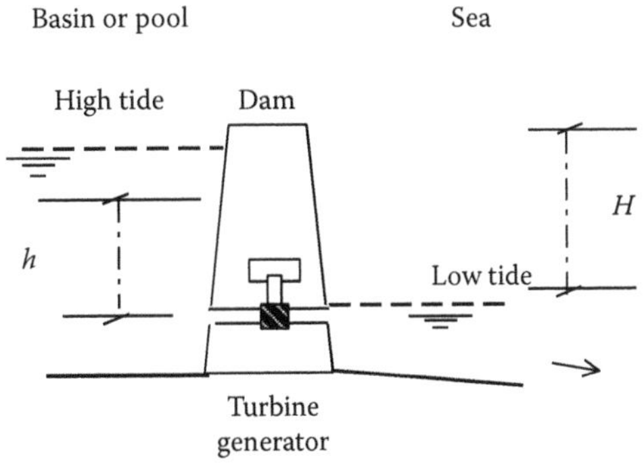

FIGURE 12.23 Barrage-based tidal power generation.

This is a simplistic model, but illustrates that tidal power of a cycle is increased by tidal basin surface area $\bar{A}$ and density, and very significantly as the square of the cycle's tidal range H.

Example 12.14

Assume you have a tidal basin of 1 km^2 with the center of mass at half the tidal range, $a = 0.5$. Take a tidal cycle with range of 6 m and lasting 12 h and 24 min. What is the tidal power available in this cycle? What is the electrical power generated if the facility is 40% efficient?

Answer: The tidal power in this cycle is $P = \dfrac{a}{T}\rho g \bar{A} H^2 = \dfrac{0.5}{(12 \times 3600 + 24 \times 60)\text{s}} 1025 \times 9.8 \times$
$1 \times (1000)^2 \times 6^2 \simeq 4.05$ MW. At 40% efficiency, we would generate 1.62 MW in a cycle corresponding to 1.6 MW × 12.4h = 20.09 MWh. Function power.barrage.cycle expedites this calculation:

```
> power.barrage.cycle(list(a=0.5,Abasin=1,nu=0.4,z=6))
$pow.tide.MW
[1] 4.05
$pow.gen.MW
[1] 1.62
$gen.MWh
[1] 20.09
>
```

This function becomes time saving when calculating power over many cycles.

When we look at plots such as the ones in Figure 12.22, we realize that the tidal range varies significantly during a month, in particular when reaching neap and spring tides. Therefore, we should account for these differences when performing power calculations such as the one we just did in Example 12.14. A comprehensive approach would be to input the values from the tidal model to the calculation, along with accounting for a decrease in surface area following the area–capacity curve of the basin. Let us assume for simplicity of illustration that surface area does not change significantly, and focus on the variation of tidal range.

Example 12.15

Function find.peaks of renpow facilitates obtaining all values of range, and power.cycle calculates power in a cycle using Equation 12.16. To demonstrate, we model a one-year time series for a hypothetical series, which is based on the constituents at Eastport, Maine, in the Bay of Fundy area (there is not a barrage there, this is just a hypothetical example).

```
x <- read.tide("extdata/ElevationTide.csv")
y <- harmonics.tide(x,days=29)
y <- harmonics.tide(x,days=365, plot=FALSE)
z <- find.peaks(y, band=c(0,1))
pp <- power.barrage.cycle(list(a=0.5,Abasin=1,z=z$xp,nu=0.4))
```

Alternatively, we could have used

```
x <- read.tide(system.file("extdata","ElevationTide.csv",package="renpow"))
```

The plots in Figure 12.24 show a monthly pattern beginning with spring tides (top panel), and the all peaks found in a year (bottom panel). Those peaks multiplied by 2 are an estimate of tidal range for those cycles. The results for power in MW and energy in MWh are

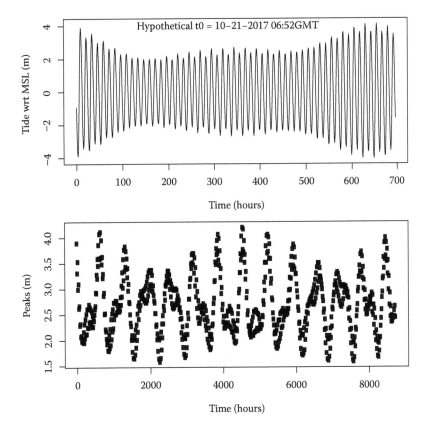

FIGURE 12.24   Hypothetical example of tides to calculate tidal power and generation over a year.

```
> mean(pp$pow.gen.MW)
[1] 0.340783
> sum(pp$gen.MWh)
[1] 2979.25
>
```

indicating an average generation power per cycle of 0.34 MW and annual generation of 2979 MWh.

Please make note of something very important that will come up again later in the chapter and in Chapter 13. It would be incorrect to average all tidal ranges in a year, and then use that average as the tidal range for the calculation of Equation 12.16, simply because the square of the average is not the same as the average of the squares.

Tidal power would increase if we could use a *double effect* to generate for both parts of the cycle, that is to say, during flood when water comes in, and during ebb when water goes out. This requires a turbine generator set that can generate in both directions of flow. This has been attempted, but for cost and practical reasons barrage-based plants operate as a single effect.

12.6.5   TIDAL CURRENTS

As mentioned earlier, the most recent emphasis is on direct use of the kinetic energy in tidal currents that occur with the ebb-and-flow of tides. This is also called *tidal in-stream energy conversion* and is based on submerged turbines. This technology is relatively new compared to other renewable conversion processes based on fluid dynamics. A common form is to use turbines similar to wind turbines consisting of a rotor with three blades (Chapter 13).

For tidal currents, we can use the power equation we have encountered before

$$P = \frac{1}{2}\rho A v^3 \qquad\qquad (12.17)$$

A difference in this case with respect to freshwater is that density of seawater is higher, say, 1025 versus 1000 kg/m^3. We will assume that the turbine can rotate to accommodate the current direction, and that therefore the negative parts of a tidal cycle can be converted to positive and used by the turbine.

Example 12.16

We will demonstrate how to calculate the average power per square meter in a tidal current with a velocity modeled by constituents given by file VelocityTide.csv. We first use harmonics.tide for 15 days and then take the absolute value with tide.current.absto produce Figure 12.25.

```
x <- read.tide("extdata/VelocityTide.csv")
y <- harmonics.tide(x,days=15,ylabel="Velocity m/s)")
z <- tide.current.abs(y, ylabel="Velocity (abs val) (m/s)")
```

Alternatively we could use

```
x <- read.tide(system.file("extdata","VelocityTide.csv",package="renpow"))
```

The effect of tide.current.abs is shown in Figure 12.25 for only 15 days; we can see how all negative parts of the cycle have been made positive. Then, applying for a year we get

```
> tidal.power(z,Aflow=1)
$Pavg
[1] 1.01
$Punits
[1] "(kW)"
```

Please realize from Figure 12.25 that velocity v has a form similar to the tidal waveforms, showing semidiurnal peaks and monthly neap and spring peaks. Consequently, when estimating power over multiple cycles, we should use values predicted by those waveforms and not just an average. Again, it would be incorrect to average all velocity values in a year and use that average for the calculation of Equation 12.17, simply because the cube of the average is not the same as the average of the cubes.

Since many of the details of tidal in-stream generation are similar to wind power, we defer those topics to the next chapter.

12.6.6 Environmental Effects of Tidal Power Generation

Building barrages to exploit tides has a variety of consequences for the coastal environment and ecosystems, some of them deemed negative and others positive. A key aspect is that the typical location of barrages is on estuaries and thus splits the estuary. A simple effect (i.e., one that works only for one direction of flow) plant reduces the water level behind the barrage, which has several negative impacts. These include reducing salinity in the basin and thus affecting the original ecosystem, increasing the area of exposed mud flats, blocking movement of marine life and fish into the estuary, and altering sediment transport and depositional patterns. The latter has been deemed positive in terms of increasing sunlight penetration in the water column and thus primary productivity. On the other hand, changing sediment transport and deposition patterns has effects on the ocean side of the barrage.

All of these potential impacts of barrages on coastal systems has prompted the interest in tidal current power generation as a more environmentally friendly approach. As we briefly discussed, the

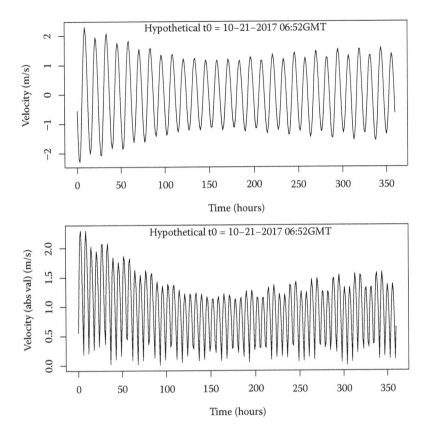

FIGURE 12.25 Tidal velocity is taken as absolute value to calculate power in the current.

technology of exploiting this resource is akin to wind turbines and it is just beginning. Therefore, it is opportune to note that installing a multitude of large tidal current turbines, in a manner similar to wind farms on land, would also have impacts on marine ecosystems, since they may interrupt marine life movements, and patterns of nutrient and sediment transport.

12.6.7 WAVE ENERGY

Wave energy is derived from wind energy, which is transferred to water by pressure and friction forces occurring at the water surface. In Chapter 13, we will study the wind resource with more detail. In this chapter we are solely concerned with the power that can be extracted from the waves. While the water stays in place, oscillating up and down, the potential energy changes and this pattern of change propagates across the ocean surface.

For a single frequency, propagation velocity c is related to frequency of oscillation f and wavelength λ by $c = f\lambda$, therefore the frequency and wavelength are related $f = \sqrt{\dfrac{g}{2\pi\lambda}}$. Now, using wave height H (i.e., total excursion from maximum to minimum height), the power is given per unit length (normal to the direction of wave propagation) as a flux of wave energy in W/m

$$P_{flux} = \frac{\rho g^2 H^2}{16(2\pi f)} = \frac{\rho g^2 H^2}{16(2\pi c/\lambda)} \tag{12.18}$$

In reality, waves are a combination of multiple heights, frequencies, and propagation wavelengths. A simplified approach is to define two parameters, the *significant height* (Hs) of a wave (in m) and the *energy period* of the wave (in s). Significant height is four times the rms of the water elevation $h(t)$

over a number of cycles: $Hs = 4\sqrt{\langle h(t)^2 \rangle}$. The energy period has been defined in several ways: One is $1.12 \times Tz$, where Tz is the average time to zero crossings in the upward direction, and another is Tp based on peak of a spectral distribution of wavelengths. We will use Tp in the following calculations.

Power flux in W/m along the wave crest is described by the simple relation

$$P_{flux} = 0.42 \times Hs^2 \times Tp \qquad (12.19)$$

The power for each pair of values Hs and Tp can be calculated from Equation 12.19 and displayed as shown in Figure 12.26 (left-hand side). This figure corresponds to a hypothetical site with Hs from 1 m to 10 m and Tp from 5 to 20 s. The result is a matrix $\mathbf{P}_{flux}$, where each entry has the value of power for the Hs and Tp values. The parameters Hs and Tp are used in data sets that contain the number of hours per year for each pair of values Hs and Tp (Figure 12.26, right-hand side). Examples for a variety of sites are provided by the Electric Power Research Institute [15]. The latter figure or its corresponding table is often referred to as a scatter diagram, but we will opt to refer to it as Hs–Tp duration diagram since that is the information contained and it relates to what we have done before with a flow–duration curve. Let us denote this as matrix $\mathbf{D}$, which has the same dimension as matrix $\mathbf{P}_{flux}$.

Example 12.17

The panels of Figure 12.26 are drawn using three functions of renpow—powflux.wave duration.wave, and wave.contour—in the following manner. Here, duration was read from file WaveHsTp.csv:

```
Pflux <- powflux.wave(Hs=seq(1,10),Tp=seq(5,20))
wave.contour(X=Pflux,label="Power flux (kW/m)")
D <- duration.wave("extdata/WaveHsTp.csv")
wave.contour(X=D,label="Duration (hrs/yr)")
```

Alternatively, we could have used

```
D <- duration.wave(system.file("extdata","WaveHsTp.csv",package="renpow"))
```

or by directly using the dataset

```
D <- duration.wave(WaveHsTp,file=FALSE)
```

Note that multiplying the power flux matrix $\mathbf{P}_{flux}$ by the Hs–Tp duration matrix $\mathbf{D}$ (entry wise multiplication or Hadamard multiplication) $\mathbf{E}_{flux} = \mathbf{P}_{flux} \circ \mathbf{D}$, we obtain a matrix of energy flux in MWh/yr/m (Figure 12.27, left-hand side). The total energy is obtained by summing all values in the resulting matrix $\mathbf{E}_{flux}$ and is shown in the figure to be 875 MWh/yr/m. Then, assuming an energy conversion device spanning a length L (m) and with efficiency η (here we assume for simplicity the same for all Hs and Tp, but in reality it varies), we can obtain electrical energy generated in a year in MWh/yr by multiplying the energy flux times L and efficiency: $\mathbf{E}_g = L \times \eta \times \mathbf{E}_{flux}$ (Figure 12.27, right-hand side).

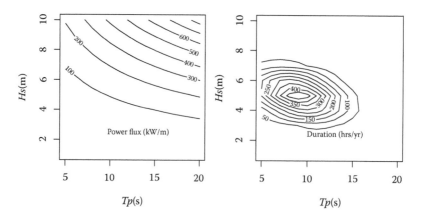

FIGURE 12.26 Wave power and duration diagram for a hypothetical site.

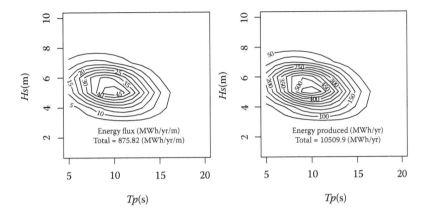

FIGURE 12.27   Wave energy and electric energy production for a hypothetical site.

Example 12.18

The panels of Figure 12.27 are drawn using two functions of renpow: energy.wave and energy.gen, plus contour.wave in the following manner. We use a structure of L = 30 m and efficiency of 40%.

```
Ew <- energy.wave(Pflux,D)
wave.contour(X=Ew,label="Energy flux (MWh/yr/m)",sum=TRUE,sumlabel="(MWh/yr/m)")
Eg <- energy.gen(Ew,L=30,nu=0.4)
wave.contour(X=Eg,label="Energy produced (MWh/yr)",sum=TRUE,sumlabel="(MWh/yr)")
```

Note that this is a substantial amount of energy generated (10.5 GW/h) in a year.

Calculations, such as the ones we have done in these examples, for coastal areas around the world have indicated that wave power generation could contribute a significant amount to the total consumption. For example, it has been estimated that the western coast of the United States has a potential of 440 TWh/y, that even with very low conversion efficiency (10%) would mean several fold more the current electric demand of the entire country [3].

Wave energy conversion is still in early stages of development. Generally, there are four categories: terminators, attenuators, point absorbers, and overtopping devices. You can find details in Boyle [1]. So far, potential environmental impacts of wave energy conversion, when deployed offshore, are deemed benign compared to tidal. These include low mechanical noise, low risk of contamination, and low risk of interrupting fish migration.

EXERCISES

12.1   A centrifugal pump works by rotating an impeller and forcing water through a volute giving water an exit velocity such that it can reach a given head h. Calculate the required exit velocity for a centrifugal pump located at point 2 of the pipe in Figure 12.7 with both sides open to atmospheric pressure. The water at point 1 has zero velocity $v_1 = 0$, and the streamline at 1 is $z_1 = 5$ m with respect to point 2, which is zero, $z_2 = 0$.

12.2   Consider a low-head hydroelectric plant with head difference of 2 m and cross-sectional area of $A = 6$ m^2. What is the power P?

12.3   Suppose we have a reservoir with $h = 10$ m that can sustain flow of $Q = 1000$ m^3/s at constant head. What is the power in the fluid assuming that loss has been designed to yield maximum power?

12.4   From a creek, we take a flow of 0.02 m^3/s to a 90% efficient turbine/generator located 30 m elevation below using 200 m of PVC pipe of $D = 0.1$ m. Calculate head loss in the pipe and power.

12.5 Compare the ratio $\dfrac{Q_{95}}{Q_{mean}}$ and $\dfrac{Q_{95}}{Q_{50}}$ for a baseflow river and for a flashy river using model.flow.

12.6 Consider a small reservoir 1 km wide and 2 km long, at 1000 m asl elevation and the conservation pool elevation is 1010 m at the dam. Assuming a simplistic bathymetry, obtain area–capacity curves. Assume we developed a micro hydroelectric plant drawing a flow of 0.01 m3/s for 8 h every day. What is maximum drought duration (in consecutive days of no inflow) that we can sustain before having to stop power generation?

12.7 Use harmonic constituents to determine tides during a 15-day and a 29-day period for Eastport, Maine, in the Bay of Fundy area. You can visit NOAA's website [14] and study the parameter values for this station. Package renpow includes one csv sample data file for this location that we can use for this exercise. Use function harmonics.tide of renpow.

12.8 Using wave power functions of renpow and file Hrs-hypothetical.csv. Estimate wave energy generation for a structure of $L = 100$ m and low efficiency of 10%.

REFERENCES

1. Boyle, G., *Renewable Energy: Power for a Sustainable Future*. Third edition. 2017: Oxford University Press. 566 pp.
2. Fay, J.A., and D.S. Golomb, *Energy and the Environment. Scientific and Technological Principles*. Second edition. MIT-Pappalardo Series in Mechanical Engineering, M.C. Boyce and G.E. Kendall. 2012, New York, Oxford: Oxford University Press. 366 pp.
3. Masters, G.M., *Renewable and Efficient Electric Power Systems*. Second edition. 2013: Wiley-IEEE Press. 690 pp.
4. Tester, J.W. et al., *Sustainable Energy: Choosing Among Options*. Second edition. 2012: MIT Press. 1021 pp.
5. World Energy Council. *Hydropower*. 2017. Accessed October 2017. Available from: https://www.worldenergy.org/data/resources/resource/hydropower/.
6. Artiola, J.F., Monitoring Surface Waters, in *Environmental Monitoring and Characterization*, J.F. Artiola, I.L. Pepper, and M.L. Brusseau, editors. 2004, Elsevier Academic Press: Burlington, MA. pp. 141–161.
7. Engineering ToolBox. Hazen-Williams coefficients for piping materials. 2017. Accessed October 2017. Available from: http://www.engineeringtoolbox.com/hazen-williams-coefficients-d_798.html.
8. Hemond, H.F., and E.J. Fechner, *Chemical Fate and Transport in the Environment*. 1994: Academic Press. 338 pp.
9. U.S. EPA. Summary of the Clean Water Act. 2014. US EPA. Accessed October 2014. Available from: http://www2.epa.gov/laws-regulations/summary-clean-water-act.
10. U.S. EPA. NPDES Home. 2014. US EPA. Accessed October 2014. Available from: http://water.epa.gov/polwaste/npdes/.
11. U.S. EPA. STORET/WQX. 2014. Accessed October 2014. Available from: http://www.epa.gov/storet/.
12. World Energy Council. Energy Resources: Marine. 2017. Accessed October 2017. Available from: https://www.worldenergy.org/data/resources/resource/marine/.
13. National Ocean Service. Currents, Tidal Currents 1. 2017. NOAA, National Oceanic and Atmospheric Administration. Accessed October 2017. Available from: https://oceanservice.noaa.gov/education/tutorial_currents/02tidal1.html.
14. NOAA. *Harmonic Constituents - Station Selection*. 2017. NOAA. Accessed October 2017. Available from: https://tidesandcurrents.noaa.gov/stations.html?type=Harmonic+Constituents.
15. EPRI, Ocean Tidal and Wave Energy. 2005, Electric Power Research Institute. p. 154.

13 Wind Resources and Wind Power

In this chapter, we start with a brief overview of atmospheric circulation patterns that will help us understand where and when we expect higher winds on the surface of the Earth. Then, we study how to calculate power in the wind taking into account air density, pressure, and temperature. Statistics of wind speed are important to calculate expected production by wind turbines, thus we will show how to analyze wind speed data using known distributions. Details of wind turbine types and performance are examined next, emphasizing the power curve. After a brief overview of large wind farms, we go into details of off-grid and microgrid wind power installations. Most textbooks on renewable power cover wind power generation [1–5] and can serve as supplementary reading to this chapter.

Early attempts to use windmills to generate electricity occurred in the late 1880s in the United Kingdom. In 1891, Poul la Cour, a Danish inventor, used electricity from a wind turbine to electrolyze water and produce hydrogen, which was used for gaslights in the local schoolhouse. It is remarkable how this pioneering effort fits well with the 21st century vision of powering electrolysis from wind and solar, and using the hydrogen for fuel cells.

In contrast to hydroelectric, wind power to generate electricity has not become an important form of generation until much more recently. Wind power generation became more readily available in the early 1970s and has continued to grow since then, experiencing a rapid increase since the 1990s. Electrical power generation using wind power is now recognized as one of the low investment, high-yield sources of power generation.

Wind power generation is mostly based on harvesting kinetic energy by means of a wind turbine that spins a generator. In a similar manner to hydroelectric, ultimately wind power is solar, because it is the differential distribution of solar radiation over the surface of the Earth that drives wind.

Recall from Chapter 1 (see Figure 1.15) that wind is not yet a significant contributor of electricity production worldwide. Lumped together with solar and other non-carbon-based renewables it does not reach but 4.2% of world production, surpassing only biomass and waste. However, installed capacity of wind, together with solar, is growing very fast. In 2016, there was about 433 GW installed capacity of wind power generation, which has increased by 54 GW in 2016 [6] to reach about 486 GW installed capacity worldwide in 2017.

In the United States, wind power accounted for ~226 TWh electricity generation or ~5.5% of the United States' electricity, with an installed capacity of 82 GW [7]. This capacity is only surpassed by China, which doubles it at 168 GW [8], representing nearly 35% of the world's capacity. Other major producers are Germany (third largest, 50 GW), India (fourth largest ~29 GW), and Spain (fifth largest with 23 GW).

The largest wind turbines range from several MW to nearly 10 MW and are used in *wind farms*, that is, hundreds of wind turbines covering an area of hundreds of km^2, and having a total capacity from hundreds of MW to tens of GW. Wind farms are developed inland and offshore. This type of arrangement has a large footprint on the landscape or the sea, which we will discuss later in the chapter. Smaller wind turbines, 100 kW to 1 MW, are used on small wind farms, wind turbine clusters, or individually and typically grid-tied. There are also smaller turbines (up to tens of kW) that are mostly used off-grid or in microgrids, as we discuss later this chapter.

13.1   WIND: DRIVING FORCES AND CIRCULATION PATTERNS

At a global scale, three important factors constitute driving forces for wind: atmospheric pressure gradient as the prime force, the Coriolis effect affecting circulation patterns, and friction with the surface of the Earth. In this section, we do a brief overview of atmospheric circulation patterns that will help us understand where and when to expect higher winds on the surface of the Earth.

13.1.1   DRIVING FORCES

We have previously encountered the concept of fluid pressure in this textbook, and in particular for gases in Chapter 4. We used pressure to understand the thermodynamics of expansion and compression of air. Please refer to that chapter to remember that 1 bar of pressure is 100 kPa. We also want to recall atmospheric concepts from Chapter 7; mainly that air temperature and pressure change with increasing altitude. Pressure steadily declines as less air exerts less weight. However, temperature changes at varying rates and signs, establishing an alternation of thermal layers, starting from the planet's surface.

In this chapter, we will focus on the troposphere, which contains most (~3/4) of the mass of the atmosphere and it extends from the surface to about 10 km, where the air pressure is about 0.1 of the pressure at sea level. In this layer, temperature decreases with altitude. The lowest part of the troposphere (~1 km) is the planetary *boundary layer*, where thermal processes interact with the Earth's surface.

Every point in the atmosphere has a value of pressure at a particular instant; this is a complicated 3D picture. It is common to slice the atmosphere at several heights and plot the pressure onto each one of these 2D planes. These are pressure maps and are an essential tool in understanding weather. Joining all points on a pressure map with equal values of pressure forms an *isobar* or a line of equal pressure. The isobars are displayed as contour lines and given in millibar (mb).

A *pressure gradient* from high pressure to low pressure is the driving force for wind. As Figure 13.1 illustrates, a gentle pressure gradient is seen as widely spaced isobars and produce light winds, whereas closely spaced isobars produce strong winds [9].

A vertical profile shows a complementary picture (Figure 13.2). At a low-pressure (L) center, we have converging wind (top view) and ascending air (side view), whereas at a high-pressure (H) center, we have diverging wind (top view) and descending air (side view). A center of high pressure is an *anticyclone*, whereas a center of low pressure is a *cyclone* [9].

The *Coriolis effect* is not a real force, but the result of the Earth's rotation that deflects trajectories of air masses in the atmosphere. As an apparent effect, it depends on the observer. It produces deflection to the right in the Northern Hemisphere and deflection to the left in the Southern Hemisphere (Figure 13.3) [9].

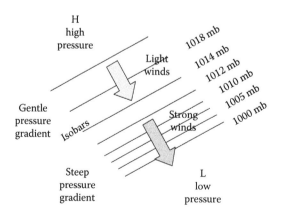

FIGURE 13.1   Barometric pressure gradient drives wind.

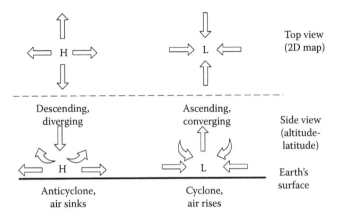

FIGURE 13.2   Top and side view of airflow around low- and high-pressure centers.

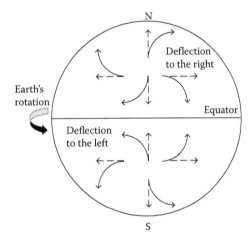

FIGURE 13.3   Coriolis effect: deflection to the right in the Northern Hemisphere and deflection to the left in the Southern Hemisphere.

In the upper atmosphere, the combination of pressure gradient and Coriolis effect generate a circular pattern of winds around centers of high pressure (H) and low pressure (L). These are *geostrophic* winds. At this high altitude, friction against the surface of the Earth is negligible. Winds flowing between L and H flow parallel to isobars (Figure 13.4). For an anticyclone (H), the wind circulates clockwise in the Northern Hemisphere and, oppositely, counterclockwise in the Southern Hemisphere. For a cyclone (L), the pattern flips: The wind circulates counterclockwise in the Northern Hemisphere and clockwise in the Southern Hemisphere (Figure 13.4).

In the lower atmosphere where we have surface winds, *friction* makes the wind change direction; it becomes convergent or divergent (Figure 13.5). The clockwise patterns are diverging around a high-pressure center in the Northern Hemisphere (anticyclone), and converging around a low-pressure center in the Southern Hemisphere (cyclone). In opposite manner, counter-clockwise patterns are converging around a low-pressure center in the Northern Hemisphere (cyclone) and diverging around a high-pressure center in the Southern Hemisphere (anticyclone) (Figure 13.5).

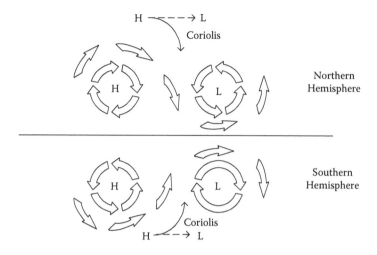

FIGURE 13.4   Circular geostrophic wind patterns in the upper atmosphere.

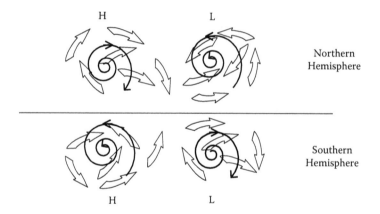

FIGURE 13.5   Surface winds spiraling patterns in the lower atmosphere.

These basic patterns, driven by pressure gradient, Coriolis, and friction, explain wind flows at any particular time in the atmosphere. For instance, Figure 13.6 illustrates a hypothetical condition of high- and low-pressure centers, and corresponding wind flow overlain on a map of the United States.

Example 13.1

Explain why Atlantic hurricanes approaching the Caribbean islands show spiraling counter-clockwise rotation. Answer: Hurricanes are cyclones with very strong pressure gradients. A hurricane approaching the Caribbean islands would be in the Northern Hemisphere, thus the rotation is counterclockwise (see Figure 13.5). It spirals inward because the lowest layers of the hurricane are near the ocean surface.

13.1.2   CONVECTION

Recall from Chapter 7 that convection is driven by a density gradient in the fluid, which will be air for the purposes of this chapter. The density gradient is due to a difference in temperature; density

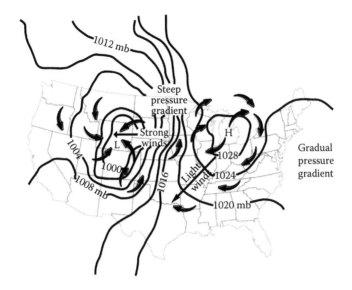

FIGURE 13.6   Hypothetical low- and high-pressure centers and winds over a map of the United States.

decreases with increasing temperature. Therefore, warmer air has lower density and is more buoyant. Convective transport in the atmosphere is vertical; lighter parcels rise against gravity and denser parcels fall.

This process creates large *cell circulation* in the atmosphere, which is a combination of up flows from a low, horizontal increasing latitude, down flows at a high, and horizontal flows back to the low (Figure 13.7). The most significant is the *Hadley cell* that rises in tropical areas and sinks at sub-tropical latitudes (Figure 13.7). The returning flows in the Hadley cell from the subtropical high to the equatorial areas are the trade winds.

Embedded in these circulation patterns, we have *jet streams* that are concentrated high velocity winds.

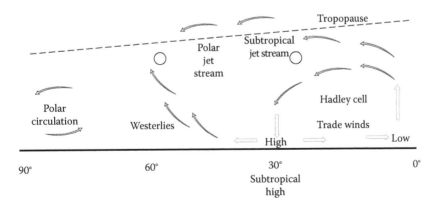

FIGURE 13.7   Vertical movement in the atmosphere, illustrated for a hemisphere. Horizontal axis is latitude and vertical axis is elevation.

13.1.3 GLOBAL CIRCULATION PATTERNS

Putting together horizontal and vertical flows, we characterize large circulation patterns on the Earth's atmosphere. See Figure 13.8 for a diagram including vertical movement and Figure 13.9 for a world map view. A prominent feature is the subtropical high (STH) pressure centers, which correspond to the descending (air sinking) arms of the Hadley cells. From these centers, we have the easterly trade winds. They are northeasterlies in the Northern Hemisphere and southeasterlies in Southern Hemisphere and converge in the *intertropical convergence zone* (ITCZ).

Higher surface temperature at the tropical areas lead to low-pressure centers at the ITCZ and correspond to air rising in the ascending arms of the Hadley cells, rising near the Equator. The westerlies are located in midlatitudes (30°–60° N and S) and flow out of the STH centers.

13.1.4 MONSOONS AND LOCAL WINDS

In addition to the global circulation patterns, we have regional and local patterns. Monsoons correspond to a regional seasonal reversal of wind patterns at some land–ocean interface areas on the

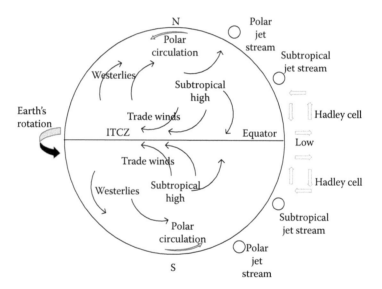

FIGURE 13.8   Schematic view of global circulation patterns.

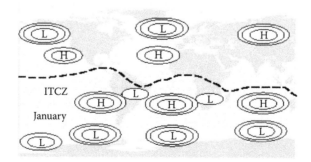

FIGURE 13.9   Global high and low pressure patterns expressed on a world map for a particular season.

Earth. In summer, winds loaded with moist air blow onshore and cause heavy rains; in winter, dry continental air blows offshore. This pattern is strong in India and West Africa.

Other local patterns are found associated with the land and ocean interface. An example is the sea–land breeze, which changes daily; it blows from sea to land at daytime, and from land to sea at nighttime. This reversal is due to the differential heating of land and sea.

13.2 WIND POWER

13.2.1 POWER IN THE WIND

Recall from Chapter 1 that for a fluid flowing with velocity v, the power is $p(t) = \frac{1}{2}\dot{m}v^2$, where $\dot{m}$ is the fluid mass flow rate, and $\dot{m} = \rho A v$ assuming a simple geometry of cross-sectional area A, normal to the direction of flow. With these elements, we find that power W available is proportional to the cube of velocity of the fluid

$$p(t) = \frac{1}{2}\rho A v^3 \tag{13.1}$$

We have used this equation in Chapter 12 for calculations of power in the flow of water. Now, in dealing with air, we use air density. The density of air changes according to elevation above sea level and temperature. To specify wind turbine performance, there are *reference conditions*, which are pressure 1 atm, temperature 15°C, and density 1.225 kg/m3.

A depends upon the rotor diameter, which depends upon the turbine design. When we divide Equation 13.1 by A, we obtain *specific power* or *power density* in W/m^2

$$p(t) = \frac{1}{2}\rho v^3 \tag{13.2}$$

which is simply *power per unit of cross-sectional area*. Since area goes with the square of diameter, doubling the rotor diameter quadruples the available power. This helps explain why use larger wind turbines: they are more economical because the turbine cost is just proportional to rotor diameter, whereas the power generated is proportional to the square of the diameter [2].

Example 13.2

The simplest calculation would be to assume density to be known and unit area. What would be the specific power or power density for wind velocity of 1 m/s at reference conditions? Answer: $p(t) = \frac{1}{2}\rho v^3 = \frac{1}{2}1.225 \times 10^3 = 612.5$ W/m2.

Function pow.rho.v3.table of renpow performs this calculation:

```
> x <- list(rho=1.225,v=10,A=1)
> pow.rho.v3.table(x)
$X
 Density(kg/m3) Vel(m/s) Area(m2) Power(W)
1        1.225   10       1  612.5
```

To visualize the effect of density and wind speed, we can use functions pow.rho.v3 and pow.v3.plot for a range of density and wind

```
x <- list(rho=c(0.9,1,1.1,1.225,1.3),v=seq(0,30),A=1)
X <- pow.rho.v3(x)
x <- list(v=X$v,y=X$rho,Pow=X$Pow,yleg="rho",ylabel="Density(kg/m3)")
pow.v3.plot(x)
```

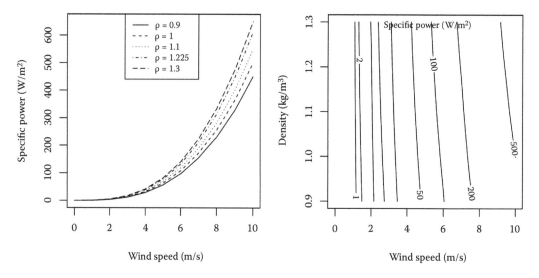

FIGURE 13.10 Variation of specific power with wind speed and density. Left: Family of curves versus wind speed with density as parameter. Right: Isopower lines as a function of density and wind speed.

to give us Figure 13.10, which illustrates the fast increase of power with wind speed and just a linear increase with density. The right-hand side represents lines of equal specific power and we can describe these as isopower lines.

13.2.2 Correcting Air Density for Temperature and Pressure

First, consider air density as mass per unit volume in kg/m^3:

$$\rho = \frac{n(\text{mol}) \times M(\text{g/mol}) \times 10^{-3}(\text{kg/g})}{V(\text{m}^3)} \tag{13.3}$$

Recall from Chapter 4 that M is molecular mass of air 28.97 g/mol (dry air, which is close to that of N_2, the most abundant gas in the atmosphere), and the number of moles per volume n/V can be found from the ideal gas law $pV = nRT$:

$$\frac{n}{V} = \frac{p}{RT} \tag{13.4}$$

where p is pressure in Pa, T is temperature in K, and R is the universal gas constant 8.314 J/(mol K). Therefore, we obtain density as a function of pressure and temperature:

$$\rho\left(\text{kg/m}^3\right) = \frac{p \times M(\text{g/mol}) \times 10^{-3}(\text{kg/g})}{RT} \tag{13.5}$$

Example 13.3

What is the air density at 1 atm (sea level) and 15°C? Use pressure 1 atm (101.325 kPa; see Chapter 4).

Answer: $\rho(\text{kg/m}^3) = \dfrac{101.325 \times 10^3 \text{Pa} \times 28.97 \text{ g/mol} \times 10^{-3} (\text{kg/g})}{8.314 \text{ J/(mol K)}(15 + 273)\text{K}} = 1.225 \text{ kg/m}^3.$

Function rho.pT.air of renpow performs this calculation (pressure can be input as "atm" or "bar").

```
> x <- list(p=1,T.C=15,punit="atm")
> rho.pT.air(x)
 Pressure(kPa) Temp(C) Density(kg/m3)
1    101.325      15    1.2259
>
>
```

Note that we can combine all constants in Equation 13.5 into one denoted by κ to facilitate a repetitive calculation and have

$$\rho = \kappa \frac{p}{T} \tag{13.6}$$

Here, if pressure is given in atm and temperature in K, we have

$$\kappa = \frac{101.325 \times 10^3 \ \mathrm{Pa/atm} \times 28.97 \ \mathrm{g/mol} \times 10^{-3} (\mathrm{kg/g})}{8.314 \ \mathrm{J/(mol \ K)}} = 353.06 \frac{\mathrm{kg}}{\mathrm{m}^3} \times \frac{\mathrm{K}}{\mathrm{atm}} \tag{13.7}$$

When pressure is given in bar and temperature in K, we have

$$\kappa = \frac{100 \times 10^3 \ \mathrm{Pa} \times 28.97 \ \mathrm{g/mol} \times 10^{-3} (\mathrm{kg/g})}{8.314 \ \mathrm{J/(mol \ K)}} = 348.45 \frac{\mathrm{kg}}{\mathrm{m}^3} \frac{\mathrm{K}}{\mathrm{bar}} \tag{13.8}$$

13.2.3 Correcting Air Density with Elevation

Now let us consider elevation above sea level. Pressure p should decrease with elevation z in m. To see this, think of an air column of height z, and the differential decrement of weight of air per unit area, or differential of pressure $dp = -\rho g dz$. Note that density depends on pressure itself and temperature according to Equation 13.6. We can equate it to $dp = p(z + dz) - p(z)$ to obtain

$$\frac{dp}{dz} = -\kappa g \frac{p}{T} = -\beta \frac{p}{T} \tag{13.9}$$

We have lumped the two constants κg into one β by dividing by 101.325 or 100 kPa according to whether we are using bar or atm $\beta = \kappa g = 348.45 \times 9.8/10^5$ Pa/bar = 0.0341. The ordinary differential equation (ODE) in Equation 13.9 is easy to solve analytically if we assume the temperature at the elevation z is constant and equal to T throughout the column [2]. Separating for integration $\frac{dp}{p} = -\beta \frac{dz}{T}$, and integrating $\ln(p/p_0) = -\beta \frac{(z-0)}{T}$, where $p_0 = p(0)$, and using exp on both sides

$$p = p_0 \exp\left(-\frac{\beta}{T}z\right) = \exp\left(-\frac{\beta}{T}z\right) \tag{13.10}$$

We can assume $p(0) = 1$ bar, or alternatively 1 atm, and use value of β accordingly. Now, we can insert this pressure equation into Equation 13.6

$$\rho = \kappa \frac{\exp\left(-\frac{\beta}{T}z\right)}{T} \tag{13.11}$$

Example 13.4

Compare air density at two conditions: A, 1000 m asl and 10°C; B, 100 m asl and 30°C. Answer: We use pressure in bar and assume pressure at sea level is 1 bar. Condition A is

$$\rho = 348.45 \times \frac{\exp(-0.0341 \times 1000/(10+273))}{(10+273)} = 1.091 \text{ kg/m}^3$$

Condition B is

$$\rho = 348.45 \frac{\exp(-0.0341 \times 100/(30+273))}{(30+273)} = 1.137 \text{ kg/m}^3$$

We can do these calculations much quicker with function rho.zT.air:

```
> x <- list(z=1000, T.C=10, punit="bar")
> rho.zT.air(x)
 Elevation (m) Pressure (bar) Temp(C) Density(kg/m3)
1      1000        0.8863          10          1.0913
> x <- list(z=100, T.C=30, punit="bar")
> rho.zT.air(x)
 Elevation (m) Pressure (bar) Temp(C) Density(kg/m3)
1       100        0.9888       30       1.1371
>
```

However, recall from Chapter 7 the concept of dry adiabatic rate in the troposphere, which means air cools at a rate of ~6°C/km. We can write a linear approximation $T(z) = T(0) - \alpha z$, where $\alpha \sim 6 \times 10^{-3}$ °C/m. Now denoting $T_0 = T(0)$ and inserting into the denominator of Equation 13.9:

$$\frac{dp}{dz} = -\beta \frac{p}{T_0 - \alpha z} \tag{13.12}$$

Separating for integration $\dfrac{dp}{p} = -\beta \dfrac{dz}{T_0 - \alpha z}$, and integrating $\ln(p/p_0) = -\beta(-1/\alpha) \ln\left(\dfrac{T_0 - \alpha z}{T_0}\right) = \ln\left\{\left(\dfrac{T_0 - \alpha z}{T_0}\right)^{\beta/\alpha}\right\}$ using exp on both sides

$$p = p_0 \left(\frac{T_0 - \alpha z}{T_0}\right)^{\beta/\alpha} \tag{13.13}$$

We can assume $p(0) = 1$ bar, or alternatively 1 atm, and use value of β accordingly. Now we can insert this pressure equation into Equation 13.6

$$\rho = \kappa \frac{\left(\dfrac{T_0 - \alpha z}{T_0}\right)^{\beta/\alpha}}{T_0 - \alpha z} \tag{13.14}$$

The function rho.zT.air calculates pressure and density using this expression when the argument x includes a value for lapse.

Example 13.5

Use function rho.zT. air to estimate pressure and density at 1000 m elevation starting with 20°C at the ground and a lapse rate of ~6°C/km. Answer:

```
>x <- list(z=1000, T0.C=20, punit="bar", lapse=6)
> rho.zT.air(x)
  Elevation (m) Pressure (bar) Temp(C) Density(kg/m3)
1     1000      0.8889      14      1.0792
  >
  >
```

Note that if we were to consider a constant temperature of 14°C using the simpler expression 13.11, we would get

```
> x <- list(z=1000, T.C=14, punit="bar")
> rho.zT.air(x)
 Elevation (m) Pressure (bar) Temp(C) Density(kg/m3)
1     1000      0.8878      14      1.0779
>
```

which is close to the value obtained by using the lapse rate. So as long as we adjust the temperature to be expected at a higher elevation, we can apply the simpler expression 13.11.

13.2.4 Specific Power as a Function of Elevation, Temperature, and Wind Speed

Now that we know how to calculate air density as a function of elevation and temperature (via pressure), we can adjust specific power (or power density) calculations to those factors. We combine several functions into one function named pow.wind, which we can call with several values of elevation and wind speed.

Example 13.6

Plot specific power for four elevations—0, 500, 1000, 2000 m asl—and wind speed from 0 to 30 m/s. Use ground temperature of 20°C and lapse rate of 6°C/km. Answer:

```
x <- list(z=c(0,500,1000,2000), T.C=20, v=seq(0,30),punit="bar",
     lapse=6,yleg="z",ylabel="Elevation(m)")
pow.wind(x)
```

The result is shown in Figure 13.11.

13.2.5 Variation of Wind Speed with Height

As we move away from the ground, wind speed increases because the friction decreases. Since power in the wind increases with v^3, a taller tower means higher wind speed, and thus more power. The increase in velocity due to height is calculated from the empirical relation [2]

$$\frac{v}{v_0} = \left(\frac{H}{H_0}\right)^\alpha \tag{13.15}$$

where H_0 is a *reference height* (10 m), v is wind speed at height H, v_0 is wind speed at the reference height, and α is a friction coefficient, which characterizes the terrain conditions taking into account whether this is on water (the lowest friction) or on land with various types and heights of vegetation and land use. There are several tables available for this coefficient. We will use a variation of these with the following α values: water 0.05, smooth bare ground 0.1, rural with grass or crops 0.15, rural with shrubs 0.2, suburban with shrubs 0.25, small town with trees 0.3, small town with low buildings 0.35, and urban including buildings 0.4.

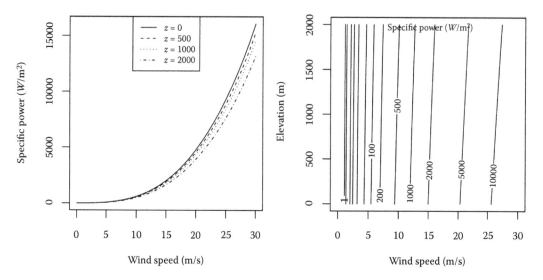

FIGURE 13.11   Variation of specific power with wind speed and elevation. Left: Family of curves versus wind speed with elevation as parameter. Right: Isopower lines as a function of elevation and wind speed.

An alternative relation, inspired on aerodynamics in the atmospheric boundary layer, is based on the *roughness length l* in m and assumes that the air flow varies logarithmically with elevation [2]

$$\frac{v}{v_0} = \frac{\ln(H/l)}{\ln(H_0/l)} \tag{13.16}$$

The parameter roughness length is the height above the ground at which wind speed is zero for atmosphere thermal conditions with lapse rate of $-9.8°C/km$. Roughness length varies from 0.0002 for water surface to 1.6 for dense urban areas or forests. Intermediate values are 0.03 (for open areas), 0.1 (for crop areas with few windbreaks), and 0.4 (for urban and rural areas with windbreaks).

Since power increases as the cube of wind speed, assuming equal air density, we see that the ratio of specific power goes as the cube of speed ratio, $\frac{P}{P_0} = \left(\frac{v}{v_0}\right)^3$, and therefore we can get the ratio of specific power for both models by raising them to the third power.

Example 13.7

Compare wind speed ratios $\frac{v}{v_0}$ and power ratio $\frac{P}{P_0}$ obtained from the empirical exponential model of Equation 13.15 with the aerodynamics model of Equation 13.16 for three terrain conditions: A, open areas with short crops; B, rural to suburban areas with trees; and C, dense urban areas. Answer: To represent these conditions we select α to be 0.1, 0.25, and 0.4, and *l* to be 0.1, 0.4, and 1.6. Function v.H of renpow allows quick calculation and plotting:

```
x <- list(alpha=c(0.1,0.25,0.4),rough=c(0.1,0.4,1.6))
v.H(x)
```

The results are shown in Figure 13.12. We can see that for condition A (open areas), the logarithmic model predicts faster wind speed and increases higher up than the exponential model. Conversely, for denser urban areas, the exponential model predicts faster wind speed, and increases higher up than the logarithmic model.

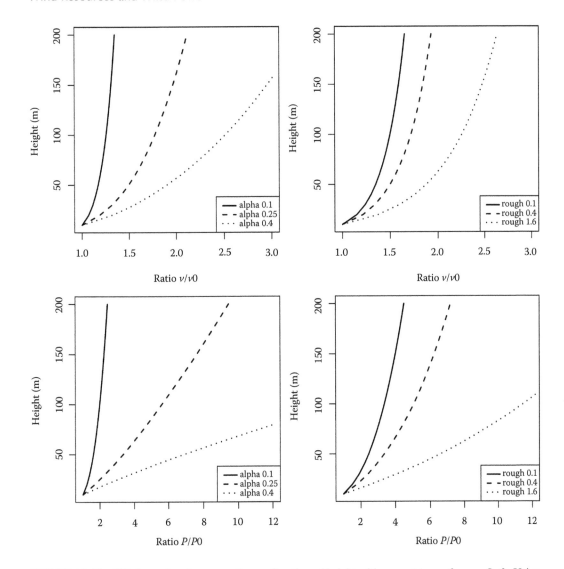

FIGURE 13.12 Wind speed and power ratio as a function of height with respect to a reference. Left: Using empirical exponential function. Right: Using logarithmic aerodynamics function.

Reasonable estimates of α and l based on land cover are good for survey or scoping work. However, a better approach is to calibrate the coefficients α and l from wind speed data at various heights at the same site. For instance, suppose we have data v_1 and v_2 from two anemometers, one at height H_1 and another at height H_2, such that $H_2 > H_1$. A zero-intercept linear regression of v_2 versus v_1 yields a slope of a.

Then the exponential model would be $\dfrac{v_2}{v_1} = a = (H_2/H_1)\alpha$, and we can solve for α to get $\alpha = \dfrac{\ln a}{\ln(H_2/H_1)}$. The logarithmic model would be $a = \dfrac{\ln(H_2/l)}{\ln(H_1/l)}$, which we can rearrange $a\ln(H_1/l) = \ln(H_2/l)$, or $(H_1/l)^a = (H_2/l)$. Solving for the roughness length, we get $H_1^a/H_2 = l^{a-1}$ and finally $l = (H_1^a/H_2)^{1/(a-1)}$.

Package renpow includes a calibration function to facilitate these calculations.

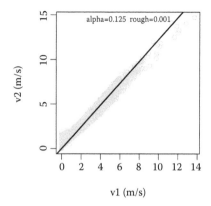

FIGURE 13.13   Calibration of wind speed versus height function from data collected at two different heights.

Example 13.8

Consider data on wind speed for two anemometers on the same tower. The area is relatively open with grass cover and several small buildings nearby. The lowest gauge is at 2 m above the ground, and the other at 10 m above the ground. The data consists of wind speed averaged every 15 min for 3 months from mid May to mid August 2017.

```
Read the data file by using
# reading file
tt.xx<-read.table("extdata/TEODP2017.csv", header=TRUE, sep=",", skip=1)
```

Alternatively, we can use the system file
```
tt.xx <- read.table(system.file("extdata","TEODP2017.csv", package="renpow"),
header=TRUE, sep=",", skip=1)
```

```
Alternatively load it from renpow data.
tt.xx <-TEODP2017
```

```
Select the average wind speed at two heights which are variables 8 and 4.
ws.2m.10m <- tt.xx[,c(1,8,4)]
v1.v2<- ws.2m.10m[,2:3]
> x <- list(v1.v2=v1.v2,H1=2,H2=10)
> cal.vH(x)
$alpha
[1] 0.125
$rough
[1] 0.001
>
```

The function produces the plot shown in Figure 13.13. We have $\alpha = 0.125$ and $l = 0.001$. These are reasonable values for open grassy areas. Note from Figure 13.13 that the zero-intercept regression is better once the lower level wind speed is above 2 m/s. In fact for low wind speed, you would see slow speed at the top anemometer when the lower anemometer is not even moving.

13.3   STATISTICS OF WIND SPEED

In the same manner as other forms of renewable power, the wind resource varies randomly and we must understand this variability to estimate how much power we may harvest during a particular week, month, or season. Very importantly, average wind speed for a location does not alone indicate the energy a wind turbine could produce; frequency of wind speeds at various intervals is also needed. All of this information is determined for a location from a probability distribution function fitted to the long-term observed data.

13.3.1 RAYLEIGH AND WEIBULL DISTRIBUTIONS

The *Weibull distribution* is often used to model wind speed v, since many wind speed data fit this distribution well. The Weibull probability density function (pdf) has two parameters: *scale* (c) and *shape* (k), and it is given by [2]

$$p(v) = \frac{k}{c} \left(\frac{v}{c}\right)^{k-1} \exp\left[-\left(\frac{v}{c}\right)^k\right] \tag{13.17}$$

Figure 13.14 shows three examples where we have varied the shape to be $k = 1, 2, 3$, while keeping the scale fixed at $c = 6$. The left-hand side shows the pdf while the right-hand side shows the cumulative distribution function (cdf). You can see that for $k = 1$, the distribution reduces to an *exponential*

$$p(v) = \frac{1}{c} \left(\frac{v}{c}\right)^0 \exp\left[-\left(\frac{v}{c}\right)^1\right] = \frac{1}{c} \exp\left(-\frac{v}{c}\right) \tag{13.18}$$

Also, note that for $k = 2$, the distribution is skewed to the left

$$p(v) = \frac{2}{c} \left(\frac{v}{c}\right)^1 \exp\left[-\left(\frac{v}{c}\right)^2\right] = \frac{2v}{c^2} \exp\left[-\left(\frac{v}{c}\right)^2\right] \tag{13.19}$$

A Weibull with $k = 2$ has a special name of *Rayleigh distribution*. Note that for $k = 3$, the density starts narrowing and acquiring more symmetry. The graphs in Figure 13.14 were obtained using function weibull.plot of renpow, which makes use of dweibull and pweibull of R.

```
weibull.plot(xmax=20,scale=6,shape=c(1,2,3))
```

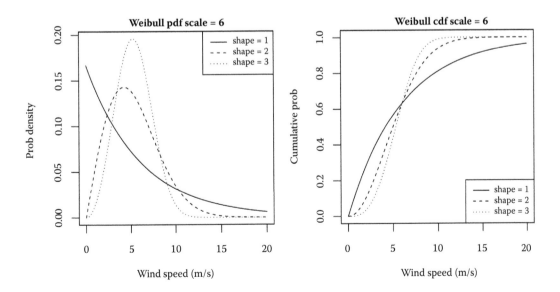

FIGURE 13.14   Weibull probability density and cumulative distribution.

A good fit to wind speed is often found for $k = 2$, or a Rayleigh distribution as given in Equation 13.19. The mean or expected value of a Rayleigh pdf is related directly to the scale c [2].

$$\mu_v = E(v) = \int_0^\infty vp(v)dv = \frac{\sqrt{\pi}}{2}c \qquad (13.20)$$

A first step to find a fit to Rayleigh is to assume shape $k = 2$, use Equation 13.20 to determine the scale from the mean wind speed, and estimate the mean of wind speed by its average

$$c = \frac{2\mu_v}{\sqrt{\pi}} \simeq \frac{2\bar{v}}{\sqrt{\pi}} \qquad (13.21)$$

where $\bar{v}$ is the sample mean or arithmetic average of all values. Once we have a model, we can calculate the probability of having a range of given wind speed.

Example 13.9

Estimate the parameters of a Rayleigh distribution for a location with average wind of 2.5 m/s. Plot the pdf and cdf using weibull.plot. Determine the probability that wind speed is above 2 m/s and above 3 m/s. Answer: The shape is $k = 2$, then use $c = \dfrac{2\bar{v}}{\sqrt{\pi}} = \dfrac{2 \times 5}{1.77} = 5.65$. To find the probability that wind speed is above 2 m/s, we use the pweibull function:

```
> weibull.plot(xmax=10,scale=2*2.5/sqrt(pi),shape=2)
> 1 - pweibull(c(2,3),scale=2*2.5/sqrt(pi),shape=2)
[1] 0.6049226 0.3227190
> abline(v=2,lty=2);abline(v=3,lty=2)
>
```

The plot is shown in Figure 13.15, including the two vertical lines for $v = 2$ and $v = 3$ on the cdf to emphasize the points at which we seek the probability. These are $\Pr[v > 2] = 1 - \Pr[v \leq 2] = 0.605$ and $\Pr[v > 3] = 1 - \Pr[v \leq 3] = 0.323$. Concluding that the wind will exceed 2 m/s 60% of the time and 3 m/s 32.3% of the time.

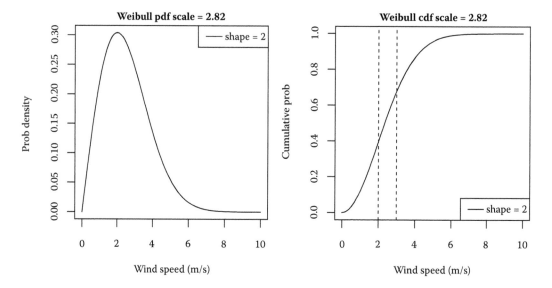

FIGURE 13.15   Example of Rayleigh distribution.

13.3.2 Fit to Real Data

Once we have an initial estimate for scale and shape, we can verify the fit to the data by calculating the histogram or its density approximation and the empirical cumulative distribution (ECDF), and compare the theoretical values to the data values. The fit can be then improved by trial and error or by optimization methods.

Example 13.10

We consider here data from the top anemometer given in Example 13.8 and apply function fit.wind that produces histograms, ECDF plots, and plots comparing fitted versus data for both pdf and cdf.

```
x <- v1.v2[,2]
fit.wind(x)
```

The results are shown in Figure 13.16. We can see that at least by visual inspection, we have a reasonable fit to the pdf and cdf.

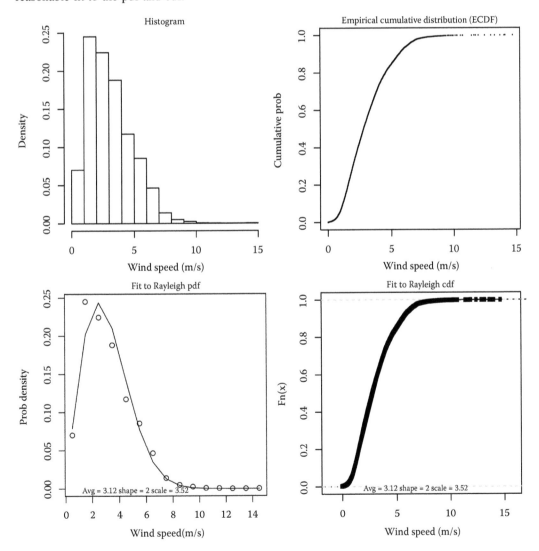

FIGURE 13.16 Wind speed data fit to a Rayleigh distribution.

13.3.3 Average Wind Speed

It should be evident to you by now that you cannot apply the specific power equation $p(t) = \frac{1}{2}\rho v^3$ to the average wind speed because the cube term is nonlinear. However, to emphasize further, compare the specific energy of two wind regimes with the same average at standard conditions.

Example 13.11

Suppose we have wind of 3 m/s and 9 m/s each 50% of the time. Compare the specific power of average wind speed to the average of the specific power. Answer: The average power is $\bar{P} =$ 0.5 × 1.225 × $(3^3 + 9^3)/2$ = 231.5 W/m². The average wind speed is $(9 + 3)/2$ = 6 m/s. If we calculate power of the average wind, we get $P(\bar{v})$ = 0.5 × 1.225 × 6^3 W/m² = 132 W/m². We can see how the results are completely different $\bar{P} > P(\bar{v})$. In this case, the average power is almost double the power of the average wind.

More generally, the mean of power is the expected value

$$\mu_P = E[P] = \frac{1}{2}\rho \int_0^\infty v^3 p(v)dv \tag{13.22}$$

which is not the same as the power of the mean wind speed

$$P(\mu_v) = P(E[v]) = \frac{1}{2}\rho \left(\int_0^\infty v p(v)dv \right)^3 \tag{13.23}$$

However, if wind speed fits a Rayleigh distribution, we can use the average wind speed times a factor 1.91 to calculate power. Let us see why we can make this simplification. We will first substitute Equation 13.21 in Equation 13.19 to obtain the Rayleigh pdf as a function of mean wind speed:

$$p(v) = \frac{2v}{c^2} \exp\left[-\left(\frac{v}{c}\right)^2\right] = \frac{2v}{\left(\frac{2\mu_v}{\sqrt{\pi}}\right)^2} \exp\left[-\left(\frac{v}{\frac{2\mu_v}{\sqrt{\pi}}}\right)^2\right] = \frac{\pi v}{2\mu_v^2} \exp\left[-\frac{\pi v^2}{4\mu_v^2}\right]$$

That is to say, the Rayleigh pdf is reduced to the following function of the mean:

$$p(v) = \frac{\pi v}{2\mu_v^2} \exp\left[-\frac{\pi}{4}\left(\frac{v}{\mu_v}\right)^2\right] \tag{13.24}$$

Now use Equation 13.24 in Equation 13.22 to calculate the mean of power

$$E[P] = \frac{1}{2}\rho \int_0^\infty v^3 \left(\frac{\pi v}{2\mu_v^2} \exp\left[-\frac{\pi}{4}\left(\frac{v}{\mu_v}\right)^2\right]\right) dv \tag{13.25}$$

which is a complicated integration that we will not develop here. We will simply use the known result [2] that

$$\mu_P = E[P] = \frac{1}{2}\rho \frac{6}{\pi}\mu_v^3 \text{ W/m}^2 \tag{13.26}$$

This a remarkable result in as much as it gives us a rule to calculate mean wind power from mean wind speed! Of course, this is true as long as wind speed fits a Rayleigh pdf. In practical terms we us the sample average $\bar{v}$ of wind speed as an estimator of the mean wind speed and $\dfrac{6}{\pi} \approx 1.91$

$$\mu_P \simeq \frac{1}{2}\rho \times 1.91 \times \bar{v}^3 \ \text{W/m}^2 \tag{13.27}$$

Example 13.12

Consider the 3-month summer data used in Example 13.10, which yields a Rayleigh pdf with $\bar{v} = 3.12$ m/s and $c = 3.52$. What is the estimated average power in the wind at standard conditions for that season? Answer: Use Equation 13.27 and substitute values $\mu_P \simeq \frac{1}{2}1.225 \times 1.91 \times 3.12 \ \text{W/m}^2 = 35.53 \ \text{W/m}^2$. This is a relatively low power density, but it corresponds to the low-wind season at that location.

13.3.4 WIND CLASSES

A practical tool used to map out wind power potential for a region is to use average power of Rayleigh winds with long-term annual average wind speed at 10 m height and standard conditions. What this means is to apply Equation 13.27 with $\rho = 1.225$ kg/m^3, and let $\bar{v}$ be an annual average over many years of wind speed measured at 10 m. The resulting μ_P is used to distinguish *classes* of wind power density by using interval breakpoints of 0, 100, 150, 200, 250, 300, 400, and 1000 W/m^2. Therefore, there are seven classes: class 1 has power density between 0 and 100 W/m^2, class 2 between 100 and 150 W/m^2, class 3 between 150 and 200 W/m^2, and so on [10].

A visualization tool for this classification is shown in Figure 13.17 obtained using function pow.class of renpow, which is called using a wind speed as an argument. The function displays where this point is on the chart and returns its power density and class.

```
> pow.class(8)
$Pow.x
[1] 598.98
$class.x
[1] 7
```

Wind classes were used to map wind resources in the United States as exemplified in the popular maps of the *Wind Energy Resource Atlas* [10]. More updated mapping tools for wind resources use raster type (pixel-by-pixel)–based maps and wind speed from models at various elevations. Examples are found on the National Renewable Energy Laboratory (NREL) website [11].

13.4 WIND TURBINES

13.4.1 WIND TURBINE TYPES

According to the axis of rotation, wind turbines can be a *horizontal axis* wind turbine (HAWT), or a *vertical axis* wind turbine (VAWT). See Figure 13.18. HAWT is the most common type employed for large turbines; it consists of a *rotor* and *nacelle* placed at the top of a *tower*. Typically, the rotor has only a few blades, thus providing faster rotation speed while minimizing turbulence that one blade creates for its following blade. Modern large turbines use *three blades*, which sacrifice rotation speed with respect to two-blade designs but produce a smoother rotation. The gearbox, drives, generator, and controllers of a HAWT are placed on the nacelle behind the rotor. The nacelle is made accessible from the ground via the hollow tower.

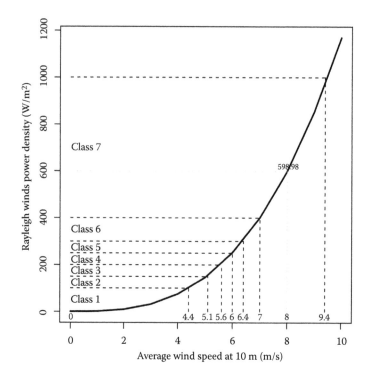

FIGURE 13.17 Wind power classes based on annual average wind speed.

Figure 13.19 shows the major components at the top of a HAWT. The rotor converts wind kinetic energy to rotation of the *low speed shaft*. A gearbox increases the rotational speed, which is needed for the *high-speed shaft* to drive the electrical generator. The *yaw drive* and *yaw motor* are used to make the rotor face upwind to generate, or to turn away from upwind when the wind speed is too strong (this condition is also aided by a *brake*). The blades can also change pitch to adjust to wind conditions.

A VAWT has an advantage in that the heavy generator and gearbox can be placed on the ground; therefore, it is not top heavy as in a HAWT. This simplifies the structure and weight of the tower. The main disadvantage is slower wind and more turbulence near the ground. VAWTs are not used in large wind farms; however, they play a role producing several kW in off-grid and microgrid installations.

13.4.2 BETZ LAW AND MAXIMUM EFFICIENCY

Power available in the wind is given by Equation 13.1, but how much of this can be extracted by a wind turbine? It turns out that there is an upper limit to the conversion efficiency in a fashion similar to the Carnot limit in thermal processes. More precisely, the limit is given by *Betz's law*, which was developed in 1919 by the German physicist Albert Betz. This law states that a turbine cannot capture more than 59.3% of the power from the wind.

This value holds for an ideal or maximum rotor efficiency, but there are additional practical factors. In other words, Betz's law defines the upper limit imposed by nature on wind power generation schemes, but frictional losses and other nonideal conditions will result in a lower conversion efficiency. Modern wind turbines have efficiency of 45% to 50%.

How did Betz arrive at this result? Please refer to the schematic of the interaction between wind and the turbine rotor (Figure 13.20). Now, we can make the following assumptions or idealizations: negligible drag force, wind velocity is axial at the inlet as well as outlet, flow is incompressible (density ρ is constant), and heat transfer effects are negligible. We will follow the presentation in Masters [2].

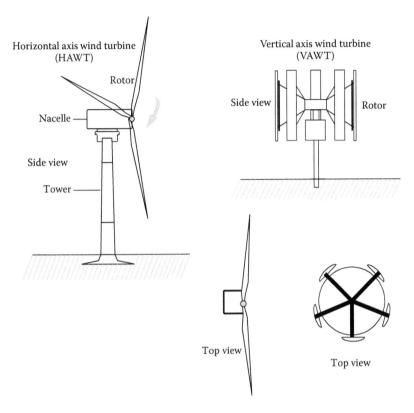

FIGURE 13.18 Types of turbines according to axis of rotation: horizontal axis wind turbine (HAWT) and vertical axis wind turbine (VAWT).

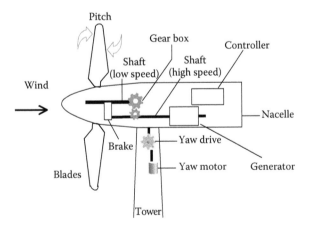

FIGURE 13.19 Typical components at the top of a large HAWT.

Next, we apply mass conservation law, that is, mass flow rate should be the same at the inlet (cross section 1 with area A_1) and outlet (cross section 2 with area A_2), and at the rotor (cross section of area A_r)

$$\dot{m} = \rho A_1 v_1 = \rho A_2 v_2 = \rho A_r v_r \tag{13.28}$$

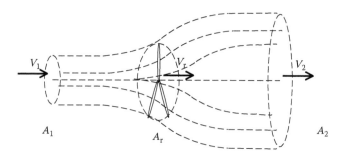

FIGURE 13.20 Schematic of wind flow over rotor to derive Betz's law.

where v_1 is wind velocity upwind of the rotor, v_2 is wind velocity downwind of the rotor, v_r is wind velocity at the rotor, and ρ is air density.

Using the concepts from Chapter 1, according to Newton's second law, the force F exerted by the wind on the rotor may be written as the rate of change of momentum $F = \dot{m}(v_1 - v_2)$, which from Equation 13.28 is $F = \rho A_r v_r (v_1 - v_2)$. Also from Chapter 1, power is force times velocity $P = F \times v$, therefore

$$P = \rho A_r v_r (v_1 - v_2) v_r = \rho A_r v_r^2 (v_1 - v_2) \tag{13.29}$$

Now, derive an expression for power based on change in kinetic energy. Recall from Chapter 1 that $p(t) = \dfrac{d}{dt}\left(\dfrac{1}{2}mv^2\right) = \dfrac{1}{2}\dfrac{dm}{dt}v^2 = \dfrac{1}{2}\dot{m}v^2$. Thus, the power output should be the difference in rate of kinetic energy coming in and rate of kinetic energy going out

$$P = \frac{1}{2}\dot{m}v_1^2 - \frac{1}{2}\dot{m}v_2^2 = \frac{1}{2}\dot{m}\left(v_1^2 - v_2^2\right) \tag{13.30}$$

Again, use Equation 13.28 to rewrite Equation 13.30 as

$$P = \frac{1}{2}\rho A_r v_r \left(v_1^2 - v_2^2\right) \tag{13.31}$$

Equate expressions Equation 13.29 and Equation 13.31 $P = \dfrac{1}{2}\rho A_r v_r (v_1^2 - v_2^2) = \rho A_r v_1^2 (v_1 - v_2)$, and then simplify

$$\frac{1}{2}\left(v_1^2 - v_2^2\right) = v_r (v_1 - v_2) \tag{13.32}$$

We can expand $v_1^2 - v_2^2 = (v_1 - v_2)(v_1 + v_2)$, then substitute in Equation 13.32 and simplify

$$v_r = \frac{1}{2}(v_1 + v_2) \tag{13.33}$$

which tells us the wind velocity at the rotor is the average of the upwind and downwind velocities! This is not very intuitive; it is the result of conservation of mass and energy, ignoring frictional loss.

Now that we know v_r, we can go back and use one of our power equations, say, Equation 13.31

$$P = \frac{1}{2}\rho A_r v_r \left(v_1^2 - v_2^2\right) = \frac{1}{4}\rho A_r (v_1 + v_2)\left(v_1^2 - v_2^2\right) \tag{13.34}$$

which is a nonlinear expression solely on the upwind and downwind velocities

$$P = \frac{1}{4}\rho A_r \left(v_1^3 - v_1 v_2^2 + v_2 v_1^2 - v_2^3 \right) \tag{13.35}$$

Power must reach a maximum for a pair of these velocities. Use the dimensionless ratio $\zeta = \frac{v_2}{v_1}$ for optimization, and rewrite Equation 13.35

$$P = \frac{1}{4}\rho A_r v_1^3 \left(1 - \zeta^2 + \zeta - \zeta^3 \right) \tag{13.36}$$

What is the value of ζ for maximum power? Take derivative $\frac{dP}{d\zeta} = \frac{1}{4}\rho A_r v_1^2 \left(-2\zeta + 1 - 3\zeta^2 \right)$ and make it equal to zero, solving for $\hat{\zeta}$ yields $3\hat{\zeta}^2 + 2\hat{\zeta} - 1 = 0$. A plot of this equation is displayed in Figure 13.21. The quadratic equation has roots $\hat{\zeta} = \frac{-2 \pm \sqrt{4 - 4 \times 3 \times (-1)}}{2 \times 3} = \frac{-2 \pm \sqrt{16}}{6} = 1/3$ and -1. Since ζ is a ratio of velocities it must be positive. Therefore, the result is $\hat{\zeta} = \left(\frac{v_2}{v_1} \right)_{\max P} = 1/3$. In an ideal wind turbine, maximum power occurs when the wind speed leaving the turbine is one-third of its value approaching the turbine (Figure 13.21).

Now, substitute $\hat{\zeta} = 1/3$ in Equation 13.36 to get $P_{\max} = \frac{1}{4}\rho A_r v_1^3 (1 - 1/9 + 1/3 - 1/27) = \frac{1}{4}\rho A_r v_1^3 (32/27) = \frac{1}{2}\rho A_r v_1^3 (16/27)$ and recognize $P_{in} = \frac{1}{2}\rho A_r v_1^3$ as the power in the wind coming into the rotor (not the power in the upwind section), thus the maximum *power coefficient* (C_P), is

$$C_{P\max} = \frac{P_{\max}}{P_{in}} = \frac{16}{27} = 0.593 \tag{13.37}$$

This is how we calculate the 59.3% maximum efficiency of Betz's law. Note that the reference (denominator) was arbitrarily (but reasonably) assumed to have an area equal to the rotor area at upwind velocity. A different result would be obtained if the reference P_{in} were upwind velocity v_1 at the upwind cross section A_1, or the velocity v_r at the rotor cross section and A_r.

Example 13.13

How much power per m^2 would you ideally extract from a wind turbine at standard conditions of density 1.225 kg/m3 and 15°C when the wind speed at 10 m is 5 m/s and the turbine is on a tower

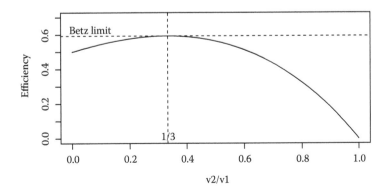

FIGURE 13.21 Efficiency versus wind speed ratio showing maximum.

20 m above the ground? What is the downwind wind speed? Assume terrain is open grass. Answer: Extrapolate wind speed at 20 m from $v = v_0 \left(\dfrac{H}{H_0}\right)^\alpha = 5 \times (20/10)^{0.12} = 5.43$ m/s. Calculate specific power for this wind speed $p(t) = \dfrac{1}{2}\rho v^3 = \dfrac{1}{2} \times 1.225 \times 5.43^3 = 98.1$ W/m2 and take Betz's efficiency $P_{max} = 0.593 \times 98.1 = 58.17$ W/m2. The downwind speed is one-third of the upwind $5.43/3 = 1.81$ m/s.

13.4.3 Tip Speed Ratio

Other factors decrease efficiency below the ideal Betz limit. A major factor is the aerodynamics impact of rotational speed and its relation to wind speed. When v is given, the efficiency of the rotor depends on how fast it rotates. If it rotates too slowly, it does not harvest enough kinetic energy and efficiency is low. If it turns too fast, it creates turbulence that affects the movement of the blades and thus the efficiency. Therefore, there must be an optimum rotation speed.

Denote by t_w the time it takes for the wind to settle after a blade has gone through and use $t_w = d/v$, where d is the distance that wind is affected (upwind and downwind), and v is wind speed (m/s). It has been empirically determined that $d \simeq r/2$, where r is the radius (in m) of the rotor [11]. The time it takes for the net blade to arrive is $t_b = \dfrac{2\pi}{n\omega}$, where n is the number of blades and ω is rotor angular speed (rad/s). When $t_b > t_w$, some energy is not harvested, and when $t_b < t_w$ there is interference. An optimum could be specified when $t_b \approx t_w$, which means $\dfrac{2\pi}{n\omega_{opt}} = \dfrac{r}{2v}$ and solving for angular velocity

$$\omega_{opt} = \frac{2v}{r}\frac{2\pi}{n}.$$

Example 13.14

What is the optimum angular frequency for a three-blade rotor of 10 m radius in 10 m/s wind?
Answer: $\omega_{opt} = \dfrac{2v}{r}\dfrac{2\pi}{n} = \dfrac{2 \times 10 \times 2\pi}{10 \times 3} = 4\pi/3$ rad/s

A parameter to specify this aspect of performance is the speed at the rotor tip. Define *tip speed ratio* (TSR) as the quotient of rotor tip tangential speed v_{tr} to wind speed

$$\lambda = \frac{v_{tr}}{v} = \frac{\omega r}{v} \tag{13.38}$$

Equivalently, using rotational speed f in rotation per sec (Hz), N in rpm, and 60 s/min, we could also write

$$\lambda = \frac{2\pi f \times r}{v} = \frac{2\pi r \times N}{60 \times v} \tag{13.39}$$

Example 13.15

Consider radius of 10 m wind speed of 10 m/s and angular speed of $4\pi/3$ rad/s. What is the TSR?
Answer: $\lambda = \dfrac{\omega \times r}{v} = \dfrac{(4\pi/3) \times 10}{10} = 4\pi/3 = 4.19$.

Efficiency or power coefficient C_P varies with TSR and this dependency is typically given as a C_P-λ curve. Three-blade rotors have optimum efficiency at about 4 to 6 TSR and thus show a peak in the C_P-λ curve around these values. Before we leave this section, another impact on efficiency is the whirlpool losses, as the wind is affected by the blades. This was not accounted for in the Betz limit, since it assumes the wind does not change direction as it goes through the rotor. As mentioned earlier, after taking into account all of these other losses, modern wind turbines have an efficiency of 45% to 50%.

13.4.4 WIND TURBINE GENERATORS

Recall from our discussion of generators in previous chapters (8, 10, and 11) that the frequency is fixed to that of the grid. We have dealt with synchronous generators, such that the rotational speed is fixed to match the electrical frequency of the grid. That is how most electricity is generated in the world from coal-fired power plants, natural gas-fired plants, nuclear, and hydroelectric. Denote by N_s the synchronous speed, by f the frequency, and p the number of poles we have

$$N_s = \frac{120 \times f}{p} \qquad (13.40)$$

For instance, for 60 Hz and 6 poles, the rotation frequency is $N_s = \dfrac{120 \times 60}{4} = 1800$ rpm.

However, wind power generation uses *asynchronous generators*, meaning that they are not rotating at a fixed speed. Most wind turbines use *induction generators*, which at low speeds behave as a motor and at higher speeds as a generator. A common induction motor employed in wind turbines is a *squirrel cage generator* (SCIG), so named because the rotor of the generator resembles a cage made out of rings and bars; the bars are shorted at the rings (Figure 13.22). The magnetic field is created by the stator, not the rotor, which is fed with three-phase alternating current (AC). An advantage of this design is that there is no need for brushes and slip rings to connect to the rotor.

When the high-speed shaft of the wind turbine rotates the generator rotor, it starts turning at a slower speed N_r than N_s, and behaves as a motor. However, once the machine is turning faster than the synchronous speed, it behaves as a generator and feeds three-phase power back to the stator windings. The slip s is defined as a relative difference between the two rotation frequencies

$$s = \frac{Ns - Nr}{Ns} = 1 - \frac{Nr}{Ns} \qquad (13.41)$$

For instance, take 60 Hz and $p = 4$. We know $N_s = 1800$ rpm from Equation 13.40. For grid-tie, we want the slip to be low, say ±1%. When $s = 1\%$, then N_r is $Nr = (1 - s)Ns = 0.99 \times 1800 = 1782$ rpm, and the machine behaves as a motor. However, for negative slip $s = -0.01$ or -1%, the rotational frequency $Nr = (1 - s)Ns = (1 - (-0.01)) \times 1800 = 1.01 \times 1800 = 1818$ rpm behaves as a generator.

Since the generator is driven by the high-speed shaft, its gear ratio affects the turbine rotor speed. Suppose the gear ratio was $G{:}1$; then the turbine will have to rotate at $1800/G$ rpm. At this point, you need to link this concept to the one that relates turbine efficiency to rotational speed as a function of wind speed. You can see that matching requirements of turbine speed for efficiency to the generator

Squirrel cage rotor

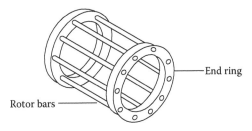

End ring

Rotor bars

FIGURE 13.22 Squirrel-cage rotor of an induction generator.

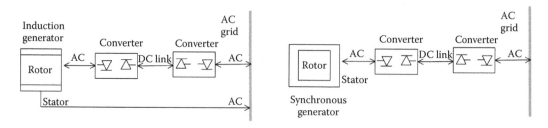

FIGURE 13.23 Left: Doubly-fed induction generator. Right: Variable speed synchronous generator.

needs can be a complicated problem. One approach is to build the generator so that the number of poles can be adjusted.

Example 13.16

Suppose the wind speed is 5 m/s and a 10 m radius turbine spins at 19 rpm to have TSR = 4.19 and optimum efficiency at 5 m/s winds. How close would that rpm value be to that required by a 4-pole generator if the gear ratio is 100:1? Answer: The 4-pole generator Ns is 1800 rpm, and the high-speed shaft would be at 1900 rpm. The slip is $s = \dfrac{1800 - 1900}{1800} = \dfrac{-100}{1800} = -0.055$ or 5% of required N_s. The machine generates because the negative slip.

As we can appreciate from these simple calculations, an SCIG does not really offer a great flexibility to adapt speed as function of wind speed variability. Two generators that can offer better performance in this regard are the wound-rotor *doubly-fed induction generator* (DFIG) and the *variable-speed synchronous generator* (Figure 13.23) [2].

The DFIG uses coils in the rotor and therefore it requires slip rings, but the stator is as in the SCIG and sends power back to the grid when the machine is generating. To feed power to the rotor, the grid-side AC to DC converts grid power to DC, which serves as input to the rotor-side DC to AC converter, feeding the rotor with AC at the right frequency. This power flow can be bidirectional, meaning power is sent back to the grid at the right frequency using the converters (Figure 13.23) [2].

The variable speed synchronous generator exerts speed control of a synchronous generator by using converters as in the rotor of the DFIG. Note that the converters need to have the same power capacity as the turbine. When the generator's rotor is a permanent magnet type, the required rotational speed of the generator becomes independent of the turbine rotation. This design allows for removal of the gearbox, but requires a permanent magnet generator (Figure 13.23) [2].

13.4.5 Performance or Power Curves

Now we put together all of the processes we have seen to be involved in converting wind speed to turbine rotation (including tower height and efficiency calculations), and turbine rotation to electric power, which depends on the generator technology. The power curve is a plot of electric power generated versus wind speed. In principle, this curve is part of the specifications provided by a manufacturer so that the user can understand how much power to expect given the winds at the location. A hypothetical example is illustrated in Figure 13.24 (left-hand side). Denote the power curve as $P(v)$ and realize that this is a nonlinear function with sharp changes in three important ranges of wind speed, which are characterized by three parameters.

First, *cut-in wind speed* (v_C) is that value of wind speed below which there is no electrical generation. In other words, it is the minimum wind speed required to overcome friction and acquire synchronous velocity. Therefore, wind power at speed lower than v_C is not usable.

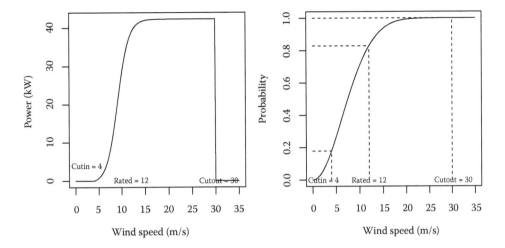

FIGURE 13.24 Left: Hypothetical idealized power curve. Right: Cumulative distribution function in conjunction with power curve.

Once the turbine goes above cut-in, power increases with cube of velocity (at least ideally if that wind power is harvested at a given efficiency) until it reaches *rated wind speed*. This is the wind speed v_R, at which generation acquires its rated capacity P_R. Above this v_R value, wind power is excessive and must be shed. Large turbines shed wind in a variety of ways, including changing the pitch and stall control. As wind-speed continues to increase, it reaches the *cut-out wind speed* (v_F) or *furling* wind speed at which the turbine must shut down to avoid damage.

Under the most idealized assumptions, the power curve could be described by the following equation with discontinuous breakpoints:

$$P(v) = \begin{cases} 0 \text{ when } v < v_C \\ \eta \times 0.5 A \rho v^3 \text{ when } v_C < v \leq v_R \\ \eta \times 0.5 A \rho v_R^3 \text{ when } v_R < v \leq v_F \\ 0 \text{ when } v \geq v_F \end{cases} \qquad (13.42)$$

where η is an overall efficiency and A is rotor area. Of course, in reality, there is a gradual transition at those breakpoints, particularly in the approach to rated power $P_R = \eta \times 0.5 A \rho v_R^3$. Function power.curve of renpow calculates a hypothetical power curve based on cut-in, rated, and cut-out parameters for a sequence of values of wind speed v and for a given rotor area. Efficiency is assumed to be $\eta = 0.4$. For simplicity, this function approximates the power by a sigmoid curve with limits v_C and v_R, and is not based on a model accounting for pitch, TSR, and other aerodynamic considerations.

Example 13.17

Assume cut-in, rated, and cut-out wind speeds are 4, 12, and 30 m/s, respectively. The rotor is 100 m². Wind speed will be varied from 0 to 35 m/s in steps of 0.1 m/s. With these parameter values we can calculate the hypothetical power curve:

```
x <- list(cutin=4,vrated=12,cutout=30, A=100,v=seq(0,35,0.1))
Pow <- power.curve(x)
```

which produces the plot of Figure 13.24 (left-hand side).

13.4.6 COMBINING WIND SPEED STATISTICS WITH POWER CURVES

As you can see from the right-hand side of Figure 13.24, we can determine the probability that the winds would have the speed corresponding to the cut-in, rated, and cut-out speeds of the power curve. This is done using a Rayleigh cdf for the average wind speed of the location. With these probabilities, we can determine the number of hours in a year that the generator would not run, or wind speed is below cut-in

$$\Pr[v < vc] \times 8760$$

Similarly, the number of hours in the year that the generator is stopped because of winds above cut-out

$$(1 - \Pr[v < v_F]) \times 8760$$

and the number of hours in the year that the generator operates at rated power or faster than rated speed but slower than cutout

$$(1 - \Pr[v < v_R]) \times 8760 - (1 - \Pr[v < v_F]) \times 8760$$

and finally the number of hours in a year generating below rated capacity, that is, faster than cut-in, but slower than rated speed

$$(\Pr[v < v_R] - \Pr[v < v_C]) \times 8760$$

Function prob.power.curve uses those parameters (cutin, rated, and cutout) to calculate the number of hours in a year for each condition as described earlier for a given average wind speed.

Example 13.18

Assume cut-in, rated, and cut-out wind speeds are 4, 12, and 30 m/s, respectively, as in the previous example. Wind speed will be varied from 0 to 35 m/s in steps of 0.1 m/s. Second, assume that in addition to these parameters, we have Rayleigh winds with 8 m/s average.

```
x <- list(cutin=4,vrated=12,cutout=30, A=100,v=seq(0,35,0.1))
prob.power.curve(x,avg=8)
```

which produces the plots of Figure 13.24 (right-hand side). The number of hours is

```
> prob.power.curve(x,avg=8)
$h.cutin
[1] 1561.689
$h.cutout
[1] 0.1399164
$h.rated.nstop
[1] 1496.242
$h.run.below.rated
[1] 5701.929
>
```

We can see that the machine would generate 5702 + 1496 = 7198 hours in the year, but only 1496 hr (or equivalent to 2 months) correspond to rated capacity.

In general, we can calculate the probability, and hence the expected number of hours for any wind speed value v by using a slice of incremental speed dv and the Rayleigh cdf

$$h(v) = \Pr(v < x \le v + dv) \times 8760 = (\Pr[x \le v + dv] - \Pr[x \le v]) \times 8760 \qquad (13.43)$$

Multiplication of expected number of hours of operation at given wind speed (from the Rayleigh cdf) to the corresponding generated power $P(v)$ from the power curve

$$e(v) = h(v)P(v) \qquad (13.44)$$

would yield energy in kWh. Integrating over the entire wind speed range (assume max is v_F)

$$E = \int_0^{v_F} h(v)P(v)dv \qquad (13.45)$$

would yield expected energy production. This divided by rated power times 8760 would be capacity factor (CF)

$$CF = \frac{\int_0^{v_F} h(v)P(v)dv}{8760 \times P_R} \qquad (13.46)$$

Function wind.energy facilitates this lengthy calculation. Arguments are x, as in the power.curve arguments; Pow, which is the output of power.curve; and the average wind speed.

Example 13.19

Assume the parameters of the previous two examples. Calculate expected energy production and capacity factor.

```
x <- list(cutin=4,vrated=12,cutout=30, A=100,v=seq(0,35,0.1))
wind.energy(x,Pow,avg=8)
> wind.energy(x,Pow,avg=8)
$energy
[1] 135007.3
$unit
[1] "kWh/yr"
$CF
[1] 0.36
>
```

We expect 135 MWh/yr and a capacity factor of 36%.

13.5   WIND FARMS

A group or array of large wind turbines at a good wind site is a *wind plant* or most often called a *wind farm*. This arrangement makes sense in terms of efficient sharing of development, transmission, and other infrastructure for maintenance and operation. However, clustering turbines in wind farms requires a substantial amount of surface area whether on land or offshore. The primary reason for this need is that large rotors are more cost efficient, and tall towers allow harvesting more wind. Increasing rotor diameter determines the turbine footprint. Due to interference effects on wind, array efficiency drops as the distance between turbines decreases, so spacing (see Example 13.20) between turbines is limited. A common spacing is 5 D (across wind direction) by 10 D (downwind direction), where D is the turbine diameter. As a first approximation, the footprint of one machine is $\frac{A}{S} =$ $\frac{\frac{\pi}{4}D^2}{10D \times 5D} = \frac{\pi}{4 \times 50} = 0.015$, where A is the rotor area and S is surface area. For a windy site, say middle of class 7, and winds at power density of 600 W/m², we would have about

$$P_w \frac{A}{S} = 0.015 \times 600 \times \frac{W}{m^2} = 9 \frac{W}{m^2}$$

where P_w is power in the wind. Taking into account individual turbine efficiency, say, at 40% (but not accounting for potential array inefficiencies due to interference), we would have $P_G \frac{A}{S} = 9 \times 0.4 \frac{W}{m^2} = 3.6 \frac{W}{m^2}$. This number is lower to the comparable expected output from a solar plant based on photovoltaic panels (Chapter 14).

Example 13.20

Approximately how much total land is needed for a 100 MW wind plant if we assume 25% additional land is needed for buffers around the production area and easements? Answer: The area needed for generation is $S_{gen} = \frac{100 \times 10^6}{3.6} = 27.77 \, km^2 = 2777$ ha. The extra land required is $S_{tot} = S_{gen}(1 + 0.25) = 34.71 \, km^2 = 3471$ ha, which is a significant amount of surface area.

Offshore wind farms offer several advantages over land-based wind farms, because primarily the wind is steady, has less surface friction, and is stronger in the afternoon when demand is higher. When located far away from shore, it has lesser impact in terms of noise and visual impairment. It is advantageous in terms of distribution for countries where the load is concentrated near coastal areas. On the other hand, construction, installation, and maintenance are more costly. The bathymetry of the area has a great influence on cost, since water depth determines the type of structure needed. Transmission requires submarine cables, and for three-phase transmission, this represents a loss due to reactive power. It is then preferable to use HVDC transmission, particularly for installations very distant from shore. In these cases, DC/AC converters are needed onshore to connect the wind turbines to the AC grid [2].

13.6   OFF-GRID AND MICROGRIDS, DISTRIBUTED GENERATION

Small wind turbines (several or tens of kW) play a role in distributed generation, where it can be tied to the grid, or in off-grid installations and microgrids, alone or in combination with PV panels. A major consideration for this type of arrangement is energy storage, typically in batteries, because of the variability of wind speed. These installations require a *charge controller* to control the power flow and charge the batteries (Figure 13.25).

13.6.1   SMALL WIND TURBINE

Small wind turbines typically have a three-phase asynchronous generator with an output voltage that varies with wind speed. Increasing the wind speed increases the output voltage magnitude and frequency. A full three-phase rectifier (Chapter 8) converts the three-phase voltage to DC, whose amplitude also varies with wind speed. This DC voltage is used to charge the batteries of the battery bank and tie to a DC bus, which can feed other loads and connect to a DC microgrid, be inverted to connect to an AC microgrid, or tie directly to the grid (Figure 13.25).

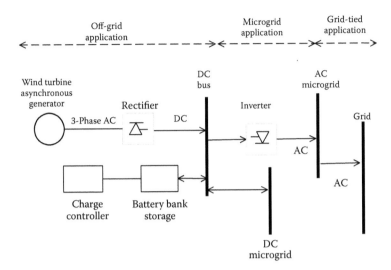

FIGURE 13.25 Small wind turbine and asynchronous generator used for off-grid or microgrid applications and optional connection to the grid.

The DC bus voltage for a small system is typically 12, 24, or 48 V. Higher voltage is advantageous because it reduces current, while maintaining the same power. Lower current reduces wire ohmic losses and allows for smaller wires, breakers, and fuses. This makes it simpler and reduces costs. For instance, for 1.2–2.4 kW systems, using 24 V would keep current below 100 A, but a 2.4–4.8 kW system would require 48 V to keep the current below 100 A.

The wind turbine has specifications (cut-in, rated, and cut-out wind speed), a power curve, and efficiencies as we previously discussed. The rectifier, charge controller, and inverter have their own efficiencies. For instance, typical values are 80%–90% for the inverter and 95%–98% for the charge controller.

Example 13.21

Suppose we have an off-grid system composed of a small 1 kW wind turbine of rotor area 4 m^2 and 0.4 efficiency with cut-in 2 m/s, rated 10 m/s, and cut-out 30 m/s. The winds are Rayleigh with average of 5 m/s. The daily AC load energy demand is 3 kWh with peak power demand of 1.5 kW. The inverter efficiency is 85% and the charge controller efficiency is 97%. We will not consider the batteries in this example. Determine suitable voltage, the peak current, DC load, and CF. Does the expected energy from the turbine cover the demand? Answer: The DC load is

$$E_{DC} = \frac{E_{AC}}{\eta_{inv} \times \eta_{cc}} = \frac{3 \text{kWh/d}}{0.85 \times 0.97} \simeq 3.64 \text{ kWh/d} \tag{13.47}$$

For a peak load of 1.5 kW we could use 24 V, which will keep current below its peak of $I_{peak} = \frac{1500}{24} = 62.5$ A. Now, use the functions of renpow with specs and winds above

```
x <- list(cutin=2.5,vrated=10,cutout=30, A=4,v=seq(0,35,0.1))
Pow <- power.curve(x)
> wind.energy(x,Pow,avg=6)
$energy
[1] 2675.07
```

```
$unit
[1] "kWh/yr"
$CF
[1] 0.31
>
```

We see that we expect 2675 kWh/yr or 7.33 kWh/d with CF = 0.31. This expected production covers the demand.

13.6.2 BATTERIES

Recall from Chapter 3 that *electrochemical cells* generate electricity from chemical reactions, and that the difference in half-cell potential between anode E_a and cathode E_c is the cell voltage or electromotive force (emf) $V_{cell} = E_a - E_c$. We then have a power source, and this cell can form a *battery* by combining it with other cells in series to increase the voltage. For instance, 6 cells of 2.12 V cells make a 12.7 V battery, which is a *nominal* 12 V battery. Conversely, providing electric current to a cell can promote chemical reactions. For example, a current can produce electrolysis or split water into hydrogen and oxygen. Hydrogen can be used in an electrochemical cell or a fuel cell to generate electric power (Chapter 6).

Lead-acid batteries are still the most used for small wind and PV power systems. Lithium-ion batteries have greater *specific energy* (Wh/kg) or *energy density* (Wh/L), but they are expensive. These batteries are typically used in electric vehicles, for which higher energy density is necessary. However, energy density is not that critical in off-grid or microgrid installations. Future use of lithium-ion batteries in off-grid and microgrid systems is likely when their costs decrease.

In brief, *maximum battery capacity* is the amount of charge Q_0 it can hold, but this simple concept is specified with more detail later using current and time. *Depth of discharge* (DOD) is the discharge expressed as percent of maximum battery capacity. You are likely familiar with batteries employed in internal combustion engine vehicles, which are used for starting, lighting, and ignition (*SLI*). Starting a vehicle consumes only a few percent of battery charge and the alternator quickly recharges it. Once the car is running, most power is supplied by the alternator. Therefore, this type of battery does not normally go into deep discharge, thus their allowable DOD is small. In fact, you may already know that SLI batteries will not last long if undergoing deep discharges.

In contrast, *deep cycle batteries* allow for greater discharge (DOD up to 25%) and are more suitable for renewable power storage applications. This type of battery is made with thicker plates, and more space above and below the plates, therefore they are large and heavy batteries. Deep cycle batteries can undergo several thousands of charge–discharge cycles not exceeding an allowable DOD (~25%). However, discharges of 50% or more may shorten life of the battery.

A lead–acid cell consists of a positive electrode (lead dioxide PbO_2) and a negative electrode (a porous metallic lead Pb structure), immersed in a sulfuric acid electrolyte. There is outgassing in a lead–acid cell when it approaches full charge because the voltage rises enough to cause electrolysis, releasing hydrogen and oxygen, and consuming water. *Flooded batteries* are vented to allow release of gases and provide access to refill the water. This arrangement is in contrast with *sealed batteries*, which have valves to minimize gas release and recycling gases within the battery.

During discharge, the sulfate ion SO_4^{2-} accumulates in the electrodes as lead sulfate $PbSO_4$, and therefore the specific gravity of the electrolyte decreases while the resistance increases. When charging, the sulfate ion SO_4^{2-} returns to the electrolyte, consequently increasing specific gravity, and then resistance drops. Each cycle leaves a little bit of sulfate at the electrodes, thus limiting the lifetime; it is good practice to charge the batteries to full often, in order to prolong their life.

State of charge (SOC) is the battery capacity at a given time $S(t)$ given in percent of maximum capacity Q_0

$$S(t) = \frac{Q_0 - \int_{t_0}^{t} i(x)dx}{Q_0} \times 100 \tag{13.48}$$

where the integral of current $i(x)$ is the charge supplied by the battery up to time t [12]. Battery voltage varies with time t and it is specified under two conditions: $v_{oc}(t)$ when in open circuit or $v(t)$ when connected to a load. The open circuit battery voltage varies according to SOC by an approximately linear relation

$$v_{oc}(t) \simeq aS(t) + b \tag{13.49}$$

Note that when SOC is 0, we have $v_{oc}|_{S=0} = b$ and that when SOC is 100% $v_{oc}|_{S=100} = a \times 100 + v_{oc}|_{S=0}$, which allows determining a. SOC could be estimated by measuring the open circuit voltage and applying a relationship [12]

$$S(t) = \frac{v_{oc}(t) - b}{a} \tag{13.50}$$

Example 13.22

Suppose that for a nominal 12 V battery, the manufacturer specifies that v_{oc} is 8 V when the SOC is 0% and v_{oc} to be 12.77 when the SOC is 100%. Determine coefficients a and b. Using these values, estimate SOC if the battery has discharged to $v_{oc} = 12$ V. Answer: b is 8 V, and using $v_{oc}|_{S=100} = a \times 100 + v_{oc}|_{S=0}$ we have $a = (12.77 - 8)/100 = 0.0477$. Then, after discharge to 12 V we estimate $S(t) = \dfrac{v_{oc}(t) - b}{a} = \dfrac{12 - 8}{0.0477} \simeq 83.86\,\%$.

In this example, we assume that we disconnect the load to measure v_{oc}. Of greater interest, is to estimate SOC from the battery voltage $v(t)$ when loaded. This requires a battery model relating $v(t)$ to $v_{oc}(t)$. A possible battery model specifies internal resistances R_c and R_d when charging and discharging and one capacitance (Figure 13.26, left-hand side) [12]. The diodes are not real parts of the battery and are just used for modeling. For simplicity, if we ignore the effect of capacitance and the terminal resistance R_b, we can relate $v(t)$ to $v_{oc}(t)$. When charging $v(t) = v_{oc}(t) - i(t)R_c$ and when discharging $v(t) = v_{oc}(t) - i(t)R_d$. Another common model is based on the generic Randles model shown on the right-hand side of Figure 13.26, and is based on an internal series resistance, two RC circuits accounting for charge transfer and diffusion [13,14].

Now let us look at capacity with more detail. Most often, *capacity* is given in ampere-hours (Ah) at a nominal voltage, at room temperature (25°C), and at a given discharge rate in hours. In this manner, the Ah capacity is specified as a ratio C/T. For instance, C/20 is for 20 h; in this case C/20 of 200 Ah battery means the battery has a capacity to provide 10 A over 20 h.

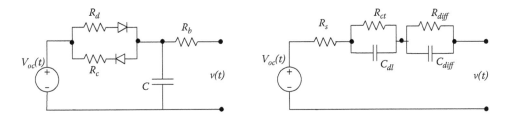

FIGURE 13.26 Two battery models.

Example 13.23

Nominally, how many amps can a battery provide for 20 hours if the C/20 is 300 Ah? How long will it take to discharge if the battery continuously provides 10 A to a load? Answer: The nominal 20 h current is 300/20 = 15 A. It would discharge in 300/10 = 30 h.

There are three types of efficiency measures for batteries. *Voltage efficiency* is the ratio of voltage after discharge (V_{out}) to voltage before discharge (V_{in})

$$\eta_V = \frac{V_{out}}{V_{in}} \tag{13.51}$$

For instance, if a battery discharges from 14 V to 12 V, we have $\eta_V = \dfrac{V_{out}}{V_{in}} = \dfrac{12}{14} = 86\%$. *Coulomb efficiency* is the ratio of capacity Ah_{out} after discharge to Ah_{in} before discharge

$$\eta_C = \frac{Ah_{out}}{Ah_{in}} \tag{13.52}$$

This efficiency is typically high (90%–95%). The inefficiency 5%–10% is by electrolysis out-gassing at higher voltage. *Energy efficiency* is the product of voltage efficiency and coulomb efficiency

$$\eta_E = \eta_V \times \eta_C \tag{13.53}$$

For instance, $\eta_E = 86\% \times 90\% = 77\%$.

Temperature affects batteries. Ideal performance is at room temperature, often considered ideally at 20°C. Capacity decreases with decreasing temperature. Rated capacity, for example, C/20 is given at 25°C; it derates to 90% at 5°C, and to 85% at 0°C. Thus, a 300 Ah would be 270 Ah at 5°C. Moreover, a battery should not discharge below a threshold at freezing temperature. At –20°C, allowable depth of discharge is reduced ~20%, at –30°C to about 40%.

13.6.3 BATTERY BANKS

Batteries are connected in series to obtain a battery *string* of higher voltage since voltage in series adds up. For a 24 V system, we can use a string of two 12 V batteries or a string of four 6 V batteries. It is best to use all identical batteries in the string. Note that capacity in Ah for the string would be the same as the capacity of one of the batteries in the string, since current is the same for elements in series.

Battery banks are formed by connecting strings of the same voltage in parallel. The resulting voltage is the same as the string voltage since voltage is the same for parallel circuits. Banks have increased capacity; Ah will add up since current of elements in parallel add up. For instance, suppose we have six 12 V batteries of 150 Ah. A 24 V bank could be formed by the parallel of three 2-battery strings, yielding a capacity of 150 Ah × 3 = 450 Ah (Figure 13.27).

A battery bank capacity is designed starting with the demand from the load, and the number of days the bank energy should be available in case of sustained low-wind conditions. Once the voltage and required bank capacity are known, the circuit configuration is designed taking into consideration cost and availability. We can explain this process by means of an example [2].

Example 13.24

Consider the off-grid system given in Example 13.21. Design a battery bank that would provide enough power for 3 days of wind of speed below cut-in with a DOD not to exceed 80%. Assume batteries are kept at room temperature. Answer: We know from the previous example that the DC bus is 24 V and the DC load is 3.64 kWh/d. Thus, the bank must have a capacity of

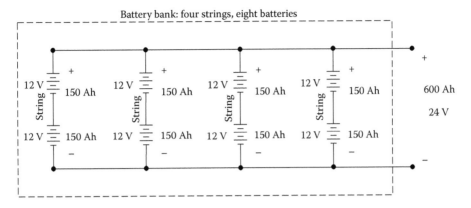

FIGURE 13.27 Battery bank composed of eight 12 V batteries of 150 Ah each. The bank is four strings, 24 V, and 600 Ah.

$$C = \frac{E_{DC}}{V} \times \frac{\text{days}}{DOD_{\max}} = \frac{3640 \quad \text{Wh/d}}{24V} \times \frac{3d}{0.8} \simeq 568 \quad \text{Ah} \qquad (13.54)$$

We can use four strings of two batteries each in series 12 V of 150 Ah each to get 24 V and 150 Ah × 4 = 600 Ah, which means we need 2 × 4 = 8 batteries.

13.6.4 CHARGE CONTROLLERS

The simplest charge controller for a small wind turbine and battery bank is a device that will charge the battery avoiding both overcharging and excessive DOD. A simple scheme to avoid overcharging is to disconnect the wind turbine generator from the battery bank when the latter achieves a predefined voltage threshold. When disconnecting the generator from the battery, and there is excess power not consumed by the load, the charge controller connects the generator output to a *diversion load* or *dump load*. The excess power generated is then just dissipated by this load, often as wasted heat. It is of value to use this excess power for some useful purpose. For example, the dump load could be water heater elements that use this excess power to heat water.

13.7 ENVIRONMENTAL CONSIDERATIONS OF WIND POWER GENERATION

When considering environmental impacts of wind power, as with other renewable sources, we must think of a balance between positive and negative aspects. Prominent positive aspects are its contribution to reducing greenhouse gas emission and interference with water resources, both inland surface water and coastal. On the other hand, negative aspects include increased land use, noise, landscape aesthetics, and interference with avian wildlife [1].

As discussed earlier in the chapter (see Example 13.20), wind farms require approximately 0.3 km²/MW, which is a substantial amount of land. However, it is possible to use this land for other purposes. In fact, many wind farms lease land from ranchers who can still conduct ranching activities with minimal interference from the wind farm, representing an extra income for the rancher.

Wind turbines are very noticeable on the landscape and are considered by many individuals to affect the *visual* aspects and *aesthetics* of areas that otherwise would be scenic. There are many factors involved in the perception of landscape aesthetics and attitudes toward changes of familiar views. These factors are similar to those related to urban expansion (e.g., tall buildings) and other large development projects (e.g., bridges), and structures that are also part of electric power systems (e.g., transmission lines). Wind turbines include additional factors of movement, noise, and of intermittent lights on top of the towers to prevent aviation accidents.

Perceived public concerns with wind turbines include noise. Noise from wind turbines is mainly from the rotating blades and this increases with wind speed. However, for large turbines, relative noise level has been reported to be about 40 dB(A) at 300 m away from a turbine. Just for comparison, this is about the same level as experienced in a typical home or office. For small turbines, noise can be higher closer to the turbine and at higher wind speeds, but decreases with the number of blades.

A major environmental concern has been the impact on wildlife, particularly interference with avian and bat species habitats. Elevated counts of bird strikes have been reported at some locations but has not been a reported problem in many other locations. This indicates that when evaluating sites for wind farms or offshore facilities, accounting for avian habitat and migration routes could help reduce these negative impacts. Less research has been conducted on the impact of small wind power generating stations on avian populations and bats.

EXERCISES

13.1. What type of rotation (clockwise or counterclockwise) would you expect for typhoons in the Northwest Pacific and for cyclones in the South Pacific?

13.2. What would be the specific power or power density for wind velocity of 3 m/s at reference conditions? What would be the air density at 2000 m asl and 5°C?

13.3. Plot specific power for four elevations—0, 200, 500, 1000 m asl—and wind speed from 0 to 30 m/s. Use ground temperature of 25°C and lapse rate of 6°C/km.

13.4. Compare wind speed ratios $\frac{v}{v_0}$ and power ratio $\frac{P}{P_0}$ obtained from the empirical exponential model with the aerodynamics model for $\alpha = 0.25$ and $l = 0.4$.

13.5. Estimate the parameters of a Rayleigh distribution for a location with average wind of 3.5 m/s. Plot the pdf and cdf. Determine the probability that the wind speed is above 3 m/s and below 5 m/s.

13.6. Suppose we have Rayleigh winds with $\bar{v} = 3.12$ m/s and $c = 3.52$. What is the estimated average power in the wind at standard conditions?

13.7. How much power per m^2 would you ideally extract from a wind turbine at standard conditions of density 1.225 kg/m^3 and 15°C when the wind speed at 10 m is 5 m/s and the turbine is on a tower 20 m above the ground? What is the downwind wind speed? Assume the terrain is shrub and small trees.

13.8. What is the optimum angular frequency for a three-blade rotor of 15 m radius in 5 m/s wind? What is the TSR for this angular speed? What is the gear ratio required to match the rpm of a 4-pole generator producing 60 Hz?

13.9. Assume cut-in, rated, and cut-out wind speeds are 3, 10, and 30 m/s, respectively. We have Rayleigh winds with 5 m/s average. Calculate number of hours with expected generation at rated power, expected generation, and CF.

13.10. Consider an off-grid wind power system with a DC load 4 kWh/d and peak DC load of 1.5 kW. The cut-in, rated, and cut-out wind speeds are 3, 10, and 30 m/s, respectively. We have Rayleigh winds with 5 m/s average. Design a battery bank that would provide enough power for 2 days of wind of speed below cut-in with a DOD not to exceed 70%. Assume batteries are kept at room temperature.

REFERENCES

1. Boyle, G., *Renewable Energy: Power for a Sustainable Future*. Third edition. 2017: Oxford University Press. 566 pp.
2. Masters, G.M., *Renewable and Efficient Electric Power Systems*. Second edition. 2013: Wiley-IEEE Press. 690 pp.
3. Fay, J.A., and D.S. Golomb, *Energy and the Environment. Scientific and Technological Principles*. Second edition. MIT-Pappalardo Series in Mechanical Engineering, M.C. Boyce and G.E. Kendall. 2012, New York: Oxford University Press. 366 pp.
4. Tester, J.W. et al., *Sustainable Energy: Choosing Among Options*. Second edition. 2012: MIT Press. 1021 pp.
5. Vanek, F., L. Albright, and L. Angenent, *Energy Systems Engineering: Evaluation and Implementation*. Third edition. 2016: McGraw-Hill. 704 pp.
6. International Energy Agency. *Renewables 2017*. 2017. Accessed October 2017. Available from: https://www.iea.org/renewables/.
7. US EIA. *Electric Power Monthly*. 2017. US Energy Information Administration. Accessed October 2017. Available from: https://www.eia.gov/electricity/monthly/.
8. GWEC. Global Wind Reports. 2017. Global Wind Energy Council. Accessed October 2017. Available from: http://gwec.net/publications/global-wind-report-2/.
9. Christopherson, R.W., and G.E. Birkeland, *Geosystems: An Introduction to Physical Geography*. 10th edition. 2018: Pearson. 610 pp.
10. Elliott, D.L., C.G. Holladay, W.R. Barchet, H.P. Foote, and, W.F. Sandusky, *Wind Energy Resource Atlas of the United States*. 1986. National Renewable Energy Laboratory, DOE, US. Accessed November 2017. Available from: http://rredc.nrel.gov/wind/pubs/atlas.
11. NREL. *Wind Maps*. 2017. Accessed November 2017. Available from: https://www.nrel.gov/gis/wind.html.
12. Chiasson, J., and B. Vairamohan, Estimating the State of Charge of a Battery. *IEEE Transactions on Control Systems Technology*, 2005. **13**(3): 465–470.
13. Alavi, S.M.M., A. Mahdi, S.J. Payne, and D.A. Howey, Identifiability of Generalized Randles Circuit Models. *IEEE Transactions on Control Systems Technology*, 2017. **25**(6): 2112–2120.
14. Nejad, S., D.T. Gladwin, and D.A. Stone, Sensitivity of Lumped Parameter Battery Models to Constituent Parallel-RC Element Parameterisation Error. In *IECON 2014: 40th Annual Conference of the IEEE Industrial Electronics Society*. 2014.

14 Solar Power

Together with hydroelectric and wind turbines, conversion of solar radiation received on Earth to electricity is one of the most notorious of the renewable conversion processes. Solar energy received on Earth is abundant and drives the ocean currents, pressure systems, winds, and weather patterns. Thus, most other renewable resources used to generate electricity are indirectly a form of solar energy (Chapter 1). We start this last chapter of the book examining the solar resource and its variation according to latitude, atmospheric conditions, and relative movement of Earth and Sun. Then, we study photovoltaic (PV) technology and various models to predict performance of solar cells. For this purpose, we integrate the knowledge of solar resource to predict production by PV cells. We examine solar farms and smaller off-grid and microgrid installations. We end the chapter by studying concentrating solar power, which is based on producing electricity by generating and storing heat from solar radiation. Most textbooks on renewable power systems include chapters in solar energy and can serve as supplementary reading [1–9].

14.1 SOLAR RESOURCE

Solar radiation received by Earth outside the atmosphere, measured as *radiation flux*, which is power per unit area or *power density* (Chapter 2), is transmitted through the atmosphere and reaches the Earth's surface. Incoming solar radiation is also called *irradiance* and *insolation*. We will have more to say about atmospheric processes later in the chapter, but for now let us focus on the distribution of this power density over the planet's surface and its variation over time.

14.1.1 ORBITAL ECCENTRICITY AND EXTRATERRESTRIAL SOLAR RADIATION

Extraterrestrial solar radiation or irradiance received above Earth's atmosphere varies daily during the year because of Earth's revolution around the sun following an elliptical orbit, which is inscribed to a plane called the *plane of the ecliptic*. Being elliptical, the eccentricity of the orbit exhibits two contrasting sun–Earth distances. The *perihelion* is when the Earth is closest to sun (currently ~July 4), whereas the *aphelion* occurs when the Earth is farthest from sun (currently ~January 3). Currently, the eccentricity of the Earth's orbit is 0.0167.

An often-used model for the variation of extraterrestrial power density I_0 or radiation received above the atmosphere over a year is to add a sinusoidal function of amplitude related to the eccentricity

$$I_0(n) = S_0 \times \left(1 + 0.033 \times \cos\left(\frac{2\pi}{365} \left(n - n_p \right) \right) \right) \tag{14.1}$$

where S_0 is the *solar constant* or average extraterrestrial radiation, and currently has a value of ~1.377 kW/m²; n is the day of the year; and n_p is the day at which the perihelion occurs, which should be maximum I_0 according to the cosine of $n - n_p = 0$.

As you can infer from Equation 14.1, the variation $2 \times 0.033/1.377 = 0.048$ represents only 4.8% of change due to orbit eccentricity. Therefore, this is less important (because it only amounts to a small difference in insolation) than the effect of the position and tilt of the axis of rotation with respect to the plane of the ecliptic, as we will discuss in the next section. For practical purposes, the analysis of solar radiation resource for solar energy conversion to electricity can often assume that I_0 is constant and equal to S_0.

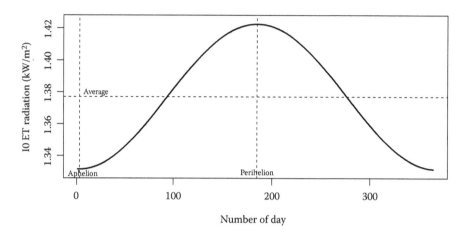

FIGURE 14.1 Example of daily extraterrestrial solar radiation during a year.

However, the eccentricity of the orbit changes cyclically over ~100,000 years from nearly circular to highly elliptical, reaching values as high as 0.05. In the latter case, the difference in insolation between aphelion and perihelion could be ~20% to 30%, and therefore more substantial. This is one of the *Milankovitch cycles* [10–12] and relates to climate changes over long time scales and Earth's geologic history, as discussed in Chapter 2.

Example 14.1

Use function I0.orbit of renpow to display extraterrestrial I0 over the course of a year. Answer: I0.orbit uses Equation 14.1, default arguments are leap = FALSE and plot = TRUE.

```
I0.orbit()
```

Using all default values for the arguments produces the graph of Figure 14.1 which applies to a non-leap year.

14.1.2 Spectral Density

Recall from Chapter 1 that the frequency ν of electromagnetic (EM) waves multiplied by the wavelength λ equals to the speed of light c (3×10^8 m/s), $c = \nu\lambda$, and that the energy E of a photon is directly related to the frequency ν of the wave by Planck's constant ($h = 6.6 \times 10^{-34}$ Js) $E = h\nu = \dfrac{hc}{\lambda}$. The spectrum of extraterrestrial solar irradiance can be modeled using blackbody radiation

$$p(\lambda) = \frac{A\lambda^{-5}}{\exp(hc/kT\lambda) - 1} \tag{14.2}$$

where $p(\lambda)$ is a flux density or irradiance (per differential of wavelength $d\lambda$), T is absolute temperature, and k is the Boltzmann constant. The constant A is

$$A = 2\pi h \left(c\frac{R}{r}\right)^2 \tag{14.3}$$

where R is radius of the sun and r is radius of the Earth.

Values for I_0 at the average sun–Earth distance are available from detailed models together with Earth's surface data as part of the ASTM G173-03 standard [13–15].

Example 14.2

Plot spectral density of I_0 in the λ range from 150 to 2500 nm assuming the sun temperature is 5800 K. Plot together with data from ASTM G173 at the average sun–Earth distance. Answer: We use I0.blackbody and spectral functions of renpow:

```
I0.blackbody(T.sun=5800,wl.nm=seq(150,2500))
X <- read.table("extdata/ASTMG173.csv",skip=1,sep=",",header=TRUE)
# or directly X <- ASTMG173 by using renpow data
spectral(X,label="G173-03",wl.lim.nm=c(150,2500),
        T.sun=5800,plot.surf=FALSE)
```

Alternatively we could have read the system file using

```
X <- read.table(system.file("extdata","ASTMG173.csv",package="renpow"),skip=1,
sep=",",header=TRUE)
```

This density is shown in Figure 14.2. The top graph illustrates the blackbody power density by itself, whereas the bottom graph compares to the ASTM G173 reference data.

From short to long waves, or high to low energy, the graph shows ultraviolet (UV; ~10–400 nm), visible (VIS; ~400–750 nm), and near infrared (NIR; ~700 nm–1400 nm). Recall from Chapters 1 and 2 that the various wavelength ranges of solar radiation determine important types of radiation in regard to Earth's climate. Incoming shortwave includes UV, VIS, and NIR, and *outgoing longwave* (low energy) emitted by the Earth's surface is in the longer infrared (IR; e.g., 4–50 μm), not shown in Figure 14.2. The balance of incoming and outgoing radiation determines Earth's temperature (Chapter 2).

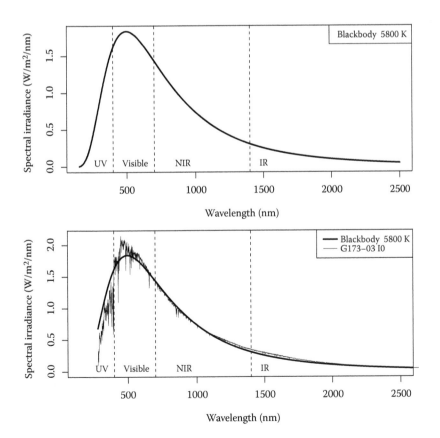

FIGURE 14.2 Blackbody model of spectral density of extraterrestrial solar radiation.

14.1.3 Latitudinal Distribution

Recall that *latitude* is the angle α (in degrees) that a hypothetical line going through the center of the Earth makes with respect to the equatorial plane. As shown in Figure 14.3, lower latitudes correspond to smaller angles and higher latitudes correspond to larger angles. One of the most important aspects of solar radiation received at the Earth's surface is its *latitudinal distribution*. As you can see from the simple diagram of Figure 14.4, the sunrays are more concentrated toward low latitudes and less concentrated toward the higher latitudes, because the surface area over which the same amount of power is received increases toward higher latitudes.

This latitudinal distribution of solar power density drives the ocean currents, pressure systems, winds, and weather patterns. Thus, this radiation differential is behind converting solar energy to other forms of resources we have discussed, such as wind. In addition, latitude combined with other factors determines where on the surface of the planet we have more availability of solar radiation resources (Figure 14.5).

Example 14.3

What would be the insolation variation from 0° to 80° latitude? Answer: Using the simple tangential approximation shown on the left-hand side of Figure 14.4, we can see that the length y of the spread of the rays is related to the ray length d by the cosine of latitude. This is $y \approx \dfrac{d}{\cos \alpha}$. Surface area for a width x would be $A(\alpha) = xy = xd/\cos\alpha$. Insolation is $S(\alpha) = P/A(\alpha) \approx \cos\alpha$.

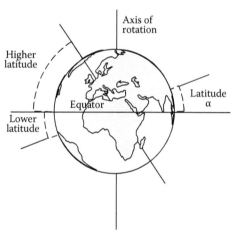

FIGURE 14.3 Latitude is an angle with respect to equatorial plane.

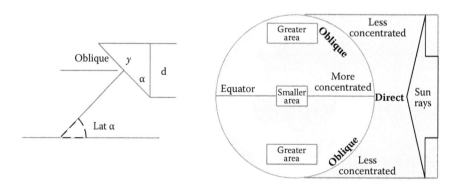

FIGURE 14.4 Solar radiation density received at the Earth's surface depends on latitude.

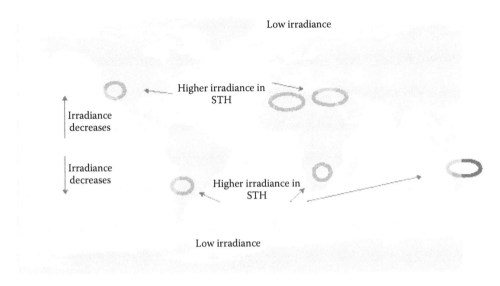

FIGURE 14.5   Average solar radiation density world distribution. Irradiance decreases with increasing latitude, but there is higher irradiance around the sub-tropical centers of high pressure (STH).

Therefore, at 80°, $S(80°) \approx 0.17 \times S(0°)$ or 17% of the radiation at the equator. This is just a simple approximation to illustrate the changes.

14.1.4 AXIAL OBLIQUITY AND DAILY VARIATION

The Earth's axis of rotation is currently tilted ~66.45° with respect to the plane of the ecliptic (Figure 14.6) in such a way that the axis of rotation points toward the star Polaris at all times during a revolution year, that is, the axis vectors are always parallel to each other. By observing Figure 14.6, you should realize that the tilt of the axis of rotation with respect to the normal to the plane of the ecliptic is the complement angle, which is 90° − 66.45° = 23.45°. Consequently, the equatorial plane is tilted 23.45° with respect to the plane of the ecliptic. Actually, the tilt angle varies with time, between 22.1° and 24.5° in a cycle that lasts ~40,000 years, which is another one of the Milankovitch cycles [10,11].

The *obliquity* and *parallelism* of the axis, causes the amount of radiation to vary unequally with latitude, as can be seen in the diagram of Figure 14.7. As this diagram shows, the circle of illumination changes with latitude during the year, and *solstices* occur when the sunrays are perpendicular

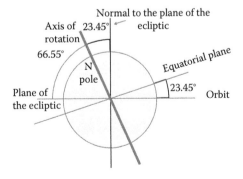

FIGURE 14.6   Planes of the equator and ecliptic. Tilt of the axis of rotation.

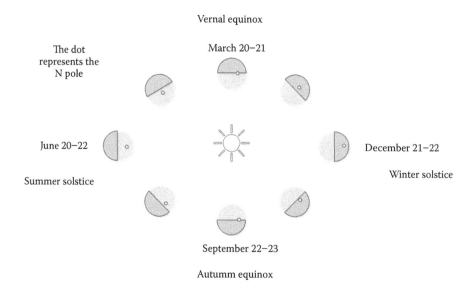

FIGURE 14.7   Revolution around the sun, solstices, and equinoxes. Illustrated for the Northern Hemisphere.

to the northernmost position (Tropic of Cancer, 23.45° latitude north), or to the southernmost position (Tropic of Capricorn, 23.45° latitude south). *Equinoxes*, on the other hand, correspond to the times when the sunrays are perpendicular to the equator.

In the Northern Hemisphere, you have the *summer solstice* on a day between June 20 and 22 (corresponding to the Southern Hemisphere winter solstice). On that day, there are 24 hours of daylight within the Arctic Circle (corresponding to 24 hours of darkness within the Antarctic Circle). Conversely, in the Northern Hemisphere, you have the *winter solstice* on a day between December 21 and 22 (corresponding to the Southern Hemisphere summer solstice), and the sun-rays are perpendicular to southernmost position (Tropic of Capricorn). On that day, there are 24 hours of daylight within the Antarctic Circle (corresponding to 24 hours of darkness within the Arctic Circle). The *vernal equinox* and the *autumnal equinox* occur on a day between March 20–21 and September 22–23, respectively, for the Northern Hemisphere, and vice versa for the Southern Hemisphere.

Note that because axis obliquity varies cyclically over 40,000 years, seasonality would vary as well; less tilt would mean milder seasons, whereas greater tilt would cause large seasonal differences. The axis had a maximum tilt of 24.5° in the Holocene, or about 10,000 years ago [10,11].

14.1.5   DECLINATION AND SUN ELEVATION ANGLE

Declination δ is the angle of the sun with respect to a plane parallel to the equator and placed at the latitude α. The maximum of this angle for each day is a function of the day number n in the year according to a simple sinusoidal model with a delay n_e given by the March equinox (vernal in the Northern Hemisphere and autumnal in the Southern Hemisphere) for which declination must be zero

$$\delta(n) = 23.45° \times \sin\left[\frac{2\pi}{365}(n - n_e)\right] \tag{14.4}$$

For instance, we can use $n_e = 80$ corresponding to the equinox on March 21 (Figure 14.8). This figure is produced using function declination of renpow, which has a nonleap year and $n_e = 80$ as default values.

```
declination()
```

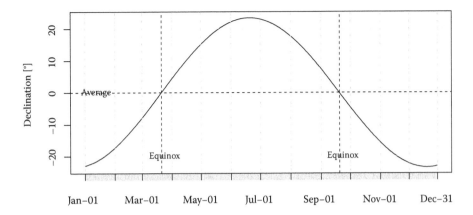

FIGURE 14.8   Example of declination during the year for a nonleap year and using $n_e = 80$ or March equinox on March 21.

Sun elevation angle is the angle that the sun is above the horizon, that is, the angle of a line from an observer to the sun with respect to the horizon. This angle varies during the day, and reaches a maximum when the sun is in the zenith. To an observer, the maximum daily sun elevation angle (sun in the zenith) reaches a maximum in the summer solstice and a minimum in the winter solstice, going through the same intermediate value on the vernal and autumnal equinox (Figure 14.9). An observer located on the equator would see an unchanged position of 90°.

As we can see from Figure 14.10, sun elevation angle β is the angle that the sun makes with respect to the tangential plane normal to latitude. The maximum elevation angle in a day is a function of the day number in the year. Therefore, declination, maximum sun elevation, and latitude for a given day number in the year are related by

$$\beta(n) = \delta(n) + (90° - \alpha) \tag{14.5}$$

Example 14.4

Determine sun elevation angle for latitude of 32.90° (decimal degrees) or 32°54′0″ (deg, min, sec) (this latitude corresponds approximately to the Dallas and Fort-Worth airport (DFW), in Texas). Answer: We can use function sun.elev of renpow. First, call declination and use its output for sun.elev:

```
x <- declination(plot = FALSE)
# latitude DFW 32.90
sun.elev(x,lat=32.9)
```

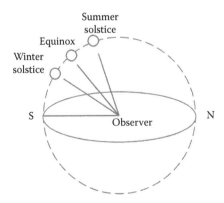

FIGURE 14.9   Sun elevation angle at noon for solstices and equinoxes, Northern Hemisphere.

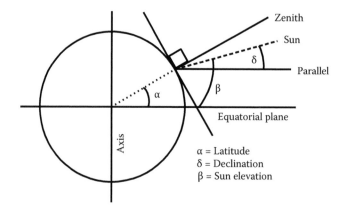

FIGURE 14.10   Latitude, declination, and sun elevation angle.

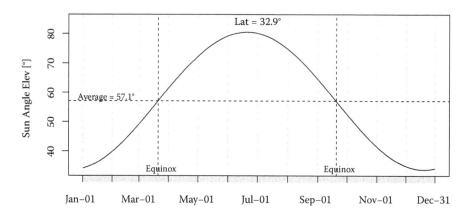

FIGURE 14.11   Example of sun elevation for a given latitude in the Northern Hemisphere.

The result is shown in Figure 14.11.

14.1.6   ATMOSPHERIC EFFECTS: DIRECT AND DIFFUSE IRRADIANCE

As the solar radiation flux goes through the atmosphere, a good part of it is absorbed by atmospheric gases and scattered by particles. Direct radiation reaching the surface of the Earth can be as high as ~70% of the extraterrestrial solar radiation I_0, or solar radiation received by Earth outside the atmosphere. There are two major components of the radiation reaching the Earth's surface: direct and diffuse radiation. A model for radiation flux through the atmosphere follows the Bougher-Lambert-Beer exponential attenuation

$$I_{bn} = I_0 \exp\left(-\tau_b \times m^{a_b}\right) \tag{14.6}$$

$$I_d = I_0 \exp\left(-\tau_d \times m^{a_d}\right) \tag{14.7}$$

where I_{bn} is the *direct beam* normal portion, I_d is the diffuse horizontal portion of clear-sky radiation reaching the Earth's surface, m is the *air mass*, τ_b, τ_d are atmosphere pseudo *optical depths*, and a_b and a_d are coefficients. Values of τ_b, τ_d are location specific and vary through the year [16]. The ASHRAE handbook and other resources provide coefficients for each month for thousands of locations [17]. The power coefficients a_b and a_d also vary, but they are related to the optical depths by empirically derived equations

$$a_b = k_b - k_{bb}\tau_b - k_{bd}\tau_d - k_{bbd}\tau_b\tau_d \qquad (14.8)$$

$$a_d = k_d - k_{db}\tau_b - k_{dd}\tau_d - k_{dbd}\tau_b\tau_d \qquad (14.9)$$

The air mass ratio is a ratio of two path lengths as the sunrays go through the atmosphere. The simplest calculation assumes a flat Earth; denote by h_1 path length if the sun were overhead or an elevation angle of 90°, and h_2 path length when the sun is at elevation angle β (Figure 14.12)

$$m = \frac{h_2}{h_1} = \frac{1}{\sin\beta} \qquad (14.10)$$

Note that h_2 is larger than h_1, and therefore the air mass ratio will be larger than 1, except when the sun is overhead for which $m = 1$. This special condition is denoted as AM1. For other values of m, say, $m = 1.5$, the notation is AM1.5. The extraterrestrial air mass ratio is considered to be zero and denoted as AM0.

Equation 14.10 works fine for low to mid latitude, but for higher latitudes, a more detailed equation for air mass ratio accounts for spherical shape of Earth [4]

$$m = \sqrt{(708\sin\beta)^2 + 1417} - 708\sin\beta \qquad (14.11)$$

We can use equations to calculate the effect of optical depth and air mass for each month at a particular location. This is quickly done using function beam.diff of renpow, which includes several sample locations from published sources.

Example 14.5

Determine direct and diffuse radiation using parameter values for the Golden, Colorado, radiometric station maintained by the National Renewable Energy Laboratory (NREL) [16].

```
tau <- read.tau("extdata/tauGolden.csv")
# or directly tau <- tauGolden
Ibd <- beam.diffuse(tau)
```

to obtain Figure 14.13. We can query the proportion of diffuse with respect to direct.

```
> Ibd$Id.Ib
 [1]  1.41  3.21  6.21  8.00  9.17  9.50 10.02  8.90  5.43  2.43  1.25  0.86
>
```

Alternatively we could have read the system file using

```
tau <- read.tau(system.file("extdata","tauGolden.csv",package="renpow"))
```

You can see that in this example, diffuse irradiance is between 1% and 10% of direct beam irradiance. Therefore, contribution of diffuse radiation to radiation received by a photovoltaic (PV) module may be important when estimating expected production. As we explain later, a PV module is made from wiring PV cells in series.

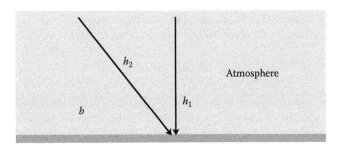

FIGURE 14.12   Air mass ratio.

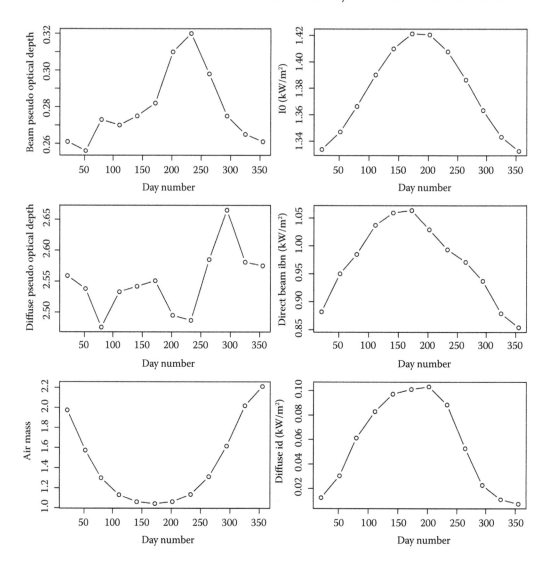

FIGURE 14.13 Example of modeled atmospheric effects (Golden, Colorado, USA).

14.1.7 Atmospheric Effects: Spectral Density

Another important aspect of atmospheric effects on solar irradiance reaching Earth's surface, relates to the changes of spectral distribution as the radiation is absorbed and scattered at different wavelengths λ by various gases and aerosols. Radiation is absorbed by ozone in the UV, oxygen in parts of the VIS, water vapor in the NIR and IR, and carbon dioxide in the IR.

Absorption is described by the Beer-Lambert exponential relation

$$I(\lambda) = I_0(\lambda) \exp(-m\sigma(\lambda)c) \tag{14.12}$$

where $I(\lambda)$ and $I_0(\lambda)$ are irradiance at the Earth surface (sea level) and outside the atmosphere, respectively, at the specific wavelength λ; and $\sigma(\lambda)$ is the *absorption cross-section* of a particular gas with concentration c. For several absorbers we add the cross section and concentration of each gas

$$I(\lambda) = I_0(\lambda) \exp\left(-m \sum_i (\sigma_i(\lambda)c_i)\right) \qquad (14.13)$$

The absorption cross-section for many trace gases are available in the HITRAN (high-resolution transmission molecular absorption) database [18].

In addition to absorption, radiation is scattered by particles in different proportions according to particle size. *Rayleigh scattering* $\varepsilon_R(\lambda)$ corresponds to reflection from particles smaller than the wavelength in consideration, whereas *Mie scattering* $\varepsilon_M(\lambda)$ refers to reflection from particles about the same size as the wavelength. These processes can be added to Equation 14.13

$$I(\lambda) = I_0(\lambda) \exp\left(-m\left(\sum_i (\sigma_i(\lambda)c_i) + \varepsilon_R(\lambda) + \varepsilon_M(\lambda)\right)\right) \qquad (14.14)$$

As mentioned in Section 14.1.2, values for spectral irradiance are available from detailed models as part of the ASTM G173-03 standard [13–15]. We will use the direct plus circumsolar surface values for an average U.S. latitude (37°) and AM1.5.

Example 14.6

Plot spectral density at sea level direct plus circumsolar together with I0 using data from ASTM G173. Plot together with black body radiation assuming the sun's temperature is 5800 K in the λ range from 150 to 2500 nm. Answer: We use spectral function of renpow.

```
X <- read.table("extdata/ASTMG173.csv",skip=1,sep=",",header=TRUE)
# or directly X <- ASTMG173
spectral(X,label="G173-03",wl.lim.nm=c(150,2500),
      T.sun=5800,plot.surf=FALSE)
```

Alternatively we could have read the system file using

```
X <- read.table(system.file("extdata","ASTMG173.csv",package="renpow"),skip=1,
sep=",",header=TRUE)
```

This density is shown in Figure 14.14. We can observe the bands of absorption of various gases; ozone in the UV, oxygen in parts of the VIS, water vapor in the NIR and IR, and carbon dioxide in the IR.

14.1.8 SUN PATH

Figure 14.15 shows Earth rotating at 15°/h. We define hour angle H in *hn* (hours before noon) as

$$H = 15 \times hn \qquad (14.15)$$

Trigonometric equations relate sun azimuth ϕ (Figure 14.16) and sun elevation β to hour angle:

$$\sin\phi = \frac{\cos\delta\sin H}{\cos\beta} \qquad (14.16)$$

$$\sin\beta = \cos L\cos\delta\cos H + \sin L\sin\delta$$

Example 14.7

Calculate sun elevation and azimuth at the latitude of DFW, Texas, for day 20 of the year. Answer:

```
# latitude DFW
sun.path(lat=32.90,nday=20)
```

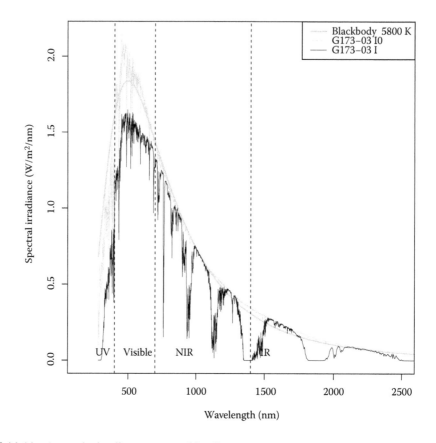

FIGURE 14.14 Atmospheric effects on spectral irradiance.

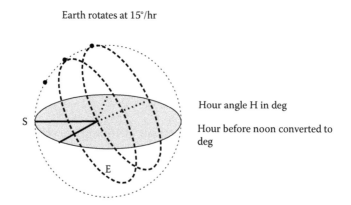

FIGURE 14.15 Sun elevation angle changes during the day.

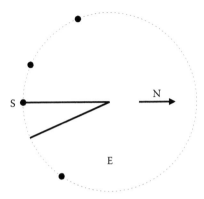

FIGURE 14.16 Projection is azimuth angle.

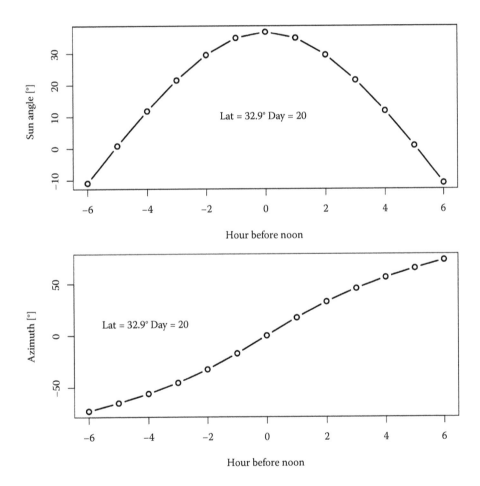

FIGURE 14.17 Sun angle and azimuth as a function of hour angle.

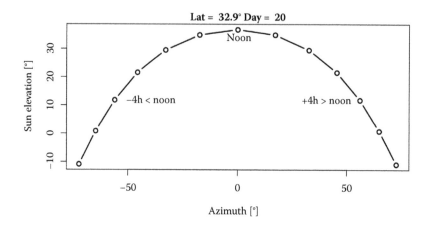

FIGURE 14.18 Position of sun in azimuth–elevation plane.

and obtain results for the sun angle (Figure 14.17, top) and azimuth (Figure 14.17, bottom). These results are combined in an azimuth–elevation plane, where the hour is implicit by the graph marker (Figure 14.18).

Now, for selected days of the year, we obtain a sun path diagram (Figure 14.19) with the following function call:

```
# latitude DFW
sun.diagram(lat=32.90)
```

This diagram will be used later to analyze potential shading of a PV module.

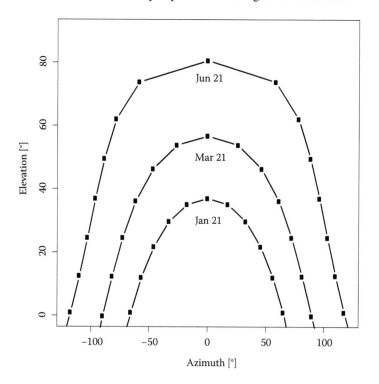

FIGURE 14.19 Example of a sun path diagram for selected days of the year.

14.1.9 DAILY SOLAR RADIATION AND 1-SUN

We can combine the sun path calculations and the atmospheric calculations to derive direct and diffuse radiation received by a flat collector at any particular time of the day. The sun elevation forms an angle of incidence θ with the normal to the surface of the collector, which is tilted with angle γ with respect to the horizontal (Figure 14.20, left-hand side).

The direct beam projected onto the normal would be

$$I_{bc} = I_{bn} \cos \theta \tag{14.17}$$

and observe that when $\gamma = 0$ (horizontal collector), the projection onto the normal is just $I_{bh} = I_{bn} \cos (90° - \beta) = \sin \beta$. Now consider the top view of the right-hand side of Figure 14.20 to see that the collector has an orientation of azimuth ϕ_c (relative to the south). At any particular time of the day, the sun has elevation β and azimuth ϕ. The cosine of the angle of incidence θ of the sun elevation with the normal to the surface of the collector can be calculated from all the aforementioned angles using [4]

$$\cos \theta = \cos \beta \cos(\phi - \phi c) \sin \gamma + \sin \beta \cos \gamma \tag{14.18}$$

The diffuse component incident on the collector can be assumed to come from all angles; when horizontal ($\gamma = 0$) it should be the same ($I_{dh} = I_d$), and when vertical it should be 1/2; and therefore have a value in between when tilted. Thus,

$$I_{dc} = I_d \left(\frac{1 + \cos \gamma}{2} \right) \tag{14.19}$$

Part of the radiation reflected from surrounding surfaces would be incident on the collector, thus

$$I_{rc} = f_r (I_{bh} + I_{dh}) \left(\frac{1 - \cos \gamma}{2} \right) \tag{14.20}$$

where f_r is a coefficient representing a fraction of reflected incident on the collector. The total radiation incident on the collector is then the sum of direct beam, diffuse, and reflected

$$I_c = I_{bc} + I_{dc} + I_{rc} \tag{14.21}$$

All of these calculations are quickly performed by calling function collector of renpow that takes as arguments (1) the direct and diffuse components calculated by beam.diff for a particular location; (2) the sun elevation and azimuth calculated by sun.path for the site latitude and a day number of interest; and (3) the collector data tilt, azimuth, and reflectivity of surroundings.

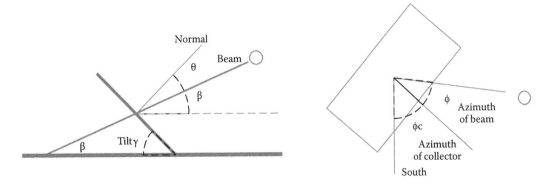

FIGURE 14.20 Collector positioning and angles.

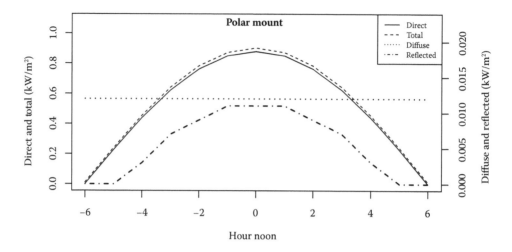

FIGURE 14.21 Radiation incident on a collector.

Example 14.8

Use the file with optical path for Golden, Colorado, and day 20 to calculate radiation on a collector tilted at the latitude, facing south ($\phi c = 0°$), and 20% contribution of reflected radiation from surrounding surfaces

```
Ibd <- beam.diffuse(tauGolden)
sunpath <- sun.path(lat=Ibd$lat,nday=20)
Ic <- collector(Ibd, sunpath, tilt= Ibd$lat,azi.c=0,fr=0.2,label="Polar Mount")
```

The result is shown in Figure 14.21, where we can see that most of the radiation is direct, but there is some contribution from diffused and reflected. The last few items in the output returned by function collector, are the kWh/m² total of the day for direct, diffuse, reflected, and total. We can query these to obtain

```
> Ic$I.h
[1]  6.281 0.011 0.064 6.488
>
```

We can then say we have 6.488 kWh total irradiance or almost ~6.5 kWh/m² that day.

Irradiance reaching the Earth surface in a clear day, when the sun is at the zenith, can be slightly above 70% of the solar constant or 1 kW/m² (($\approx 0.7 \times 1.377 \approx 1$ kW/m²). This is such a convenient number that has been defined as *1-sun* equivalent of irradiance, or 1-sun insolation. Therefore, it is easy to think of daily energy density in 1 kWh/m² as the number of hours per day of full sun or 1-sun insolation. It is then equivalent to say, for example, 6.5 kWh/m² per day as 6.5 *hours of full sun*, or 6.5 hours of 1-sun. We will see later that this expresses the capacity factor of a PV solar installation.

14.2 PHOTOVOLTAIC (PV) BASICS

The photovoltaic (PV) effect is the phenomenon of converting energy of photons of light into electrical current or voltage. The PV effect, first reported in 1839 by Becquerel, was used in cells made from selenium as early as 1880, but the conversion had very low efficiency (~1%), and the process was not well understood. In the 1940s and 1950s, research demonstrated how to make cells from silicon crystals. Demand for space exploration applications in the 1960s advanced the technology considerably. In the last couple of decades, we have seen an increased use in residential, commercial, and utility-scale applications.

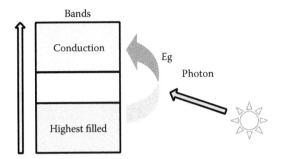

FIGURE 14.22 Semiconductor band model.

A semiconductor PV cell is made from a p-n junction using semiconductor material, mostly crystalline silicon (Si), but also germanium (Ge). These elements are combined with boron (B), phosphorous (P), gallium (Ga), and arsenic (As), among others.

We can better understand the PV effect by using the energy band model of semiconductors (Figure 14.22). This model assumes that electrons can move from the last filled energy band to the conduction band, and that this jump requires an amount of energy at least equal to the "band-gap energy" or E_g, which varies from material to material. For example, in silicon $E_g = 1.12$ eV. The charge of one electron is 1.6×10^{-19}C. One eV is an amount of energy equivalent to 1.6×10^{-19} J or the energy of one electron.

Light provides the energy to move electrons to the conduction band, or in other words to jump the E_g gap. Of course, for this to happen, light has to be of a frequency such that a photon has to have energy of at least E_g to move the electrons.

Recall that the photon energy E in J is inversely related to the wavelength by the Planck's constant $h = 6.6 \times 10^{-34}$ J-sec $E = \dfrac{hc}{\lambda}$. Using this relationship, we can restate the condition for PV effect as saying that to overcome the band-gap, the photon energy should exceed the band-gap energy $E = \dfrac{hc}{\lambda} \geq E_g$. For example, in silicon

$$\lambda \leq \frac{hc}{Eg} = \frac{\overbrace{6.6 \times 10^{-34}}^{\text{J-s}} \times \overbrace{3 \times 10^{8}}^{\text{m/s}}}{\underbrace{1.12}_{\text{eV}} \times \underbrace{1.6 \times 10^{-19}}_{\text{J/ev}}} = 1.11 \times 10^{-6} \quad \text{m} = 1.11\,\mu\text{m} = 1110 \text{ nm}$$

We need a wavelength shorter than 1110 nm to overcome the band-gap. However, a shorter wavelength with more photon energy above this level is wasted (Figure 14.23). In other words, the required level is equal to the usable [19]. The percent of the energy available from 1110 nm to 150 nm that is wasted represents about 40%.

14.2.1 Solar Radiation and Efficiency

As we have discussed, solar radiation reaching Earth is distributed by wavelength according to Planck's law of blackbody radiation increasing for short wavelengths (150–500 nm), reaching a peak at ~500 nm, and then decreasing as wavelength increases (Figure 14.24). As we can observe from this figure, the right tail of this density (above 1100 nm) has energy lower than that required to excite the electrons; the area under the curve for this tail amounts to about 20% of the total area. The energy corresponding to wavelengths shorter than 1100 nm represents the remaining 80% and can excite the electrons, but as we know from the previous section, only 60% is usable. Therefore, 0.6 × 80% = 48% is usable. Consequently, PV maximum efficiency is approximately 50%.

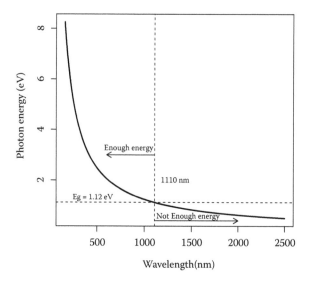

FIGURE 14.23 Photon energy required to bridge the gap.

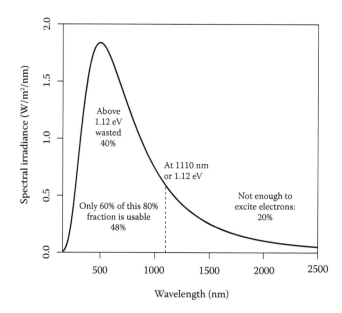

FIGURE 14.24 Solar radiation spectrum with required and usable photon energy.

One manner to increase efficiency is to increase band-gap to maximize power, ideally in the 1.2–1.8 eV by adding elements to silicon, thus making cells such as GaAs and CdTe. See b table 5.2 of Masters [19].

However, the actual efficiency is lower because of other factors: PV cells get hot and radiate energy following the blackbody radiation law (~7% loss), or there is a hole saturation effect causing hole–electron pairs to recombine (~10% loss). Therefore, the 50% theoretical efficiency is reduced to ~33%. The Shockley-Queisser limit states that the maximum efficiency is 33.7%.

14.2.2 SOLAR CELL MODEL

A p-n junction is formed by adjoining semiconductor material that has been differentially doped (Figure 14.25) and constitutes a diode. In Chapter 3, we cover the basic diode model and a basic PV cell model that consists of a diode, a current source I_{sc} that produces a current when illuminated by solar radiation, and two resistances Rp and Rs (Figure 3.19). The I_{sc} value at full sun conditions, or fully illuminated, becomes a specification of a solar cell. Also, recall from Chapter 3 that we can calculate the *I-V* characteristics as shown in Figure 3.20 using function PVcell:

```
x <- list(I0.A=1, Isc.A=40, Area=100, Rs=0.05, Rp=1, Light=1)
# units: I0.A pA/cm2 Isc.A mA/cm2 Area cm2 Rs ohm Rp ohm
X <- PVcell(x)
ivplane(X, x0=T,y0=T)
```

Since power is the product of current and voltage, we see that power increases as voltage increases, but then quickly decreases due to the decrease in current. As light decreases, the I_{sc} goes down, and therefore the cell produces less current for the same voltage (recall Example 3.14).

I-V and power as a function of light can be visualized more generally using an array for the light argument in function PVcell.

Example 14.9

Display *I-V* and power curves for light values of 1, 0.75, and 0.5 of full sun. Answer:

```
x <- list(I0.A=1, Isc.A=40, Area=100, Rs=0.05, Rp=1, Light=c(1,0.75,0.5))
# units: I0.A pA/cm2 Isc.A mA/cm2 Area cm2 Rs ohm Rp ohm
X <- PVcell(x)
PVcell.plot(X)
```

We can see that current decreases for the same voltage with elbow shifting to the right (Figure 14.26). Power decreases for the same voltage as well, but also the maximum power decreases and shifts to the right as light decreases (Figure 14.27). We conclude that to optimize power output from the cell, we have to modify the operating conditions of voltage and current as light changes. This is the subject of a subsequent section.

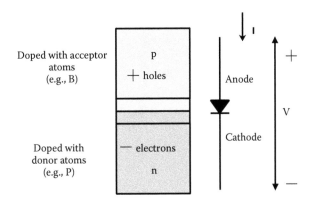

FIGURE 14.25 An illustration of a p-n junction and diode.

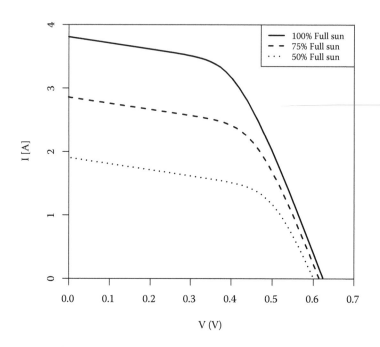

FIGURE 14.26   *I-V* as a function of light.

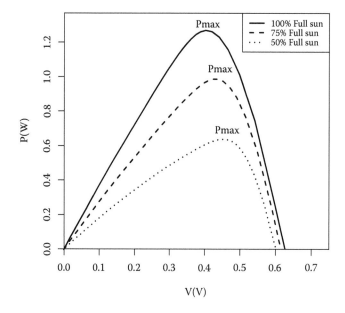

FIGURE 14.27   Power as a function of light.

14.2.3   CELLS, MODULES, AND ARRAYS

One PV cell outputs about 0.5 V. We make a *module*, often called a *panel*, by wiring cells in series; 36 cells would output about 18 V, which is referred to as a nominal 12-V module. Likewise, 72 cells would output about 36 V, referred to as a 24-V module. For example, multiplying the curve of each cell by 36, we obtain Figure 14.28 for the *I-V* curves and Figure 14.29 for power.

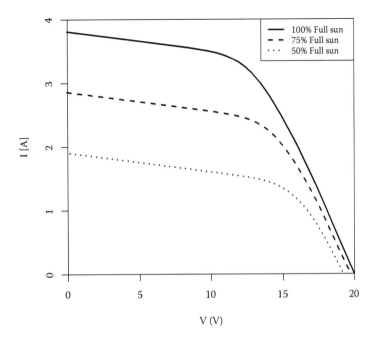

FIGURE 14.28   *I-V* for a module made from 32 cells.

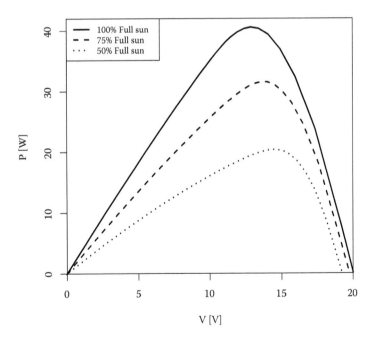

FIGURE 14.29   Power for a module made from 32 cells.

Arrays are made wiring modules in series or parallel. When in series, we form a *string* and thus we increase voltage, that is, *I-V* curves are added by voltage axis. When adding modules in parallel, we increase current, and *I-V* curves are added by current axis. It is common to form strings and then connect strings in parallel.

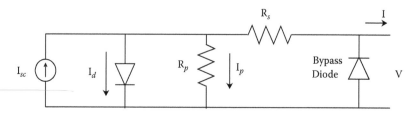

FIGURE 14.30 Bypass diode.

14.2.4 SHADING AND BYPASS DIODE

A shaded cell in a module affects the entire module because the cells are in series. Assuming we have n cells and that some are shaded, we can calculate the decrease in voltage due to shading using the cell model. When we shade one cell

$$V_{sh} = \frac{n-1}{n} V - I \left(R_p + R_s \right) \tag{14.22}$$

and therefore the reduced voltage is

$$\Delta V = V - V_{sh} = V \left(1 - \frac{n-1}{n} \right) + I \left(R_p + R_s \right)$$

or rearranging

$$\Delta V = \frac{V}{n} + I \left(R_p + R_s \right) \sim \frac{V}{n} + I R_p$$

To avoid this issue we use a bypass diode. This is a diode placed in parallel with the cell and shunting the shaded cell (thus decreasing the voltage drop across R_p) (Figure 14.30). We can also connect a bypass diode across a module. When using a parallel combination of strings, we can add a blocking diode to each string to avoid sending current to a malfunctioning string.

14.3 PV PERFORMANCE

14.3.1 LOAD AND POWER

A load is represented as a curve on the I-V plane, and when superimposed to the I-V curve for the PV module, it will determine operating points at each sun condition. A resistive load with resistance R has a straight-line I-V characteristic with slope equal to $1/R$. A charging or discharging battery at steady state have straight-line I-V characteristics with slope that varies from charging to discharging conditions.

Example 14.10

Assume a 10-Ω resistive load. What are the values of voltage corresponding to sunlight of 20%, 40%, 60%, and 80% of full sun? What is the power for these voltage values? How do they compare with maximum power values? Answer: As shown in Figure 14.31, we obtain the plot for a 10-Ω load as a straight line with slope 1/10, and then find voltages for each intersection as shown in the figure. These are 5.6, 11.3, 14.3, and 14.8 volts, respectively. Use $V^2/10$ to obtain the power for each sun condition, 3.1, 12.8, 20.4, and 21.9 W. These are not optimal power values, as we can see on the power versus voltage graph (Figure 14.32). Indeed the voltage at which the maximum power point (MPP) is established is 12.6, 13.8, 12.7, and 11.6 volts, which correspond to power 4.6, 14.0, 21.2, and 25.8 watts, respectively.

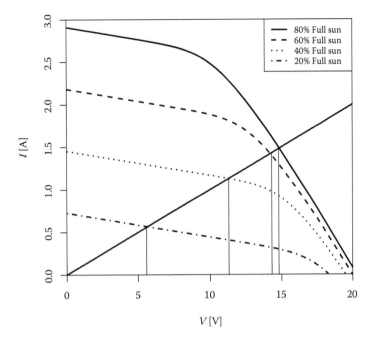

FIGURE 14.31 Resistive load and intersection points.

Similarly, we can look at the *I-V* lines for a battery charging and discharging, and use these to determine the operating point by superimposing on *I-V* curves of the PV module at various levels of sunlight.

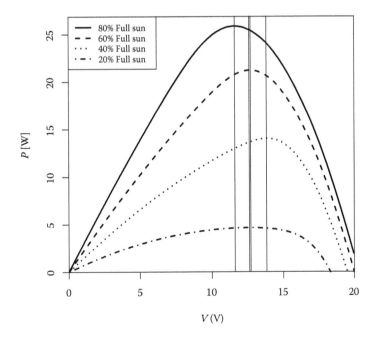

FIGURE 14.32 Power curves and maximum power points.

14.3.2 MAXIMUM POWER POINT TRACKING

As already mentioned, the peak power in the P versus V curve will vary as sunlight varies. To adjust the maximum power point (MPP), we use a device to track this maximum power point. This device is called a MPP tracker (or MPPT). The device is in essence a DC-DC converter or buck-boost converter (Chapter 5). The output voltage is proportional to the input voltage $V_{out} = V_{in} \times \left(\dfrac{D}{1 - D} \right)$, where the factor is depending on the duty cycle D of PWM signal controlling the switch.

Example 14.11

We want the MPPT to change from $V_{in} = 11$ V to $V_{out} = 15$ V. What is D? Answer: We should adjust D such that $15 = 11 \times \left(\dfrac{D}{1 - D} \right)$, therefore $\left(\dfrac{D}{1 - D} \right) = 15/11 = 1.36$, and solving for D we get $D = \dfrac{1.36}{2.36} = 0.57$.

14.3.3 EFFICIENCY AND PERFORMANCE

Other factors lower a PV module's power efficiency by 20% to 40%. These include temperature, dirt and dust, inverter efficiency (if converted to AC), and mismatched modules.

Recall from Section 14.1.9 that for a given site, insolation S can be given in kWh/m^2 per day, or equivalently as hours/day of 1-sun, or hours of peak sun. For example, if the average solar radiation is 7.0 kWh/m^2 per day, then we have the equivalent of 7 hours of 1-sun per day. We can then use area A of the module (in m^2) and system efficiency η to evaluate energy produced by the system at a given site

$$E = S \times A \times \eta \qquad (14.23)$$

either as kWh or hours of 1-sun.

Example 14.12

Assume we have 10 m^2 of panels with efficiency of 40% at a site that has 7.0 kWh/m^2 per day on day 20 of the year when sunny. Calculate energy production assuming the panels capture all sunlight available. Answer: $E = S \times A \times \eta = 7 \times 30 \times 0.4$ kWh = 84 kWh for that day.

14.3.4 IMPACT OF TEMPERATURE ON SOLAR PANELS

The temperature of a PV cell depends on ambient temperature and solar radiation; the cell gets hotter as the air gets warmer and radiation increases. The cell temperature T_c in °C can be calculated using the equation

$$T_c = T_a + \left(\frac{NOCT - 20°C}{0.8} \right) S \qquad (14.24)$$

where T_a is ambient temperature (°C); S solar irradiation (in kW/m^2); and NOCT is the nominal operating cell temperature, which is the expected cell temperature when the ambient temperature is 20°C, the irradiation is 0.8 kW/m^2, and the wind speed 1 m/s. The NOCT is given by the cell manufacturer, and for many panels this number is in the 45°C to 47°C range (see, for example, Masters [4]).

Other temperature effect specifications include nominal Voc at 25°C (NV_{oc}), a negative voltage temperature coefficient TC_V in %/°C, and a negative power temperature coefficient TC_P in %/°C. For example, for voltage

$$V_{oc} = NV_{oc} \times (1 + TC_V \times (T_c - 25)) \tag{14.25}$$

Using the NOCT specification, we can estimate the cell temperature, and with this value we can calculate the change in open circuit voltage, and maximum power as ambient temperature and irradiation conditions change.

Example 14.13

Assume a module with NOTC = 45°C, NV_{oc} of 38 V, rated power 100 W, TC_V of –0.30%/C, and TC_P of –0.4%/C. At full sun (say, 1-sun) and ambient 32°C, we get the cell temperature $T_c = 32 + \left(\dfrac{46 - 20°C}{0.8} \right) \times 1 = 64.5°C$. With this increased temperature, we calculate the effect on open circuit voltage $V_{oc} = 38 \times (1 - 0.003 \times (64.5 - 25)) = 33.5$ V, and power $P = 100 \times (1 - 0.004 \times (64.5 - 25)) = 84.2$ W.

14.3.5 SUN PATH AND SHADING

The diagram shown on Figure 14.19 can be used to analyze potential shading of the solar panel at the installation site using simple tools such as a clinometer and a compass. Locate potential shading features, and note their azimuth using a compass and height using the clinometer. Then, place the features on the sun path diagram to determine potential shading by these features. Other tools are available, such as shadow diagrams for your location that can be generated from a script or downloaded from the web once you provide the latitude [4].

14.3.6 CAPACITY FACTOR

Capacity factor (CF) can be evaluated by dividing insolation S in kWh/d or h/day of peak sun by 24 h/day. Thus, CF for PV installations depends on location, and it is a way of expressing insolation Annual energy E (in kWh/yr) can be estimated from installed capacity P (in kW), and CF using the number of hours in a year as

$$E = P \times CF \times 8760 \tag{14.26}$$

14.4 TILTING THE PANEL AND SUN TRACKING

14.4.1 POLAR MOUNT

When a PV panel tilt is fixed, this tilt is selected such that the panel receives as much sunlight as possible, which means looking in the direction of the sun or perpendicular to the maximum sun elevation for the day, which varies according to latitude (Figure 14.33). Note from this figure the tilt must be $\gamma(n) = 90° - \beta(n)$, because sun elevation is related to declination and latitude, as given in Equation 14.5

$$\gamma(n) = 90° - \beta(n) = 90° - \delta(n) - (90° - \alpha) = \alpha - \delta(n) \tag{14.27}$$

In other words, the tilt for day n is the latitude minus the declination for the day. To select a fixed tilt, we can use an average value. Since average declination is zero, then average tilt is just equal to latitude. This is called a *polar mount*, and because it is fixed, it would not be optimal for most days, and we would observe large deviations for extreme values of the declination during the year.

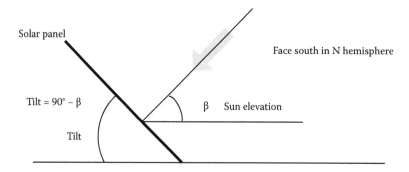

FIGURE 14.33   PV panel tilt.

Example 14.14

Assume a polar mount installation for the Golden, Colorado, site and compare energy produced on a clear day at either equinox with the energy produced on clear days at the winter and summer solstices. Answer: Call beam.diff with file for this site. This site is in the northern hemisphere, use the vernal equinox day of March 21, summer solstice June 21, and winter solstice December 21. Calculate the day number that corresponds to these dates

```
Ibd <- beam.diffuse(tauGolden)
days <- days.mo(21)$day.mo[c(3,6,12)]
> days
[1]  80 172 355
>
```

Now apply sun.path and collector functions for each one of these days and query the energy produced (total) in position 4 of the I.h array

```
E <- array()
for(i in 1:3){
sunpath <- sun.path(lat=Ibd$lat,nday=days[i])
Ic <- collector(Ibd, sunpath, tilt=Ibd$lat,azi.c=0,fr=0.2)
E[i] <- Ic$I.h[4]
}
> E
[1] 8.330 8.815 6.087
>
```

We see that expected production is 8.33 kWh/m^2 on the vernal equinox day of March 21, 8.82 kWh/m^2 on the summer solstice June 21, and 6.09 kWh/m^2 on the winter solstice December 21.

For greater efficiency, the tilt can be varied manually for small installations (just a few PV panels for off-grid or microgrid applications) according to the season or month of the year or even day of the year. Function tilt.adj of renpow utilizes Equation 14.27 to calculate the tilt required at a given latitude for a set of days using the declination for those days.

Example 14.15

For latitude 33.64°N (Hartsfield-Jackson Atlanta International Airport, Georgia), calculate tilt required for representative days of each month (say middle of the month) and season. Answer: The following lines of code will produce the top plot of Figure 14.34:

```
days <- days.mo(15)$day.mo # mid day of the month
labels <- c("Jan","Feb","Mar","Apr","May","Jun","Jul","Aug",
```

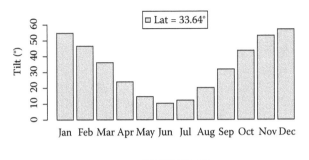

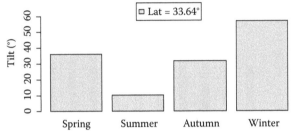

FIGURE 14.34 PV panel tilt adjustment by month and season.

```
        "Sep","Oct","Nov","Dec")
  tilt<- tilt.adj(lat=33.64,days,labels)
```

The results can be queried from object tilt:

```
> tilt
$days
 [1]  15  46  74 105 135 166 196 227 258 288 319 349
$tilt
 [1] 54.74 46.59 36.06 23.86 14.61 10.29 12.29 20.19 31.83 43.61 53.02 57.01
```

We can pick representative days for the season using months 3, 6, 9, and 12 and write

```
days.season <- tilt$days[c(3,6,9,12)]
labels <- c("Spring","Summer","Autumn","Winter")
tilt.adj(lat=33.64,days.season,labels)
```

to produce the bottom plot of Figure 14.34. Spring and Autumn can be set to the same value. They are slightly different here because we selected a representative day.

Evaluation of the impact of tilt adjustment on electric energy production requires calculating direct beam and diffuse radiation (using optical depths for the site), together with the sun path for selected days (e.g., day 21 of each month or days that optical depth parameters are given by the ASHRAE data sets). These steps involve applying functions direct.diffuse, sun.path, and collector of renpow. For user friendliness, a single call to function month.prod of renpow using a filename for the data for the site (or simply the name of the corresponding dataset), performs all of these lengthy calculations, plots graphs, and summarizes the result using bar graphs.

Example 14.16

Calculate monthly tilt for the Hartsfield-Jackson Atlanta International Airport, Georgia, and energy produced on a clear day for each month. Compare to energy produced each month for a polar mount installation. Answer: Call function

```
month.prod(tauAtlanta)
```

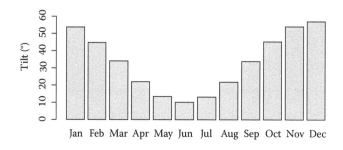

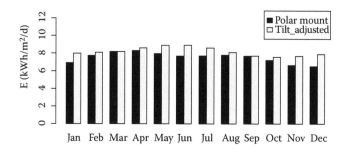

FIGURE 14.35 PV panel tilt adjustment by month and energy produced compared to polar mount.

The result shown in Figure 14.35 displays both the tilt required and the difference in production. We can also query the percent gain in production achieved by adjusting the tilt. It can be as high as 15% and 21% around the winter solstice months.

```
$Eadd
 [1] 15.31 4.59 0.07 3.84 11.89 16.07 11.91 4.17 0.00 4.99 15.80 21.16
>
```

More efficiency is achieved by tracking the sun automatically during its daily ascent and descent for each day, as we discuss in the next two sections. Sun trackers can be *two-axis* (or dual-axis), both solar azimuth and elevation, which adjust the panel's azimuth and its tilt to track the curves given in Figure 14.19, or *one-axis* (or single-axis), which adjust either the azimuth or the elevation, but not both.

14.4.2 TWO-AXIS SUN TRACKING DURING THE DAY

To understand the impact of two-axis tracking on production, we can use Equation 14.18 with panel tilt adjusted to be $\gamma = 90° - \beta$ every hour and panel azimuth adjusted to $\phi_c = \phi$ every hour, substituting

$$I_{bc} = I_{bn} \cos \theta = I_{bn}(\cos \beta \cos(0°) \sin(90° - \beta) + \sin \beta \cos(90° - \beta)) \qquad (14.28)$$

Therefore, the equation for direct beam reduces to

$$I_{bc} = I_{bn} \cos \theta = I_{bn}\left(\cos^2(\beta) + \sin^2(\beta)\right) = I_{bn} \qquad (14.29)$$

Equation 14.19 for diffuse reduces to

$$I_{dc} = I_d\left(\frac{1 + \cos \gamma}{2}\right) = I_d\left(\frac{1 + \sin \beta}{2}\right) \qquad (14.30)$$

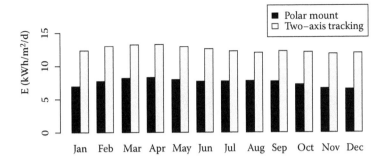

FIGURE 14.36 Example of two-axis tracking percent gain in production by month.

and Equation 14.20 for reflected reduces to

$$I_{rc} = f_r(I_{bh} + I_{dh}) \left(\frac{1 - \cos \gamma}{2} \right) = f_r(I_{bh} + I_{dh}) \left(\frac{1 - \sin \beta}{2} \right) \tag{14.31}$$

In a similar way to monthly tilt adjustment, evaluation of the impact of two-axis tracking on electric energy production requires selecting a set of days (e.g., day 21 of each month), calculating the sun path for each hour of each day, and calculating direct beam and diffuse radiation using optical depths for the site for each day. The latter entails using the panel's tilt and azimuth; these are adjusted as required by the two-axis tracking Equations 14.29 to 14.31. These steps involve repeatedly applying functions direct.diffuse, sun.path, and collector of renpow. For user friendliness, a single call to function two.axis.tracking of renpow using a filename for the data for the site (or directly the name of the dataset), calls all other functions to perform these lengthy calculations, plots graphs, and summarizes the result using bar graphs.

Example 14.17

Calculate impact of two-axis tracking on production for the Hartsfield-Jackson Atlanta International Airport, Georgia, and energy produced on a clear day for each day of the evaluation. Compare to energy produced each month for a polar mount installation. Answer: Call function

```
two.axis.tracking (tauAtlanta)
```

The result is shown in Figure 14.36 that displays the difference in production. We can also query the percent gain in production achieved by two-axis tracking. It can be as high as ~80% around the winter months, and as low as ~50%–60% during later summer months.

```
$Eadd
[1] 78.42 68.16 61.54 59.92 62.26 63.83 58.77 54.28 58.58 67.51 78.17 83.41
```

14.4.3 One-Axis Sun Tracking

While there are at least four ways of implementing one-axis tracking [4], we will discuss only two practical setups. One of these is to rotate a south-facing panel around a horizontal axis, that is, tracking sun elevation while the panel's azimuth is fixed at $\phi_c = 0°$. This is a horizontal north-south (HNS) tracker. The second one is a polar mount north-south (PNS) tracker, where the tilt is fixed at the angle of a polar mount and is rotated around the polar axis tracking the sun's azimuth [4]. Equations for direct, diffuse, and reflected radiation for one-axis tracking are provided in Masters [4]. For examples and exercises, we will work here with PNS, for which the equations are

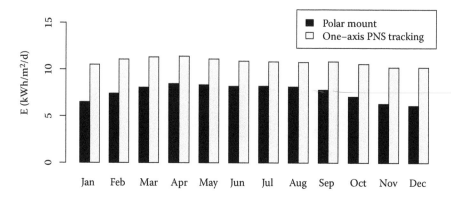

FIGURE 14.37 Example of PNS one-axis tracking percent gain in production by month.

$$I_{bc} = I_{bn} \cos \theta = I_{bn} \cos \delta \tag{14.32}$$

$$I_{dc} = I_d \left(\frac{1 + \sin(\beta - \delta)}{2} \right) \tag{14.33}$$

$$I_{rc} = f_r (I_{bh} + I_{dh}) \left(\frac{1 - \sin(\beta - \delta)}{2} \right) \tag{14.34}$$

We can use function one.axis.tracking of renpow with argument mode to specify the type of tracking. By default mode = 'PNS'.

Example 14.18

Evaluate PNS one-axis tracking impact on production for Hartsfield-Jackson Atlanta International Airport, Georgia, and energy produced on a clear day for each day of the evaluation. Compare to energy produced each month for a polar mount installation. Answer: Call function

```
one.axis.tracking(tauAtlanta)
```

to produce figures for each day and a summary bar plot shown in Figure 14.37. The added energy can be queried to be

```
$Eadd
 [1] 61.43 49.47 40.29 34.69 33.09 32.80 31.71 32.26 38.68 49.41 61.62 67.17
>
```

We can see that it can be as high as ~70% around the winter months and as low as ~30%–40% during later summer months.

14.5 SOLAR FARMS, GRID-TIE, OFF-GRID, AND MICROGRIDS

14.5.1 CONNECTING TO THE GRID

Since PV cells produce DC power, the major equipment needed to connect to the grid are inverters (Chapter 8). In small systems, such as a residential installation, one inverter may suffice to convert all the DC produced to AC (Figure 14.38). In larger systems, several or many inverters are needed. It is also possible to have microinverters assigned to each module together with a MPPT. In this case, the module is an AC source and can be combined with other modules as AC units.

Small PV systems connected to the grid should have the capability to disconnect from the grid (islanding) when the grid goes down in order to avoid hazard to grid personnel during grid

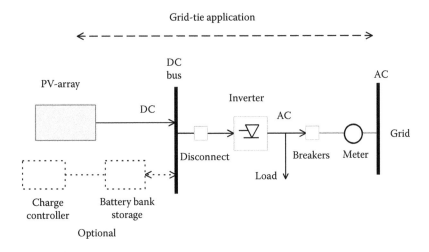

FIGURE 14.38 Grid-connected small PV installation.

maintenance. It is common now to have power flow to and from the grid, or *net metering*, in such a way that surplus PV power is sold to the utility to compensate for the user's bill (Figure 14.38). When converting to AC, it is important to include inverter efficiency to account for potential production using $P_{ac} = P_{dc} \times \eta_{inv}$. For instance, if $\eta_{inv} = 0.8$, then a DC capacity of 1 kW could only yield $P_{ac} = 0.8$ kW.

14.5.2 SOLAR FARMS

In a manner similar to wind farms formed by arrays of many large wind turbines, a large number of PV panels can be arranged in a "solar farm" or "sun farm" to form a *utility-scale solar plant* with capacity from tens of MW to hundreds of MW, and more recently reaching the GW range in plants built in China. Very large solar farms include millions of individual panels.

Utility-scale power plants benefit from economy of scale, but require land. In general, the better the irradiance at a site, the lower the requirements for land for the same production. Other factors, like latitude and topography, influence how much of the land can be covered by the panels. Take, for instance, a good site with 7 kWh/m^2 per day. Estimating PV panel efficiency at 40% and an occupancy of 50% of the land surface area, we can estimate $\frac{7000}{24} \times 0.4 \times 0.5 \approx 60\,\mathrm{W/m^2}$. The reported capacity and area of major solar farms in the world have lower numbers ranging from 30 to 40 MW/ Km^2, which is equivalent to 30 to 40 W/m^2. Note that this number is ten times larger than that typical for wind farms at 3 to 4 W/m^2 (Chapter 13).

14.5.3 OFF-GRID AND MICROGRIDS

As we studied in Chapter 13, a small off-grid power system based on wind requires a battery bank for storage. A similar consideration applies for PV-based small systems (Figure 14.39). Required battery bank capacity and solar panel power are determined through summing the average power consumption of each load.

Example 14.19

Suppose we have an off-grid system composed of four 320 W PV modules (nominal 24 V) connected in parallel. The site insolation is 6 h/d of peak sun. The daily AC load energy demand is 4 kWh, the inverter efficiency is 85%, and the charge controller efficiency is 97%. Does the

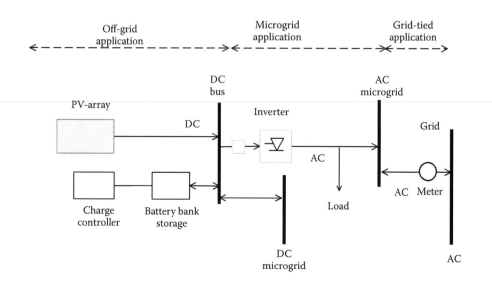

FIGURE 14.39 Off-grid, microgrid, and grid-tie.

expected energy from the PV cover the demand? Design a battery bank to provide 3 days of power of consecutive low insolation of only 1 h/d of full sun. DOD not to exceed 80%. Assume batteries are kept at room temperature. Answer: The DC load is

$$E_{DC} = \frac{E_{AC}}{\eta_{inv} \times \eta_{cc}} = \frac{4kWh/d}{0.85 \times 0.97} \simeq 4.85 kWh/d \qquad (14.35)$$

The PV installation would provide $E = P \times CF = 4 \times 320$ W $\times$ 6h/d = 7.68 kWh/d. It covers the demand. For 3 days of only 1 h/d of 1 sun, the PV produces $E_{low} = 4 \times 320$ W $\times$ 1 h/d = 1.28 kWh/d. Thus, the bank must have a capacity of

$$C = \frac{E_{DC} - E_{low}}{V} \times \frac{days}{DOD_{max}} = \frac{(4850 - 1280) \text{ Wh/d}}{24V} \times \frac{3d}{0.8} \simeq 558 \text{ Ah} \qquad (14.36)$$

We can use four strings of two batteries each in series 12 V of 150 Ah each to get 24 V and 150 Ah $\times$ 4 = 600 Ah, which means we need 2 $\times$ 4 = 8 batteries.

A PV-based microgrid is formed by having several generation units, each using their own inverter and charge controller to feed a load, and possibly their own storage. Then, all units are connected to the microgrid bus (Figure 14.39). The principle of operation is that surplus power from any generating unit will be supplied to the microgrid bus, and consumed by any of the loads, or possibly stored in any of the storage units. When the microgrid is connected to the grid, overall surplus from the microgrid is supplied to the grid.

14.6   CONCENTRATING SOLAR POWER (CSP)

Solar energy can be used as an alternative to PV collectors to concentrate radiation and provide heat to a working fluid in a reservoir, thus raising its temperature and operating a heat engine using a Rankine cycle or a Stirling cycle (Figure 14.40). This conversion technology is called *solar thermal* in general and more specifically *concentrating solar power* (CSP). This takes us back to producing electricity by a heat engine (Chapters 4, 6, 7, and 9), but in this case, we do not burn a fuel to produce heat but use concentrated solar radiation to do so.

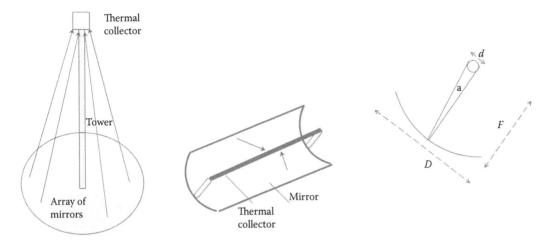

FIGURE 14.40 Solar thermal, concentrated solar power (CSP). Left: Field of mirrors, thermal collector on top of tower. Center: Parabolic trough. Right: Geometrical model.

Most solar thermal plants have been built in the United States and Spain. Large (200–400 MW) solar thermal plants have been built in the southwestern deserts of the United States. In Spain, most plants range from tens of MW to 100 MW, with some plants reaching the 200 MW capacity.

A great advantage of solar thermal or CSP is that it is possible to store thermal energy in hot fluids or other materials (e.g., molten salt), instead of chemical energy in a battery, to compensate for intermittency of the resource. For large thermal storage, say, 8 hours, it would be possible to run the plant during nighttime and provide power continuously.

New working fluids and other thermodynamic cycles are being researched. A promising technology is to employ supercritical carbon dioxide (sCO_2) and a Brayton cycle (Chapter 9) to improve CSP efficiency [20]. In its supercritical state (above the temperature and pressure critical point, Chapter 6), CO_2 acts like a gas but with the density of a liquid. Small changes in temperature or pressure cause large changes in density.

There are two major types of CSP plants. One type is an *array of mirrors* focusing sunlight on the thermal collector (e.g., a tank with the heat transfer fluid) located on top of a tower (Figure 14.40, left-hand side), and the other type is an array of cylindrical mirrors or *troughs* with linear thermal collectors (e.g., a tube with heat transfer fluid) (Figure 14.40, center).

The surface of the mirrors and the troughs can be *parabolic* or *spherical*, and simple geometrical considerations allow to calculate how much radiation is concentrated [21]. A lens or mirror surface of size D, with an axis pointing in the direction of the sun, forms an image of size d at the collector located at focal length F (Figure 14.40, right-hand side). The angle subtended by looking at the sun from Earth is $\alpha = 0.0093°$. For a parabolic surface, the *concentration ratio* (CR) is given by a linear relation between D and d

$$CR = \frac{D}{d} = \frac{D}{\alpha F} \qquad (14.37)$$

whereas for a spherical surface, the CR is given by a ratio of area

$$CR = \left(\frac{D}{d}\right)^2 = \left(\frac{D}{\alpha F}\right)^2 \qquad (14.38)$$

Example 14.20

Calculate CR for a parabolic trough of $D = 1$ m, $F = 25$ cm. Calculate minimum size d of collector.

Answer: $CR = \dfrac{D}{\alpha F} = \dfrac{1}{0.0093 \times 0.25} = 430.11$. The minimum thermal collector diameter would be

$d = \dfrac{D}{CR} = \dfrac{1}{430.11} \simeq 2$ mm.

The simplest model for the net heat input to the thermal collector would be

$$dQ = Q_{in} - Q_{loss} = \zeta \times I_{bc} \times CR - C(T_c - T_a) \tag{14.39}$$

where the input heat Q_{in} is determined by the absorbed fraction ζ of concentrated direct irradiance $I_{bc} \times CR$, and the heat loss Q_{loss} is determined by the heat exchange capacity C and the temperature difference of the collector T_c, with respect to ambient temperature T_a. Only direct irradiance is included because the solar collector would not focus the diffuse and reflected irradiance [8,21].

Thermal efficiency could be estimated by the ratio of thermal input to irradiance

$$\eta_{th} = \frac{dQ}{I_{bc}} \tag{14.40}$$

whereas overall efficiency could be estimated by the ratio of electrical power to irradiance

$$\eta = \frac{P}{I_{bc}} \tag{14.41}$$

CSP systems exhibit overall efficiency in the order of 25% to 30%. Solar thermal farms are more efficient in the use of area than PV solar farms and much more than wind farms. Some CSP achieve 50 W/m2, which is ~50% higher than PV-based farms. Axis tracking is often implemented in CSP plants to achieve higher production.

14.7 ENVIRONMENTAL CONSIDERATIONS OF SOLAR POWER GENERATION

When considering environmental impacts of solar power, as with other renewable sources, we must think of a balance between positive and negative aspects. Prominent positive aspects are its contribution to reducing greenhouse gas emission and interference with water resources. On the other hand, negative aspects may vary according to the type and size of installation. Small PV systems installed on rooftops or spaces that are otherwise unusable in the urban environment do not typically have an impact on land use, but may have an impact on visual appearance of towns and aesthetics. Large utility scale solar farms have impacts that may include increased land use, perceived impact on landscape aesthetics, and interference with ecosystems and wildlife habitat. Large installations, of course, have secondary impacts related to area used for transmission lines and substations. Some large solar farms are colocated with other installations, such as hydropower plants, and this can mitigate some of those impacts. PV-based solar power in particular has many positive aspects since it has no moving parts (except when sun tracking is implemented) and does not produce noise.

Manufacturing of PV panels involve primarily silicon, which is abundant and nontoxic, and secondarily a small amount of other chemicals that are less abundant, such as indium and tellurium, and in some cases toxic materials, such as cadmium. These chemicals are used in much less proportion than silicon. PV panel manufacturing is energy intensive, but the energy employed in their production is recovered in about 1 to 2 years of production depending on insolation [1]. On balance with what is known now, solar power production, PV-based in particular, has fewer environmental impacts than other renewable conversion processes.

EXERCISES

14.1. Assuming the sun temperature is 6000 K, plot spectral density of I_0 in the λ range from 150 to 4000 nm. Plot together with data from ASTM G173 at the average sun–Earth distance for both I_0 and sea level.
14.2. What would be the insolation variation from 30° to 70° latitude?
14.3. Determine sun elevation angle for latitude of 39.74° (decimal degrees) (this latitude corresponds to Golden, Colorado).
14.4. Calculate the sun elevation angle β that will yield an air mass ratio of m = 1.5 (AM1.5).
14.5. Calculate sun elevation and azimuth at the latitude of 39.74° (decimal degrees) for day 50 of the year.
14.6. Use the file with optical path for Hartsfield-Jackson Atlanta International Airport, Georgia, and day 50 to calculate radiation on a collector tilted at the latitude, facing south ($\phi c = 0°$), and 10% contribution of reflected radiation from surrounding surfaces.
14.7. Display *I-V* and power curves for light values of 1, 0.8, and 0.4 of full sun. Calculate the voltage and current for a 10-Ω load for each curve. What is the power given to the load for each one of these curves? Calculate the voltage and current at which the maximum power point (MPP) is established for all these curves. What would be the power for each curve if a MPP tracker adjusts the voltage output to the above values?
14.8. Calculate tilt required for a solar panel at four distinct days: spring equinox (March 21), fall equinox (September 22), winter solstice (December 21), and summer solstice (June 21) for Dallas, Texas (latitude 32.9N).
14.9. Assume a module with NOTC = 45°C, NV_{oc} of 38 V, rated power 100 W, TC_v of −0.30%/C, and TC_p of −0.4%/C.
14.10. Calculate impact of two-axis tracking and PNS one-axis tracking on production for the Golden, Colorado, site and energy produced on a clear day for each day of the evaluation. Compare to energy produced each month for a polar mount installation.
14.11. Calculate CR for a spherical collector of $D = 1$ m, $F = 25$ cm. Calculate minimum size d of collector.

REFERENCES

1. Boyle, G., *Renewable Energy: Power for a Sustainable Future*. Third edition. 2017: Oxford University Press. 566 pp.
2. Demirel, Y., *Energy: Production, Conversion, Storage, Conservation, and Coupling*. Second edition. Green Energy and Technology. 2016: Springer. 616 pp.
3. Keyhani, A., *Design of Smart Power Grid Renewable Energy Systems*. 2017: Wiley IEEE Press. 566 pp.
4. Masters, G.M., *Renewable and Efficient Electric Power Systems*. Second edition. 2013: Wiley-IEEE Press. 690 pp.
5. Nelson, V., and K. Starcher, *Introduction to Renewable Energy*. Second edition. Energy and the Environment, A. Ghassemi. 2016: CRC Press. 423 pp.
6. Sørensen, B., *Renewable Energy: Physics, Engineering, Environmental Impacts, Economics and Planning*. Fourth edition. 2011: Academic Press. 954 pp.
7. Tester, J.W., *et al.*, *Sustainable Energy: Choosing Among Options*. Second edition. 2012: MIT Press. 1021 pp.
8. Vanek, F., L. Albright, and L. Angenent, *Energy Systems Engineering: Evaluation and Implementation*. Third edition. 2016: McGraw-Hill. 704 pp.
9. da Rosa, A., *Fundamentals of Renewable Energy Processes*. Third edition. 2013: Elsevier, Academic Press. 884 pp.
10. Graedel, T.E., and P.J. Crutzen, *Atmospheric Change: An Earth System Perspective*, 1993: Freeman. p. 446.
11. NASA. *Earth Observatory*. 2017. Accessed November 2017. Available from: https://earthobservatory.nasa.gov/Features/Milankovitch/milankovitch_2.php.

12. Berger, A., M.F. Loutre, and M. Crucifix, The Earth's Climate in the Next Hundred Thousand Years (100 kyr). *Surveys in Geophysics*, 2003. 24(2): 117–138.

13. ASTM. *ASTM G173 - 03(2012), Standard Tables for Reference Solar Spectral Irradiances: Direct Normal and Hemispherical on 37° Tilted Surface.* 2017. Accessed November 2017. Available from: https://www.astm.org/Standards/G173.htm.

14. NREL. *Reference Solar Spectral Irradiance: Air Mass 1.5.* 2017. Available from: http://rredc.nrel.gov /solar/spectra/am1.5/.

15. Gueymard, C.A., The Sun's Total and Spectral Irradiance for Solar Energy Applications and Solar Radiation Models. *Solar Energy*, 2004. 76(4): 423–453.

16. Gueymard, C.A. and D. Thevenard, Revising ASHRAE Climatic Data for Design and Standards— Part 2: Clear-Sky Solar Radiation Model. *ASHRAE Transactions*, 2013. 119: 194–209.

17. ASHRAE. *ASHRAE Climate Data Center.* 2017. Accessed November 2017. Available from: https:// www.ashrae.org/resources–publications/bookstore/climate-data-center#std169.

18. HITRAN. *The HITRAN Database.* 2017. Accessed November 2017. Available from: https://www.cfa .harvard.edu/hitran/.

19. Masters, G.M., *Renewable and Efficient Electric Power Systems.* Second edition. 2013, Hoboken, NJ: Wiley-IEEE Press. 712 pp.

20. DOE. SCO2 Power Cycles. 2017. *Office of the Under Secretary for Science and Energy.* Accessed November 2017. Available from: https://www.energy.gov/under-secretary-science-and-energy/sco2 -power-cycles.

21. Fay, J.A., and D.S. Golomb, *Energy and the Environment. Scientific and Technological Principles.* Second edition. MIT-Pappalardo Series in Mechanical Engineering, M.C. Boyce and G.E. Kendall. 2012, New York: Oxford University Press. 366 pp.

Appendix: Introduction to R

This appendix is a very brief tutorial guide on how to install and use R. Its purpose is to help those readers not familiar with R get started with the examples given in the book. This tutorial illustrates R installation, basic data structure and calculations, installing and loading packages, reading data from files, plotting graphics, and programming loops. This tutorial is intended only as a starting point. There are many resources online to learn and use R. You can also refer to other books where I have written detailed tutorials on using R, and other references therein [1,2].

A.1 INSTALLING R

Download R from the Comprehensive R Archive Network (CRAN) repository http://cran.us.r -project.org/ by looking for the *precompiled binary distribution* for your operating system (Linux, Mac, or Windows). In this book, we assume a Windows installation, but all examples and exercises can be done using Mac or Linux. For Windows, select **base** and then the executable download for the current release; at the time of this writing, the release was R-3.4.2. Save and run this program following the installation steps, which take just a few minutes.

A good practice to facilitate working with R is to set up a *working directory* or folder to write and read files. This directory can be typically located in a local hard disk **c:**, a network home drive **h:**, or in a removable drive (flash memory), **e:**. For example, say the working directory is **c:\ labs**. Then folders or directories could be created within the working directory to contain a variety of files, containing data as well as programs.

You can establish a working directory as default by including it in the **Start In** option of the R shortcut, and avoid changing it every time. It takes setup time, but then it will be worthwhile saving time in the long term. To do this, right click on **Shortcut**, go to **Properties**, then type the name of your launch or work directory (Figure A.1). Once you have done this, just double click the shortcut to run R, which will launch from this folder. You need to create this folder beforehand if it does not yet exist. Note, that now when you select **File|Change dir**, the working folder will be the one selected as Start In for the shortcut, and therefore there is no need to set it every time.

When working on machines that are shared by many individuals (e.g., a university computer lab), a more permanent solution is to create a new personalized shortcut in your working directory. Find the R program on the desktop; right click, select **Create Shortcut**, and browse to the desired location (say, folder **labs**), as shown in Figure A.2. Then right click on this new shortcut and select properties, edit **Start In** as shown before. Thereafter, run R by a double click on this new personalized shortcut.

A.2 RUNNING R

You can just double click on the shortcut created during installation, and preferably modified as explained in the last section. Once the R system loads, we get the R Graphical User Interface (RGui) that opens up with the ">" prompt on the R Console (Figure A.3). If you did not edit the Shortcut, use the **File|Change Dir** option in the RGui menu, then type the path or browse to your working directory. Another manner to avoid repeating this step every time you start the R system is to save the workspace file **.Rdata** in the working folder, and double click it to launch the program from that folder. We will explain more on this later when we describe the workspace.

A.2.1 BASIC R SKILLS

The R console is where you type commands upon the > prompt and where you receive text output (Figure A.3). We can also enter scripts or programs in a separate text editor, and then copy and paste

FIGURE A.1   Modifying Start in folder of R shortcut properties.

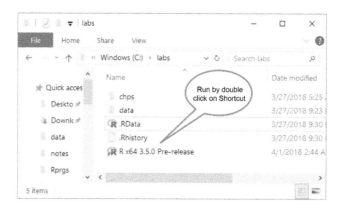

FIGURE A.2   New R shortcut to reside in your working folder C:\labs.

to the R console for execution. Better, you can use the **Script** facility of the RGui described in the next section.

R Manuals are available in pdf and in html formats via the Help menu item. PDF are Portable Document Files and can be viewed and printed using the Acrobat Reader. HTML (Hypertext Markup Language) can be viewed using a web browser. You can obtain help on specific functions from the

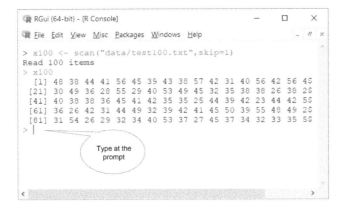

FIGURE A.3 Start of R GUI and the R Console.

Help menu; select **R functions (text)**, then type the function name on the dialog box. Equivalently at the console, type **help(*name of function*)** at the prompt in the console; example, **help(plot)** to get help on function **plot**. Also typing question mark followed by the function name would work, for example **?plot**. You can also launch the help with **help.start** function.

A.2.2 SCRIPTS

From the **File** menu, select **New Script** to type a new set of commands. This will generate an editor window where you can type your set of commands. In this window, you can type and edit commands, and then right click to run one line or a selection of lines by highlighting a section (Figure A.4). Then, save the script by using **File|Save As** followed by selecting a folder and a file name. For easy access later, it is convenient to save your scripts in your working directory, (possibly under separate folders to keep them organized). Scripts are saved as files with ***.R**, that is, with extension **.R**. Later the script can be recalled by **File|Open Script**, browse to your folder, select the file, and edit the script. Further edits can be saved using **File|Save**.

An alternative way of running your script (if you do not need to edit it before running it) is to use **File|Source R Code**, browse to folder, and select the script. This is equivalent of typing the function **source("myscript.R")** to execute the set of commands stored in file **myscript.R**.

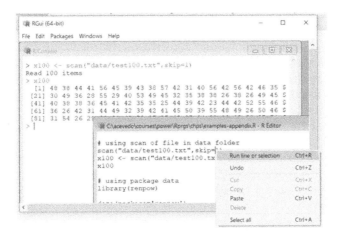

FIGURE A.4 Script Editor: right click to run a line or lines of the script.

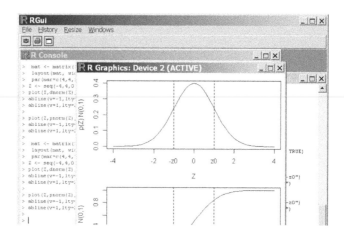

FIGURE A.5   R graphics windows.

A.2.3   GRAPHICS DEVICE

This is where you receive graphical output interactively (Figure A.5). To print a hard copy of a graph, you can just select **File|Print** from the menu, while the graphics window is selected. To save in a variety of graphics formats, use **File|Save As**. These formats include metafile, postscript, bmp, pdf, and jpeg. To capture the image on the clipboard and then paste on another application, you could simply use **Copy and Paste** from the graphics window. To do this you can also use **File|Copy to clipboard**, and then paste the clipboard contents on the selected cursor position of the file being edited in the application. Notice that one can work with windows as usual, go to **Windows** in the menu bar; here you can **Tile** the windows or **Cascade** the windows, and so on. In addition, this is where you can open the console window.

A.2.4   PACKAGE INSTALLATION AND LOADING

Packages are set up to use by following two steps: (1) installation from the Internet and (2) loading it for use. First, to install from the RGui: Go to **Packages|Install package(s)**, select a mirror site depending on your location, and then select Install package. For example, select the **renpow** package, which is the package developed for this book. The download process starts and will give you messages about progress and success or not of the package installation. You do not need to repeat the installation as long as you do not reinstall the R software or want to update the package.

The second step is to load the package. From the RGui: Go to **Packages|Load package**, and then select the package. For example, browse and select renpow. Alternatively, you can run the function **library** from the R console by typing library (renpow). Help on functions provided by renpow would now be available. For example, go to HTML help, and when you click on packages, you will see renpow. Datasets in renpow are used throughout the book as examples. You can find the available datasets by typing data(package='renpow').

These datasets are based on data files in the folder **inst\extdata** of package renpow and are used throughout this book as examples. There are three different ways of using these data files. One way, is to get the data files from the CRAN website; for this, use the link Packages on the left side navigation list, then go to the table listing the package by name. Once you click on the package name, you will have the option of downloading the package source as *.tar.gz, then you can extract the inst \extdata folder from this archive. Copy this folder under your working directory labs, so that you will find a data file using path **extdata**. Another way is to read the file from the installation directory using the function system.file("extdata",datafilename,package="renpow") where datafilename refers to the target data file. A third way is to read the file from the installation directory using the function system.file("extdata",datafilename, package="renpow") where datafilename refers to the target data file. A third way is to use command data( ) in R to load the data set corresponding to the file. Once the data set is loaded, you can use it and address its contents as a **data frame** or data set object. A shortcut is simply referring to the dataset as the file name excluding the extension. All three alternatives are explained in Section A.3.1 using a simple example.

Part of my intent writing package renpow, which supports this textbook, is to bundle many lengthy calculations into functions requiring a single call or at most a sequence of a few lines of code. Most functions of renpow have been described and used in the textbook in the context of specific examples. More options and details are available from the help of renpow.

A.3   SIMPLE TASKS: DATA FILES, GRAPHS, ANALYSIS

A.3.1   READ A SIMPLE TEXT DATA FILE

Before you write code to import or read a data file, it is a good practice to examine the file and understand its contents. Assuming that you copied the folder extdata under your working directory labs, we can use the file extdata/test100.txt as an example. For instance, use the Notepad to look at file extdata/test100.txt. You will see that it has a header x specifying the variable name and that it is just one field per line with no separator between fields (Figure A.6, left-hand side).

A more convenient text editor is Vim, an open source program that you can download from the Internet (www.vim.org). Some nice features are that you get line numbers and position within a line (Figure A.6, center), a more effective find/replace, and tool and color codes for your script, if you choose to write the scripts on this editor. Another option is the Notepad++ editor available in Windows, which provides line numbers and convenient programming setup for a variety of languages (Figure A.6, right-hand side).

Make sure you set **File|Change Dir** to the working folder labs so that the path to the file is relative to this folder. If you have personalized the R shortcut to start in folder labs, then the path to the file is relative to this folder, for example, in this case extdata/test100.txt. Therefore, you could use this name to scan the file.

```
> scan("extdata/test100.txt", skip=1)
```

On the console, we receive the response

```
Read 100 items
 [1] 48 38 44 41 56 45 39 43 38 57 42 31 40 56 42 56 42 46 35 40 30 49 36 28 55
[26] 29 40 53 49 45 32 35 38 38 26 38 26 49 45 30 40 38 38 36 45 41 42 35 35 25
[51] 44 39 42 23 44 42 52 55 46 44 36 26 42 31 44 49 32 39 42 41 45 50 39 55 48
[76] 49 26 50 46 56 31 54 26 29 32 34 40 53 37 27 45 37 34 32 33 35 50 37 74 44
```

Alternatively using the file from the installed package directory we can type

```
scan(system.file("extdata","test100.txt", package="renpow"),skip=1)
```

Next, create an object x100 by scanning an input data file

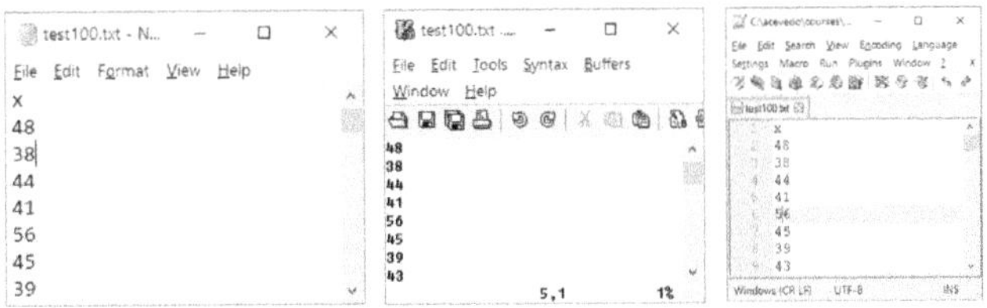

FIGURE A.6   Example of a text file. Viewed with the notepad text editor (left), the Vim text editor (center), and the Notepad++ (right).

```
> x100 <- scan("extdata/test100.txt", skip=1)
```

or its alternative

```
x100 <- scan(system.file("extdata","test100.txt", package="renpow"),skip=1)
```

The operator "<-" is used for assignment. Equivalently, you can write the same using the equal sign (=). However, the equal sign is better used for other purposes, such as giving values to the arguments of functions.

Double check that you have the newly created object by **Misc|List objects** or using **ls**().

```
> ls()
[1] "x100"
```

The object x100 is stored in the workspace **labs\.Rdata**. Double check the object contents by typing its name

```
> x100
  [1] 48 38 44 41 56 45 39 43 38 57 42 31 40 56 42 56 42 46 35 40 30 49 36 28 55
 [26] 29 40 53 49 45 32 35 38 38 26 38 26 49 45 30 40 38 38 36 45 41 42 35 35 25
 [51] 44 39 42 23 44 42 52 55 46 44 36 26 42 31 44 49 32 39 42 41 45 50 39 55 48
 [76] 49 26 50 46 56 31 54 26 29 32 34 40 53 37 27 45 37 34 32 33 35 50 37 74 44
>
```

We can see that this object is a one-dimensional array. The number given in brackets on the left-hand side is the position of the entry first listed in that row. For example, entry in position 26 is 29. Since this object is a one-dimensional array, we can check the size of this object by using function **length**():

```
> length(x100)
[1] 100
```

When entering commands at the console, you can recall previously typed commands using the up arrow key. For example, after you type

```
> x100
```

you can use the up arrow key, and edit the line to add length

```
> length(x100)
```

We now explain the alternative method using the data commands in R applied to the renpow data files that have been stored as a database in the libraries when you installed the package. For additional information you may want to look at the help for data( ) using help("data"). Using this method, typing data(package='renpow') at the console will list the data objects and files of renpow; you will see test100 in this list. Now to use it, simply type test100 at the console. This is a data frame of 100 rows and a single column with name x (which was the header x in the file. You can address the column as an array typing test100$x. Here the $ symbol followed by x means the particular column of this data frame. The result is the same that we obtained earlier for x100. In fact we could have just typed x100 <- test100$x instead of reading the file. This is a direct and shorter way of using the data of this package, but of course it is good to learn the skills of reading the data files. Keep this in mind for all chapters of the book.

A.3.2 SIMPLE STATISTICS

Now, we can calculate sample mean, variance, and standard deviation:

```
> mean(x100)
```

```
[1] 40.86
> var(x100)
[1] 81.61657
> sd(x100)
[1] 9.034189
>
```

It is good practice to round the results, for example, to zero decimals:

```
> round(mean(x100),0)
[1] 41
> round(var(x100),0)
[1] 82
> round(sd(x100),0)
[1] 9
>
```

We can concatenate commands in a single line by using the semicolon (;) character. Thus, for example we can round the above to two decimals:

```
> mean(x100); round(var(x100),2); round(sd(x100),2)
[1] 40.86
[1] 81.62
[1] 9.03
```

A.3.3 SIMPLE GRAPHS TO A GRAPHICS WINDOW

When applying a graphics command, a new graph window will open by default if none is opened; otherwise, the graph is sent to the active graph window. Plot a histogram by using function **hist** applied to a single variable or one-dimensional array object

```
>hist(x100)
```

to obtain the graph shown in Figure A.7. Bar heights are counts of how many measurements fall in the bin indicated in the horizontal axis.

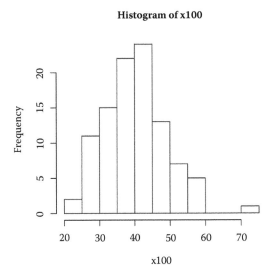

FIGURE A.7   Histogram of x100.

A.3.4 Addressing Entries of an Array

We can refer to specific entries of an array using brackets or square braces. For example, entry in position 26 of array x100

```
> x100[26]
[1] 29
```

That is to say, x100[26]=29. A colon (:) is used to declare a sequence of entries. For example, the first 10 positions of x100

```
> x100[1:10]
 [1] 48 38 44 41 56 45 39 43 38 57
```

Entries can be removed. For example, x100[-1] removes the first entry

```
> x100[-1]
 [1] 38 44 41 56 45 39 43 38 57 42 31 40 56 42 56 42 46 35 40 30 49 36 28 55 29
> length(x100[-1])
[1] 99
```

Using a blank or no character in the bracket means that all entries are used

```
> x100[]
 [1] 48 38 44 41 56 45 39 43 38 57 42 31 40 56 42 56 42 46 35 40 30 49 36 28 55
[26] 29 40 53 49 45 32 35 38 38 26 38 26 49 45 30 40 38 38 36 45 41 42 35 35 25
[51] 44 39 42 23 44 42 52 55 46 44 36 26 42 31 44 49 32 39 42 41 45 50 39 55 48
[76] 49 26 50 46 56 31 54 26 29 32 34 40 53 37 27 45 37 34 32 33 35 50 37 74 44
> length(x100[])
[1] 100
```

A.3.5 Example: Two Variables

Next, we work with an example of a data file containing more than one value in a line. As discussed in the book, specific heat is an important parameter in thermodynamic calculations (Chapter 4). Look at the extdata/CpCvT.txt file. It consists of two columns; the first is temperature in °C and the second is the ratio $\gamma = \dfrac{c_p}{c_v}$ of specific heat for air with blank separations (Figure A.8).

```
TC     cpcv
-40    1.401
-20    1.401
0      1.401
5      1.401
10     1.401
15     1.401
```

... more records like this until values for 500 and 1000

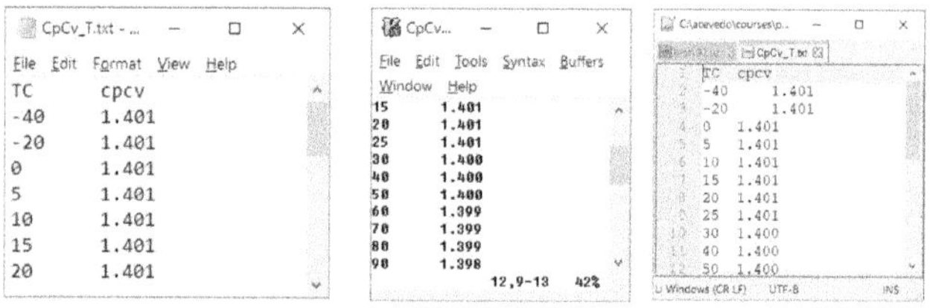

FIGURE A.8 Example of a text file. Viewed with the notepad text editor (left), the Vim editor center, and the Notepad++ (right).

```
500    1.357
1000   1.321
```

We will see how to read a file like this, and create a data set or data frame object containing this data. For this purpose, we use R function **read.table** in the following manner, which specifies that we have a header line and the values are separated by blanks.

```
X <- read.table("extdata/CpCvT.txt",header=TRUE,sep="")
```

or alternatively

```
X <- read.table(system.file("extdata","CpCvT.txt", package="renpow"),
header=TRUE,sep="")
```

Using the third alternative, we could have obtained the object directly by typing X <- CpCvT at the console.

The header strings become the names of the variables in the data set and each row is assigned a number, as we can see when we query X:

```
> X
   TC cpcv
1  -40 1.401
2  -20 1.401
3   0 1.401
4   5 1.401
5  10 1.401
6  15 1.401
7  20 1.401
...
```

We can refer to one of the variables by using the name of the data set followed by $ and the name of the variable. For instance, X$TC is temperature.

```
> X$TC
 [1]  -40 -20   0   5  10  15  20  25  30  40  50  60  70  80  90
[16]  100 200 300 400 500 1000
>
```

You can use **dim** to check the dimensions of the data set to be 21 rows and two columns

```
> dim(X)
[1] 21 2
>
```

For brevity, the names of the components can be addressed directly once we **attach** the data frame.

```
> attach(X)
```

For instance, to plot variable cpcv as a function of TC (Figure A.9, left-hand side) we can refer to TC and cpcv directly instead of X$TC and X$cpcv

```
> plot(TC,cpcv)
```

You can use a line graph by adding an argument **type="l"** to the plot function. Careful, this is a lowercase letter "l" for line not the number one "1" (Figure A.9, right-hand side).

```
> plot(TC,cpcv, type="l")
```

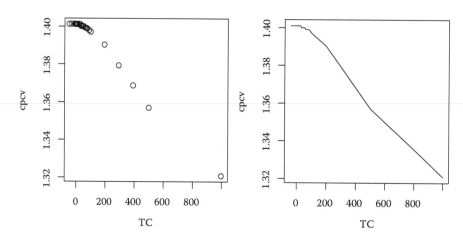

FIGURE A.9 Simple plots: x-y using points and x-y using lines.

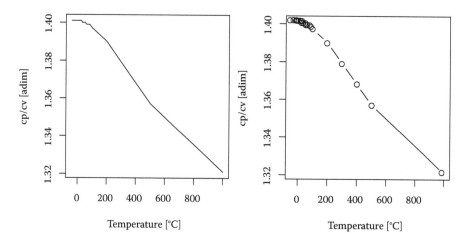

FIGURE A.10 Simple plots. Left: x-y using lines and axis labels. Right: x-y using both points and lines plus axis labels.

By default, the x- and y-axis are labeled with the variable name. These can be changed using xlab= and ylab= arguments. For example, we obtain Figure A.10 (left-hand side) by applying

```
> plot(TC,cpcv, type="l", xlab="Temperature [°C]", ylab="cp/cv [adim]")
```

We can plot both points and lines using type="b"

```
> plot(TC,cpcv, type="b", xlab="Temperature [°C]", ylab="cp/cv [adim]")
```

See Figure A.10 (right-hand side).

When using attach, it is a good idea to the use detach when you are done to avoid confusing variables:

```
>detach(X)
```

You will see that TC is no longer available after you use detach:

```
> TC
Error: object 'TC' not found
```

A.3.6 MULTIPLE GRAPHS TO A PAGE

You may be wandering how I produced two graphs on the same page. There are several ways to do this. Most of the figures you see in this book that have more than one panel were produced using function **panels** of renpow, which in turn uses function layout of R and several graphics parameters given by function par.

For instance, to produce Figure A.9, we can write attach(X) and then

```
panels(7,7/2,1,2,pty="m")
plot(TC,cpcv)
plot(TC,cpcv, type="l")
```

which precedes the calls to plot by calling the panel function with width 7 inches and height 7/2 inches. As another example, we can use panels to produce four plots on a page as shown in Figure A.11:

```
panels(7,7,2,2,pty="m")
plot(TC,cpcv)
plot(TC,cpcv, type="l")
```

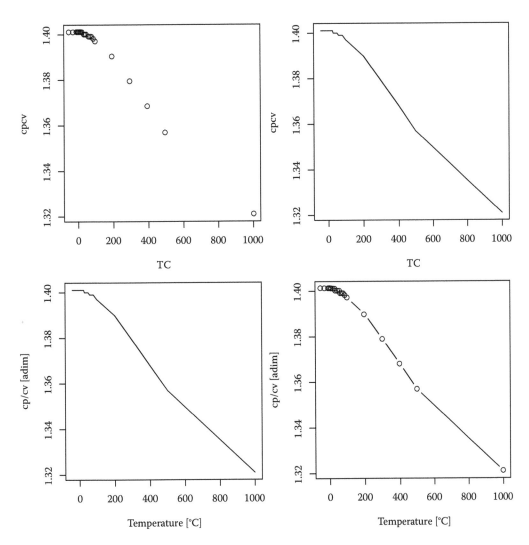

FIGURE A.11   Four plots on the same graphics device.

```
plot(TC,cpcv, type="l", xlab="Temperature [°C]", ylab="cp/cv [adim]")
plot(TC,cpcv, type="b", xlab="Temperature [°C]", ylab="cp/cv [adim]")
```

Remember to detach the data set by using detach(X).

A.4 WORKSPACE

A.4.1 STORE YOUR OBJECTS

The workspace containing objects you create can be stored in a file **.Rdata**. Use **File|Save workspace** menu to store an image of your objects. File **.Rdata** is created in the launch folder specified in the **Start In** field of the R shortcut, or you can browse to find a desired working folder to store. For example, c:/labs, the console will inform you of the save operation.

```
> save.image("C:/labs/.RData")
```

A convenient option is to store the workspace in your working folder. Once you save the workspace, right after opening the commands window, you will see a message like this:

```
[Previously saved workspace restored]
```

When done this way, **.Rdata** resides in your working drive and you can use this **.Rdata** file in order to facilitate access to functions created by various scripts. After launching the program, you could load the workspace using **File|Load Workspace** and browse to find the desired **.Rdata** file. Alternatively, you can also double click on the **.Rdata** file to launch the program and load the workspace.

To list objects in your workspace, type **ls()** or using **Misc|List objects** of the RGui menu. Check your objects with **ls()** or with **objects()**; if this is your first run, then you will not have existing objects and you will get **character(0)**.

```
> ls()
character(0)
> objects()
character(0)
>
```

A.4.2 COMMAND HISTORY AND LONG SEQUENCES OF COMMANDS

When editing through a long sequence of commands by using the arrow keys, one could use **History** to visualize all the previous commands at once. However, it is typically more convenient (especially if writing functions) to type the commands first in a script using the script editor or a text editor (say Vim). Then, copy and paste from the text editor to the R console to execute.

A.4.3 EDITING DATA IN OBJECTS

For example to edit object x100, use

```
>edit(x100)
```

to invoke the data editor or from **Edit** menu item GUI use **data editor**.

A.4.4 CLEAN UP AND CLOSE R SESSION

Many times, we generate objects that may not be needed after we use them. In this case, it is good practice to clean up after a session by removing objects using the **rm** function. A convenient way of doing this is to get a list of objects with **ls()** and then see what we need to remove.

For example, suppose at this point we may want to keep object **x100** because they contain data we may need later, but remove object **x**:

```
> ls()
[1] "x100" .."x"
> rm(x)
```

You can also confirm that objects were indeed removed and get an update of the list to see if there is some more cleanup required:

```
> ls()
[1]    "x100"
```

The objects can also be listed using **Misc|List Objects** in the RGui menu.

You can use **q()** or **File|Exit** to finalize R and close the session. When you do this, you will be prompted for saving the workspace. Reply yes; this way you will have the objects created available for the next time you use R.

A.5   MORE SKILLS

A.5.1 DEFINING FUNCTIONS

We define a function in R by declaring a name assigned to the word function, and in parenthesis the arguments followed by lines of code. As a simple example, define an R function to calculate a hyperbolic mathematical function $X(t) = \dfrac{a}{t+b}$. The following line assigns f as a function with arguments t, a, and b:

```
f <- function(t, a, b) X <- a/(t+b)
```

Now to call the function, first give values to the arguments and use the name of the function:

```
t <- seq(0,10,0.1)
a <- 2; b <-2
X <- f(t,a,b)
plot(t,X, type="l")
```

Once we store a function in the workspace, we can write a call to it with proper arguments. It is available whenever we need it as long as we save the workspace before we exit the session. You can verify that the function f is stored in your workspace using ls().

As another example of how to define a function, take the conversion to rectangular from polar used in this book (Chapter 5). Study the following, and observe that the lines of code that constitute the function are fenced by curly braces; starting with { and ending with }. Also, note that the result is provided by a return. For practice, type and store this function

```
recta <- function(pol){
mag <- pol[1]; ang <- pol[2]
a <- mag * cos(ang*pi/180)
b <- mag * sin(ang*pi/180)
return(round(c(a,b),3))
}
```

Then for instance, to convert polar 2 ∠ 45° to rectangular, call this function by its name recta, and argument is an array with values 2 and 45:

```
recta(c(2,45))
```

As another example, the function panels mentioned in Section A.3.6 is implemented by the following code:

```
panels <- function (wd,ht,rows,cols,pty,int="r"){
np <- rows*cols # number of panels
mat <- matrix(1:np,rows,cols,byrow=T) # matrix for layout
layout(mat, widths=rep(wd/cols,cols), heights=rep(ht/rows,rows), TRUE)
par(mar=c(4,4,1,.5),xaxs=int,yaxs=int,pty=pty)
}
```

A.5.2   CSV Text Files

A CSV file is a text file with fields separated by commas. CSV stands for the format of comma separated values. As an example, we will use extdata/ERCOT2010.01.csv. In windows, a default program to open the CSV files is Excel (Figure A.12). To see more numbers you would have to scroll to the right. A CSV file is just a text file, and therefore it opens with the notepad, Notepad++, and Vim. Right click on the file name, select Open with, and then select Vim. As you can see, commas separate the numbers, and the lines are "word wrapped" (Figure A.13). If using the notepad, you can choose the option **Format|Word wrap** to show all the numbers.

One way to read a CSV file into R is to use read.table with with **sep=","** argument:

```
Y <- read.table("extdata/ERCOT2010.01.csv",header=TRUE,sep=",")
```

or alternatively

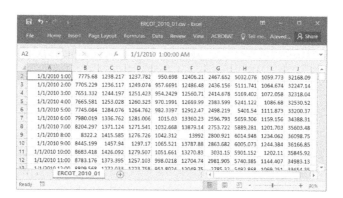

FIGURE A.12   An example of CSV file opened in MS Excel.

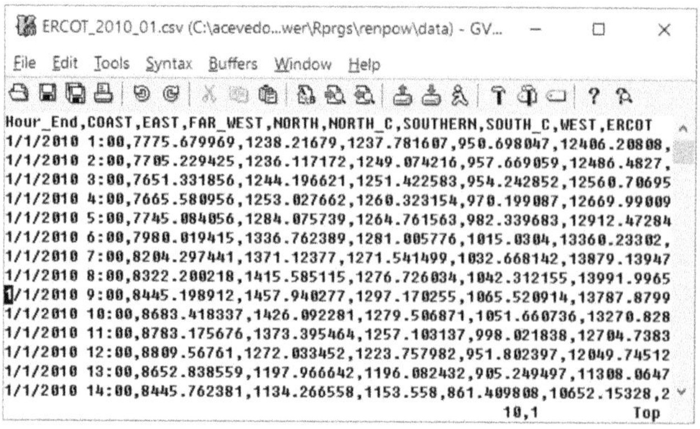

FIGURE A.13   A CSV file opened in Vim.

```
Y <- read.table(system.file("extdata","ERCOT2010.01.csv", package="renpow"),
header=TRUE,sep=",")
```

Then, operate on object Y as before

```
> dim(Y)
[1] 743 10
> names(Y)
 [1] "Hour_End" "COAST"    "EAST"     "FAR_WEST" "NORTH"    "NORTH_C"
 [7] "SOUTHERN" "SOUTH_C"  "WEST"     "ERCOT"
>
```

We could have directly used Y <- ERCOT2010.01.

A.5.3 PLOTTING MORE THAN ONE VARIABLE

Let us use the preceding example to see how to plot more than one variable using function **matplot**, and add a legend to identify the curves. We will plot variables 3 through 5, using line types from 1 to 3, all in color black (col is color, col=1 will be black color for all lines). We can place a legend using the **legend** command.

```
matplot(Y[,3:5],type="l",lty=1:3,col=1, xlab="Hours",ylab="Demand (MW)")
legend("topright",leg=names(Y[,3:5]),col=1,lty=1:3)
```

The first two arguments of legend are the x,y coordinates to place the legend, then **leg** argument is an array with the labels; here we have used the names of the data set (Figure A.14).

A.5.4 TIME SERIES

An important type of data set is organized by a time stamp. An example is the data set used in the previous two sections. The first column Y$Hour_End consists of a day and hour. We can convert to time and use a package to analyze time series such as **xts**. Install and load this package using the menu or commands

```
tt <- strptime(Y$Hour_End, format="%m/%d/%Y
# create time series
x.t <- xts(round(Y[,3],2),tt)
plot(x.t,ylab="Demand(MW)")
```

which produces Figure A.15.

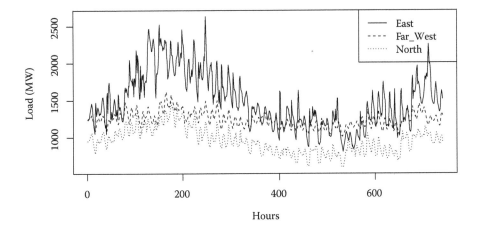

FIGURE A.14   Plot with a family of lines.

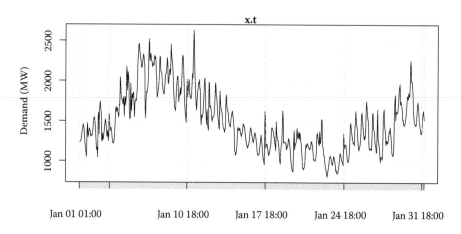

FIGURE A.15   Time series plot.

A.5.5   MORE ON GRAPHICS WINDOWS

You could also direct graphics to specific formats. For example, the plots we have done earlier can be
directed to a pdf using

```
wd=7; ht=7
pdf("appendix-ERCOT.pdf",wd,ht)
 matplot(Y[,3:5],type="l",lty=1:5,col=1, xlab="Hours",ylab="Demand (MW)")
 legend("topright",leg=names(Y[,3:5]),col=1,lty=1:5)
 plot(x.t,ylab="Demand(MW)",main="")
dev.off()
```

Because we have two graphics functions between the pdf() and the dev.off() lines, each graph
function produces one page of the pdf file. When opened, the pdf file will show two pages and one
graph per page. You can add a path to the pdf filename to direct it to a specific directory:

```
wd=7; ht=7
outfile <- paste(path, "appendix-ERCOT.pdf",sep="")
pdf(outfile,wd,ht)
 matplot(Y[,3:5],type="l",lty=1:5,col=1, xlab="Hours",ylab="Demand (MW)")
 legend("topright",leg=names(Y[,3:5]),col=1,lty=1:5)
 plot(x.t,ylab="Demand(MW)",main="")
dev.off()
```

Alternatively, we could produce two plots on the same page using function panels of renpow after
the pdf function and before the first graphics function (Figure A.16).

```
wd=7; ht=7
outfile <- paste(path, "appendix-ERCOT-twopanels.pdf",sep="")
pdf(outfile,wd,ht)
 panels(wd,ht,2,1,pty="m")
 matplot(Y[,3:5],type="l",lty=1:5,col=1, xlab="Hours",ylab="Demand (MW)")
 legend("topright",leg=names(Y[,3:5]),col=1,lty=1:5)
 plot(x.t,ylab="Demand(MW)",main="")
dev.off()
```

You can use this method to send the graphics output of most of the examples in this book to pdf,
and organize it by pages and panels, as you deem necessary and useful.

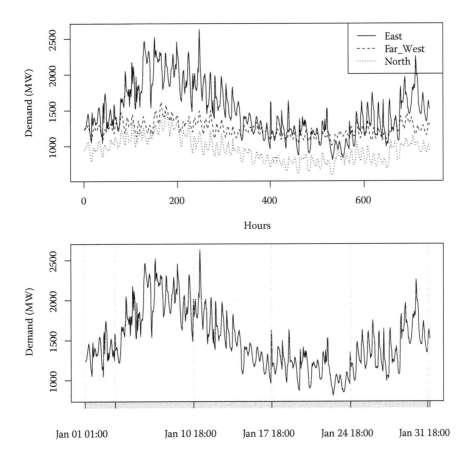

FIGURE A.16 Two graphs on the same page that were sent to a PDF.

A.5.6 PROGRAMMING LOOPS

We will use the simple example of evaluating a function of one variable for different values of a parameter. For instance, evaluate the exponential function $x = exp(rt)$ for several values of the coefficient r. First, declare the sequence of values of t:

```
> t<-seq(0,10,0.1)
```

Then, store values of r in an array:

```
>r <- c(-0.1,0,0.1)
```

Now declare a two-dimensional array (matrix) to store the results; each column corresponds to the values of x for a given value of r:

```
> x <- matrix(nrow=length(t), ncol=length(r))
```

Here nrow and ncol denote the number of rows and columns, respectively. Now use function **for()** to execute a loop:

```
> for(i in 1:3) x[,i]<-exp(r[i]*t)
```

We can ask to see the results; the first few lines are

```
> x
```

```
        [,1] [,2]    [,3]
[1,] 1.0000000  1 1.000000
[2,] 0.9900498  1 1.010050
[3,] 0.9801987  1 1.020201
[4,] 0.9704455  1 1.030455
[5,] 0.9607894  1 1.040811
[6,] 0.9512294  1 1.051271
```

We can round to two decimal places:

```
> round(x,2)
     [,1] [,2] [,3]
[1,] 1.00  1 1.00
[2,] 0.99  1 1.01
[3,] 0.98  1 1.02
[4,] 0.97  1 1.03
[5,] 0.96  1 1.04
[6,] 0.95  1 1.05
[7,] 0.94  1 1.06
[8,] 0.93  1 1.07
[9,] 0.92  1 1.08
[10,] 0.91  1 1.09
```

A more effective and preferred way of applying loops is using functions **sapply()**, **vapply()**, and **mapply()**. For example, the preceding calculation can be done using

```
x <- sapply(r, function(r) exp(r*t))
```

This simple line yields the same results in just one statement without having to predefine the matrix x.

REFERENCES

1. Acevedo, M.F., *Data Analysis and Statistics for Geography, Environmental Science & Engineering: Applications to Sustainability*. 2013, Boca Raton, Florida: CRC Press, Taylor & Francis Group. 535 pp.
2. Acevedo, M.F., *Simulation of Ecological and Environmental Models*. 2012, Boca Raton, Florida: CRC Press, Taylor & Francis Group. 464 pp.

Index

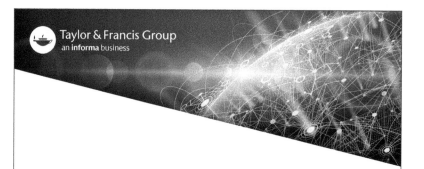

Taylor & Francis eBooks

www.taylorfrancis.com

A single destination for eBooks from Taylor & Francis
with increased functionality and an improved user
experience to meet the needs of our customers.

90,000+ eBooks of award-winning academic content in
Humanities, Social Science, Science, Technology, Engineering,
and Medical written by a global network of editors and authors.

TAYLOR & FRANCIS EBOOKS OFFERS:

A streamlined
experience for
our library
customers

A single point
of discovery
for all of our
eBook content

Improved
search and
discovery of
content at both
book and
chapter level

REQUEST A FREE TRIAL
support@taylorfrancis.com

Milton Keynes UK
Ingram Content Group UK Ltd.
UKHW051942071024
449327UK00026B/2126